Masses of Fundamental Particles

Cargèse 1996

NATO ASI Series

Advanced Science Institutes Series

A series presenting the results of activities sponsored by the NATO Science Committee, which aims at the dissemination of advanced scientific and technological knowledge, with a view to strengthening links between scientific communities.

The series is published by an international board of publishers in conjunction with the NATO Scientific Affairs Division

A Life Sciences **B Physics**	Plenum Publishing Corporation New York and London
C Mathematical and Physical Sciences **D Behavioral and Social Sciences** **E Applied Sciences**	Kluwer Academic Publishers Dordrecht, Boston, and London
F Computer and Systems Sciences **G Ecological Sciences** **H Cell Biology** **I Global Environmental Change**	Springer-Verlag Berlin, Heidelberg, New York, London, Paris, Tokyo, Hong Kong, and Barcelona

PARTNERSHIP SUB-SERIES

1. Disarmament Technologies	Kluwer Academic Publishers
2. Environment	Springer-Verlag
3. High Technology	Kluwer Academic Publishers
4. Science and Technology Policy	Kluwer Academic Publishers
5. Computer Networking	Kluwer Academic Publishers

The Partnership Sub-Series incorporates activities undertaken in collaboration with NATO's Cooperation Partners, the countries of the CIS and Central and Eastern Europe, in Priority Areas of concern to those countries.

Recent Volumes in this Series:

Volume 362—Low-Dimensional Applications of Quantum Field Theory
edited by Laurent Baulieu, Vladimir Kazakov, Marco Picco, and Paul Windey

Volume 363—Masses of Fundamental Particles: *Cargèse 1996*
edited by Maurice Lévy, Jean Iliopoulos, Raymond Gastmans, and Jean-Marc Gérard

Volume 364—Quantum Fields and Quantum Space Time
edited by Gerard t'Hooft, Arthur Jaffe, Gerhard Mack, Pronob K. Mitter, and Raymond Stora

Series B: Physics

Masses of Fundamental Particles

Cargèse 1996

Edited by

Maurice Lévy

Université Pierre et Marie Curie
Paris, France

Jean Iliopoulos

Ecole Normale Supérieure
Paris, France

Raymond Gastmans

Katholieke Universiteit Leuven
Leuven, België

and

Jean-Marc Gérard

Université Catholique de Louvain
Louvain-la-Neuve, Belgique

Plenum Press
New York and London
Published in cooperation with NATO Scientific Affairs Division

Proceedings of a NATO Advanced Study Institute on
Masses of Fundamental Particles,
held August 5 – 17, 1996
in Cargèse, France

NATO-PCO-DATA BASE

The electronic index to the NATO ASI Series provides full bibliographical references (with keywords and/or abstracts) to about 50,000 contributions from international scientists published in all sections of the NATO ASI Series. Access to the NATO-PCO-DATA BASE is possible in two ways:

—via online FILE 128 (NATO-PCO-DATA BASE) hosted by ESRIN, Via Galileo Galilei, I-00044 Frascati, Italy

—via CD-ROM "NATO Science and Technology Disk" with user-friendly retrieval software in English, French, and German (©WTV GmbH and DATAWARE Technologies, Inc. 1989). The CD-ROM contains the AGARD Aerospace Database.

The CD-ROM can be ordered through any member of the Board of Publishers or through NATO-PCO, Overijse, Belgium.

Library of Congress Cataloging-in-Publication Data

Masses of fundamental particles : Cargèse 1996 / edited by Maurice Lévy ... [et al.].
p. cm. -- (NATO ASI series. B, Physics ; v. 363)
"Published in cooperation with NATO Scientific Affairs Division."
"Proceedings of the NATO Advanced Study Institute on Masses of Fundamental Particles, held August 5-17, 1996 in Cargèse, France"--T.p. verso.
ISBN 0-306-45694-X
1. Quarks--Congresses. 2. Atomic mass--Congresses. 3. Particles (Nuclear physics)--Congresses. I. Lévy, Maurice, 1922- . II. North Atlantic Treaty Organization. Scientific Affairs Division. III. NATO Advanced Study Institute on Masses of Fundamental Particles (12th : 1996 : Cargèse, France) IV. Series.
QC793.5.Q2526M38 1998
539.7'25--dc21 97-31221
CIP

ISBN 0-306-45694-X

A Division of Plenum Publishing Corporation
233 Spring Street, New York, N. Y. 10013

http://www.plenum.com

10 9 8 7 6 5 4 3 2 1

Printed in the United States of America

PREVIOUS CARGÈSE SYMPOSIA PUBLISHED IN THE NATO ASI SERIES B: SERIES

Volume 350 FRONTIERS IN PARTICLE PHYSICS: *Cargèse 1994*
edited by Maurice Lévy, Jean Iliopoulos, Raymond Gastmans, and Jean-Marc Gérard

Volume 311 QUANTITATIVE PARTICLE PHYSICS: *Cargèse 1992*
edited by Maurice Lévy, Jean-Louis Basdevant, Maurice Jacob, Jean Iliopoulos, Raymond Gastmans, and Jean-Marc Gérard

Volume 261 Z° PHYSICS: *Cargèse 1990*
edited by Maurice Lévy, Jean-Louis Basdevant, Maurice Jacob, David Speiser, Jacques Weyers, and Raymond Gastmans

Volume 223 PARTICLE PHYSICS: *Cargèse 1989*
edited by Maurice Lévy, Jean-Louis Basdevant, Maurice Jacob, David Speiser, Jacques Weyers, and Raymond Gastmans

Volume 173 PARTICLE PHYSICS: *Cargèse 1987*
edited by Maurice Lévy, Jean-Louis Basdevant, Maurice Jacob, David Speiser, Jacques Weyers, and Raymond Gastmans

Volume 156 GRAVITATION IN ASTROPHYSICS: *Cargèse 1986*
edited by B. Carter and J. B. Hartle

Volume 150 PARTICLE PHYSICS: *Cargèse 1985*
edited by Maurice Lévy, Jean-Louis Basdevant, Maurice Jacob, David Speiser, Jacques Weyers, and Raymond Gastmans

Volume 130 HEAVY ION COLLISIONS: *Cargèse 1984*
edited by P. Bonche, Maurice Lévy, Phillipe Quentin, and Dominique Vautherin

Volume 126 PERSPECTIVES IN PARTICLES AND FIELDS: *Cargèse 1983*
edited by Maurice Lévy, Jean-Louis Basdevant, David Speiser, Jacques Weyers, Maurice Jacob, and Raymond Gastmans

Volume 85 FUNDAMENTAL INTERACTIONS: *Cargèse 1981*
edited by Maurice Lévy, Jean-Louis Basdevant, David Speiser, Jacques Weyers, Maurice Jacob, and Raymond Gastmans

Volume 72 PHASE TRANSITIONS: *Cargèse 1980*
edited by Maurice Lévy, Jean-Claude Le Guilliou, and Jean Zinn-Justin

Volume 61 QUARKS AND LEPTONS: *Cargèse 1979*
edited by Maurice Lévy, Jean-Louis Basdevant, David Speiser, Jacques Weyers, Raymond Gastmans, and Maurice Jacob

Volume 44 RECENT DEVELOPMENTS IN GRAVITATION: *Cargèse 1978*
edited by Maurice Lévy and S. Deser

Volume 39 HADRON STRUCTURE AND LEPTON–HADRON INTERACTIONS: *Cargèse 1977*
edited by Maurice Lévy, Jean-Louis Basdevant, David Speiser, Jacques Weyers, Raymond Gastmans, and Jean Zinn-Justin

Volume 26 NEW DEVELOPMENTS IN QUANTUM FIELD THEORY AND STATISTICAL MECHANICS: *Cargèse 1976*
edited by Maurice Lévy and Pronob Mitter

Volume 13 WEAK AND ELECTROMAGNETIC INTERACTIONS AT HIGH ENERGIES: *Cargèse 1975* (Parts A and B)
edited by Maurice Lévy, Jean-Louis Basdevant, David Speiser, and Raymond Gastmans

PREFACE

The 1996 Cargèse Summer Institute on Frontiers in Particle Physics was organized by the Université Pierre et Marie Curie, Paris (M. Lévy), the Ecole Normale Supérieure, Paris (J. Iliopoulos), the Katholieke Universiteit Leuven (R. Gastmans), and the Université Catholique de Louvain (J.-M. Gérard), which, since 1975, have joined their efforts and worked in common. It was the twelfth Summer Institute on High Energy Physics organized jointly at Cargèse by three of these universities.

The Standard Model for fundamental interactions is constructed on two essential ingredients: the gauge symmetry and the mass generation mechanism. Now that the gauge theory aspect has been firmly established, the new challenge for the young researchers in elementary particle physics is the understanding of the origin of the masses.

The standard Higgs mechanism is believed to be responsible for generating the masses of ALL fundamental particles. Professor D. Treille discussed the prospects for Higgs boson search and described the experimental determinations of the gauge boson masses. The influence of the top quark mass on electroweak processes has been emphasized by Professor J.L. Rosner, while Professor M. Neubert introduced the heavy-quark effective theory which allows you to get rid of heavy-quark masses. The theoretical determinations of the light-quark masses have been critically analyzed by Professor H. Leutwyler. Professor A. Pich presented the various experimental tests on lepton universality and Professor R.L. Mössbauer reviewed our present knowledge on the neutrino masses.

The interplay between particle physics, astrophysics, and cosmology has become increasingly intensive. Professor J. Silk reviewed the various dark matter candidates responsible for the mass distribution in our Universe. Among these candidates are the supersymmetric particles. Professors G.G. Ross and C. Kounnas speculated on the origin of masses in supersymmetric extensions of the Standard Model.

The experimentalists must learn how to face the numerous difficulties associated with new colliders. Professor J. Feltesse introduced the deep-inelastic physics related to HERA (DESY). Finally, Professor J.-J. Aubert gave a detailed presentation of the future LHC collider (CERN) able to explore the TeV energy domain where the Higgs phenomena are expected to manifest themselves.

We owe many thanks to all those who have made this Summer Institute possible!

Special thanks are due to all the Scientific Committee of NATO and its President for a generous grant. We are also very grateful for the financial contribution given by the

C.N.R.S. and the Institut National de Physique Nucléaire et de Physique des Particules (IN2P3).

We also want to thank Ms. A. Touchant for her efficient organizational help, Mr. and Ms. Ariano and Ms. Cassegrain for their kind assistance in all material matters of the school, and, last but not least, the people from Cargése for their hospitality.

Mostly, however, we would like to thank all the lecturers and participants: their commitment to the school was the real basis for its success.

M. Lévy
R. Gastmans
J. Ilipoulos
J.-M. Gérard

CONTENTS

BOSON MASSES IN THE STANDARD MODEL

Daniel Treille

CERN, 1211 Geneva 23, Switzerland

INTRODUCTION

In this review I shall first describe briefly the experimental determinations, present and future, of the intermediate vector bosons' masses. These will be put in perspective with the other electroweak accurate measurements, to take stock of the Standard Model (SM) tests.

The key problem of the SM concerns the Higgs sector. Prospects for Higgs boson search and study at the various existing and foreseen machines will then be summarized for the SM and SM-like boson(s). The abundance of the literature on this topic is such that I shall focus only on a few main themes.

Another hypothetical boson has appeared in the SM, to solve the so-called strong CP problem: the axion. Without entering the debate on the legitimacy of its existence, I shall describe the experimental programmes aiming at its discovery.

Finally, following comments appearing in the Particle Data Group, I shall raise a few questions about the photon and gluon masses, from an empirical and experimental point of view, since no real theoretical motivation for a non-zero mass exists for them.

THE Z^0 MASS

Several years ago, M_Z was promoted to the status of a basic entry in the SM. As is well known, the SM Lagrangian has three fundamental quantities, g, g' and v, the $SU(2)_L$ and $U(1)$ coupling constants and the Higgs vacuum expectation value (v.e.v.) respectively. At tree level, these are sufficient to compute any observable in the SM. It is, however, natural to substitute for these abstract quantities three very accurately measured ones. The usual choice is:

$$\alpha(M_Z)\,, \quad G_\mu \text{ and } M_Z\,.$$

I shall first focus on M_Z, keeping comments on the other two for the next chapter.

M_Z has been measured at LEP during the last few years by a careful scan of the Z^0 resonance curve. This curve is a Breit–Wigner (BW), modified by well-understood

Masses of Fundamental Particles: Cargèse 1996
Edited by M. Lévy *et al.*, Plenum Press, New York, 1997

radiation effects. Whether the fitting curve is a normal BW, or a variant with an energy-dependent width, is a matter of choice which will lead to different but related values for the mass.

Given the large Z^0 hadronic cross-section and the absence of problems in isolating the signal, the statistical and detector systematic errors are very small. The only real problem is to get the absolute value of the c.m. energy of the scanning points, that we call peak, peak−2, peak+2 (the actual step was $\sim$ 1760 MeV in the '93 and '95 scans).

The LEP Energy Determination [1]

The LEP beam energy is obtained with high accuracy using the technique of resonant depolarization. As is well known, the build-up of transverse polarization in an e^+e^- ring is a natural phenomenon, linked to a slight spin dependence of spin flip synchrotron radiation emission (Sokolov–Ternov effect). One needs, however, to avoid and counteract the most offending resonant depolarization effects by choosing carefully the exact energy working point (this obviously influences the choice of the scan points) and by implementing the technique of harmonic spin matching [2]. A sufficient level of transverse polarization is thus achieved at the Z^0 energies (and even recently up to $E_b = 50$ GeV), and measured by backward Compton scattering.

The behaviour of the transverse polarization, considered here as a classical vector, is described by the BMT equation [3]. In a rotating frame following the particles, it processes ν times around the direction of the guiding magnetic field; ν, the spin tune, is $\nu \equiv \gamma a = \frac{E}{440.65\ \text{MeV}} = \text{integer} + \Delta\nu$ where $a \equiv (g-2)/2$.

Seen from an observer at rest, this becomes a precession at the frequency $\Delta\nu$. Therefore a local field oscillating at that frequency can resonantly flip the spin and the frequency measurement gives the absolute energy. The accuracy of the method is extremely good (200 keV). It may seem surprising since the energy width of the LEP beam is $\sim$ 55 MeV. However, the oscillations in energy of individual particles across that spread occur on a much shorter time-scale than the depolarization process; therefore what is actually measured is really the mean energy of the particles.

Unfortunately the depolarization measurement has only been commissioned in accelerator conditions without colliding beams, typically at the end of a physics fill. One then needs a model to follow the evolution of the energy during a fill and provide the instantaneous energy through a set of corrections to the absolute energy determined by the resonant depolarization method. One can group them in two main categories: those that affect the mean LEP energy of $e^\pm$ beams and those that cause differences between the two beams locally at experimental points.

Let me recall [4] that, for a given magnetic structure, the nominal energy E_0 is defined for a beam circulating on the central orbit C_o, in average through the centre of the quadrupoles. For a beam off the central orbit, i.e. not centred in the quads, the change in momentum is

$$\Delta p = \frac{p_o}{\alpha_C} \frac{\Delta C}{C_o}$$

where α_C is the momentum compaction factor ($1.86 \pm 0.02 \times 10^{-4}$), C_o the length of the closed orbit (26.658873 km, known at $\sim 10^{-8}$) and ΔC the variation of orbit length. The amplification factor $1/\alpha_C$, related to the strong focusing mechanism, explains why such very small distortions may have observable effects on the beam energy.

On the other hand, the horizontal displacements of the beam in the arcs, accurately measured by appropriate beam-orbit monitors, provide a parametrization of these energy variations.

In 1995 two of 3600 LEP dipoles were equipped with NMR probes capable of monitoring the magnetic field behaviour between the RD measurements. These allowed for instance to (re)-discover the phenomenon of vagabond currents, due to the electric trains on the surface, and causing a drift in the magnetic field through a monodirectional walk of the magnets along their hysteresis curve.

This recently discovered effect is to be added to the list of previously understood causes of drift, either on a very long time basis, like the influence of the level of water in the Lake of Geneva, or on a daily or fill basis, like the tide effect and the magnet temperature effect.

A parametrization of the rise of energy, as a function of the time in the fill, time of day and ring temperature, has thus been developed to reproduce the 1995 data. This model can be applied to the datasets of previous years in a coherent way.

Besides the determination of the LEP mean energy one has also to evaluate the corrections to the c.m. energy which depend on the location around the ring and the interaction point. These are corrections linked to the RF system and requiring its detailed modelling, and to the dispersion effects at the collision point, well-monitored and minimized during the scan. These careful procedures led to a substantial improvement in the final precision obtainable on the Z^0 parameters. Details can be found in Ref. [5]. At the moment the common LEP errors are estimated to be $\Delta M_Z = \pm 1.5$ MeV and $\Delta\Gamma_Z = \pm 1.7$ MeV, and the result is

$$M_Z = 91.1865 \pm 0.0020 \text{ GeV} ,$$

a 20 ppm measurement.

REMARKS ON THE OTHER SM INPUTS

The muon decay constant G_μ is also known at the level of 20 ppm:

$$G_\mu = 1.16639\,(2)\,10^{-5} \text{ GeV}^{-2} .$$

The fine structure constant α is extremely accurately known at zero energy (Thomson limit). What matters here, however, is the value it takes at the scale M_Z, $\alpha(M_Z)$, and unfortunately the accuracy on $\alpha(0)$ is spoilt by the uncertainty in the evolution of the coupling with energy.

At the moment $\delta(\Delta\alpha) = \pm 0.0007$ [6]. This uncertainty is linked to the imperfect knowledge of the $e^+e^- \to$ hadrons total cross-section at low energies, from threshold to the Υ region. One can therefore conceive the possibility of an improvement, although practically such a coordinated programme at several low-energy machines is difficult to set up.

The tree-level relation giving the W mass in the SM is

$$M_W^2 \left(1 - \frac{M_W^2}{M_Z^2}\right) = \frac{\pi}{\sqrt{2}} \frac{\alpha}{G_\mu} .$$

The uncertainty on α corresponds to

$$\Delta M_W = \pm 14 \text{ MeV} .$$

In terms of $\sin^2\theta_W$, it represents:

$$\delta \sin^2\theta_W = \pm 0.00023 .$$

Since the present experimental uncertainty on $\sin\theta_W$ is

$$\delta \sin^2\theta_W = \pm 0.00024 \ ,$$

one sees that the $\Delta\alpha$ problem is on the verge of becoming the leading uncertainty for some sectors of the SM tests.

The Top Mass Determination [7]

While the three quoted basic inputs are enough to compute all other SM observables at tree level, the knowledge of the top mass becomes crucial at the loop level, since the virtual effects due to this heavy quark dominate all others and must therefore be accurately evaluated and subtracted.

From LEP and other e.w. measurements, the top mass, within the SM and for a Higgs mass of 300 GeV, is extracted as:

$$m_t = 177 \pm 7^{+16}_{-19} \text{ GeV} \ ,$$

where the second uncertainty corresponds to the possible excursion of the Higgs mass (between the LEP present limit and 1 TeV).

From Fermilab experiments, the combined value for the direct determination is:

$$m_t = 175 \pm 6 \text{ GeV} \ .$$

Fermilab, with the new injector (> 1999), should reach $\Delta m_t = \pm 2$–3 GeV. The LHC experiments quote ±3 GeV, and may reach ±1–2 GeV.

One would need a linear collider to get, by a threshold measurement, $\Delta m_t = \pm 0.5$ GeV. To guide the reader, the sensitivities of the various observables to physics inputs are summarized in Table 1 [8] which shows their respective merits and limitations.

M_W MEASUREMENT

A basic observable of the SM, M_W is, at the loop level, predicted as [9]:

$$M_W^2\left(1-\frac{M_W^2}{M_Z^2}\right) \equiv \frac{\pi\alpha}{\sqrt{2}G_\mu}\,\frac{1}{1-\Delta r} \ .$$

The correction factor can be expressed as:

$$1-\Delta r = (1-\Delta\alpha)\left(1+\frac{c_W^2}{s_W^2}\Delta\bar{\rho}\right) - (\Delta r)_{\text{remainder}}$$

where $\Delta\alpha \equiv -\text{Re}\,\hat{\Pi}^\gamma(M_Z^2) = 0.0593 \pm 0.0007$ is the already quoted evolution of α, while

$$\Delta\bar{\rho} = N_C \frac{\alpha}{16\pi s_W^2 c_W^2}\,\frac{m_t^2}{M_Z^2}$$

and

$$(\Delta r)^{\text{top}}_{\text{remainder}} = -\frac{\alpha}{4\pi s_W^2}\left(\frac{c_W^2}{s_W^2}-\frac{1}{3}\right)\log\frac{m_t}{M_Z} + \ldots$$

$$(\Delta r)^{\text{Higgs}}_{\text{remainder}} \simeq \frac{\alpha}{16\pi s_W^2}\,\frac{11}{3}\left(\log\frac{M_H^2}{M_W^2}-\frac{5}{6}\right)$$

express the virtual effects of the top and the Higgs boson. The sensitivity of M_W to the various inputs appears numerically in Table 1.

Table 1. Summary of the present experimental errors, physics sensitivity, and theoretical uncertainties in SM predictions for the main LEP observables. Courtesy of M. Martinez.

Observable	Value	Error	Sensitivity to:			Theory uncertainties in MSM				
	Value	Error	M_t (GeV)	M_H (GeV)	α_s	$\Delta\alpha$	$\Delta\rho_H$	$\Delta\delta_{vb}(m_b)$	Higher orders	Techn.
	(Warsaw 1996)		175 ± 6	60–1000	± 0.003	128.89 (9)		4.7 ± 0.3		
Γ_z (MeV)	2494.6	2.7	+1.4	−9.6	+1.7	0.7	0.5	0.2	0.3	0.1
Γ_ℓ (MeV)	83.91	0.11	+0.06	−0.26	0.02	–	0.02	–	0.01	0.003
$\sigma_{had}^{peak,0}$ (pb)	41508	56	+3	8	−16	2	–	2	–	2
$R_\ell \times 10^3$	20778	29	−1.8	−29	+21	4	1	2	1	4
$\sin^2\theta_w^{eff} \times 10^4$	2316.5	2.4	−2	+16	0.05	2.3	0.5	–	0.8	1.0
$R_b \times 10^4$	2178	11	−2	+0.7	0.05	0.2	–	0.8	0.1	1
M_W (MeV)	80356	125	+35	−204	1	14	9	–	5	2

Within the SM one can infer a value for M_W from LEP e.w. measurement [10]: $M_W^{LEPI} = 80.292 \pm 0.048 \pm^{0.010}_{0.018}$.

This same is true from $\nu\mathcal{N}$ data: $M_W^{\nu\mathcal{N}} = 80.22 \pm 0.21$.

For direct M_W measurements, the two machines which are giving information at present are the Fermilab Tevatron Collider and LEP 200 at CERN.

Hadron Collider [11]

Access to the W mass is obtained there exclusively through the $W \to \ell\nu$ leptonic decay. Since the longitudinal momentum of the ν is unknown, one has to consider transverse quantities. Instead of the invariant mass which cannot be reconstructed, one focuses on the transverse mass

$$M_T^2 = 2E_T^e E_T^\nu (1 - \cos\varphi^{e\nu}) .$$

This quantity exhibits a Jacobian peak which contains the mass information.

Both M_T and p_T^e are invariant under longitudinal boosts. $M_T = 2p_T^e$ if $p_T^W = 0$. Under a transverse boost, $p_T^e|_{max}$ depends on p_T^W, while $M_T|_{max}$ is still M_W: M_T is thus less sensitive to W transverse motion.

The downside of using M_T is that the required $p_T^\nu \equiv \not{E}_T$ has to be determined as:

$$\sum_i \vec{p}_{T_i} = -\vec{p}_T^{\,e} - \vec{p}_T^{\,rec} - \vec{u}(\mathcal{L}) .$$

where $\vec{u}$ reflects the underlying momentum flow which turns out to be a quantity depending on the number of interactions per crossing.

The careful determinations of the energy scale, for CDF and D^0, and of the momentum scale, for CDF, are described in Ref. [11] and use J/ψ, Z, Y, W and π^0 decay modes.

The effect of the choice of structure functions needed to model the measurement is controlled by reproducing the observed W charge asymmetry.

The underlying momentum flow increases M_T by a quantity varying between 70 MeV (run 1a) and 270 MeV, as predicted for run II in 1999 and beyond, foreseen with a luminosity 20 times higher, but six times more bunches.

Table 2 gives as an example the present uncertainties in CDF and Table 3 summarizes the situation.

M_W at LEP 200

This is a well-documented subject [12]. I shall focus on a few new developments, and since LEP has recently crossed the W-pair threshold, I shall give a status report on the first and still modest results on M_W.

The channel of interest is $e^+e^- \to W^+W^-$, studied theoretically in great detail but only recently observed. LEP 200, which has already achieved 172 GeV c.m. this autumn, will reach 186–188 GeV in 1997, and 192 GeV or so a year later. Higher energies, up to 200–205 GeV, are conceivable, but this is not an approved programme. Although this luminosity figure may now look somewhat high, we shall keep for the integrated luminosity per experiment the traditional quantum of $\sim$ 500 pb^{-1}. Since most of the foreseeable results are statistically dominated, it is easy to scale down at will the quoted accuracies.

Ignoring the non-competitive method resting on the lepton end-point energy of W leptonic decay, I shall focus on the reconstruction method, underlining some possible problems with the totally hadronic channel, and on a recently proposed threshold cross-section measurement [13]: the latter has already been performed in summer with a modest integrated luminosity.

Table 2. Components of the uncertainty in the CDF W mass measurement from run 1a, in MeV/c^2.

Source	$W \to e\nu$	$W \to \mu\nu$	Common
Statistics	145	205	–
Lepton scale	120	50	50
Momentum	50	50	50
E/p (energy)	110	–	–
Other	130	120	90
Lepton Res.	80	60	–
Recoil model	60	60	60
e/μ towers/ID	25	10	5
Trigger	0	25	–
Background	10	25	–
W_{p_T}	45	45	25
PDF	50	50	50
QED	20	20	20
QCD	20	20	20
Fitting	10	10	–
Overall (180)	230	240	100

Table 3. Status of m(W) measurements outside LEP.

Experiment	m(W) (MeV)	Stat.	Syst.	Scale	Total error
CDF 90	79 910	292	227		390
UA2 92	80 360	220	297		370
CDF 95 (1A)	80 410	118	136		180
*D0 96 (1A)	80 350	140	160	160	270
*D0 96 (1B)	80 380	70	130	80	170
νN WA	80 220				210

*Preliminary

The Direct Reconstruction Method: We can distinguish three categories of W-pair events:

- all hadronic decay (a)
- hadronic–leptonic decay (b)
- all leptonic decay (c).

Invariant masses are reconstructed from the decay products using the energy-momentum conservation and the equality of the two masses as constraints. Neutrinos are undetected and the energy of taus is not measured. One should therefore ignore $W \to \tau\nu_\tau$ decays, as well as purely leptonic events. On the other hand, jet directions and e, μ energy and directions are, in principle, correctly measured.

(a) For hadronic events the procedure is the following:

- One rejects events with too small total energy or too much missing momentum.
- One reconstructs events as 4 jets: one way is to maximize a generalization of thrust to 4-jet axes, the 4-thrust T4. One can thus eliminate broad events associated with hard gluon radiation.
- Associating the 4 jets 2 by 2, one performs a constrained fit for each of the 3 combinations: the jet energies and momenta are corrected using the available constraints.
- If $E_{\rm min}$ is the energy of the lowest energy jet and $\theta_{\rm min}$ is the smallest interjet angle, one can for instance reject events with

$$E_{\rm min} \times \theta_{\rm min} < 10 \text{ GeV} \cdot \text{rad} ,$$

which are likely to correspond to QCD background.

- One selects the favoured combination of jets on the basis of χ^2.

Figure 1 gives the results for the mass distributions before and after fit, showing the improvement of the resolution. The background comes mostly from bad combinations, but also from other channels. The reconstruction efficiency is typically 2/3. The statistical error, for 500 pb^{-1}, is $\sim \pm 70$ MeV/c^2.

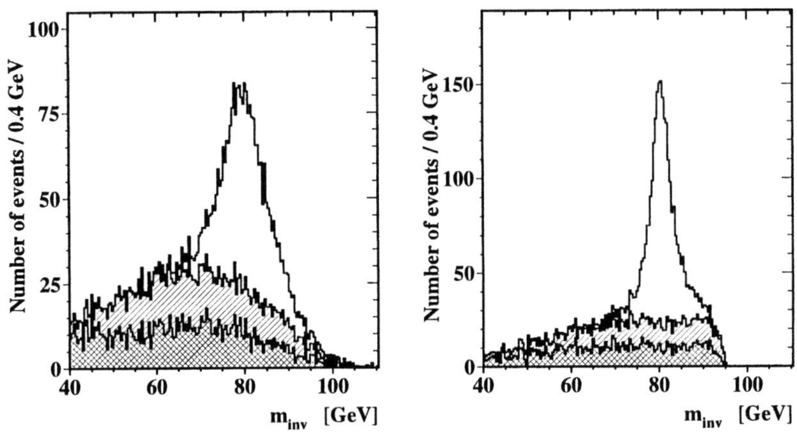

Figure 1. The 2-jet invariant mass from W-pair 4-jet decay before and after a kinematical 5C fit (Monte Carlo at 192 GeV).

From the beam energy constraint an average mass shift of +400 MeV (mass found minus mass generated) is expected. This fact, due mostly to the effect of initial-state radiation (ISR), exhibits the main difficulty of the reconstruction method.

To check biases in this method one can run the Monte Carlo without initial-state radiation. The net result is that $M_{\rm W}$ is reconstructed $\sim$ 130 MeV/c^2 lower than the generated value.

As for the effects of ISR, they can be decomposed in two components:

i) the photon, if detected, is assigned to a jet: this leads to a mismeasurement of the jet direction and decreases $M_{\rm W}$.

ii) if the photon is not detected, the most frequent case, the fit assumes a nominal energy, while the effective energy is in fact lowered by the radiation of E_γ; this leads to a positive offset

$$\Delta M = E_\gamma \frac{M_{\rm W}}{\sqrt{s}} .$$

At $\sqrt{s} = 175$ GeV, the mean radiated energy is 1.2 GeV and $\Delta M = +550$ MeV/c^2.

These various contributions lead to the expected shift. A fit to the mass distribution gives more weight to the peak, yielding a shift of only ~ 200 MeV at this energy.

(b) Hadronic-leptonic events have been studied similarly. The neutrino momentum is defined as:

$$P_\nu = -P_{\text{miss}} \ .$$

Angles are kept as they are measured. Jet energies come from a fit with the constraint

$$(E_{\text{J1}} + E_{\text{J2}} + E_{\text{l}} - \sqrt{s})^2 - E_\nu^2 = 0 \ .$$

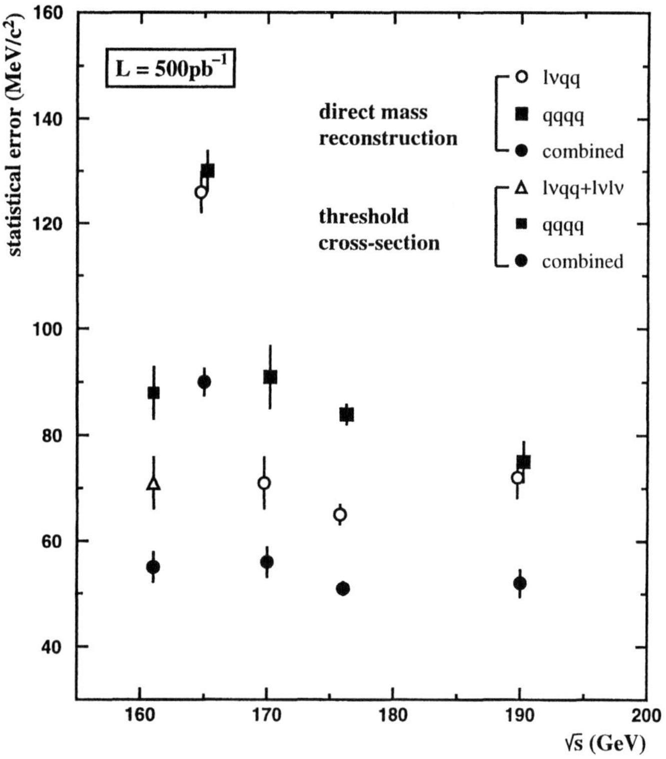

Figure 2. The expected statistical error of the LEP M_{W} measurement.

The masses M_{q1q2} and $M_{\text{e}\nu}$ are calculated separately and their average is taken. Only e and μ are used in this exercise.

The statistical error on the mass, from that method, is 72 MeV/c^2 for 500 pb^{-1}.

Combining hadronic events and hadronic-leptonic ones, one finds for $\sqrt{s} = 175$ GeV and 500 pb^{-1} a statistical error of ± 50 MeV/c^2.

The dependence of the statistical error on E_{beam} is not strong, as Fig. 2 shows.

The choice of energy has thus to be made by considering other physics topics in addition.

Systematic Errors of the Reconstruction Method:

The Beam Energy. An accurate knowledge of the beam energy is necessary since a procedure of constraint is used. At LEP I this is possible through the already described resonant depolarization method. Unfortunately it is likely that polarization will not be obtained at LEP 200, although it was recently observed at $E_b = 50$ GeV, as we said. The expected energy spread of the beam is $\Delta E \simeq 150$ MeV, a value dangerously comparable to the interval of 440 MeV between polarization resonances. From magnetic measurements only, one can expect an uncertainty of $\sim \pm 26$ MeV.

With a calibration near 50 GeV through resonant depolarization and an extrapolation to 90 GeV using the flux loop method one can obtain $\sim \pm 17$ MeV.

The best way would be to install a H_2 gas jet target in the machine and detect Moeller scattering of the beam electrons by the gas jet electrons. From the measurement of the minimum opening angle between the two outgoing electrons, one can expect an accuracy of $\sim \pm 2$ MeV. This demanding possibility has not been retained up to now.

The Radiative Correction. Much effort has been devoted to the study of radiative corrections and their implementation.

For instance, the Dubna–Zeuthen group has now written a program to calculate initial-state radiative corrections, taking into account the finite width of the W and higher order terms. The biggest uncertainty in the calculation comes from the non inclusion of the $0(\alpha^2)$ non universal t-channel contribution. This is evaluated to be:

$$\Delta M_W \sim \pm 5 \text{ MeV}/c^2 .$$

However, an ideal Monte Carlo, needed to estimate the shift, should include a correct description of initial and final-state radiation, in particular:

- the correct angular and energy distribution of the radiated photon,
- the correlation between the angle and energy of the photon,
- the correlation between the radiated photon and the W directions,
- higher order effects.

No Monte Carlo has all these effects implemented yet: to take this into account, the systematic error resulting from ISR and FSR has been increased to:

$$10 \text{ MeV}/c^2 .$$

Control of Systematics. Once the mass is reconstructed from real data, the result will have to be shifted by the offset found in the Monte Carlo. The problem is to check the Monte Carlo.

A method using mixed Lorentz-boosted Z^0 events, from real data, to simulate W pairs, has been proposed by DELPHI: the mass shift in the reconstructed Z^0 mass is: -140 MeV$/c^2$. The similarity of this result with the shift found in the Monte Carlo generated W^+W^- events means that one can trust the shift prediction to ~ 10 MeV$/c^2$.

The exact knowledge of detector performances is not critical. Results on hadronic events are essentially independent of the energy and momentum resolution of the detectors, because of the beam energy constraint. Systematic shifts in energy will not lead to more than a 10 MeV$/c^2$ systematic error. Background effects are not expected to contribute appreciably in any category of events. Tables 4 and 5 summarize the systematic errors.

Table 4. Estimated systematic errors on M_W per experiment in MeV. Colour reconnection and Bose–Einstein effects are not included in the total.

Source	$W^+W^- \to q\bar{q}q\bar{q}$	$W^+W^- \to q\bar{q}\ell\nu$
E_{beam}	12	12
ISR	10	10
Fragmentation	16	16
Backgrounds	12	6
Calibration	10	10
Monte Carlo statistics	10	10
Mass fit	10	10
Jet assignment	5	–
Interconnection	?	–
Total	31	29

Table 5. Estimated total error on M_W, in MeV, which could be obtained by combining four experiments each with an integrated luminosity of 500 pb^{-1} at $\sqrt{s} = 175$ GeV. Colour reconnection and Bose–Einstein effects are not included in the systematic error.

Source	$W^+W^- \to q\bar{q}q\bar{q}$	$W^+W^- \to q\bar{q}\ell\nu$	Combined
Statistical	36	36	25
Common systematic	25	23	23
Uncorr. systematic	9	9	6
Total	45	44	34

Potential Problems in the Purely Hadronic Channel: In the hadronic decay mode $e^+e^- \to W^+W^- \to q_1\bar{q}_2q_3\bar{q}_4$, QCD interference effects can mix up the two colour singlets and produce hadrons that cannot be uniquely assigned to either of W^+ and W^- [14]. In plain language, the two W's could speak to each other. This obviously concerns soft particles of energy $\omega \lesssim \Gamma_W$.

Perturbative colour rearrangement (CR) can be evaluated and is found to be a small effect. However, the non-perturbative one is a more difficult problem (see J/ψ production in B decay) and requires model building, generally Lund string fragmentation. The models differ by the assumptions on the nature of the string and on the reconnection probability. We refer to Ref. [15] for a thorough review. The annoying fact is that: 1) the effect may be large, i.e. larger than the previously quoted systematic errors; 2) there does not seem to be other visible effects whose measurement could give information on the occurrence and magnitude of CR effects.

Since W decay vertices are much closer to each other than typical hadronization distances, there exists also a cross talk between W's through identical bosons [16].

A correct theory would have to use amplitude-based quantum mechanics. This is not the case for existing models of hadronization which are formulated in a probabilistic language. Therefore some 'cooking' is the usual practice to mimic BE effects, involving a local reweighting of allowed particle momentum configurations. The impossibility to conserve both E and $\vec{p}$ in the pairwise shifts calls for ad hoc rescaling procedures. Spurious mass shifts thus introduced must be separated from true BE ones. From BE interference large effects on M_W are also found to be possible.

Work is going on in this sector. In the present situation, there clearly exists a doubt about the validity of the systematic errors quoted in the preceding chapter for the totally hadronic channel. On the contrary, the mixed decay channels are free from such uncertainties.

The Threshold Measurement [13]**:** The idea is simply to perform an accurate measurement of the W pair cross-section at an energy point not far above threshold, which offers an optimum compromise between the sensitivity to the W mass value (calling for low $\sqrt{s}$) and the magnitude of the cross-section, hence the statistical accuracy.

This energy is found to be 161 GeV (Fig. 3).

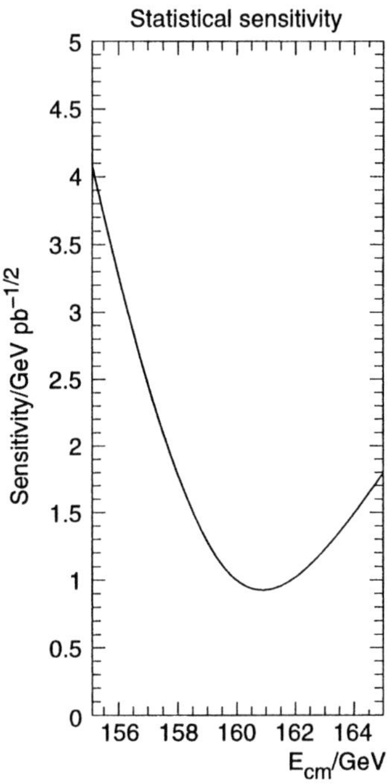

Figure 3. Sensitivity of the threshold method versus $E_{\rm cm}$.

The systematic errors on the mass come from the uncertainty on σ, L and the efficiency ϵ for the WW channel, and also from the uncertainty on the W width. If $\Delta E_{\rm b}(\cong \Delta M)$ is known to $\sim$ 17 MeV, the overall systematic will be $\Delta M_{\rm syst} \simeq 36$ MeV. For the time being the knowledge of $E_{\rm b}$ is much poorer.

As for the statistical error, if each of the four experiments receives an integrated luminosity L_0, it will be:

$$\Delta M_{\rm stat} = 67 \text{ MeV} \left[\frac{L_0}{100 \text{ pb}^{-1}}\right]^{-1/2} .$$

Figure 2 gives an overview of the accuracy on $M_{\rm W}$ expected from LEP 200. One sees that the region $\sim$ 161–170 GeV, between the threshold method and the start of a performant reconstruction one, is useless for that observable. Since the triple gauge boson coupling measurement cries for energy, it is clear that W physics, besides the threshold measurement, should be performed at maximal E.

Present Situation [17]**:** Since the time of these lectures, which closely followed the advent of the first W pairs at LEP, more data were obtained at threshold (161 GeV) and at 172 GeV. The LEP M_W determination from the threshold method is given by Fig. 4. From the second exposure at 172 GeV, preliminary mass plots of the reconstruction method were obtained (Fig. 5). The errors, statistically dominated in both cases, are similar ($\Delta M \simeq \pm 0.2$ GeV for all four experiments together). The LEP energy determination uncertainty is still of secondary importance, given the low statistics.

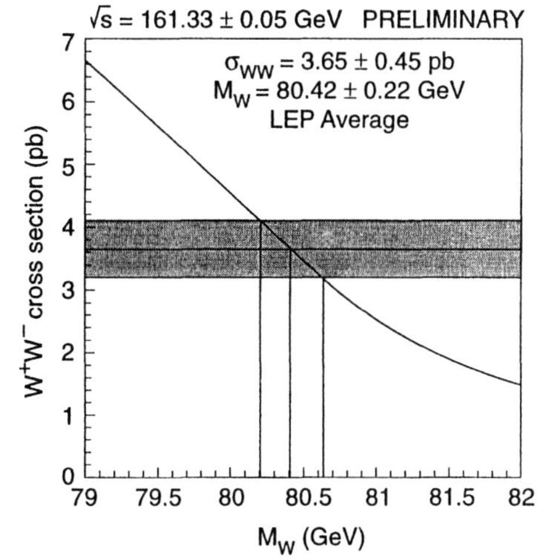

Figure 4. The LEP W mass determination from the threshold method.

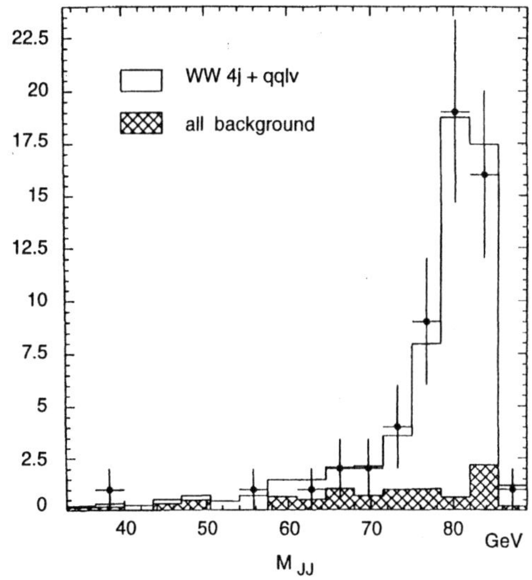

Figure 5. The 2-jet mass spectrum from W pairs, 4-jet and mixed decay modes, in the DELPHI experiment at 172 GeV.

Higgs Phenomenology

If, in spite of its shortcomings, one tries to stick to the SM it is natural to study the bounds this model announces for the Higgs mass. In particular, most relevant for LEP 200 is the lower limit that one can deduce from the requirement of stability or metastability of the vacuum [17]. This limit depends on the scale at which the perturbative character of the SM is expected to break down. If the model is supposed to be valid up to the GUT or Planck scale, one finds: $m_{\rm h} \geq 135$ GeV. If, on the contrary, the model breaks down at low energy (1–100 TeV) the limit is lower and the boson could be in the region accessible to LEP 200: however, in such a case, the upper limit is $\sim$ 600 GeV or so, and there is no particular reason to expect the Higgs boson in the 100 GeV region, nor any strong argument to try to gain 10–20 GeV of accessibility there, at any machine.

The situation is totally opposite in the case of SUSY where the lightest boson $\rm h^0$ is bound to exist in this region. The tree-level upper bound on its mass is raised by loop corrections which depend on the fourth power of $m_{\rm t}$ and, logarithmically, on the masses in the stop sector and therefore on the mixing which determines them

$$m_{\rm h}^2 \leq m_{\rm Z}^2 |\cos 2\beta|^2 + F\left(m_{\rm t}^4,\, \ln \frac{m_{\tilde{\rm t}}}{m_{\rm t}},\, ...\right) .$$

In the last few years, the value of $m_{\rm t}$ has been sharpened and the computation of $m_{\rm h}$ in the MSSM has been refined. One can summarize the results by Fig. 6 [18] which shows four extreme cases: large and small $\operatorname{tg}\beta$ (i.e. IR fixed point), large and small mixing. It is thus possible to tell which scenarios will be covered by a given mass reach. For instance reaching $m_{\rm h} \cong 110$ GeV would allow one to cover the small $\operatorname{tg}\beta$ cases, whatever the mixing, and the large $\operatorname{tg}\beta$ case with small mixing, while the large $\operatorname{tg}\beta$ situation with large mixing stays partly uncovered.

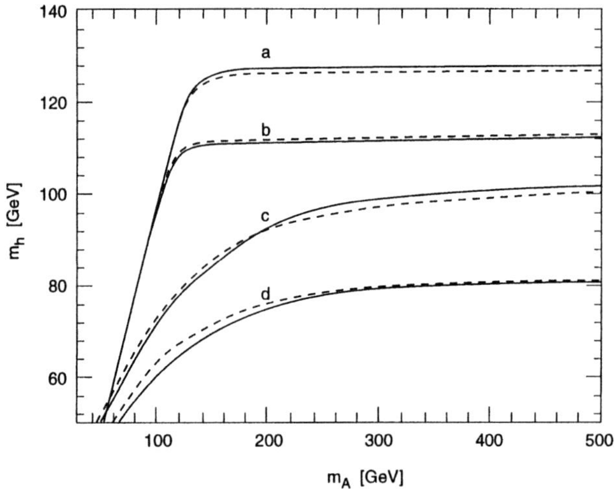

Figure 6. The lightest CP-even Higgs mass as a function of the CP-odd Higgs mass (a) and (b): large $\operatorname{tg}\beta$, with large and zero squark mixing, respectively. (c), (d) = small $\operatorname{tg}\beta$ (IR fixed point), with large and zero squark mixing, respectively.

We recall that, as soon as m_A is beyond $\sim$ 100 GeV, the h^0 boson is essentially SM-like, both for its bremsstrahlung production mode and in its decay. So the description of the SM boson search which will follow is actually done having the h^0 in mind.

A last point is the slow variation of m_h in the upper right part of the $\text{tg}\,\beta - m_A$ plane: one understands then that a small change in $\sqrt{s}$ can lead to a large one in the coverage of the plane.

Search for the SM Higgs Boson

I shall limit myself to a brief status report of the present situation, and a survey of future prospects.

Production and Decay: The features of Higgs boson production and decay are dominated by its property of coupling preferentially with the highest-mass objects available.

In e^+e^- collisions the SM Higgs production occurs through bremsstrahlung and fusion processes, the latter dominating at high energies. In hadronic collisions, the evolution of gg and qq processes when $m_H/\sqrt{s}$ increases is shown by Fig. 7.

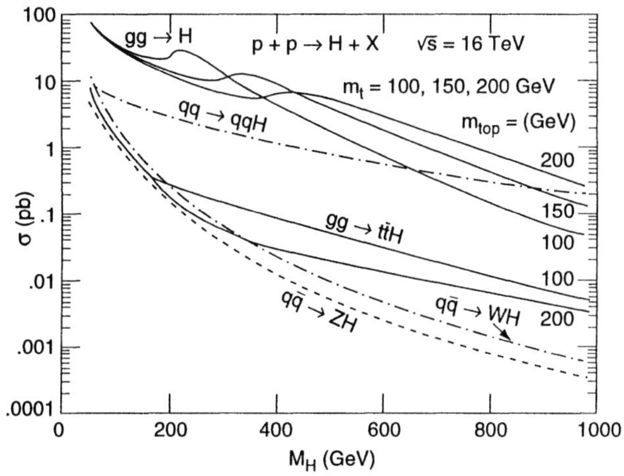

Figure 7. Various components of the Higgs hadronic production cross-section.

The expected decay modes of the SM Higgs boson as a function of its mass are well known: dominance of $b\bar{b}$ up to $\sim$ 140 GeV, and then of intermediate boson pairs.

The 2γ decay mode, vital for the exploration of the intermediate mass region (90–140 GeV) at hadron colliders, occurs through loop diagrams and has a small branching ratio ($\sim 10^{-3}$).

Higgs Search at LEP: Searching for a SM-like Higgs boson has been an important activity at LEP I [19].

LEP I. A low-mass boson was rapidly excluded in all foreseeable decay modes. Focusing on the highest mass region one can, for LEP I, draw the following conclusions:

- the present overall limit is $\sim$ 65 GeV;
- one is not far from reaching saturation and only a few GeV more are to be expected;

- while the $H\nu_e\bar{\nu}_e$ channel is still essentially background-free, the Hl^+l^- channel starts being populated and thus weakened by the expected background from four-fermions;
- the 4-jet channel will remain inaccessible for the SM boson search at LEP I in spite of the progress made in b-tagging;
- the alternative mode $e^+e^- \to H\gamma$ will also stay out of reach in the SM frame. But a level of about 10^{-5} for the branching ratio has been reached and this sets limits on various still possible anomalous couplings in the Higgs sector.

LEP 200. Figure 8 gives the cross-section of the Higgs production process versus energy and shows that above $m_h \sim 55$ GeV it is more profitable to search for it at LEP 200 [20] than on the Z^0 since the cross-section is bigger and the background smaller.

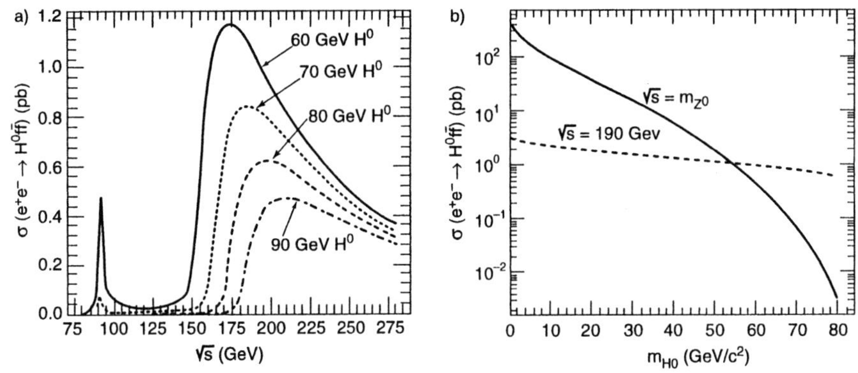

Figure 8. The Higgs boson production cross-section versus $\sqrt{s}$ (a) and versus m_H for two energies (b).

The sharp threshold occurs at $\sim m_Z + m_h$ and one needs ~ 10 GeV c.m. more to get close to the maximum of the cross-section: hence the old rule of thumb giving the reach:

$$m_h \sim \sqrt{s} - 100 \text{ GeV} .$$

For the relevant LEP 200 mass region the Higgs boson decays mostly into $b\bar{b}$ and this explains the overall importance of b-tagging in its search. The $b\bar{b}$ decay is a tree-level process, unlike the $\gamma\gamma$ mode, and its branching ratio is unambiguously calculable. The Z^0 is observed in the usual ways: $q\bar{q}$, $\nu\bar{\nu}$ and l^+l^-.

Although it is still modest, the fusion channel $e^+e^- \to \nu_e\bar{\nu}_e H$ does exist and interferes with the main one $ee \to HZ(Z \to \nu_e\bar{\nu}_e)$. This channel is not bound to the kinematical limit expressed above: the possibility that it allows the gain of a few GeV in mass reach is under study.

The background channels are mostly $q\bar{q}\gamma$ (half being a radiative return to the Z region), WW, and ZZ. Combinations of kinematics and b-tagging allow them to be reduced to manageable levels and all HZ modes turn out to be exploitable. The most difficult case is when $m_H \approx m_Z$, since the ZZ final state is then kinematically indistinguishable and one rests fully on the b tag.

The b-tag efficiency and purity needed have already been achieved in the present LEP experiments. It was shown that at LEP 200 one can keep $\sim 50\%$ efficiency to the HZ channel and reject WW by ~ 70 since there are no b's in W decay.

Even in the most difficult case, the signal/background figures are ≥ 1. Figure 9 gives for the three energies the exclusion and discovery curves which have the behaviour expected from our previous considerations. To interpret them in terms of running time, one must make tentative assumptions about the experimental procedure at LEP 200: if it is possible to combine the four experiments, as I believe, the luminosity needed per experiment to answer the Higgs question is rather modest and corresponds to at most one year of running.

The rule of thumb presented above is fairly well verified for discovery: for exclusion at 95% C.L. one still gains ~ 5 GeV or so in the mass reach.

On the other hand, the Tevatron, even after its upgrade, would not be able to produce an SM-like Higgs boson in this mass range at a sufficient rate. At least an order of magnitude increase in luminosity — or in the cross-section — would be needed.

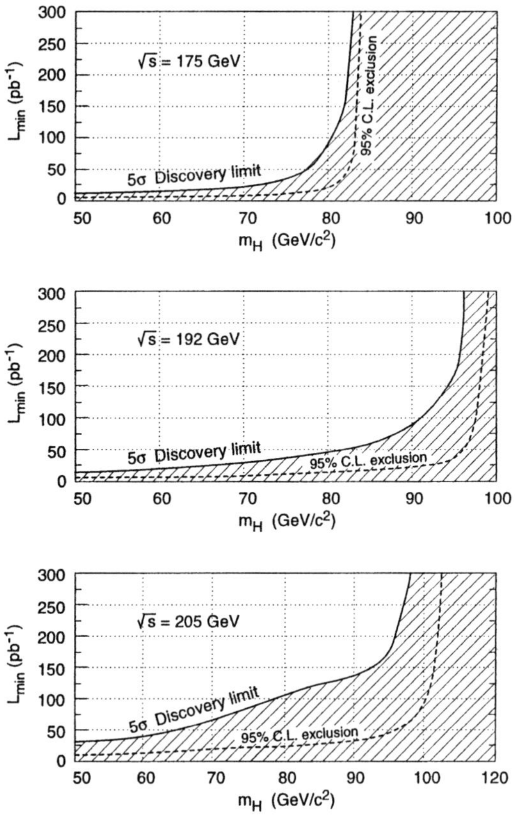

Figure 9. The exclusion and discovery curves for the SM Higgs boson at LEP 200.

Prospects at Linear Colliders: The problem is quite similar to the one at LEP 200 as long as the bremsstrahlung process dominates: in particular b-tagging will be very useful. It should, however, be remembered that for a given Higgs mass the Higgs cross-section slowly decreases, whilst various fusion backgrounds rapidly increase with $\sqrt{s}$: the optimal mode for Higgs study, once it is discovered, is therefore to bring the energy of the machine down to $\sim m_{\rm H} + 110$ GeV.

For $m_{\rm H} \sim 100$–150 GeV, at the NLC ($\sqrt{s} \sim 500$ GeV) the fusion mechanism takes over. This machine will allow the discovery of a Higgs boson up to ~ 350 GeV, but a 1.5 TeV machine is needed to cover the full mass domain. Various studies [21] have demonstrated that the visibility of a Higgs boson at an LC is guaranteed when the c.m. energy is sufficient.

As previously mentioned, an e^+e^- (or e^-e^-) collider can, in principle, be turned into a $\gamma\gamma$ collider: this would be an ideal machine to study (but not to discover) the Higgs boson as we shall see below.

Prospects at the LHC: Future large hadron colliders have a large potential for the exploration of the Higgs sector [22].

The production cross-section shown in Fig. 7 is relatively comfor up to $m_H \sim$ 800 GeV.

When the Higgs boson decays substantially into a pair of Z's and is abundantly produced, namely for $140 \lesssim m_H \lesssim 800$ GeV, the search, through four-lepton final states, is relatively straightforward.

On the other hand, the extreme regions below $\sim$ 140 GeV (a) and above $\sim$ 800 GeV (b) are certainly quite difficult to explore.

In (a) one way is to rely on the $H \rightarrow \gamma\gamma$ mode in spite of its small branching ratio. With an outstanding electromagnetic calorimeter, retaining its quality at full luminosity, the signal should be visible over the irreducible 2γ background spectrum: this assumes that the reducible background from π^0's or jets mimicking a photon, can be mastered, as well as the background from $Z^0 \rightarrow e^+e^-$, with the e confused with a γ, for masses around 90 GeV.

The possibility to observe the light Higgs boson in its dominant $b\bar{b}$ decay mode, when it is tagged by the presence of a W or a $t\bar{t}$ system, is considered as well, after the encouraging results of b identification by the CDF microvertex.

Above $\sim$ 800 GeV, various tricks such as those described in the case of a strongly interacting sector (central jet veto, forward jet tag) have to be used.

If reality conforms with Monte Carlo expectations, the hadron colliders, with several years at full luminosity, should solve the SM Higgs problem.

The MSSM Higgses

The previous section devoted to the SM Higgs describes as well the search for h^0 through the bremsstrahlung mechanism. However, in the case of SUSY, one can also exploit the Associated Production (AP) processes.

LEP 200: The relevant AP process is there: $e^+e^- \rightarrow A^0h^0$ [20].

The complementarity between the two mechanisms is well known, although the P wave AP has the handicap of a $(\text{velocity})^3$ factor. At large $\text{tg}\,\beta$ and modest m_A, the AP dominates, while it is the contrary at small $\text{tg}\,\beta$.

The AP process, leading to 4 b's, is particularly prone to b-tagging: this is welcome to allow the elimination of the much larger WW background, which is kinetically identical to the signal when $m_A \sim m_h \sim m_W$.

Let us recall that at LEP I both h and A are excluded up to 45 GeV. Searches for hA associated production in the 4b final state exploiting b-tagging were particularly efficient.

However, this result is obtained for a particular set of SUSY parameters corresponding to no or small mixing in the stop sector.

Turning now to the prospects for LEP 200, the exclusion/discovery domains in the $\text{tg}\,\beta - m_A$ plot are shown in Fig. 10. One distinguishes clearly the regions where the two search channels dominate. The coverage obtained through the bremsstrahlung channel depends critically, as we saw previously, on the available $\sqrt{s}$, and also, for a given m_{top}, on the level of mixing in the stop sector.

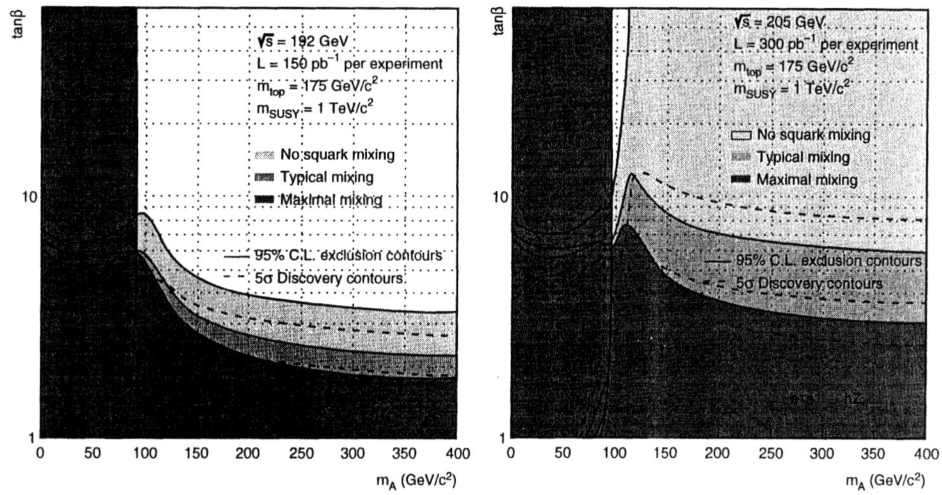

Figure 10. The exclusion/discovery domains in the $\text{tg}\,\beta - m_{\text{A}}$ plot at LEP 200 for two energies and various sets of parameters.

At $\sqrt{s} = 205$ GeV, small mixing and $1\,\text{fb}^{-1}$, the plane is nearly fully covered. This can be seen also in an alternative representation using the $(m_{\text{h}},\ \text{tg}\,\beta)$ plane.

One must remember that, in particular in the case of SUSY, one may be led to a situation where the Higgs bosons decay invisibly. This can happen if the $\chi^0\chi^0$ decay mode is open, or in various scenarios of R parity breaking.

Detecting this mode is possible at e^+e^- machines.

LHC: Here again the c.m. energy opens a large set of possibilities, described in the now familiar MSSM plot $\text{tg}\,\beta - m_{\text{A}}$. To simplify the presentation, let us consider separately the various channels. The LHC potential is summarized by Figs. 11(a)–11(d) [23].

One will search, as in the SM case, for a signal in the $\gamma\gamma$ mode. For SM-like couplings the reach is given by Fig. 11(a). The two large experiments are rather equivalent, and the apparent difference of coverage reflects the different assumptions made. A caveat: since both h production and $\text{h} \to \gamma\gamma$ decay are mediated by loops, one should in principle take into account the actual population of particles circulating in the loops, which depends on the actual SUSY scenario.

Another direct signal to be searched for is the 4-lepton one [Fig. 11(c)], which allows the coverage of the bottom region of the diagram. One may notice here that such a channel would reveal the existence of the H boson, while in the same region LEP 200 would give access to the h boson: this illustrates the complementarity between the information of different machines.

Other channels, like $\tau\tau$ or $\mu\mu$, can reveal the existence of Higgs bosons in different regions of the MSSM plane, above the lines shown in Fig. 11(b) and 11(d).

Putting together the various pieces of information one will essentially cover all the plane, with the possible exception of a 'hole' around $m_{\text{A}} \sim 100$–200 GeV, $\text{tg}\,\beta \sim 5$–10.

The size of the hole depends on the luminosity considered and on the actual physical parameters. For instance, a heavier top is favourable to the LHC potential of exploration (and unfavourable to the LEP 200 one as we noticed).

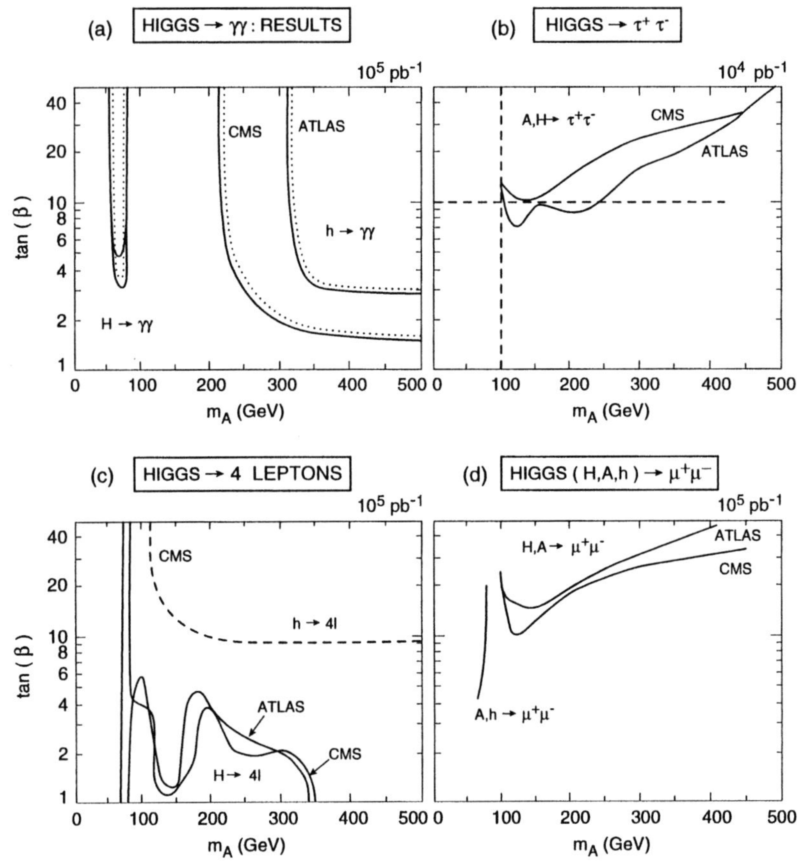

Figure 11. The discovery contours in the tg $\beta - m_{\mathrm{A}}$ plot at the LHC for four decay modes of Higgs bosons.

Linear Colliders: It is likely that a LC will be built after the bulk of the exploratory work by the LHC.

Here again the potential is large and the visibility of Higgses is guaranteed.

One can probably say that the roles of a LC will be:

1) To complement the exploratory work, in particular for special cases, like invisible modes of boson decay, which may be difficult to deal with at the LHC. Experience and simulation show that their observation should not be a problem at e^+e^- machines.
2) To distinguish between scenarios, in case of discovery, and to bring quantitative information.
   The first goal, if a single boson has been discovered by then, would be to decide whether one is dealing with SUSY or not by getting evidence for eventual partners, and/or by measuring its branching ratios with enough accuracy to draw

conclusions [24]. This is illustrated by Fig. 12 [25] which shows the domain in which such conclusions can be drawn.

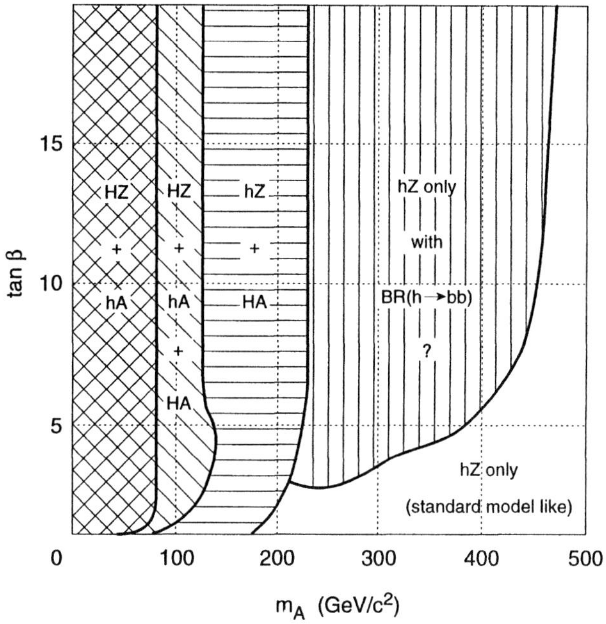

Figure 12. Regions of the tg $\beta - m_{\mathrm{A}}$ plot for which LEP 200 and a Linear Collider can distinguish SUSY from the SM.

THE AXION

The strong CP problem can be summarized as follows. To the QCD Lagrangian one should in principle add a term

$$\frac{\theta g^2}{32\pi^2} G^{\mathrm{a}}_{\mu\nu} \tilde{G}^{\mathrm{a}\mu\nu}$$

where G is the gluon tensor field and θ a parameter. Although this term can be written as a 4-divergence, and will not contribute in the perturbative mode, one can show that it plays a role through non-perturbative ones (instantons). Actually the physics of QCD depends upon θ through the combination

$$\bar{\theta} \equiv \theta - \arg \det m_{\mathrm{q}} \ ,$$

where m_{q} is the CKM matrix.

Unless $\bar{\theta} = 0$, QCD will violate C, P and CP. Experimentally the bound on the neutron electric dipole corresponds to $\bar{\theta} \lesssim 10^{-9}$.

Here lies the mystery. How is it that $\bar{\theta}$ which is the sum of two a priori unrelated factors, as in the formula above, can be fine-tuned to zero with such an accuracy?

Peccei and Quinn (PQ) [26] answered by postulating the existence of an extra global U(1) quasi-symmetry which:

– is a symmetry of the classical theory,

- is broken explicitly by non-perturbative QCD effects,
- is broken explicitly by the v.e.v. of a scalar field, the axion field

$$a(x) = v\alpha(x)$$

through which quarks get mass $m_j = K_j v e^{i\alpha}$. Then $\bar{\theta} = \theta - \arg(m_1 \dots m_n) = \theta - \arg(K_1 \dots K_n) - n\alpha$ becomes a function of the dynamical field α. One can show that the absolute minimum of the effective potential occurs for $\bar{\theta} = 0$.

For a delightful metaphoric introduction to the axion see Ref. [27].

At the time of PQ assumption it was believed that v was of the order of the electroweak scale. However, the corresponding boson did not show up. It was then realized that models with much larger scales could be built, leading to a very light and weakly coupled axion, called therefore the 'invisible' axion, which may have been produced in the early universe and appear as a relic.

Besides the past experiments, cosmological arguments set quite strong limits to the axion mass (Fig. 13) and leave only a window extending from 10^{-6} to 10^{-3} eV.

The only conceived experimental method to detect it is to induce its conversion into an electromagnetic microwave by the Primakoff effect in a static magnetic field (Fig. 14), and this is the way relic axions have been looked for up to now [28].

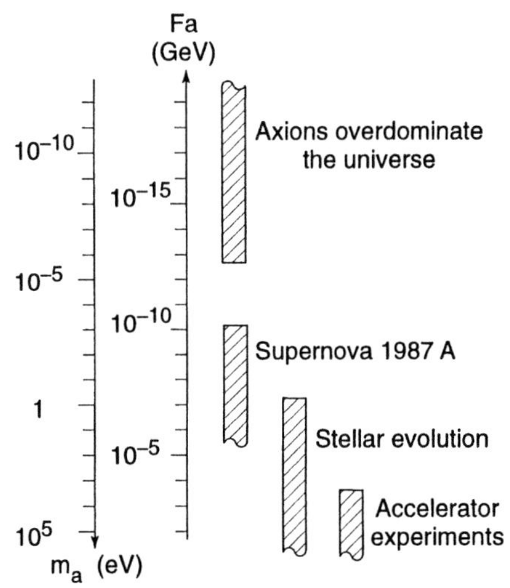

Figure 13. The present status of axion mass exclusion. 10^{-6} eV corresponds to a frequency of 2π (242 MHz).

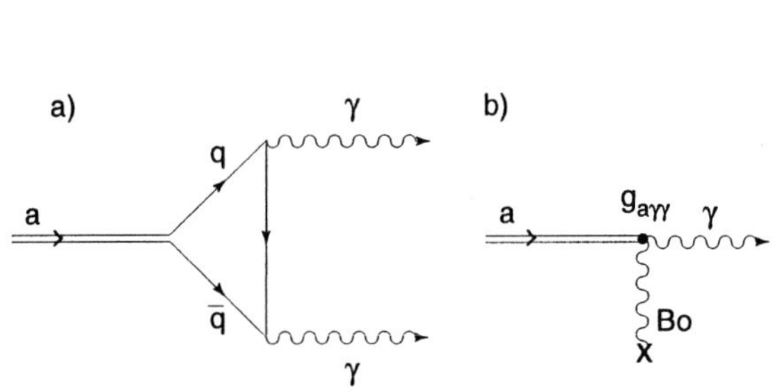

Figure 14. Axion–photon coupling (a), axion–photon conversion in an external magnetic field (b).

These axions are non-relativistic and should appear as a resonant signal in an RF cavity tuned at the frequency corresponding to the axion mass. The search apparatus is thus built around such a device (or set of devices) which can be continuously tuned in frequency by proper movable rods.

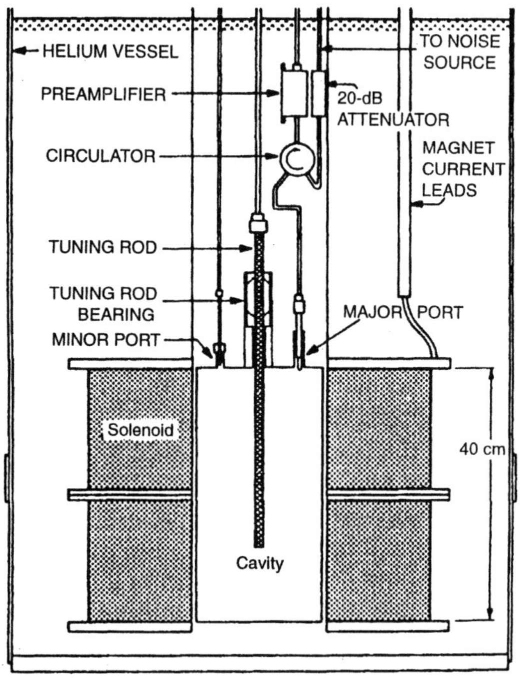

Figure 15. The set-up of the experiment of Ref. [29].

The expected output power is given by a formula like:

$$P_{a\to\gamma} \simeq 10^{-21}\,\mathrm{W}\left[\frac{B}{7.7\,\mathrm{T}}\right]^2\left[\frac{V}{200\,\ell}\right]\left[\frac{C}{0.65}\right]\left[\frac{Q}{90000}\right]\left[\frac{f}{0.7\,\mathrm{GHz}}\right]\left[\frac{\rho_a}{\rho_{\mathrm{Halo}}}\right]$$

where B is the magnetic field, V the volume of the cavity, C its mode-dependent form factor, Q its loaded quality factor, and P_a the local axion density. A most sophisticated detection set-up (Fig. 15) [29] is required to reach the sensitivity needed: as shown by Fig. 16 it works! Several experiments have reported steady progress in sensitivity and the explored range. Very recently one of them [30] has reached for a narrow mass window, around 3×10^{-6} eV (Fig. 17), the sensitivity required to test one of the cosmological axion models [31]: the authors expect, with a long scanning programme, to cover the full mass domain in the next three years or so.

Various other ideas and programs are 'worth the detour' [32].

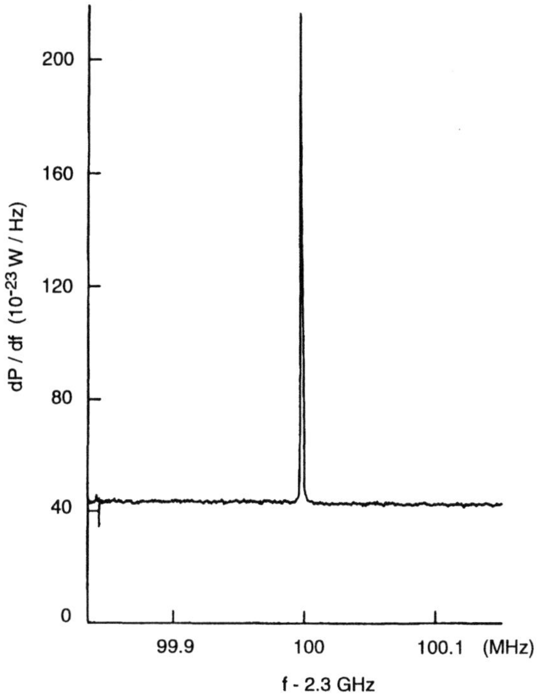

Figure 16. From Ref. [29], an injected peak with a total power of 2×10^{-18} W at 2.4 GHz. The bin width is 400 Hz.

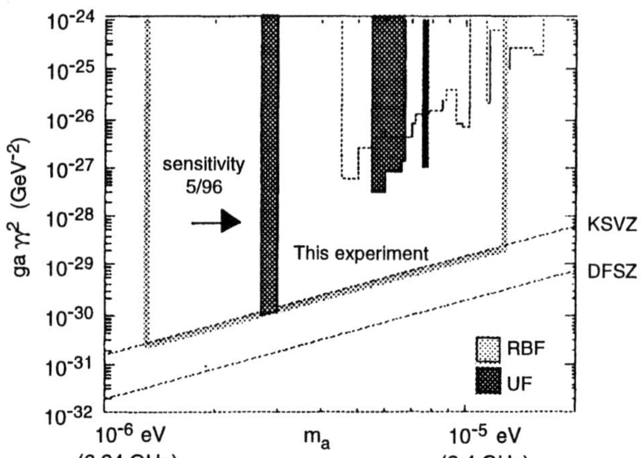

Figure 17. Exclusion plot for galactic axions. The heavy dark line indicates where the scan of Ref. [30] has already reached sensitivity to the KSVZ cosmological model.

ODD QUESTIONS ON THE PHOTON AND THE GLUON

Photon Mass

The gauge invariance of electrodynamics guarantees a zero mass for the photon and there is no good theoretical reason to envisage a departure from that situation.

One can nevertheless ask how a non-zero mass could be introduced theoretically [33] and what would be the observable consequences. The massive photon would then obey the Proca equation

$$\partial_\mu F^{\mu\nu} + M^2 A^\nu = 0 \, .$$

$\partial_\nu A^\nu = 0$, which was a choice in case of Maxwell equation, is now necessary. Gauge invariance is lost, the photon field has three degrees of freedom.

As the corresponding page of the Particle Data Group testifies, there has been much experimental activity in various domains to try to give evidence for a non-zero mass.

It would imply a departure from the inverse-square law of the electrostatic force. This can be tested in two ways:

a) One can assume it varies as $1/r^{2+\epsilon}$ and quote a value or limit for ϵ.
b) One can assume that the potential has the Yukawa form $r^{-1}e^{-\mu r}$ and quote a value or limit for μ, which is interpreted as:

$$\mu = m_\gamma c/\hbar \, .$$

Laboratory experiments give access to ϵ or μ (or m_γ). Geomagnetic experiments give access to μ (or m_γ). As a test of Coulomb law, the old Cavendish experiment, giving $|\epsilon| < 0.02$, has been progressively refined. The most accurate version of it [34] gives:

$$\epsilon = (2.7 \pm 3.1)10^{-16}$$

or

$$m_\gamma < 1.6 \times 10^{-47} g \, .$$

Speculations on the possibility that the non-zero mass would appear at very low temperature have been put forward [35], with, however, no convincing justification [36]. This led nevertheless to ingenious experiments.

A recent one [37] is a null test of Coulomb's law performed at 1.36 K and giving

$$m_\gamma \leq (1.50 \pm 1.38)10^{-42} g \, .$$

Let us briefly describe a null test of Ampere's law at 1.24 K [38] giving:

$$m_\gamma \leq (8.4 \pm 0.8)10^{-46} g \, .$$

The apparatus, shown in Fig. 18, consists of a source coil and a pick-up coil protected by a superconducting shield. The source coil is driven with a current amplitude between 0 and 10 A at a frequency of 0.11719 Hz.

From the Proca equation one derives:

$$\overrightarrow{\mathrm{rot}}\, \vec{B} = \left(\frac{4\pi}{c}\right) \vec{J} - \mu^2 \vec{A}$$

with the characteristic term proportional to the potential vector in case of non-zero photon mass. This gives a non-shielded contribution

$$\oint \vec{B}\, \vec{d\ell} = -\mu^2 \int \vec{A} \cdot d\vec{f}$$

with

$$A_z = \left(\frac{2G}{c}\right) N I \ln\left(\frac{R_0}{R_i}\right)(1 + \ldots) .$$

The observable is the spectral density of the recorded magnetic flux. The resolution is 2.86×10^{-5} Hz. At maximum current amplitude a spurious coupling, due to an imperfect shielding, appears and sets a limit to the sensitivity of the experiment (Fig. 19). The limit on the photon mass is the one quoted above.

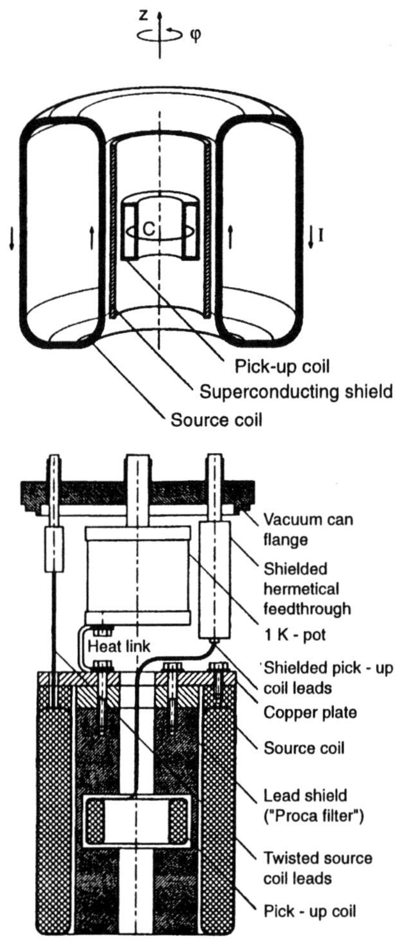

Figure 18. The set-up of the experiment of Ref. [38].

Geomagnetic or cosmomagnetic experiments exploit the large scales D offered, since the expected fractional change in the field goes like: $\sim (\mu D)^2$.

For instance in the Pioneer 10 observations of Jupiter's magnetic field [39] 134 vector field measurements are done between 2.84 and 13.1 Jupiter radii ($R_J \sim 71\ 000$ km). The field varies between ~ 150 and $\sim 18\ 000$ Gamma (10^{-5} Gauss). μ is treated as a free parameter. Fits to the data are performed with formulae containing n coefficients, corresponding to various choices of multipoles to describe the field: a minimum of 3 (internal dipole), and up to 16 (internal dipole + internal quadrupole + external dipole + external quadrupole).

One finds

$$\begin{aligned} \mu &\leq 2 \times 10^{-11} \text{ cm}^{-1} \\ &\equiv 6 \times 10^{-16} \text{ eV} \\ &\equiv 8 \times 10^{-49} \text{ g} . \end{aligned}$$

The limitation of the experiment is due to the insufficient knowledge of the plasma density in the region of the measurements.

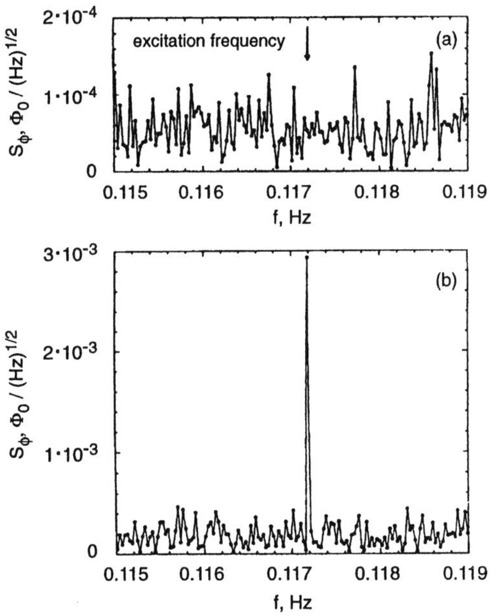

Figure 19. In the experiment of Ref. [38], spectral density of the magnetic flux (a) and evidence for a spurious coupling at full excitation (b).

Gluon Mass

Here again the standard belief is that colour symmetry is an exact symmetry and therefore that $m_g = 0$.

In the past, however, several alternatives were explored. For instance a gauge theory built on a broken $SU(3)_c \times U(1)$ gauge symmetry and integrally charged quarks (ICQ) was put forward in Ref. [40]. At large distance the mixing of the U(1) gauge field with the SU(3) octet component gives:

– a massless but 'tinted' photon

$$A_\mu = S_\mu \cos\chi + G_\mu^8 \sin\chi$$

– an orthogonal state

$$B_\mu = -S_\mu \sin\chi + G_\mu^8 \cos\chi$$

with $m_B \neq 0$.

One knows that $m_B \lesssim \Lambda_{QCD}$ and $\mathrm{tg}\,\chi < 10^{-2}$.

The electric charge felt by the photon in this model is

$$Q = Q_0 + Q_8 \frac{-m_g^2}{q^2 - m_g^2} ,$$

where Q_0 corresponds to the colour singlet and has the usual value found in the case of fractionally charged quarks. The colour octet part fades away for a virtuality larger than the gluon mass. One then needs a real photon to test the model. More exactly two of them at least are required to take part in the process, since an isolated colour octet cannot express itself in a world of singlets.

Such processes are:

$\gamma\mathcal{N} \to \gamma + \ldots$: deep inelastic QED Compton scattering [41]

$\mathrm{pp} \to \gamma\gamma + \ldots$: two real photons in hadronic collisions

$\gamma\gamma \to X$: untagged $\gamma\gamma$ DIS [42] .

This ICQ gauge theory, besides the fact that it was controversial on theoretical grounds [43], was experimentally disproven by the measurements of the processes quoted.

One can adopt a different and more pragmatic attitude and ask what upper limits can be put on m_g from some observed features of the real world [44]. The idea is to complete the QCD potential (Coulomb + linear) by a Yukawa potential e^{-rm_g}/r: actually this is approximated by a triangular potential ending sharply at $r = m_g^{-1}$. The slope is $K \sim (0.5\ \mathrm{GeV})^2$.

The fact that no quark was liberated at 200 GeV/constituent gives

$$m_g < \frac{K}{200\ \mathrm{GeV}} \approx 1.3\ \mathrm{MeV} .$$

The proton stability, interpreted through tunnelling arguments, gives

$$m_g < 20\ \mathrm{MeV} .$$

Finally the fact that apparently no quark was left free in the cosmological condensation leads to:

$$m_g < 2 \times 10^{-10}\ \mathrm{MeV} .$$

For the graviton I refer to the Particle Data Group and references therein.

Acknowledgements

I warmly thank the Desktop Publishing Service for their fast and excellent work.

References

[1] B. Assmann et al., *Z. Phys.* C66:567 (1995).

[2] Polarization at LEP, CERN 88–06.

[3] V. Bargmann et al., *Phys. Rev. Lett.* 2:435 (1959).

[4] M. Placidi, CERN/SL/96–15.

[5] T. Camporesi, contribution to Minneapolis DPF 96.

[6] H. Burkhardt and B. Pietrzyk, LEPP/EXP/95–05.
[7] S. Willenbrock, HEP-ph/9608418.
[8] A. Blondel, Plenary talk at the 1996 International Warsaw Conf.
[9] W. Hollik, hep-ph/96 02 380, G. Altarelli, CERN–TH/96–265, hep-ph/96 11 239.
[10] LEP EWWG/96–02.
[11] M. Demarteau, in Snowmass, June 95, 1996.
[12] G. Altarelli et al., eds., *Physics at LEP2*, CERN 96–01, (2 vols.), CERN, Geneva (1996).
[13] W.J. Stirling, DTP/95/24, HEP-PH-9503320.
[14] T. Sjostrand and V.A. Khoze, CERN-TH/7043–93.
[15] V.A. Khoze et al., in: *Physics at LEP2*, G. Altarelli et al., eds., CERN 96–01, Vol. 2, CERN, Geneva (1996) p. 190.
[16] L. Lönnblad and T. Sjostrand, CERN-TH/95–17.
[17] J.A. Casas et al., SCIPP–96–04, DESY 96–021.
[18] J.R. Espinosa, IEN-FT-96-133, HEP-ph/9606316.
[19] J.F. Grivaz, LAL 95–83.
[20] Higgs physics, in: *Physics at LEP2*, G. Altarelli et al., eds., CERN 96–01, Vol. 1, CERN, Geneva (1996) p. 351.
[21] P. Janot, preprint LAL 93–28.
[22] E. Richter-Was et al., CERN TH/96–11, CMS LHCC/94–38.
[23] F. Pauss, ETHZ–IPP PR–94–8.
[24] M. Peskin, SLAC-PUB-6582 1994; P.M. Zerwas, DESY 94–001.
[25] P. Janot, preprint LAL 93–28.
[26] R. Peccei and H. Quinn, *Phys. Rev. Lett.* 38:1440 (1977).
[27] P. Sikivie, UFIFT–HEP–95–9
[28] P. Sikivie, *Phys. Rev.* D32:2988 (1985).
[29] W. Wuensch et al., *Phys. Rev.* D40:3153 (1989).
[30] C. Hagmann et al., astro-ph/9607022.
[31] J. Kim, *Phys. Rev. Lett.* 43:103 (1979);
M. Shifman et al., *Nucl. Phys.* B166:49 (1980).
[32] S. Matsuki et al., ICRKU–9402, P.Sikivie et al., UFIFT–HEP 93 -2
[33] A.S. Goldhaber and M.M. Nieto, *Rev. Mod. Phys.* 43:277 (1979).
[34] E.R. Williams et al., *Phys. Rev. Lett.* 26:721 (1971).
[35] J.R. Primack and M.A. Sher, Nature 288:680 (1980).
[36] L.F. Abbot and M.B. Gavela, Nature 299:187 (1982).
[37] J.J. Ryan et al., *Phys. Rev.* D32:801 (1985).
[38] M.A. Chernikov et al., *Phys. Rev. Lett.* 68:3383 (1992).
[39] L. Davis et al., *Phys. Rev. Lett.* 35:1402 (1975).
[40] A.V. Efremov et al., *Sov. J. Nucl. Phys.* 34(5):754 (1981).
[41] P. Astbury et al., *Phys. Lett.* B152:419 (1985).
[42] R.M. Godbole et al., *Phys Lett.* 142:91 (1984)
[43] L.B. Okun et al., ITEP 79–79, July 1979.
[44] F.J. Yndurain, *Phys. Lett.* B345:524 (1995).

NONCOMMUTATIVE GEOMETRY AND THE INTERNAL SPACE OF GAUGE THEORIES

Thomas Krajewski [1]

Centre de Physique Théorique
UPR 7061
F-13288 Marseille cedex 09

ABSTRACT

In this seminar, we outline how one can use noncommutative geometry to describe the internal space of gauge theories with spontaneous symmetry breaking. This can be done only for a tiny subset of Yang-Mills-Higgs theories. We give some features of this subset, focusing on mild extensions of the standard model.

1. INTRODUCTION

From the most general point of view, noncommutative geometry is the geometry of spaces whose coordinates fail to commute. The simplest way to get such a geometry lies in deforming an ordinary one. For instance, one can consider the quantum plane, whose coordinates satisfy $xy = qyx$ instead of $xy = yx$, and whose symmetries are described by a quantum group. Although they can appear as purely algebraic constructions, these spaces occur naturally in quantum physics. Indeed, to quantise a system, we have to replace the *commutative* algebra of classical observables by a *noncommutative* algebra of operators in a Hilbert space. This is the framework of the theory developped by A. Connes: one first formulates some notions of differential geometry using the *commutative* algebra of coordinates and then, one lets the algebra of coordinates be a *noncommutative* algebra of operators on a Hilbert space. Provided suitable extensions of the usual notions of geometry are chosen, some of the results of differential geometry remain true even if the algebra of coordinates is no longer commutative. Within this framework, one can develop Riemannian geometry and gauge theory. The resulting theories turn out to be powerful enough to fully describe the standard model as a pure gauge theory on a noncommutative space. However, this cannot be done for any Yang-Mills theory with spontaneous symmetry breaking, and in this seminar we aim to outline this construction and to briefly describe the class of such theories, with special emphasis on mild extensions of the standard model.

It is organized as follows. We first give a short reminder of the way one builds a

[1] and Université de Provence and Ecole Normale Superieure de Lyon, tkrajews@cpt.univ-mrs.fr

gauge theory in a noncommutative setting. Since the Higgs field appears as a Yang-Mills connection on a finite space, we shall then study such geometries, described by the so-called *finite spectral triples*. Afterwards we will be able to give some applications to particle physics.

2. GAUGE THEORY IN NONCOMMUTATIVE GEOMETRY

In this section, we shall first briefly describe the gauge fields in a noncommutative setting and then study finite spectral triples which are the basic building blocks of such theories. Obviously it is far from being complete and we refer to the lecture of A. Connes for a much more detailed study. Noncommutative geometry is fully developped in the references [1]-[7].

2.1. General Aspects

To simplify our discussion, let us start with a finite set X of N elements. The algebra $\mathcal{A}$ of complex valued functions on X is the finite dimensional commutative and involutive algebra $\mathbb{C}^N$. Obviously, the converse also holds: any finite dimensional commutative and involutive algebra over $\mathbb{C}$ can be considered as the algebra of complex valued functions on a finite set X. Furthermore, this equivalence between a set of points X and the commutative algebra of coordinates $\mathcal{A}$ on X remains true in the more general setting of topological spaces (Gelfand- Naimarck theorem). Accordingly, the suitable definition of a noncommutative space X is that it is given by an involutive but not necessarily commutative algebra $\mathcal{A}$. The algebra $\mathcal{A}$ can be thought as the algebra of "coordinates" on X. However, as in quantum mechanics, the noncommutativity of the algebra leads us to the impossibility to locate a point.

This is the first step towards the notion of spectral triple. The latter has been introduced by A. Connes to generalize the Riemannian geometry with spin to noncommutative spaces. A spectral triple is a triple $(\mathcal{A}, \mathcal{H}, D)$ where $\mathcal{A}$ is involutive algebra together with a $\mathbb{R}$-linear representation π of $\mathcal{A}$ as operators on a complex Hilbert space $\mathcal{H}$. D is a hermitian operator on $\mathcal{H}$ called Dirac operator. These objects have to satisfy some relations called axioms of noncommutative geometry that are stated such that, in the commutative case, one recovers the ordinary notions of geometry. In this case, the algebra is simply the algebra $C^\infty(\mathcal{M})$ of smooth complex valued function on a compact and euclidean manifold $\mathcal{M}$. The Hilbert space is the space $L^2(\mathcal{M}, S)$ of square integrable spinors on $\mathcal{M}$ and D is the ordinary Dirac operator $i\gamma^\mu\partial_\mu$, γ^μ being the usual hermitian Dirac matrices. Note that we assume that the manifold is endowed with a positive definite metric so that we are in fact dealing with an euclidean field theory.

In the commutative case, a gauge field is simply a vector potential of Maxwell's theory. Using covariant notations, it can be written as a 1-form,

$$A = A_\mu dx^\mu.$$

To introduce such an object in the noncommutative setting, let us use the analogy with quantum mechanics. We know that in the quantum case, time derivatives must be replaced by commutators with the hamiltonian. Following this idea, it appears natural to replace the differential dx^μ of the coordinates by a commutator. Thus we define a noncommutative 1-form as

$$A = \sum_i \pi(a_0^i)\left[D, \pi(a_1^i)\right],$$

with $a_0^i, a_1^i \in \mathcal{A}$. A gauge field is a hermitian 1-form A, i.e. satisfying $A = A^*$ as operator on $\mathcal{H}$. It then easy to see that, in the commutative case, we recover the usual definition of a 1-form if we identify $i\gamma^\mu$ and dx^μ.

For A to be a good gauge field in the framework of noncommutative geometry, we have to specify its transformation law under a gauge transformation. To proceed, let us notice that in the commutative case, symmetries of the space X are in one-to-one correspondence with automorphisms of the algebras $\mathcal{A}$. In the noncommutative case, among all these automorphisms, internal automorphisms are of special interest. They are defined, for each unitary element u of $\mathcal{A}$ (i.e. that fulfills $uu^* = u^*u = 1$), as the map that takes $x \in \mathcal{A}$ to $uxu^* \in \mathcal{A}$. As in quantum mechanics, these symmetries have to be represented on the Hilbert space $\mathcal{H}$. To any unitary element u of $\mathcal{A}$ we associate the operator $\pi(u)\mathcal{J}\pi(u)\mathcal{J}^{-1}$, where $\mathcal{J}$ is an antiunitary operator involved in the formulation of the axioms of noncommutative geometry. In the standard model, this operator is simply the charge conjugation. Therefore, the gauge transformation associated to u maps $\psi \in \mathcal{H}$ to $\pi(u)\mathcal{J}\pi(u)\mathcal{J}^{-1}\psi \in \mathcal{H}$. Since the theory must be gauge invariant, the transformation law of the gauge field A is

$$A \rightarrow \pi(u)A\pi(u^*) + \pi(u)\left[D, \pi(u^*)\right].$$

These is similar to the usual transformation law of Yang-Mills theory

$$A_\mu \rightarrow gA_\mu g^{-1} + g\partial_\mu g^{-1},$$

if we identify the commutator with the partial derivatives.

To complete the construction, we have to say how one can get an action functional for the gauge field A. This can be achieved in two ways. On the one hand, one can stay in the framework of Yang-Mills theory and define a suitable notion of 2-form. This allows us to define the curvature $C = dA + A^2$ of the gauge field and then, to recover the action as the square norm of the curvature C [8]. On the other hand, a closer look at the axioms of noncommutative geometry exhibits a deep connection between gauge fields and the Dirac operator. Using theses ideas, one can write an action functional that is of purely gravitational nature and that yields the action of the Yang-Mills fields coupled to gravity [9]-[10]. In both cases, the fermionic action is simply given by

$$< \psi, \left(D + A + \mathcal{J}A\mathcal{J}^{-1}\right) \psi >,$$

for any spinor $\psi \in \mathcal{H}$.

To end this section, we have to specify what class of spectral triples $(\mathcal{A}, \mathcal{H}, D)$ one uses to build a Yang-Mills theory with spontaneous symmetry breaking. Since we want the group of unitary elements of $\mathcal{A}$ to be the group of local gauge transformations, it must be of the form $C^\infty(\mathcal{M}) \otimes G$, where G is a semi-simple compact Lie group. Therefore, the suitable spectral triples are tensor products of the commutative spectral triple $(C^\infty(\mathcal{M}), L^2(\mathcal{M}, S), i\gamma^\mu\partial_\mu)$ by a finite spectral triple $(\mathcal{A}_F, \mathcal{H}_F, D_F)$. The latter is defined as a spectral triple whose algebra $\mathcal{A}_F$ and Hilbert space $\mathcal{H}_F$ are both finite dimensional. More precisely, one has

$$\begin{aligned}
\mathcal{A} &= C^\infty(\mathcal{M}) \otimes \mathcal{A}_F, \\
\mathcal{H} &= L^2(\mathcal{M}, S) \otimes \mathcal{H}_F, \\
D &= i\gamma^\mu\partial_\mu \otimes I_F + \gamma^5 \otimes D_F,
\end{aligned}$$

where I_F is the identity operator of $\mathcal{H}_F$. Note that in this case, $\mathcal{H}_F$ is the Hilbert space of all fermions and antifermions and that D_F is the mass matrix of all fermions and antifermions. From now on, the Riemannian manifold $\mathcal{M}$ will be fixed and we will be concerned only with finite spectral triples, so that we can drop the index F and write a finite spectral triple as $(\mathcal{A}, \mathcal{H}, D)$.

2.1. Finite Spectral Triples

Let us now study the finite spectral triples. Recall that such a triple is a spectral triple $(\mathcal{A}, \mathcal{H}, D)$ where both $\mathcal{A}$ and $\mathcal{H}$ are assumed to be finite dimensional. Note that it has been called finite spectral triple instead of finite dimensional spectral triple since in the commutative case it corresponds to a finite set, that is to say, a manifold of dimension 0, and not to a finite dimensional manifold.

In the case of finite spectral triples, there is a great oversimplification in the formulation of the axioms of noncommutative geometry. Indeed, all the assumptions related to functional analysis, such as for instance boundedness or compactness of some operators, are obviously satisfied for finite dimensional Hilbert spaces. Besides, it can be shown that the algebra $\mathcal{A}$ is necessarily a direct sum of matrix algebras whose entries are real numbers, complex numbers or quaternions. Therefore, all the axioms of noncommutative geometry reduce to some algebraic relations involving matrices, and their general study can be performed in a very easy manner.

We now give these relations without proving their connection with the general case. They should be thought of as rules that select the class of allowed Yang-Mills-Higgs theories in the framework of noncommutative geometry.

To proceed, let us write the algebra as

$$\mathcal{A} = \bigoplus_{i=1}^{N} \mathcal{A}_i,$$

where $\mathcal{A}_i = M_{n_i}(\mathbb{K})$ is the algebra of $n_i \times n_i$ matrices with entries in the fields $\mathbb{K} = \mathbb{R}$ (real numbers), $\mathbb{C}$ (complex numbers) or $\mathbb{H}$ (quaternions). Then the axioms reduce to the four following assumptions.

Reality

There is an antiunitary involution $\mathcal{J}$ on $\mathcal{H}$ that commutes with D and such that, for any elements x and y of $\mathcal{A}$ one has $[\pi(x), \mathcal{J}\pi(y)\mathcal{J}^{-1}] = 0$.

In the commutative case $\mathcal{J}$ is simply the charge conjugation of Dirac spinors and for the standard model it is the antilinear operator that maps a fermion to the corresponding antifermion. Therefore $\mathcal{J}$ is called charge conjugation.

Orientability

There is hermitian involution χ on $\mathcal{H}$ that commutes with $\mathcal{J}$, anticommutes with D and can be written as $\sum_i \pi(a^i)\mathcal{J}\pi(b^i)\mathcal{J}^{-1}$ with $a^i, b^i \in \mathcal{A}$.

In the most general case, this axiom requires that χ can be written as the image of a Hochschild cycle of dimension n for a geometry of dimension n. In the 4-dimensional commutative case, it reduces to the relation

$$\chi = \gamma^5 = \frac{-1}{4!}\epsilon_{\mu\nu\rho\sigma}\, \gamma^\mu\gamma^\nu\gamma^\rho\gamma^\sigma,$$

where $\epsilon_{\mu\nu\rho\sigma}$ is the completely antisymmetric tensor of rank 4 such that $\epsilon_{0123} = 1$. Using the analogy between $i\gamma^\mu$ and dx^μ, χ thus corresponds to the volume form $dx^0 \wedge dx^1 \wedge dx^2 \wedge dx^3$, so that this axiom is called orientability. For the finite spectral triple of the standard model, it is the operator whose value on left-handed fermions is -1 and +1 on right-handed fermions. Therefore for finite spectral triples χ is refered to as chirality operator.

Poincaré duality

The matrix $\cap_{ij}$ of the intersection form is non degenerate. The latter is defined by

$$\cap_{ij} = \mathrm{Tr}\left[\chi\pi(p_i)\mathcal{J}\pi(p_j)\mathcal{J}^{-1}\right],$$

where $p_i \in \mathcal{A}_i$ is a minimal hermitian projection and Tr *is the ordinary trace of a matrix*

The analogue of this assumption in the commutative case is simply the isomorphism, for a manifold of dimension n, between p-forms and (n-p)-forms. For the standard model it has no obvious significance but it has an important consequence concerning the existence of right handed neutrinos.

First order condition

For any $x, y \in \mathcal{A}$, the Dirac operator satisfies $[[D, \pi(x)], \mathcal{J}\pi(y)\mathcal{J}^{-1}] = 0$

For a commutative geometry, this means that the Dirac operator is a differential operator of order one.

Contrary to its supersymmetric extensions, the standard model contains no Majorana particle. This simple fact is formulated in the framework of noncommutative geometry by the notion of S^0-real spectral triple [8]. The latter are particular cases of spectral triples that fufill an additional axiom called S^0-reality formulated as follows.

S^0-reality:

There is an operator ϵ such that $\epsilon^2 = 1$, $\epsilon^* = \epsilon$ and that commutes with π, χ and D and anticommutes with $\mathcal{J}$.

In others words, particles are eigenstates of ϵ with eigenvalue +1 and antiparticles eigenstates with eigenvalue -1. Since the charge conjugation $\mathcal{J}$ anticommutes with ϵ there are no Majorana particles. The Hilbert space of a finite S^0-real spectral triple can always be written as a direct sum of four Hilbert spaces,

$$\mathcal{H} = \mathcal{H}_L \oplus \mathcal{H}_R \oplus \mathcal{H}_L \oplus \mathcal{H}_R.$$

The spaces $\mathcal{H}_L$ and $\mathcal{H}_R$ are respectively the spaces spanned by left-handed and right-handed fermions. They both appear twice in the decomposition of $\mathcal{H}$ since we have particles and antiparticles. With respect to this decomposition of $\mathcal{H}$, π and χ block diagonal,

$$\begin{aligned} \pi &= \mathrm{diag}\left(\pi_L, \pi_R, \pi_L^c, \pi_R^c\right), \\ \chi &= \mathrm{diag}\left(-I_{N_L}, I_{N_R}, -I_{N_L}, I_{N_R}\right), \end{aligned}$$

where N_L and N_R are the dimension of $\mathcal{H}_L$ and $\mathcal{H}_R$. The Dirac operator D and the charge conjugation J are

$$D = \begin{pmatrix} 0 & M & 0 & 0 \\ M^* & 0 & 0 & 0 \\ 0 & 0 & 0 & \overline{M} \\ 0 & 0 & \overline{M}^* & 0 \end{pmatrix}, \quad J = \begin{pmatrix} 0 & 0 & I_L & 0 \\ 0 & 0 & 0 & I_R \\ I_L & 0 & 0 & 0 \\ 0 & I_R & 0 & 0 \end{pmatrix} C,$$

where C denotes complex conjugation and $M \in M_{N_L \times N_R}$ is a complex matrix that can be viewed as an endomorphism from $\mathcal{H}_R$ to $\mathcal{H}_L$. Thus, it is a coupling between left-handed and right-handed fermions i.e. their mass matrix.

We end this section by giving the spectral triple of the standard model. The algebra $\mathcal{A}$ is

$$\mathcal{A} = \mathbb{H} \oplus \mathbb{C} \oplus M_3(\mathbb{C}),$$

so that one recovers, up to a $U(1)$ factor, the gauge group $SU(2) \times U(1) \times SU(3)$ of the standard model. If we denote by N_f the number of families, the Hilbert spaces $\mathcal{H}_L$ and

$\mathcal{H}_R$ are $\mathbb{C}^{8N_f}$ and $\mathbb{C}^{7N_f}$. For three families and if we omit the colour index for quarks, a basis of $\mathcal{H}_L$ is

$$\begin{pmatrix} u \\ d \end{pmatrix}_L, \begin{pmatrix} c \\ s \end{pmatrix}_L, \begin{pmatrix} t \\ b \end{pmatrix}_L, \begin{pmatrix} \nu_e \\ e \end{pmatrix}_L, \begin{pmatrix} \nu_\mu \\ \mu \end{pmatrix}_L, \begin{pmatrix} \nu_\tau \\ \tau \end{pmatrix}_L,$$

and

$$(u)_R, (d)_R, (c)_R, (s)_R, (t)_R, (e)_R, (\mu)_R, (\tau)^-,$$

is a basis of $\mathcal{H}_L$. Within these basis (generalized to N_f families),

$$\begin{aligned}
\pi_L(a) &= \operatorname{diag}\left(a \otimes I_{3N_f}, a \otimes I_{N_f}\right), \\
\pi_R(b) &= \operatorname{diag}\left(b\, I_{3N_f}, \bar{b}\, I_{3N_f}, \bar{b}\, I_{N_f}\right), \\
\pi_L^c(b,c) &= \operatorname{diag}\left(I_{2N_f} \otimes c, \bar{b}\, I_{2N_f}\right), \\
\pi_R^c(b,c) &= \operatorname{diag}\left(I_{2N_f} \otimes c, \bar{b}\, I_{N_f}\right),
\end{aligned}$$

where $(a,b,c) \in \mathbb{H} \oplus \mathbb{C} \oplus M_3(\mathbb{C})$. Finally, the mass matrix is

$$M = \begin{pmatrix} \begin{pmatrix} M_u \otimes I_3 & 0 \\ 0 & M_d \otimes I_3 \end{pmatrix} & 0 \\ 0 & \begin{pmatrix} 0 \\ M_e \end{pmatrix} \end{pmatrix},$$

where, for three families,

$$\begin{aligned}
M_u &= \operatorname{diag}(m_u, m_c, m_t), \\
M_d &= V_{CKM} \operatorname{diag}(m_d, m_s, m_b), \\
M_e &= \operatorname{diag}(m_e, m_\mu, m_\tau),
\end{aligned}$$

where m_p stands for the mass of particle p and V_{CKM} is the Cabibbo-Kobayashi-Maskawa mixing matrix.

3. APPLICATIONS TO PARTICLE PHYSICS

Let us now analyse the constraints on our models arising from the assumption that the internal space of a gauge theory is described by a finite spectral triple. Our approach is pedestrian and we do not give the detailed proofs of our statements. Full details of this analysis are available in the standard literature listed in the bibliography.

3.1. The Gauge Group

In the previous section we have already sketched the relation between gauge transformations and $*$-automorphisms of the algebra $\mathcal{A}_F$. The latter are fully described by the group G_F of the unitary elements of $\mathcal{A}_F$. Moreover, we recall that the algebra $\mathcal{A}_F$ is a direct sum of matrix algebras $M_n(\mathbb{K})$ where $\mathbb{K}$ stands for the fields of real numbers $\mathbb{R}$, complex numbers $\mathbb{C}$ and quaternions $\mathbb{H}$. Therefore G_F is a direct product of the simple Lie groups $SO(n)$, $SU(n)$ and $SP(2n)$. These groups arise respectively from the unitary elements of the matrix algebras $M_n(\mathbb{R})$, $M_n(\mathbb{C})$ and $M_n(\mathbb{H})$. Following these ideas, it is tempting to identify the gauge group G with the group of unitary elements G_F of $\mathcal{A}_F$.

However, in the general case G is contained in G_F and the equality $G = G_F$ only holds up to some $U(1)$ factors. Indeed, these $U(1)$ may disappear for two reasons. First of all, we have seen that the gauge transformation associated to $u \in G_F$ maps $\psi \in \mathcal{H}$ to $\pi(u)\mathcal{J}\pi(u)\mathcal{J}^{-1}\psi$. Since $\mathcal{J}$ is antiunitary, the kernel of the map $u \mapsto \pi(u)\mathcal{J}\pi(u)\mathcal{J}^{-1}$ is a direct product of some $U(1)$. To obtain the gauge group, we have to divide G_F by these $U(1)$. Secondly, in some particular cases, we are obliged to rule out by hand some $U(1)$ for physical reasons such as gravitational or gauge anomaly cancellation. This is the case for the standard model where we have to remove one $U(1)$ because of anomaly cancellation, or equivalently, to select the right hypercharges. Note that this condition can be achieved without changing the metric aspect of the noncommutative setting of the standard model and is usually called a *unimodularity condition.* In general, the latter should be a set of rules that determine all the linear combinations of the $U(1)$'s (at the Lie algebra level) that have to be eliminated. For the standard model, one can check that anomaly cancellation is equivalent to $\mathrm{Tr}_{\mathrm{p}}\left[(\pi(a)+\mathcal{J}\pi(a)\mathcal{J}^{-1})\right]=0$, for any a in the Lie algebra of gauge group. Tr_{p} denotes the trace restricted on particle space [11].

As a conclusion, we can say that the gauge group can be any direct product of the simple Lie groups $SO(n)$, $SU(n)$ or $SP(2n)$. At the Lie algebra level, we obtain all direct sums of the real Lie algebras A_n, B_n, C_n and D_n appearing in Cartan's classification. Thus the Lie algebra of the gauge group can be any classical Lie algebra apart from those built with the five exceptional ones [12]. For instance, we cannot construct the $E_8 \times E_8$ gauge theory in our framework.

3.2. The Fermionic Representation

As already pointed out, the gauge transformations take the spinor ψ into $\pi(u)\mathcal{J}\pi(u)\mathcal{J}^{-1}\psi$. To analyse the fermionic representation of the gauge group, we first have to understand the underlying algebraic structure of the map $u \mapsto \pi(u)\mathcal{J}\pi(u)\mathcal{J}^{-1}$. i.e. the representation of the associative algebra $\mathcal{A}$. Recall that $\mathcal{A}$ is a direct sum of N matrix algebras

$$\mathcal{A} = \overset{N}{\underset{i=1}{\oplus}} \mathcal{A}_i,$$

$\mathcal{A}_i = M_{n_i}(\mathbb{K})$ being the algebra of $n_i \times n_i$ matrices with entries in $\mathbb{K} = \mathbb{R}$, $\mathbb{C}$ or $\mathbb{H}$. Moreover, the only irreducible representation of $\mathcal{A}_i$ is the fundamental one for matrices with entries in $\mathbb{R}$ or $\mathbb{H}$. In the complex case, $\mathcal{A}_i$ has two irreducible representations: the fundamental one and its complex conjugate. According to the reality axiom, the two representations of the associative algebra $\mathcal{A}$, $x \mapsto \pi(x)$ and $y \mapsto \mathcal{J}\pi(y)\mathcal{J}^{-1}$, commute. Therefore, we can simultaneously decompose these two representations of $\mathcal{A}$ into irreducible representations of the $\mathcal{A}_i$'s. The Hilbert space $\mathcal{H}$ can be written as a direct orthogonal sum

$$\mathcal{H} = \underset{i,j}{\oplus} \mathcal{H}_{ij}.$$

where $\mathcal{H}_{ij} = \pi(1_i)\mathcal{J}\pi(1_j)\mathcal{J}^{-1}\mathcal{H}$, 1_i denoting the unit of $\mathcal{A}_i$ considered as element of $\mathcal{A}$. Let us write any element x of $\mathcal{A}$ as $(x_1, x_2, ..., x_N)$ with $x_i \in \mathcal{A}_i$. The two representations of $\mathcal{A}$ on $\mathcal{H}$ induce two representations of $\mathcal{A}_i$ and $\mathcal{A}_j$ on $\mathcal{H}_{ij}$ that are direct sums of the fundamental representations of $\mathcal{A}_i$ and $\mathcal{A}_j$.

To simplify our discussion, let us assume that we are in the real case, i.e. all matrices are real and that there is no multiplicity, i.e. the two previous representations are the fundamental ones. We denote an unitary element u of $\mathcal{A}$ by $(u_1, ..., u_N)$ where $u_i \in SO(n_i)$. Then, it is easy matter to see that any spinor $\psi_{ij} \in \mathcal{H}_{ij}$ carries two internal indices $\alpha \in \{1, ..., n_i\}$ and $\beta \in \{1, ..., n_j\}$ and that under gauge transformation it becomes

$$\psi_{ij}^{\alpha\beta} = u_i^{\alpha\gamma}\psi_{ij}^{\gamma\delta}u_j^{\beta\delta},$$

where $u_i^{\alpha\gamma}$ and $u_j^{\beta\delta}$ are the entries of the matrices $u_i \in SO(n_i)$ and $u_j \in SO(n_j)$. Since all spinors of $\mathcal{H}$ can be written as sums of such ψ_{ij}, the fermionic representation is a direct sum of tensor products of two fundamental representations of the groups $SO(n_i)$. If we denote as usually the fundamental representation of $SO(n_i)$ by n_i, the fermionic representation can be written as

$$\bigoplus_{(i,j)\in I\times J} n_i \otimes n_j,$$

where I and J are subsets of $\{1, ..., N\}$.

These results remain valid in the complex and quaternionic cases and if we allow some multiplicities. We simply have to incorporate the complex conjugate $\overline{n_i}$ of the fundamental representation of $U(n_i)$ and we have to repeat some of the factors of the previous decomposition to take into account multiplicities. Note that for each simple factors of the gauge group, the fermions are either in the fundamental representation if $i \neq j$, or in the adjoint one if $i = j$. Consequently, it is not possible to built the grand unified $SU(5)$ theory in our framework since the representation 10 of $SU(5)$ cannot be obtained in that way.

Let us end this section with two remarks. First, we did not take care of the unimodularity conditions that change the fermionic representation. However, they only modify the representations of the $U(1)$ factors and thus do not modify drastically the previous result. In particular, they allow us to obtain every representations of $U(1)$ and not only those corresponding to charge ± 1. This is necessary to get the right hypercharges of the quarks in the standard model. Secondly, we did not worry about all other axioms of noncommutative geometry. In particular, some of these representations are not compatible with Poincaré duality as we shall see in the next section.

3.3. Poincaré Duality and Right-Handed Neutrinos

Although the axiom of Poincaré duality seems to be a rather abstract mathematical requirement, it has a deep physiscal significance in the standard model case. Indeed, it is at loggerheads with the addition of a right-handed Dirac neutrino to the standard model. To check that Poincaré duality holds for the standard model, let us define three hermitian and minimal projections of $\mathcal{A}$ by

$$p_1 = (I_2, 0, 0) \;\; p_2 = (0, 1, 0) \;\; p_3 = (0, 0, e)\,,$$

where e is the elementary matrix of $M_3(\mathbb{C})$ whose only non vanishing entry e_{11} is set equal to 1. The matrix $\cap$ of the intersection form is defined by $\cap_{ij} = \mathrm{Tr}\,[\chi\pi(1_i)\mathcal{J}\pi(1_j\mathcal{J}^{-1})]$. We can easily compute it, and we obtain

$$\cap = 2N_f \begin{pmatrix} 0 & -1 & -1 \\ -1 & 1 & 1 \\ -1 & 1 & 0 \end{pmatrix}.$$

Thus, it has non vanishing determinant and Poincaré duality holds for the standard model.

Let us now add to the spectral triple a right-handed neutrino. To proceed, we leave π_L and π_L^c unchanged and we incorporate to π_R and π_R^c a right-handed Dirac neutrino. Now π_R and π_R^c are given by

$$\begin{aligned} \pi_R(b) &= \mathrm{diag}\left(b\, I_{N_f} \otimes I_3, \bar{b}\, I_{N_f} \otimes I_3, \bar{b}\, I_{N_f}, d\, I_{N_f}\right), \\ \pi_R^c(b, c) &= \mathrm{diag}\left(I_2 \otimes I_{N_f} \otimes c, \bar{b}\, I_{N_f}, d\, I_{N_f}\right), \end{aligned}$$

for any $(a,b,c) \in \mathbb{H} \oplus \mathbb{C} \oplus M_3(\mathbb{C})$ and where d denotes indifferently b or $\bar{b}$. Similarly, χ and $\mathcal{J}$ are now written as

$$\chi = \begin{pmatrix} -I_{8N_f} & 0 & 0 & 0 \\ 0 & I_{8N_f} & 0 & 0 \\ 0 & 0 & -I_{8N_f} & 0 \\ 0 & 0 & 0 & I_{8N_f} \end{pmatrix}, \quad J = \begin{pmatrix} 0 & 0 & I_{8N_f} & 0 \\ 0 & 0 & 0 & I_{8N_f} \\ I_{8N_f} & 0 & 0 & 0 \\ 0 & I_{8N_f} & 0 & 0 \end{pmatrix} C.$$

Using the same set of projections of $\mathcal{A}$ as above, we compute the matrix of the intersection form and we get

$$\cap = 2N_f \begin{pmatrix} 0 & -1 & -1 \\ -1 & 2 & 1 \\ -1 & 1 & 0 \end{pmatrix}.$$

This matrix has vanishing determinant so that Poincaré duality is not fulfilled. Therefore, we have shown that the existence of right-handed Dirac neutrinos in every families is not compatible with the set of axioms of noncommutative geometry.

3.4. Noncommutative Geometry and the Strong Force

We already pointed out that the standard model is described by a S^0-real spectral triple. In this case, the axiom of reality and the first order condition reduce to the three following relations,

$$[\pi_L(x), \pi_L^c(y)] = 0,$$
$$[\pi_R(x), \pi_R^c(y)] = 0,$$
$$(M\pi_R(x) - \pi_L(x)M)\,\pi_R^c(y) = \pi_L^c(y)\,(M\pi_R(x) - \pi_L(x)M),$$

that must be satisfied for any $x, y \in \mathcal{A}$.

Let us now assume that the the electroweak sector, i.e. the two representations π_L and π_R of the algebra $\mathbb{H} \oplus \mathbb{C}$, and the mass matrix M are given. We can then consider the three previous relations as equations for the strong sector π_L^c and π_R^c. Moreover, if we assume that the algebra of the standard model is $\mathcal{A} = \mathbb{H} \oplus \mathbb{C} \oplus X$, where X is a simple matrix algebra, these equations are sufficient to determine the two representations π_L^c and π_R^c of the strong force. In particular, the latter are vectorial and gluons remain massless. Indeed, if Φ is the Higgs field, according to the previous relations we have

$$[\pi^c, \Phi] = 0,$$

so that the gluons are not coupled to the Higgs field and therefore do not become massive after spontaneous symmetry breaking [13].

3.5. $U(1)$ Extensions of the Standard Model

Motivated by some experimental results that have been interpreted as the existence of an additional heavy $U(1)$ gauge boson Z' (its mass should be at least 1 TeV), one can try to incorporate it in the framework of noncommutative geometry. More precisely, the question is how can we change the description of the standard model in noncommutative geometry such that the resulting Yang-Mills-Higgs theory is a $U(1)$ extension of the standard model? To proceed, we shall assume that we do not add new fermions to the theory, i.e. we leave the Hilbert space $\mathcal{H}$ and the Dirac operator unchanged D. Therefore, the previous question has three different answers. First of all, recall that the group of unitaries of the algebra of standard model is $SU(2) \times U(1) \times U(3)$ and that we used an unimodularity condition to rule out one $U(1)$. Accordingly, the most natural way to get an extra $U(1)$ gauge boson is to give up the unimodularity condition. The resulting boson is a linear combination of two $U(1)$ gauge bosons. The first one arises from the local decomposition $U(3) = U(1) \times SU(3)$ of the unitary elements of $M_3(\mathbb{C})$ and the second one is due to those of $\mathbb{C}$. According to the arguments developed in the previous section, we know that the first of these gauge bosons is massless. Furthermore, the second one is a linear combination of this massless field and of the ordinary gauge bosons γ and Z of the standard model. Thus the spectrum of the mass matrix of gauge bosons remains the same so that it is clearly impossible to get a heavy gauge field.

The two others answers lie in changing the algebra $\mathcal{A} = \mathbb{H} \oplus \mathbb{C} \oplus M_3(\mathbb{C})$ of the spectral triple of the standard model. One can first replace the quaternions $\mathbb{H}$ by the two by two matrices $M_2(\mathbb{C})$. After application of a suitable unimodularity condition, the resulting gauge group is $U(1) \times U(1) \times SU(2) \times SU(3)$ and we get an extra $U(1)$. This, together with a complete description of the associated Higgs fields, has been worked out in [14]. In particular, the resulting Z' comes out too light, 80 GeV. Let us mention that the goal was not to get an $U(1)$ extension of the standard model, but to study a model derived from a quantum symmetry. Indeed, the algebra $\mathbb{C} \oplus M_2(\mathbb{C}) \oplus M_3(\mathbb{C})$ is related to the quantum group $SL_2(q)$ with $q^3 = 1$ and the standard model based on $\mathbb{C} \oplus M_2(\mathbb{C}) \oplus M_3(\mathbb{C})$ is likely to have a quantum symmetry, in a way outlined by A. Connes in his lecture.

Finally, one can add to the algebra of the standard model an other summand $\mathbb{C}$. The resulting algebra $\mathcal{A} = \mathbb{H} \oplus \mathbb{C} \oplus \mathbb{C} \oplus M_3(\mathbb{C})$ provides us, after application of a suitable unimodularity condition, with the gauge group $U(1) \times U(1) \times SU(2) \times SU(3)$. Therefore, we have to modify the representation π of $\mathcal{A}$ as operators on $\mathcal{H}$. This is achieved by writing down the most general π leaving the $SU(2)$ and $SU(3)$ content of the theory unchanged. We select some of these representations by imposing geometrical and physical conditions. The geometrical constraints are the axioms of noncommutative geometry that rule out many of these representations. Then, we impose to the remaining representations some physical requirements such that we obtain the right hypercharges and that the photon remain massless. We end up with only a few of them that are good candidates for incorporating a extra $U(1)$ gauge boson into the standard model. Nevertheless, we still have two major troubles. First of all, all these extra $U(1)$ are anomalous. This is well known and it also remains true if we look for these $U(1)$ extensions of standard model in the more general setting of the Yang-Mills-Higgs theory The second point is that the mass of the extra gauge boson does not exceed the mass of the top quark so that it cannot be considered as a good candidate for the explanation of the anomalous experimental values [15].

3.6. Constraints on the Bosonic Sector

In the framework of noncommutative geometry, one starts with a spectral triple and then, one can construct a Yang-Mills-Higgs theory and an action for both bosonic and fermionic fields. In other words, one starts with a few basic objects concerning the fermionic sector (fermionic representation of the gauge group, mass terms,...) and then constructs the bosonic sector out of these objects. Accordingly, we then get constraints on the bosonic sector which are of two different kinds.

Since the scalar fields of the theory are obtained as 1-form computed with the finite spectral triple, we can completely determine them as functions of it. Indeed, assuming implicit space-time dependance, all the scalar fields belong to a hermitian matrix Φ that can be written as

$$\Phi = \sum_i \pi(a_0^i) \left[D, \pi(a_1^i)\right],$$

with $a_0^i, a_1^i \in \mathcal{A}$. Furthermore, we know the transformation law of Φ under a gauge transformation and its coupling to fermions. This leads us to some algebraic constraints on the bosonic sector that are under current investigation [16].

Besides, starting with a spectral triple, one can compute an action for the fields. This action looks like the action of a Yang-Mills-Higgs model (eventually coupled to gravity) and the coefficients appearing in the action can be identified with the masses and coupling constants. This gives rise to some relations between the coupling constants and the masses that are analysed, for the standard model in [17] and for the standard model coupled with gravity, in [9]. Note that it has been achieved at the tree level and the stability of these relations under renormalisation are investigated in [11].

4. CONCLUSION

In this seminar, we have outlined some basic features of the Yang-Mills-Higgs theories that can be constructed in the framework of noncommutative geometry. This construction is based on the notion of finite spectral triple which turns out to be a noncommutative geometry on a finite space. The latter can be thought as the internal space of the Yang-Mills-Higgs theory under consideration. A general theory whose internal degrees of freedom are described by a finite spectral triple leads us to some constraints on the model. We have given some of these constraints, with special emphasis on models that are close to the standard model, obtained for instance by adding a extra $U(1)$ gauge boson or a right-handed Dirac neutrino to the standard model.

Acknowledgements

It is a pleasure to thank B. Iochum and T. Schücker for their helpful advices and the organizers of the Cargese Summerschool.

References

[1] Connes A. *Géométrie non commutative*, Intereditions (1990).

[2] Connes A. *Noncommutative Geometry*, Academic Press (1994).

[3] Connes A., Lott J. *The metric aspect of noncommutative geometry*, in the Proceedings of the 1991 Cargese Summer Conference, eds: J. Frölich and al, Plenum Press.

[4] Connes A. *Gravity coupled with matter and the foundation of non commutative geometry*, hep-th/9603053 (1996).

[5] Connes A. *Noncommutative geometry and physics*, in the Proceedings of the 1992 Les Houches Summerschool in Theoretical Physics, North Holland publishing.

[6] Kastler D., Mebkhout M. *Lectures on Noncommutative Differential Geometry*, World Scientific (to be published).

[7] Madore J. *An Introduction to Noncommutative Differential Geometry and its Physical Applications*, Cambridge University Press (1995).

[8] Connes A. *Noncommutative geometry and reality*, J. Math. Phys. **36**, 6194, (1995).

[9] Chamseddine A., Connes A. *The Spectral Action Principle*, hep-th 9606001, (1996).

[10] Iochum B., Kastler D., Schücker T. *On the universal Chamseddine-Connes action 1.Details of the action computation*, hep-th 9607158, (1996).

[11] Martin C.P., Gracia-Bondía J.M. and Varilly C.: The Standard Model as a non-commutative geometry: the low energy regime, *preprint* FT/UCM-12-96, UCR-FM-6-96, hep-th/9605001 (1996).

[12] Iochum B. and Schücker T. *Yang-Mills-Higgs versus Connes-Lott*, Comm. Math. Phys. **178**, 1-26, (1996).

[13] Asquith R. *Noncommutative Geometry and the Strong Force*, phys. Lett. B 336, 220, (1995).

[14] Pris I., Schücker T. *Non-commutative geometry beyond the Standard Model*, hep-th/9604115 (1996)

[15] Krajewski T., Pris I. *Towards a Z' Gauge Boson in Noncommutative Geometry*, hep-th 9607005, (1996), to appear in Letters in Math. Phys.

[16] Krajewski T. *Classification of Finite Spectral Triples*, in preparation.

[17] Carminati L., Iochum B., Schücker T. *The noncommutative constraints on the standard model à la Connes*, hep-th 9604169 (1996), to appear in Journ. Math. Phys.

TOP QUARK MASS

Jonathan L. Rosner

Enrico Fermi Institute and Department of Physics
University of Chicago
5640 South Ellis Avenue, Chicago, IL 60637

1. INTRODUCTION

In the present article, based on a series of three lectures, we describe ways in which the large top quark mass influences a number of processes, and some ideas about how it might arise. After a brief introduction to the pattern of masses and couplings of quarks and leptons and an overview of the top quark's properties, we discuss in Section 2 the role of the top quark in mixing processes, such as particle-antiparticle mixing in the neutral kaon and B meson systems, and in other flavor-changing charge-preserving processes. Section 3 is devoted to precision electroweak experiments and the impact upon their interpretation of the precise measurements of the top quark's mass which have recently been achieved at Fermilab (Section 4). Some brief remarks about top quark production in electron-positron collisions occupy Section 5, while Section 6 contains various speculations about the source of the top quark's mass. (See also Graham Ross' lectures at this Institute [1].) Section 7 concludes.

A. Quark and lepton mass and coupling patterns

The top quark is the heaviest known elementary particle, nearly 200 times as heavy as a proton. Its mass is compared with that of the other known quarks and leptons in Figure 1. Also shown are the patterns of couplings in charge-changing weak transitions. The main coupling of the top (t) appears to be to the bottom (b) quark [2], while charm (c) couples mainly to strange (s) and up to down. Each charged lepton, in turn, appears to couple to its own neutrino. However, the pattern for quarks known at present is much richer, including weaker couplings of t to s and d, of c to b and d, and of u to b and s. This pattern is encoded in a unitary 3×3 matrix V known as the *Cabibbo-Kobayashi-Maskawa* [3, 4] matrix.

In these lectures we shall explore a number of aspects of the pattern of masses and couplings in Figure 1 and Table 1 [5]. The large mass of the top quark and its pattern of couplings to the b, s, and d quarks lead to a number of consequences for processes in which the top quark participates as a virtual particle. Not the least of these is the ability of loop diagrams involving the top quark to generate the observed CP violation in the neutral kaon system [4]. At the same time, the fact that the top quark is about

Masses of Fundamental Particles: Cargèse 1996
Edited by M. Lévy *et al.*, Plenum Press, New York, 1997

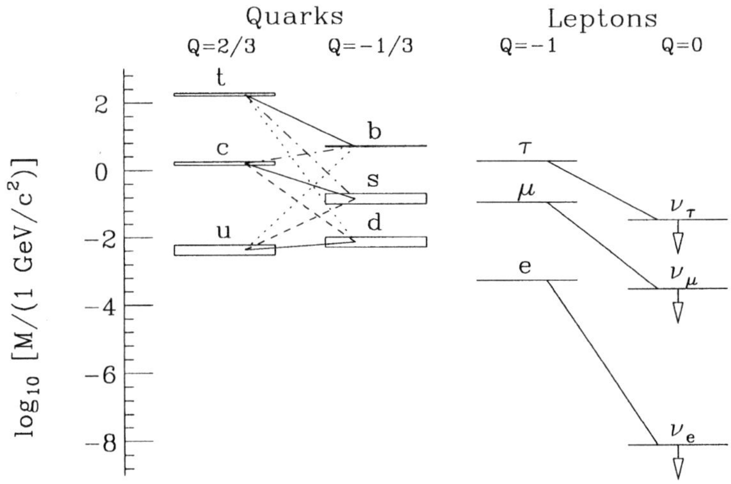

Figure 1. Patterns of charge-changing weak transitions among quarks and leptons. Direct evidence for ν_τ does not yet exist. The strongest inter-quark transitions correspond to the solid lines, with dashed, dot-dashed, and dotted lines corresponding to successively weaker transitions.

twice as heavy as the gauge bosons W and Z of the electroweak interactions [6] offers a hint that it may play a crucial role in the understanding of all the quark and lepton masses.

A note of caution: if its mass is viewed on a logarithmic scale, as in Figure 1, the top quark is not all that exceptional. Although it is more than 100 times as heavy as the next lightest quark of charge 2/3, the charmed quark, we have not yet explored similar logarithmic mass intervals above the bottom quark or τ lepton. But the Z decays to just three pairs of light neutrinos [7]. so the pattern of any fermions heavier than those in Figure 1 must at least change to a significant extent.

Table 1. Relative strengths of charge-changing weak transitions.

Relative amplitude	Transition	Source of information (example)
~ 1	$u \leftrightarrow d$	Nuclear β-decay
~ 1	$c \leftrightarrow s$	Charmed particle decays
~ 0.22	$u \leftrightarrow s$	Strange particle decays
~ 0.22	$c \leftrightarrow d$	Neutrino prod. of charm
~ 0.04	$c \leftrightarrow b$	b decays
~ 0.003	$u \leftrightarrow b$	Charmless b decays
~ 1	$t \leftrightarrow b$	Dominance of $t \to Wb$
~ 0.04	$t \leftrightarrow s$	Only indirect evidence
~ 0.01	$t \leftrightarrow d$	Only indirect evidence

B. Quark and lepton families

The case for a quark-lepton analogy is suggested by the doublets

$$\begin{pmatrix} \nu_e \\ e^- \end{pmatrix} : \text{leptons} \quad ; \quad \begin{pmatrix} u \\ d \end{pmatrix} : \text{quarks} \quad . \tag{1}$$

In both cases the states in the upper and lower rows are connected by emission or absorption of a W boson. This pattern was extended to the second family by several authors [8], who inferred the existence of a fourth quark (charm) coupling mainly to the strange quark from the existence of the $\mu - \nu_\mu$ doublet:

$$\begin{pmatrix} \nu_\mu \\ \mu^- \end{pmatrix} : \text{leptons} \Rightarrow \begin{pmatrix} c \\ s \end{pmatrix} : \text{quarks} \quad . \tag{2}$$

The suppression of processes involving weak decays of strange particles as compared with strangeness-preserving weak decays could be made compatible with the universal strength of the weak interactions if the u quark coupled to an appropriately normalized linear combination of d and s:

$$u \leftrightarrow d \cos\theta_c + s \sin\theta_c \quad . \tag{3}$$

This universality property was proposed by Gell-Mann and Lévy [9] and used by Cabibbo [3] (hence the subscript on the angle) to relate successfully a number of beta-decay processes of non-strange and strange particles. The strength of the charge-changing weak current could be normalized by considering it and its hermitian adjoint to be members of an SU(2) triplet. The neutral member of this triplet contained strangeness-changing terms and thus had to be interpreted as unphysical, since no strangeness-changing neutral-current processes had been seen. However, if one then considered the charmed quark to couple to the orthogonal combination,

$$c \leftrightarrow -d \sin\theta_c + s \cos\theta_c \quad , \tag{4}$$

the corresponding neutral current would couple to the combination $u\bar{u} + c\bar{c} - d\bar{d} - s\bar{s}$, and would preserve flavor [8, 10]. This property was preserved in higher orders of perturbation theory, and neutral strangeness-changing processes were induced only to the extent that the charm and u quark masses differed from one another [10].

A further motivation for the quark-lepton analogy was noted by Bouchiat, Iliopoulos, and Meyer [11] in 1972. In a gauge theory of the electroweak interactions, triangle anomalies associated with graphs of the type shown in Figure 2 have to be avoided. This cancellation requires the fermions f in the theory to contribute a total of zero to the sum over f of $Q_f^2 I_{3L}^f$. Such a cancellation can be achieved by requiring quarks and leptons to occur in complete families of the type mentioned above, so that the terms

$$\text{Leptons}: \quad (0)^2\left(\frac{1}{2}\right) + (-1)^2\left(-\frac{1}{2}\right) = -\frac{1}{2} \tag{5}$$

$$\text{Quarks}: \quad 3\left[\left(\frac{2}{3}\right)^2\left(\frac{1}{2}\right) + \left(-\frac{1}{3}\right)^2\left(-\frac{1}{2}\right)\right] = \frac{1}{2} \tag{6}$$

sum to zero for each family.

Some history of how the quark and lepton families were pieced together may be of interest [12]. In the family (1), the first particle to be discovered was the electron, whose unique properties were clinched by J. J. Thomson's measurement of its enormous

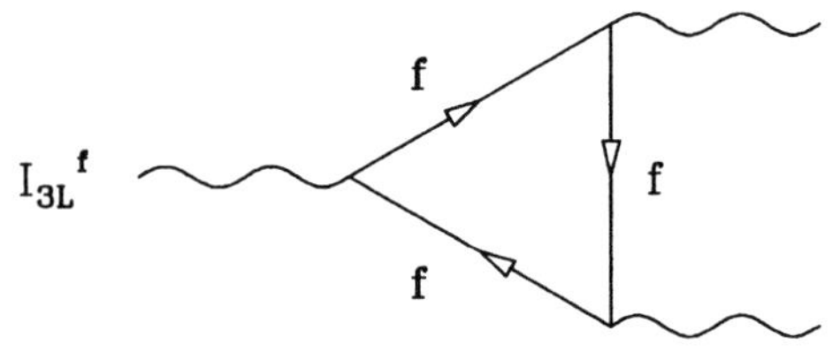

Figure 2. Example of triangle diagram for which leading behavior must cancel in a renormalizable electroweak theory.

charge-to-mass ratio (nearly 2000 times that of the proton) in 1897. The corresponding neutrino was proposed by W. Pauli [13] in 1931. As for the u and d quark, the need for two constituents of hadrons differing in charge by a unit but with very similar behaviour under the strong interactions dates back to the recognition of isotopic spin as a good symmetry [14], in the 1930's.

Evidence for the second family (2) began with the discovery in 1937 by S. Neddermeyer and C. Anderson of the muon in cosmic radiation [15]. The corresponding neutrino ν_μ was proved distinct from the electron neutrino ν_e in 1962 by a group led by L. Lederman, M. Schwartz, and J. Steinberger [16]. Meanwhile, evidence for particles containing one or more strange (s) quarks, initially seen in cosmic radiation and later produced at accelerators, began accumulating over the period 1946-53. The charmed (c) quark was discovered initially in the form of a $c\bar{c}$ bound state, the J [17] or ψ [18] (the dual name J/ψ has survived). Particles containing a single charmed quark ("bare charm") were identified a bit later [19].

At the same time evidence for the charmed quark was accumulating in electron-positron annihilations, the first member of a third family:

$$\begin{pmatrix} \nu_\tau \\ \tau^- \end{pmatrix} : \text{leptons} \quad ; \quad \begin{pmatrix} t \\ b \end{pmatrix} : \text{quarks} \quad . \tag{7}$$

was making its appearance. This was the τ lepton, identified by M. Perl and collaborators [20]. If this were to be a lepton like the e and μ, having its own neutrino, the anomaly-cancellation mechanism mentioned earlier [11] then required there to be a third pair of quarks: the "top" and "bottom" [21], our present names for the quarks proposed by Kobayashi and Maskawa [4] as candidates for explaining CP violation.

The bottom quark made its appearance in the form of $b\bar{b}$ bound states discovered at Fermilab in 1977 by a group led by L. Lederman [22] and in "bare bottom" (otherwise known as "beauty") mesons containing single b quarks seen at the Cornell Electron Storage Ring (CESR) somewhat later [23]. Eagerly awaited, the top quark finally made its appearance 17 years after the bottom, in proton-antiproton collisions at Fermilab at a center-of-mass energy of 1.8 TeV [24].

Even up to the present day, in the absence of direct evidence for the ν_τ, we have never had a complete set of quark and lepton families without hints of the next family. By the time the ν_τ is discovered, perhaps at a forthcoming experiment at Fermilab [25], who knows what other particles may appear?

C. Top quark properties

The most recent measurements of the top quark mass, reported at the 1996 Warsaw

Conference, are [26]

$$m_t = 176.8{\pm}4.4{\pm}4.8\ \mathrm{GeV}/c^2\ (\mathrm{CDF})\ ;\ 169{\pm}8{\pm}8\ \mathrm{GeV}/c^2\ (\mathrm{D0})\ ;\ 175{\pm}6\ \mathrm{GeV}/c^2\ (\mathrm{avg.})\ . \tag{8}$$

The top is expected to decay essentially 100%($\simeq |V_{tb}|^2$) to $W + b$, with branching ratios of only about $|V_{ts}|^2 \simeq 1.5 \times 10^{-3}$ to $W + s$ and $|V_{td}|^2 \simeq 10^{-4}$ to $W + d$. We shall see in Sec. 4 that recent data [2] support this expectation.

The predicted rate for the decay $t \to W + b$ is

$$\Gamma(t \to W + b) = \frac{G_F}{8\pi\sqrt{2}}|V_{tb}|^2 m_t^3 \Phi_K K_{\mathrm{QCD}} \quad , \tag{9}$$

where, in the limit in which the b quark mass can be neglected,

$$\Phi_K \equiv \left(1 - \frac{M_W^2}{m_t^2}\right)^2 \left(1 + 2\frac{M_W^2}{m_t^2}\right) = 0.885 \tag{10}$$

and K_{QCD} is a QCD correction factor which is about 0.917 [27]. For $m_t = 175$ GeV, one thus predicts $\Gamma = (1.76\ \mathrm{GeV})(0.885)(0.917) \simeq 1.43$ GeV, so that the top quark decays before it can form hadrons. Thus, sad to say, we will not be able to study the properties of $t\bar{t}$ or $t\bar{q}$ states (where q stands for one of the lighter quarks). Nonetheless, some crude information about the $t\bar{t}$ interaction can be recovered, as we shall mention in Sec. 5.

The spin of the top quark is expected to be 1/2, just like that of the other quarks. Direct evidence for this is lacking, but if we are prepared to admit that $J(t) = J(b)$, we can use all of the bottom quark's rich spectroscopy to conclude that $J(b) = 1/2$ and hence $J(t) = 1/2$. The spectra of $c\bar{c}$, $b\bar{b}$, charm – light quark and bottom – light quark bound states are shown in Figs. 3, 4, 5, and 6 respectively. We shall use some of the terminology in these figures in what follows, so they are worth getting acquainted with.

The charmonium ($c\bar{c}$) spectrum in Fig. 3 bears strong evidence that the spin of the charmed quark is 1/2, as one might have expected from its partnership with the strange quark. (Spins of strange particles have been directly measured for a number of years and support the notion that $J(s) = 1/2$.) The S-wave ($L = 0$) levels have total angular momentum J, parity P, and charge-conjugation eigenvalue C equal to $J^{PC} = 0^{+-}$ and 1^{--} as one would expect for ${}^1S_0$ and ${}^3S_1$ states, respectively, of a quark and antiquark. The P-wave ($L = 1$) levels have $J^{PC} = 1^{+-}$ for the ${}^1P_1$, 0^{++} for the ${}^3P_0$, 1^{++} for the ${}^3P_1$, and 2^{++} for the ${}^3P_2$. The $J^{PC} = 1^{--}$ levels are identified as such by their copious production through single virtual photons in e^+e^- annihilations. The 0^{-+} level η_c is produced via single-photon emission from the J/ψ (so its C is positive) and has been directly measured to have J^P compatible with 0^- [28]. Numerous studies have been made of the electromagnetic (electric dipole) transitions between the S-wave and P-wave levels [29] and they, too, support the assignments shown.

The $b\bar{b}$ (upsilon) levels shown in Fig. 4 have a very similar structure, aside from an overall shift. The similarity of the $c\bar{c}$ and $b\bar{b}$ spectra is in fact an accident of the fact that for the interquark distances in question (roughly 0.2 to 1 fm), the interquark potential interpolates between short-distance Coulomb-like and long-distance linear behavior in a manner roughly compatible with $V \sim \log r$ [30] or with an effective power near zero [31]. The copious production of 1^{--} candidiates in e^+e^- annihilations and the pattern of electric dipole transitions between S- and P-wave levels again supports the assignments shown.

States consisting of a single charmed quark and light (u, d, or s) quarks or antiquarks, shown in Fig. 5, again support $J(c) = 1/2$. The lightest mesons have $J^P = 0^-$

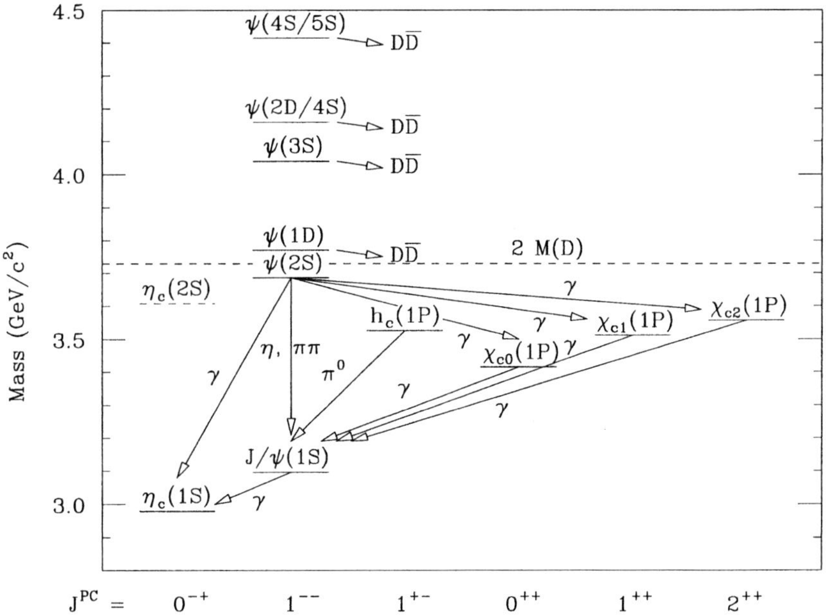

Figure 3 Charmonium ($c\bar{c}$) spectrum. Observed and predicted levels are denoted by solid and dashed horizontal lines, respectively. Arrows denote electromagnetic transitions (labeled by γ) and hadronic transitions (labeled by emitted hadrons).

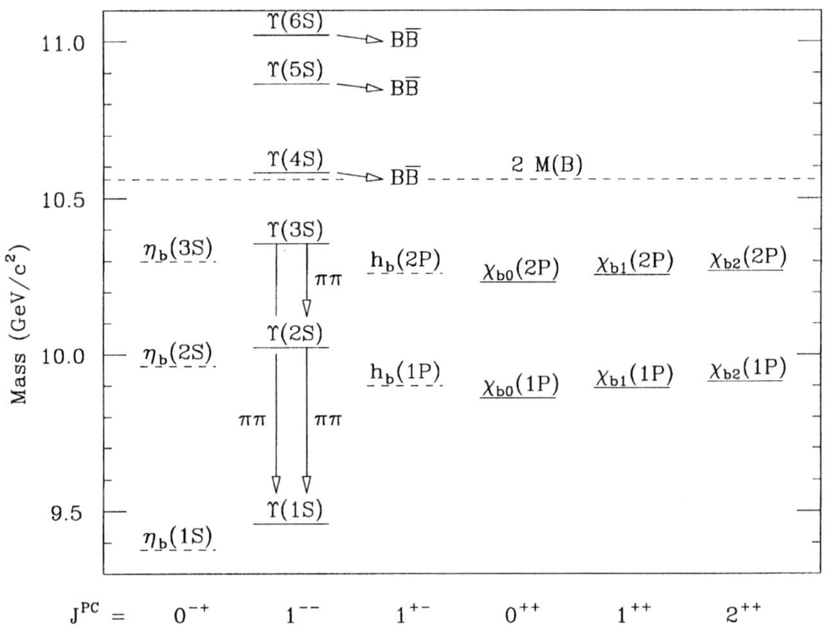

Figure 4. Spectrum of $b\bar{b}$ states. Observed and predicted levels are denoted by solid and dashed horizontal lines, respectively. In addition to the transitions labeled by arrows, numerous electric dipole transitions and decays of states below $B\bar{B}$ threshold to hadrons containing light quarks have been seen.

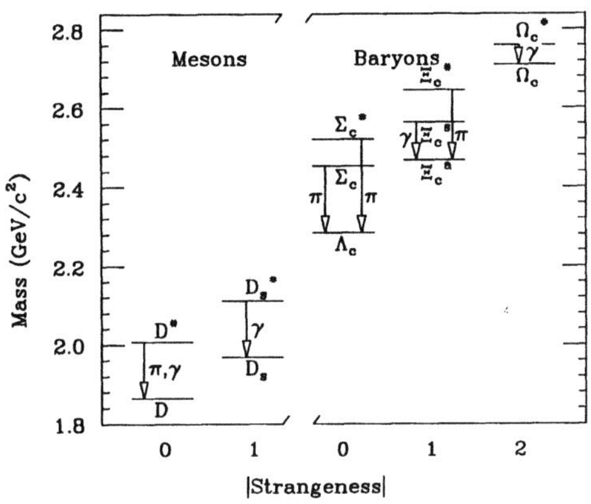

Figure 5. Spectrum of lowest-lying states containing one charmed and one light quark. Observed and predicted levels are denoted by solid and broken horizontal lines, respectively.

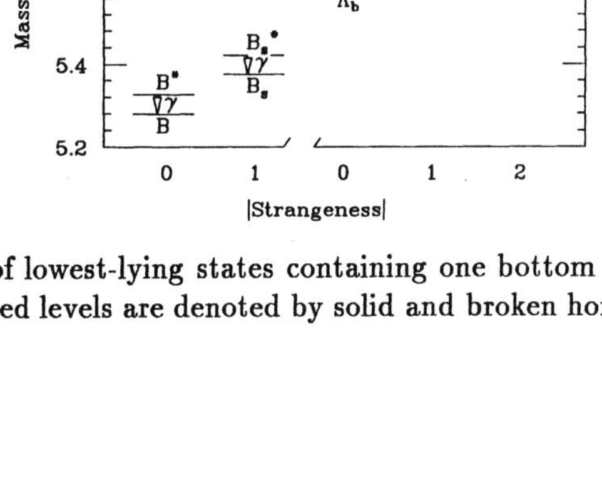

Figure 6. Spectrum of lowest-lying states containing one bottom and one light quark. Observed and predicted levels are denoted by solid and broken horizontal lines, respectively.

($^1S_0$) and 1^- ($^3S_1$). Two charmed baryons discovered fairly recently, both strange [32] and nonstrange [33], are candidates for $J = 3/2$. One expects such states as bound states of three spin-1/2 quarks.

Finally, the pattern of states containing a single b quark (Fig. 6) is very similar to that for singly-charmed states, though not as well fleshed-out. In many cases the splittings between states containing a single b quark is less than that between the corresponding charmed states by roughly a factor of $m_c/m_b \simeq 1/3$ as a result of the smaller chromomagnetic moment of the b quark. (This feature has also been noted by Neubert at this Institute [34].)

The top quark's mass, just like every other mass in the standard electroweak theory, arises as a result of Yukawa interactions of one or more Higgs bosons with the fermions. For the t and b quarks, the relevant part of the Lagrangian can be written as

$$\mathcal{L}^{t,b}_{\text{Higgs}} = -g^b_Y[\bar{Q}_L\phi_1 b_R + \text{H.c.}] - g^t_Y[\bar{Q}_L\phi_2 t_R + \text{H.c.}] \tag{11}$$

where

$$Q_L \equiv \begin{pmatrix} t \\ b \end{pmatrix}_L \ , \quad \phi_1 \equiv \begin{pmatrix} \phi_1^+ \\ \phi_1^0 \end{pmatrix} \ , \quad \phi_2 \equiv \begin{pmatrix} \phi_2^0 \\ \phi_2^- \end{pmatrix} \tag{12}$$

if there are two Higgs doublets, or $\phi_2 = i\sigma_2\phi_1^c$ if there is only one. If the neutral member pf each Higgs doublet acquires a vacuum expectation value $\langle\phi_i^0\rangle = v_i/\sqrt{2}$, then $m_b = g^b_Y v_1/\sqrt{2}$, $m_t = g^t_Y v_2/\sqrt{2}$. Here

$$2^{-1/4}G_F^{-1/2} = v_1^2 + v_2^2 = (246 \text{ GeV})^2 \quad , \tag{13}$$

while the right-hand side is replaced by $v_1^2 \equiv v^2$ if $\phi_2 = i\sigma_2\phi_1^c$.

Suppose there were just one Higgs doublet, with $v \simeq 246$ GeV. Then

$$g^t_Y = \frac{246 \text{ GeV}}{\sqrt{2}m_t} = \frac{174 \text{ GeV}}{175 \pm 6 \text{ GeV}} \simeq 1 \ ! \tag{14}$$

The significance of this relation may be accidental, however. Quark masses "run" as a result of their interactions with gauge fields (particularly with gluons), and so their values depend on the scale at which the mass is probed. As one example, the top quark's "pole" mass (what one would measure in a physical process of top quark production) and the value $\bar{m}_t(\mu)$ in the modified-minimal-subtraction scheme at a mass scale μ are related [35] by

$$\frac{m_t^{\text{pole}}}{\bar{m}_t(\mu)} = 1 + \frac{4}{3}\frac{\alpha_s(\mu)}{\pi} + \ldots \quad . \tag{15}$$

For example, with a pole mass of 175 GeV, and $\mu = M_W$ (which will frequently be appropriate for the loop calculations we shall perform), so that $\alpha_s(M_W) \simeq 0.12$, one has $m_t^{\text{pole}}/\bar{m}_t(M_W) \simeq 1.05$. Taking into account higher-order effects as well, we shall take $\bar{m}_t(M_W) = 165 \pm 6$ GeV/c^2 [36] in a number of subsequent calculations.

2. THE TOP QUARK IN MIXING PROCESSES

A. Cabibbo-Kobayashi-Maskawa matrix parameters

The electroweak Lagrangian, before electroweak symmetry breaking, may be written in flavor-diagonal form as

$$\mathcal{L}_{\text{int}} = -\frac{g}{\sqrt{2}}[\overline{U'}_L\gamma^\mu W_\mu^{(+)}D'_L + \text{H.c.}] \quad , \tag{16}$$

Table 2. Parameters of KM matrices for n doublets of quarks.

	$n =$	2	3	4
Number of parameters	$(n-1)^2$	1	4	9
Number of angles	$n(n-1)/2$	1	3	6
Number of phases	$(n-1)(n-2)/2$	0	1	3

where $U' \equiv (u', c', t')$ and $D' \equiv (d', s', b')$ are column vectors decribing *weak eigenstates.* Here g is the weak $SU(2)_L$ coupling constant, and $\psi_L \equiv (1-\gamma_5)\psi/2$ is the left-handed projection of the fermion field $\psi = U$ or D.

Quark mixings arise because mass terms in the Lagrangian are permitted to connect weak eigenstates with one another. Thus, the matrices $\mathcal{M}_{U,\ D}$ in

$$\mathcal{L}_m = -[\overline{U'}_R \mathcal{M}_U U'_L + \overline{D'}_R \mathcal{M}_D D'_L + \text{H.c.}] \tag{17}$$

may contain off-diagonal terms. One may diagonalize these matrices by separate unitary transformations on left-handed and right-handed quark fields:

$$R_Q^+ \mathcal{M}_Q L_Q = L_Q^+ \mathcal{M}_Q^+ R_Q = \Lambda_Q \quad . \tag{18}$$

where

$$Q'_L = L_Q Q_L; \quad Q'_R = R_Q Q_R \quad (Q = U, D) \quad . \tag{19}$$

Using the relation between weak eigenstates and mass eigenstates: $U'_L = L_U U_L$, $D'_L = L_D D_L$, we find

$$\mathcal{L}_{\text{int}} = -\frac{g}{\sqrt{2}}[\overline{U}_L \gamma^\mu W_\mu V D_L + \text{H.c.}] \quad , \tag{20}$$

where $U \equiv (u, c, t)$ and $D \equiv (d, s, b)$ are the mass eigenstates, and $V \equiv L_U^\dagger L_D$. The matrix V is just the Cabibbo-Kobayashi-Maskawa matrix. By construction, it is unitary: $V^\dagger V = VV^\dagger = 1$. It carries no information about R_U or R_D. More information would be forthcoming from interactions sensitive to right-handed quarks or from a genuine theory of quark masses. Because of the unitarity of the matrix, the neutral currents to which the Z^0 couples are flavor-diagonal: $\overline{Q'}\Gamma Q' = \overline{Q}\Gamma Q$, where Γ is any combination of γ^μ and $\gamma^\mu\gamma_5$.

For n u-type quarks and n d-type quarks, V is $n \times n$ and unitary. An arbitrary $n \times n$ matrix has $2n^2$ real parameters, but unitarity ($V^\dagger V = 1$) provides n^2 constraints, so only n^2 real parameters remain. We may remove $2n - 1$ of these by appropriate redefinitions of relative quark phases. The number of remaining parameters is then $n^2 - (2n-1) = (n-1)^2$. Of these, $n(n-1)/2$ (the number of independent rotations in n dimensions) correspond to angles, while the rest, $(n-1)(n-2)/2$, correspond to phases. We summarize these results in Table 2.

For $n = 2$, we have one angle and no phases. The matrix V then can always be chosen as orthogonal [3, 8, 10]. For $n = 3$, we have three angles and one phase, which in general cannot be eliminated by arbitrary choices of phases in the quark fields. It was this phase that motivated Kobayashi and Maskawa [4] to introduce a third quark doublet. It provides a potential source of CP violation, serving as the leading contender for the observed CP-violating effects in the kaon system and suggesting substantial CP asymmetries in the decays of mesons containing b quarks.

The CKM matrix V is then, explicitly,

$$V = \begin{pmatrix} V_{ud} & V_{us} & V_{ub} \\ V_{cd} & V_{cs} & V_{cb} \\ V_{td} & V_{ts} & V_{tb} \end{pmatrix} \quad . \tag{21}$$

We now parametrize its elements.

It is convenient to choose quark phases [37] so that the n diagonal elements and the $n-1$ elements just above the diagonal are real and positive. The parametrization we shall employ is one suggested by Wolfenstein [38].

The diagonal elements of V are nearly 1, while the dominant off-diagonal elements are $V_{us} \simeq -V_{cd} \equiv \lambda \simeq 0.22$. Thus to order λ^2, the upper 2×2 submatrix of V is already known from the four-quark pattern. One may identify $\lambda = \sin\theta_c$, where θ_c is the Gell-Mann – Lévy – Cabibbo angle mentioned earlier. It is the single parameter needed to describe the mixing in the four-quark ($n = 2$) system. More precisely, analyses of strange particle decays [39] have yielded $\sin\theta_c = 0.2205 \pm 0.0018$.

The matrix element $|V_{ud}|^2$ may be obtained by comparing the strengths of certain beta-decay transitions involving vector transitions with that of muon decay. One can also measure the neutron decay rate (which involves both vector and axial vector transitions), and extract the vector coupling strength by finding g_A from decay asymmetries. This vector coupling strength may be compared with that obtained in muon decay to learn $|V_{ud}|^2$. Finally, one can study the decay $\pi^+ \to \pi^0 e^+ \nu_e$. Overall, one finds [5] $|V_{ud}| \simeq 0.975 \pm 0.001$, so that $|V_{ud}|^2 + |V_{us}|^2 = 0.999 \pm 0.002$. One cannot deduce the need for the additional contribution of $|V_{ub}|^2$ in the relation $|V_{ud}|^2 + |V_{us}|^2 + |V_{ub}|^2 = 1$ required by the unitarity of the CKM matrix. Measurements of $|V_{cd}|$ and $|V_{cs}|$ (see Ref. [5]) are also consistent with the predictions of unitarity, but with larger errors.

The long b quark lifetime (about 1.5 to 1.6 ps [40]) and the predominance of charmed quarks among b decay products implies that $V_{cb} \simeq 0.04$, allowing one to express it as $A\lambda^2$, where $A = \mathcal{O}(1)$. A recent compilation of results [36, 41] on semileptonic $b \to c$ decays yields $V_{cb} = A\lambda^2 = 0.0393 \pm 0.0028$, or $A = 0.808 \pm 0.058$. [M. Neubert [34] quotes $V_{cb} = 0.0388 \pm 0.0020_{\text{exp}} \pm 0.0012_{\text{th}}$.] Unitarity then requires $V_{ts} \simeq -A\lambda^2$ as long as V_{td} and V_{ub} are small enough (which they are).

The magnitude of the element V_{ub} is learned by comparing charmless b decays to those with charm. One thus finds that $|V_{ub}|$ appears to be of order $A\lambda^3$. Here one must allow for a phase, so one must introduce two new parameters ρ and η: $V_{ub} = A\lambda^3(\rho - i\eta)$. The measured ratio [36] $|V_{ub}/V_{cb}| = 0.08 \pm 0.016$, whose main error is dominated by theoretical extrapolation from a small part of the lepton spectrum in semileptonic $b \to u$ decays, implies

$$(\rho^2 + \eta^2)^{1/2} = 0.363 \pm 0.073 \quad . \tag{22}$$

This constraint can be plotted in the (ρ, η) plane as a band bounded by circles centered at (0,0).

Finally, unitarity specifies uniquely the form $V_{td} = A\lambda^3(1 - \rho - i\eta)$. To summarize, the CKM matrix may be written in terms of the 4 parameters λ, A, ρ, and η as

$$V \approx \begin{bmatrix} 1 - \lambda^2/2 & \lambda & A\lambda^3(\rho - i\eta) \\ -\lambda & 1 - \lambda^2/2 & A\lambda^2 \\ A\lambda^3(1 - \rho - i\eta) & -A\lambda^2 & 1 \end{bmatrix} \quad . \tag{23}$$

The form (23) is only correct to order λ^3 in the matrix elements. For certain purposes it may be necessary to exhibit corrections of higher order to the elements. This can be done using the unitarity of the matrix.

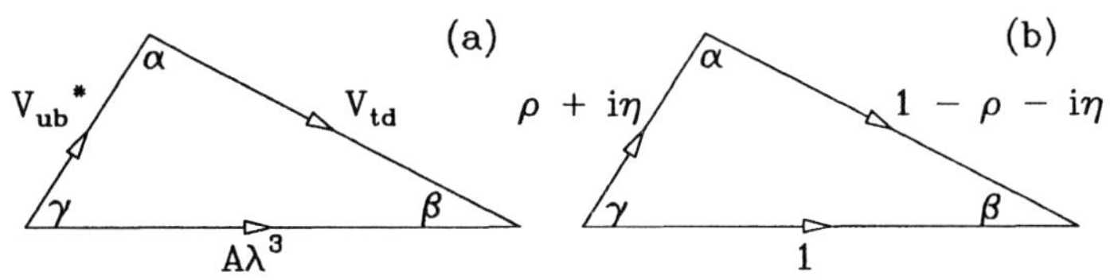

Figure 7. Unitarity triangle for CKM elements. (a) The relation (24) in the complex plane; (b) Eq. (24) divided by the normalizing factor $A\lambda^3$.

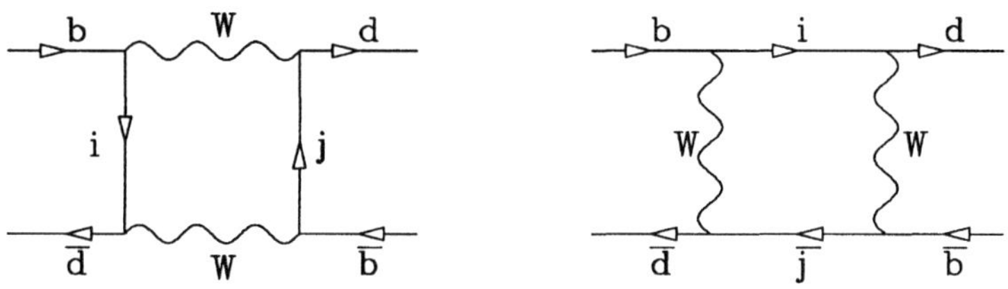

Figure 8. Box diagrams for mixing of B^0 and $\bar{B}^0$.

The unitarity of the CKM matrix implies that the scalar product of one row (or column) with the complex conjugate of any other row (or column) must vanish: for example,

$$V_{ud}^* V_{td} + V_{us}^* V_{ts} + V_{ub}^* V_{tb} = 0 \quad . \tag{24}$$

Since $V_{ud}^* \approx 1$, $V_{us}^* \approx \lambda$, $V_{ts} \approx -A\lambda^2$, and $V_{tb} \approx 1$ we have $V_{td} + V_{ub}^* = A\lambda^3$, a useful relation expressing the least-known CKM elements in terms of relatively well-known parameters. This result can be visualized as a triangle [42] in the complex plane [Fig. 7(a)]. In this figure the angles α, β, and γ are defined as in the review by Nir and Quinn [43].

Dividing (24) by $A\lambda^3$, since $V_{ub}^*/A\lambda^3 = \rho + i\eta$, $V_{td}/A\lambda^3 = 1 - \rho - i\eta$, one obtains a triangle of the form shown in Fig. 7(b). The value of $V_{ub}^*/A\lambda^3$ may then be depicted as a point in the (ρ, η) plane. The major ambiguity which still remains in the determination of the CKM matrix elements concerns the shape of the unitarity triangle. The answer depends on the magnitude of V_{td}. As we shall see, decays alone will not provide the answer. In order to learn about the elements V_{td} and V_{ts} one must resort to indirect means, which involve loop diagrams.

B. Mixing of neutral B mesons

The box diagrams of Fig. 8 dominate the mixing between B^0 and $\bar{B}^0$, leading to a splitting Δm between the mass eigenstates. (We present an abbreviated account of mixing; for more details see [5, 44, 45, 46].) The leading contributions in each of these diagrams cancel one another when one sums over all the intermediate quarks of charge 2/3, since $V_{ub} V_{ud}^* + V_{cb} V_{cd}^* + V_{tb} V_{td}^* = 0$. The remaining contributions are dominated by the top quark since the corresponding CKM products of pairs of CKM elements $V_{ib} V_{id}^*$ are each of order λ^3 and $m_t \gg m_c, m_u$. One then finds [46]

$$\Delta m = \frac{G_F^2}{6\pi^2} |V_{td}|^2 M_W^2 m_B f_B^2 B_B \eta_B S\left(\frac{m_t^2}{M_W^2}\right) \quad , \tag{25}$$

where

$$S(x) \equiv \frac{x}{4}\left[1 + \frac{3-9x}{(x-1)^2} + \frac{6x^2 \ln x}{(x-1)^3}\right] \quad . \tag{26}$$

This factor behaves as x for small x and $x/4$ for large x and equals $3/4$ for $x = 1$.

We take $m_t = 175 \pm 6$ GeV/c^2 as noted in Sec. 1 C, $M_W = 80.34 \pm 0.10$ GeV/c^2 (see Sec. 3), $m_B = 5.279$ GeV/c^2 (see [47]), and $f_B\sqrt{B_B} = 200 \pm 40$ MeV [34, 36]. Here f_B is the B meson decay constant, defined in such a way that the matrix element of the weak axial-vector current $A_\mu \equiv \bar{b}\gamma_\mu\gamma_5 d$ between a B^0 meson and the vacuum is $\langle 0|A_\mu|B^0(p)\rangle = ip_\mu f_B$. With this normalization, the decay constants of the light pseudoscalar mesons are $f_\pi = 131$ MeV and $f_K = 160$ MeV. The meson decay constants express the amplitude for the corresponding quark and antiquark (e.g., b and $\bar{d}$) to be found at a point, as they must in order to participate in the short-distance processes expressed by Figs. 8.

The factor B_B expresses the degree to which the diagrams of Fig. 8 actually provide the contribution to $B - \bar{B}$ mixing. It is sometimes known as the "vacuum saturation parameter" or, more obscurely, as the "bag parameter" since an early estimate of the corresponding quantity for kaons was performed in a "bag" model of quarks confined in hadrons. It has been estimated in lattice gauge theories [48] to be 1.16 ± 0.08. Finally, $\eta_B = 0.55$ is a QCD correction. All quantities are quoted in the same consistent renormalization scheme [49] and often appear in the literature with hats.

The first hints of $B - \bar{B}$ mixing were obtained by the UA1 Collaboration in 1986 [50]. In proton-antiproton collisions, an excess was observed of like-sign muons above background, which could be interpreted as the effects of $B - \bar{B}$ pair production, followed by oscillation of one of the (probably strange) B's to its antiparticle and semileptonic decay of both B's. (As we shall see, the mixing amplitude for strange B's is expected to be much larger than that for nonstrange ones.)

The first evidence for mixing of nonstrange B's was obtained by the ARGUS Collaboration in 1987 [51]. The reaction $e^+e^- \to B^0\bar{B}^0$ was studied at threshold; the observation of like-sign lepton pairs from the semileptonic decays of b quarks (e.g., $b \to c\ell^-\nu_\ell$, where $\ell = e$ or μ) indicated that one of the neutral B mesons had oscillated to its antiparticle. The large mixing amplitude, $\Delta m/\Gamma \simeq 0.7$ (where Δm is the mass difference between mass eigenstates and Γ is the B meson decay rate), was one early indication of a very heavy top quark.

With the rapidly moving B mesons and the fine vertex information now available at LEP [52], CDF [53], and SLD [54] (see also the review by [40]), it has become possible to directly observe time-dependent $B^0 - \bar{B}^0$ oscillations with a modulating factor $\sin(\Delta m t)$ (where t is the proper decay time). The current world average [55] is $\Delta m_d = 0.470 \pm 0.017$ ps^{-1}, where the subscript refers to the mixing between $B^0 \equiv \bar{b}d$ and $\bar{B}^0 \equiv b\bar{d}$. Using the expression (25) and the parameters mentioned above, we can then obtain an estimate of $|V_{td}|$, which leads, once we factor out a term $A\lambda^3$, to the constraint

$$|1 - \rho - i\eta| = 1.01 \pm 0.22 \quad . \tag{27}$$

This result can be plotted in the (ρ, η) plane as a band bounded by circles with centers at (1,0).

Now that the top quark mass is known so precisely, the dominant source of error in Eq. (27) is uncertainty in the B meson decay constant f_B. Pseudoscalar meson decay constants, through the axial-current matrix element mentioned earlier, govern purely leptonic processes such as $\pi \to \mu\nu$, $K \to \mu\nu$, and the recently observed $D_s \to \mu\nu$ and $D_s \to \tau\nu$. (The $D_s = c\bar{s}$ is the lowest-lying charmed-strange meson; see Fig. 5.) Information on heavy meson decay constants can be improved in several ways [5].

Direct measurements of D_s leptonic decays have been reported by the WA75, CLEO, BES, and E653 Collaborations [56], and one can also estimate D_s by assuming factorization in certain B decays in which the charged weak current produces a D_s [57]. A recent compilation of experimental values [58] yields $f_{D_s} = (241 \pm 21 \pm 30)$ MeV.

Estimates based on flavor SU(3), whereby one relates f_{D_s} to the corresponding decay constant f_D of the nonstrange charmed meson, yield ratios f_D/f_{D_s} in the range 0.8 – 0.9. Thus, one expects a value of f_D not far below the current experimental upper bound [59] $f_D < 290$ MeV (90% c.l.). This limit was placed by studying the reaction $e^+e^- \to D^+D^-$ just above threshold at the SPEAR storage ring, and looking for the distinctive decay $D \to \mu\nu$ opposite an identified D. The Beijing Electron-Positron Collider collaboration [60] has identified one $D \to \mu\nu$ event, leading to $f_D = 300^{+180+80}_{-150-40}$ MeV.

Direct measurements of f_B are possible once $|V_{ub}|$ is fairly well known, since the partial width $\Gamma(B \to \ell\nu)$ is proportional to $(f_B|V_{ub}|)^2$. The expected branching ratios are about $(1/2) \times 10^{-4}$ for $\tau\nu$ and 2×10^{-7} for $\mu\nu$ [5]. Theoretical estimates of f_B take many forms, including quark models [61] and lattice gauge theory calculations [62], leading to a rough range $f_B = 180 \pm 40$ MeV.

The same SU(3) estimates [61, 62] for the ratio of f_D/f_{D_s} also give a very similar ratio of f_B/f_{B_s}. It has been shown that the equality of these two ratios is to be expected within a few percent [63].

C. CP-violating $K^0 - \bar{K}^0$ mixing

The top quark (and its charge $-1/3$ partner, the bottom) were invented [4] to explain CP violation in the neutral kaon system. More than thirty years after its discovery [64], this remarkable phenomenon has only a candidate theory to explain it, with confirmation of the explanation still to come. (For a compendium of literature on CP violation, see [65].)

The K^0 and $\bar{K}^0$ are strong-interaction eigenstates of opposite strangeness, but the weak interactions do not conserve strangeness. Hence the weak interactions may, and do, pick out linear combinations of K^0 and $\bar{K}^0$ in decay processes [66]. As of 1957, when the weak interactions were understood to violate charge-conjugation invariance C and spatial reflection P but to preserve their product CP, one expected [67] the linear combination $K_1^0 \equiv (K^0 + \bar{K}^0)/\sqrt{2}$, with even CP, to have a much more rapid decay rate since it could decay to the CP-even final state of two pions. The orthogonal linear combination $K_2^0 \equiv (K^0 - \bar{K}^0)/\sqrt{2}$, with odd CP, would live much longer since it was forbidden by CP invariance to decay to two pions and would have to decay to three pions or a pion and a lepton-neutrino pair. In fact, a long-lived neutral kaon does exist [68], with a lifetime about 600 times that of the short-lived variety.

In 1964 J. Christenson, J. Cronin, V. Fitch, and R. Turlay reported that in fact the long-lived neutral kaon *did* decay to two pions, with an amplitude whose magnitude is about 2×10^{-3} that for the short-lived $K \to 2\pi$ decay. To reach this conclusion they constructed a beam of neutral long-lived kaons and observed their decays in a spark chamber [64].

One then can parametrize the mass eigenstates as

$$K_S \text{ (“short”)} \simeq K_1 + \epsilon K_2 \ , \quad K_L \text{ (“long”)} \simeq K_2 + \epsilon K_1 \quad , \tag{28}$$

where $|\epsilon| \simeq 2 \times 10^{-3}$ and the phase of ϵ turns out to be about $\pi/4$. The parameter ϵ encodes all that is currently known about CP violation in the neutral kaon system. But where does it come from?

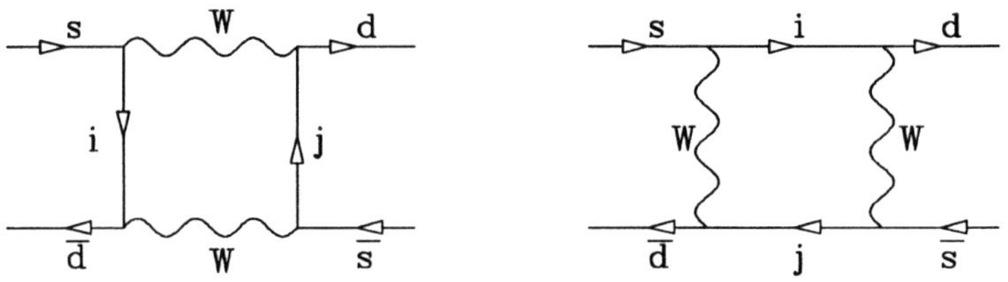

Figure 9. Box diagrams for mixing of a neutral kaon and its antiparticle.

One possibility, proposed [69] immediately after the discovery and still not excluded, is a "superweak" CP-violating interaction which directly mixes $K^0 = d\bar{s}$ and $\bar{K}^0 = s\bar{d}$. This interaction would have no other observable consequences since the $K^0 - \bar{K}^0$ system is so sensitive to it!

The presence of three quark families [4] poses another opportunity for explaining CP violation through the mixing diagram shown in Fig. 9. With three quark families, phases in complex coupling coefficients cannot be removed by redefinition of quark phases. Within some approximations [5], the parameter ϵ is directly proportional to the imaginary part of the mixing amplitude described in Fig. 9. Its magnitude (see [44] or [45] for a calculation in the limit of $m_t \ll M_W$) is [46]

$$|\epsilon| \simeq \frac{G_F^2 m_K f_K^2 B_K M_W^2}{\sqrt{2}(12\pi^2)\Delta m_K}[\eta_1 S(x_c) I_{cc} + \eta_2 S(x_t) I_{tt} + 2\eta_3 S(x_c, x_t) I_{ct}] \quad , \tag{29}$$

where $I_{ij} \equiv \mathrm{Im}(V_{id}^* V_{is} V_{jd}^* V_{js})$. In order to evaluate these expressions we need to work to sufficiently high order in small parameters in V. The application of the unitarity relation to the first and second rows tells us, in fact, that a more precise expression for V_{cd} is $V_{cd} = -\lambda - A^2\lambda^5(\rho + i\eta)$. We then find $I_{cc} = -2A^2\lambda^6\eta$, $I_{ct} = A^2\lambda^6\eta$, and $I_{tt} = 2A^2\lambda^6\eta[A^2\lambda^4(1-\rho)]$. The factors $\eta_1 = 1.38$, $\eta_2 = 0.57$, $\eta_3 = 0.47$ are QCD corrections [70], while $x_i \equiv m_i^2/M_W^2$. The function $S(x)$ was defined in Eq. (26), while

$$S(x,y) \equiv xy\left\{\left[\frac{1}{4} + \frac{3}{2(1-y)} - \frac{3}{4(1-y)^2}\right]\frac{\ln y}{y-x} + (y \leftrightarrow x) - \frac{3}{4(1-x)(1-y)}\right\} \quad . \tag{30}$$

Eq. (29) may then be rewritten (updating [71]) as

$$|\epsilon| = 4.39A^2 B_K \eta[\eta_3 S(x_c, x_t) - \eta_1 S(x_c) + \eta_2 A^2\lambda^4(1-\rho)S(x_t)] \quad . \tag{31}$$

Using the experimental values [47] $|\epsilon| = (2.28 \pm 0.02) \times 10^{-3}$, $f_K = 160$ MeV, $\Delta m_K = 3.49 \times 10^{-15}$ GeV, and $m_K = 0.4977$ GeV, the value $B_K = 0.75 \pm 0.15$ [72], and the top quark mass $\bar{m}_t(M_W) = 165 \pm 6$ GeV/c^2, we find that CP-violating $K - \bar{K}$ mixing leads to the constraint

$$\eta(1 - \rho + 0.44) = 0.51 \pm 0.18 \quad , \tag{32}$$

where the term $1-\rho$ in parentheses corresponds to the loop diagram with two top quarks, and the term 0.44 corresponds to the additional contribution of charmed quarks. The major source of error on the right-hand side is the uncertainty in the parameter $A \equiv V_{cb}/\lambda^2$. Eq. (32) can be plotted in the (ρ, η) plane as a band bounded by hyperbolae with foci at (1.44,0).

D. Summary of parameter space

The constraints (22), (27), and (32) define the allowed region of parameters shown in Fig. 10. The boundaries shown are 1σ errors, but are dominated by theoretical uncertainties in each case.

A large region centered about $\rho \simeq 0$, $\eta \simeq 0.35$ is permitted. Nonetheless, it could be that the CP violation seen in kaons is due to an entirely different source, perhaps a superweak mixing of K^0 and $\bar{K}^0$ [69]. In that case one could probably still accommodate $\eta = 0$, and hence a real CKM matrix, by going slightly outside the 1σ bounds based on $|V_{ub}/V_{cb}|$ or $B - \bar{B}$ mixing. In order to confirm the predicted nonzero value of η, we turn to other experimental possibilities.

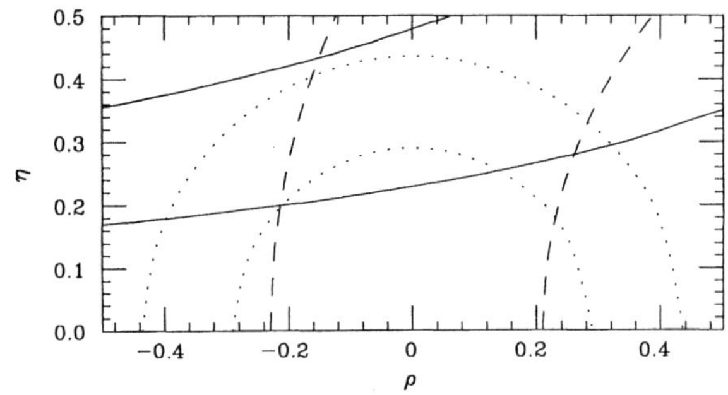

Figure 10. Region in the (ρ, η) plane allowed by constraints on $|V_{ub}/V_{cb}|$ (dotted semicircles), $B^0 - \bar{B}^0$ mixing (dashed semicircles), and CP-violating $K - \bar{K}$ mixing (solid hyperbolae).

E. CP violation in B meson decays

We have already mentioned the information provided by $B^0 - \bar{B}^0$ mixing. The large value of $\Delta m/\Gamma \simeq 0.7$ was an early hint, through graphs of the form of Fig. 8, that the top quark was very heavy. This large value also has another very important benefit: It makes possible a class of incisive studies of CP violation in neutral B meson decays [73].

By comparing rates decays for a state which is produced as B^0 and one which is produced as a $\bar{B}^0$ to final states which are eigenstates of CP, one can directly measure angles in the unitarity triangle of Fig. 7. Because of the interference between direct decays (e.g., $B^0 \to J/\psi K_S$) and those which proceed via mixing (e.g., $B^0 \to \bar{B}^0 \to J/\psi K_S$), these processes are described by time-dependent functions whose difference when integrated over all time is responsible for the rate asymmetry. Thus, if we define

$$C_f \equiv \frac{\Gamma(B_{t=0} \to f) - \Gamma(\bar{B}_{t=0} \to f)}{\Gamma(B_{t=0} \to f) + \Gamma(\bar{B}_{t=0} \to f)} \quad , \tag{33}$$

we have, in the limit of a single direct contribution to decay amplitudes,

$$A(J/\psi K_S, \pi^+\pi^-) = -\frac{x_d}{1 + x_d^2} \sin(2\beta, 2\alpha) \quad , \tag{34}$$

where $x_d \equiv \Delta m(B^0)/\Gamma(B^0)$. This limit is expected to be very good for $J/\psi K_S$, but some correction for penguin contributions (see Sec. 2 G) is probably needed for $\pi^+\pi^-$. The value $x_d \simeq 0.7$ is nearly optimum to maximize the coefficient of $\sin(2\beta, 2\alpha)$.

To see this behavior in more detail [74], we note that the time-dependent partial rates for a state which is initially B^0 ($\bar{B}^0$) to decay to a final state f may be written as

$$d\Gamma[B^0(\bar{B}^0) \to f]/dt \sim e^{-\Gamma t}[1 \mp \mathrm{Im}\lambda_0 \sin(\Delta m t)] \quad , \tag{35}$$

where in order to obtain this simple result we neglect $\Delta\Gamma/\Gamma$ in comparison with $\Delta m/\Gamma$. This step is justified for B's, in contrast to the situation for K's. The final states to which both B and $\bar{B}$ can decay are only a small fraction of those to which B or $\bar{B}$ normally decay, and so one should expect quite similar lifetimes for the two mass eigenstates. Integration of (35) gives

$$C_f = \frac{-x_d}{1+x_d^2}\mathrm{Im}\lambda_0(f) \tag{36}$$

for the total asymmetry. For the CP-eigenstate final states mentioned, $\lambda_0(J/\psi K_S) = -e^{-2i\beta}$ and $\lambda_0(\pi^+\pi^-) = e^{2i\alpha}$. The extra minus sign in the first relation is due to the odd CP of the $J/\psi K_S$ final state.

The asymmetry (36) is suppressed both when $\Delta m/\Gamma$ is very small and when it is very large (e.g., as is expected for B_s). For B_s, in order to see an asymmetry, one must not integrate with respect to time. Experiments planned with detection of B_s as their focus will require precise vertex detection to measure mixing as a function of proper time. For B^0, on the other hand, the value of $x/(1+x^2)$ for $x = 0.7$ is 0.47, very close to its maximum possible value of 1/2 for $x = 1$.

When more than one eigenchannel contributes to a decay, there can appear terms of the form $\cos(\Delta m t)$ as well as $\sin(\Delta m t)$ in results analogous to Eqs. (35) [75]. These complicate the analysis somewhat, but information can be obtained from them [76] on the relative contributions of various channels to decays.

F. Lifetime and mass differences for strange B's

The mixing between strange B's due to diagrams like those in Fig. 8 is considerably enhanced relative to that between nonstrange B's:

$$\frac{\Delta m_s}{\Delta m_d} = \frac{f_{B_s}^2 B_{B_s}}{f_B^2 B_B}\left|\frac{V_{ts}}{V_{td}}\right|^2 \simeq 17 - 52 \quad , \tag{37}$$

where we have taken the expected ranges of decay constant and CKM element ratios, and Δm_s refers to mixing between the $B_s \equiv \bar{b}s$ and $\bar{B}_s \equiv b\bar{s}$. Alternatively, we may retrace the steps of Sec. 2 B, replacing appropriate quantities in Eq. (25), to derive an analogous expression for Δm_s, which we then evaluate directly. For $V_{ts} = 0.040 \pm 0.004$, $m_{B_s} = 5.37$ GeV/c^2, $f_{B_s}\sqrt{B_{B_s}} = 225$ MeV, $\eta_{B_s} = 0.6 \pm 0.1$, and [40] $\tau_{B_s} \equiv 1/\Gamma_s = 1.55 \pm 0.10$ ps, we find $\Delta m_s/\Gamma_s = 22 \pm 6$, with an additional 40% error associated with $f_{B_s}^2 B_{B_s}$. This result implies many particle-antiparticle oscillations in a decay lifetime, requiring good vertex resolution and highly time-dilated B_s's for a measurement. The present experimental bound $\Delta m_s > 9.2$ ps^{-1} based on combining ALEPH and DELPHI results [77] begins to restrict the parameter space in an interesting manner.

The large value of Δm_s entails a value of $\Delta\Gamma_s$ between mass eigenstates of strange B's which may be detectable. After all, the short-lived and long-lived neutral kaons differ in lifetime by a factor of 600. Strong interactions and the presence of key channels (e.g., $\pi\pi$) are a crucial effect in strange particle (e.g., K^0 and $\bar{K}^0$) decays. While the b quark decays as if it is almost free, so that strong interactions here are much less important, it turns out that a corresponding difference in lifetimes for strange B's of the order of 20% is not unlikely [78].

In the ratio $\Delta m_s/\Delta\Gamma_s$, uncertainties associated with the meson decay constants cancel out, and in lowest order (before QCD corrections are applied) one finds [79, 80] $\Delta m_s/\Delta\Gamma_s \simeq \mathcal{O}(-[1/\pi][m_t^2/m_b^2]) \simeq -200$. The sign indicates that the heavier state is expected to be the longer-lived one, as in the neutral kaon system. The top quark does not contribute to the width difference associated with the imaginary part of the graphs in Fig. 8, since no $t\bar{t}$ pairs are produced in B_s decays.

Aside from small CP-violating effects, the mass eigenstates of strange B's correspond to those $B_s^{(\pm)}$ of even and odd CP. The decay of a $\bar{B}_s$ meson via the quark subprocess $b(\bar{s}) \to c\bar{c}s(\bar{s})$ gives rise to predominantly CP-even final states [81], so the CP-even eigenstate should have a greater decay rate. One calculation [78] gives

$$\frac{\Gamma(B_s^{(+)}) - \Gamma(B_s^{(-)})}{\bar{\Gamma}} \simeq 0.18\frac{f_{B_s}^2}{(200\ \text{MeV})^2} \quad , \tag{38}$$

while a more recent estimate [82] is $0.16^{+0.11}_{-0.09}$. The lifetime difference between CP-even and CP-odd strange B's thus can provide useful information on f_{B_s}, and hence indirectly on the weak interactions at short distances. One way to learn this lifetime difference [83] is to study angular distributions in $B_s \to J/\psi + \phi \to e^+e^-K^+K^-$ (or $\mu^+\mu^-K^+K^-$). The J/ψ and ϕ are both spin-1 particles and hence can be produced in states of orbital angular momenta $L = 0$, 1, and 2 from the spinless B_s decay. $L = 1$ corresponds to $P = CP = -$, while $L = 0$ or 2 corresponds to $P = CP = +$. A simple transversity analysis [84] permits one to separate the two cases.

In the rest frame of the J/ψ, let the x axis be defined by the direction of the ϕ, the $x - y$ plane be defined by the kaons which are its decay products, and the z axis be the normal to that plane. Let the e^+ (or μ^+) make an angle θ with the z axis. Then the CP-even final states give rise to an angular distribution $1 + \cos^2\theta$, while the CP-odd state gives rise to $\sin^2\theta$. In case both CP eigenstates are present in $B_s \to J/\psi\phi$, one will see a gradual increase of the $\sin^2\theta$ component relative to the $1 + \cos^2\theta$ component. More likely (if predictions [81] are correct), the CP-even state will dominate, so that one will be measuring mainly the lifetime of this eigenstate when following the time-dependence of the decay. The average decay rate $\bar{\Gamma} \equiv (\Gamma_+ + \Gamma_-)/2$ (the subscripts denote CP eigenvalues) is measured in flavor-tagged decays of $B_s = \bar{b}s \to \bar{c} + \ldots$ or $\bar{B}_s = b\bar{s} \to c + \ldots$.

A recent analysis [85] of $B \to J/\psi K^*$ has some bearing on the $B_s \to J/\psi\phi$ partial-wave structure. The two processes are related by flavor SU(3), involving a substitution $s \leftrightarrow d$ of the spectator quark. Thus, one expects the same partial waves in the two decays. The CLEO Collaboration has studied 146 $B \to J/\psi K^*$ decays in 3.36×10^6 $B\bar{B}$ pairs produced at the Cornell Electron Storage Ring (CESR).

One can decompose the amplitude A for a spinless meson to decay to two vector mesons into three independent components [86], corresponding to linear polarization states of the vector mesons which are either longitudinal (0), or transverse to their directions of motion and parallel ($\|$) or perpendicular ($\perp$) to one another. The states 0 and $\|$ are P-even, while the state $\perp$ is P-odd. Aside from the case of A_0, which is special for massive vector mesons, these arguments were advanced some time ago [87] to determine the parity of the neutral pion in its decays to two photons. Since J/ψ and ϕ are both C-odd eigenstates, the properties under P are the same as those under CP.

A suitably normalized amplitude $A_\perp$ thus can be expressed in terms of partial-wave amplitudes S, P, and D as $|A_\perp|^2 = |P|^2/(|S|^2+|P|^2+|D|^2)$. In $B_s \to J/\psi\phi$, $|P|^2$ would correspond to a CP-odd final state and hence to the decay of the odd-CP eigenstate. The CLEO results [85] for the related process $B \to J/\psi K^*$ are $|A_\perp|^2 = 0.21 \pm 0.14$ from a fit to the transversity angle, and $|A_\perp|^2 = 0.16 \pm 0.08 \pm 0.04$ from a fit to the

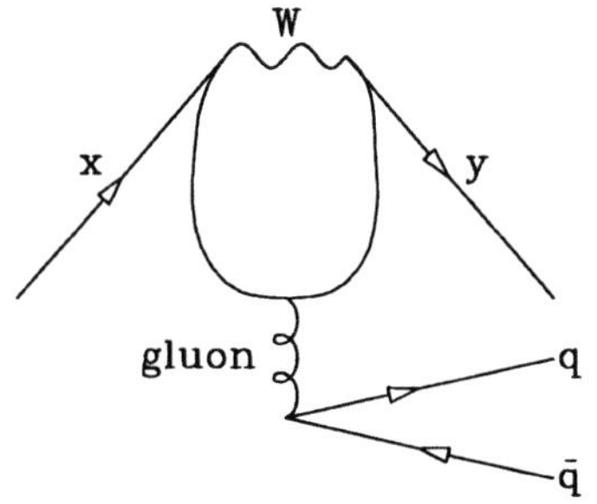

Figure 11. "Penguin" diagram describing transition of a quark x to another quark y with the same charge. The intermediate quarks have charge differing from $Q(x) = Q(y)$ by one unit. Here $q = (u, d, s)$.

full angular distribution. This implies (via flavor SU(3)) that $B_s \to J/\psi\phi$ is dominated by the CP-even final state, and thus the lifetime in this state measures approximately $\tau(B_s^{(+)})$. It will be interesting to see if any evidence for non-zero $|P|^2$ can be gathered in $B \to J/\psi K^*$, in which case $B_s \to J/\psi\phi$ should exhibit the time-variation in the transversity-angle distribution mentioned above [83].

G. Processes dominated by penguin diagrams

Although the unitarity of the CKM matrix implies flavor conservation for charge-preserving electroweak interactions in lowest order, we have seen that loop diagrams can induce flavor-changing charge-preserving interactions in higher order. Another example of this phenomenon is provided by the "penguin" diagram [88] illustrated in Fig. 11. Although the penguin's "leg" is a gluon in this illustration, it can also be a photon or Z. When the external quarks x and y have charge $-1/3$, the intermediate quarks have charge 2/3 and can include the top quark. Because of the top quark's large mass, such penguin diagrams can be very important.

An example of a predicted penguin effect in $s \to d$ transitions is a phase arising in the decays of neutral kaons to $\pi\pi$. This phase can lead to a "direct" contribution to the ratios for CP-violating and CP-conserving decays, in addition to that provided by the mixing parameter ϵ measured earlier.

One may define

$$\eta_{+-} \equiv \frac{A(K_L \to \pi^+\pi^-)}{A(K_S \to \pi^+\pi^-)} \ ; \quad \eta_{00} \equiv \frac{A(K_L \to \pi^0\pi^0)}{A(K_S \to \pi^0\pi^0)} \ ; \tag{39}$$

the effect of "direct" decays then shows up in a parameter ϵ' which causes η_{+-} and η_{00} to differ from one another:

$$\eta_{+-} = \epsilon + \epsilon' \quad ; \quad \eta_{00} = \epsilon - 2\epsilon' \quad . \tag{40}$$

Since ϵ' and ϵ are expected to have approximately the same phase (see, e.g., [5]), one expects

$$\begin{aligned} |\eta_{+-}| &\simeq |\epsilon|[1 + \mathrm{Re}(\epsilon'/\epsilon)] \quad , \\ |\eta_{00}| &\simeq |\epsilon|[1 - 2\mathrm{Re}(\epsilon'/\epsilon] \quad , \end{aligned} \tag{41}$$

and hence

$$\left|\frac{\eta_{00}}{\eta_{+-}}\right|^2 = \frac{\Gamma(K_L \to 2\pi^0)}{\Gamma(K_S \to 2\pi^0)} \Big/ \frac{\Gamma(K_L \to \pi^+\pi^-)}{\Gamma(K_S \to \pi^+\pi^-)} = 1 - 6\ \mathrm{Re}\frac{\epsilon'}{\epsilon} \quad . \tag{42}$$

Present expectations [89] are that ϵ'/ϵ could be a few parts in 10^4 (but in any case $\epsilon'/\epsilon \leq 10^{-3}$), requiring the ratio of ratios (42) to be measured to about one part in 10^3. Experiments now in progress at Fermilab and CERN should have the required sensitivity. The previous results of these experiments are:

$$\text{E731 [90]}: \quad \text{Re}(\epsilon'/\epsilon) = (7.4 \pm 6.0) \times 10^{-4} \quad , \tag{43}$$

$$\text{NA31 [91]}: \quad \text{Re}(\epsilon'/\epsilon) = (23.0 \pm 6.5) \times 10^{-4} \quad , \tag{44}$$

leading to some question about whether a non-zero value has been observed. Because of the cancelling effects of gluonic and "electroweak" penguins (in which the curly line in Fig. 11 is a photon or Z), the actual magnitude of ϵ'/ϵ is difficult to estimate, so that one's best hope is for a non-zero value within the rather large theoretical range, thereby disproving the superweak model [69] of CP violation.

The contribution of the (gluonic) penguin diagram to the effective weak Hamiltonian may be written (for x, y equal to quarks of charge $-1/3$)

$$\mathcal{H}_W^{\text{penguin}} \simeq \frac{G_F}{\sqrt{2}} \frac{\alpha_s}{6\pi} \left[\xi_c \ln \frac{m_t^2}{m_u^2} + \xi_t \ln \frac{m_t^2}{m_u^2} \right]$$

$$\left[(\bar{y}_L \gamma^\mu \lambda^a x_L)(\bar{u}\gamma_\mu \lambda^a u + \bar{d}\gamma_\mu \lambda^a d + \ldots) + \text{H.c.} \right] \quad , \tag{45}$$

where $\xi_i \equiv V_{ix} V_{iy}^*$, and λ^a are color SU(3) matrices $[a = (1, \ldots, 8)]$ normalized so that $\text{Tr}(\lambda^a \lambda^b) = 2\delta^{ab}$. The top quark is dominant in the flavor-changing processes $b \to d$ and $b \to s$ (with corrections due to charm which can be important in some cases [92]), while the charmed quark dominates the $s \to d$ process. (The top quark plays a key role, however, in the electroweak penguin contribution to this process [89].) One may imitate the effect of an infrared cutoff for the gluonic penguin graph by using a constituent-quark mass $m_u \sim 0.3$ GeV/c^2.

One can estimate the effect of Eq. (45) for the $b \to sq\bar{q}$ penguin graph; one finds it is comparable to that of the $b \to ud\bar{u}$ "tree" contribution $(G_F/\sqrt{2}) V_{ub} V_{ud}^* [\bar{u}\gamma_\mu(1 - \gamma_5)b][\bar{d}\gamma^\mu(1 - \gamma_5)u]$. The penguin contribution to $b \to dq\bar{q}$ and the $b \to us\bar{u}$ tree contribution are both expected to be suppressed by approximately one power of the Wolfenstein parameter $\lambda \sim 0.2$, as one can see by comparing CKM elements. Thus, $B^0 \to \pi^+\pi^-$ is expected to be dominated by the tree amplitude; $B^0 \to K^+\pi^-$ is expected to be dominated by the penguin amplitude; and the rates of the two processes should be similar.

In Fig. 12 we show a contour plot of the significance of detection by the CLEO Collaboration [93] of the decays $B^0 \to \pi^+\pi^-$ and $B^0 \to K^+\pi^-$. Evidence exists for a combination of $B^0 \to K^+\pi^-$ and $\pi^+\pi^-$ decays, generically known as $B^0 \to h^+\pi^-$. On the basis of 2.4 fb^{-1} of data, the most recent published result [94] is $B(B^0 \to h^+\pi^-) = (1.8\ ^{+0.6}_{-0.5}\ ^{+0.2}_{-0.3} \pm 0.2) \times 10^{-5}$. Although one still cannot conclude that either decay mode is nonzero at the 3σ level, the most likely solution is roughly equal branching ratios (i.e., about 10^{-5}) for each mode. Only upper limits exist for other modes of two pseudoscalars [95], but these are consistent with predictions [96].

Penguin diagrams play a number of roles in B decays. We enumerate several of them; others are mentioned in [5].

1. The process $B^+ \to K^0\pi^+$ is expected to be almost completely due to the penguin graph; one cannot write a corresponding tree graph for it. By comparison with the $B^0 \to K^+\pi^-$ decay, where the penguin graph is expected to be the main contribution, one expects $B(B^+ \to K^0\pi^+) \simeq 10^{-5}$. The weak phase of the process (which changes sign under charge-conjugation) thus is expected to be $\text{Arg}(V_{tb}^* V_{ts}) = \pi$,

so that the charge-conjugate process has the same weak phase. As a result, one turns out to be able to separate strong final-state interaction phases from weak phases and obtain estimates of quantities like the angle $\gamma = \text{Arg}(V_{ub}^*)$ in Fig. 7 by comparing rates for $B^+ \to (K^0\pi^+,\ K^+\pi^0,\ K^+\eta,\ K^+\eta')$ with the corresponding B^- rates [97]. One can also obtain this information by measuring the time-dependence in $B^0(\bar{B}^0) \to \pi^+\pi^-$ and the rates for $B^0 \to K^+\pi^-$, $B^+ \to K^0\pi^+$, and the charge-conjugate processes [98]. (The rates for $B^+ \to K^0\pi^+$ and $B^- \to \bar{K}^0\pi^-$ are expected to be equal because of the penguin dominance mentioned above.) The weak phases of the major amplitudes contributing

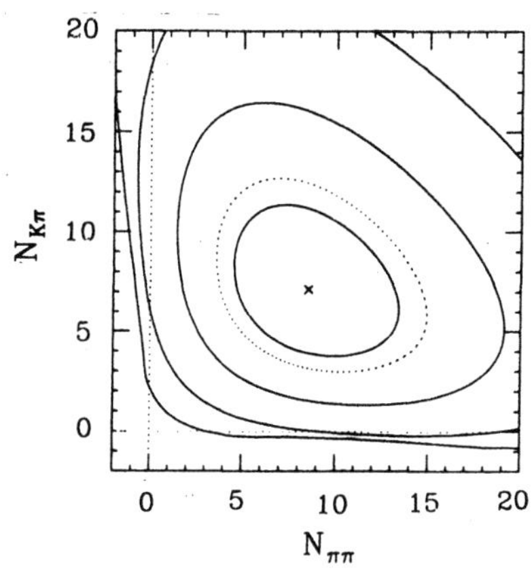

Figure 12. Contours of significance of detection of $B^0 \to \pi^+\pi^-$ and $B^0 \to K^+\pi^-$ by the CLEO Collaboration.

Table 3. Phases of amplitudes contributing to decays of B mesons to $\pi\pi$ and $K\pi$. Here ΔS refers to the change of strangeness in the process.

	"Tree"		"Penguin"	
$\lvert\Delta S\rvert$	CKM elements	Phase	CKM elements	Phase
0	$V_{ub}^* V_{ud}$	γ	$V_{tb}^* V_{td}$	$-\beta$
1	$V_{ub}^* V_{us}$	γ	$V_{tb}^* V_{ts}$	π

to these decays are summarized in Table 3. The relative weak phase of tree and penguin amplitudes for strangeness-preserving decays is $\gamma + \beta = \pi - \alpha$ (assuming the unitarity triangle to be valid), while the corresponding relative phase for strangeness-changing decays (aside from a sign) is just γ. As a result, one can measure both α and γ.

2. A number of processes (in addition to the decay $B^+ \to K^0\pi^+$ mentioned above) are dominated by penguin graphs. By comparing the rates for strangeness-preserving and strangeness-changing processes, one can measure the ratio $|V_{td}/V_{ts}|$ [99]. Examples of useful ratios are $B(B^+ \to \bar{K}^{*0}K^+)/B(B^+ \to \phi K^+)$ and $B(B^+ \to \bar{K}^{*0}K^{*+})/B(B^+ \to \phi K^{*+})$.

3. We mentioned in Sec. 2 E that the time-integrated rate asymmetry in $B \to \pi^+\pi^-$ could provide information on the angle α of the unitarity triangle. The most direct test is based on the assumption that the tree process $\bar{b} \to \bar{u}u\bar{d}$ is the only direct contribution to this decay. However, "penguin pollution" [75] (due to the $b \to d$ transition) makes the analysis less straightforward, even thought the penguin amplitude is expected to be only about 0.2 of the tree amplitude. Ways to circumvent this difficulty include the detailed study of the isospin structure of the $\pi\pi$ final state [76], and the use of flavor SU(3) to estimate penguin effects using $B \to K\pi$, where they are expected to be dominant [96, 98, 100].

H. Rare kaon decays

We give a sample of processes influenced by loop diagrams in which the top quark plays a role. The present discussion is based on a recent discussion of $K \to \pi\nu\bar{\nu}$ decays [101]. Details on other processes may be found in Refs. [5, 45, 70, 102, 103].

We are concerned with the subprocess $s \to d\nu\bar{\nu}$, which can proceed via loop diagrams involving exchange of a pair of W's (a box diagram) or one W and one Z (an electroweak penguin). In each case results will be quoted for a sum over the three neutrino species.

1. The decay $K^+ \to \pi^+\nu\bar{\nu}$ is predicted to have a branching ratio

$$B(K^+ \to \pi^+\nu\bar{\nu}) \simeq 10^{-11}A^4|T(x_t)(1-\rho-i\eta+[0.41\pm 0.09])|^2 \quad , \tag{46}$$

where $x_t \equiv m_t^2/M_W^2$, and

$$T(x) \equiv \frac{x}{4}\left[\left(\frac{3x-6}{(x-1)^2}\right)\ln x + \frac{x+2}{x-1}\right] \quad . \tag{47}$$

The term $1-\rho-i\eta$ in (46) is the contribution of the top quark, while the term in square brackets represents the contribution of charm, uncertain primarily because the charmed quark mass is not all that well known. Uncertainty in the Wolfenstein parameter A contributes a good deal to the error on the predicted branching ratio. The top quark's mass used to be a major source of error in this estimate, but is no longer. For the present value of m_t, $T(x_t) \simeq 3$.

An experimentally measured branching ratio will define a band in the (ρ, η) plane whose boundaries are circles with centers at $(1.41 \pm 0.09, 0)$. Given the region of parameters already allowed in Fig. 10, such a measurement essentially specifies ρ. The predicted branching ratio is 10^{-10}, give or take a factor of 2. The present experimental limit from an experiment (E787) at Brookhaven National Laboratory [104] is $B(K^+ \to \pi^+\nu\bar{\nu}) < 2.4 \times 10^{-9}$ (90% c.l.).

2. The decay $K_L \to \pi^0\nu\bar{\nu}$ is expected to be purely CP-violating [105]. The predicted branching ratio is

$$B(K_L \to \pi^0\nu\bar{\nu}) \simeq 4.4 \times 10^{-11}\eta^2 A^4|T(x_t)|^2 \quad . \tag{48}$$

For the allowed range of A and η, one expects a branching ratio of 2×10^{-11}, give or take a factor of 2. While the present upper limit [106] of 5.8×10^{-5}(90% c.l.) is quite far from this, it represents a considerable advance from previous limits, and much improvement is expected in the next few years.

3. PRECISION ELECTROWEAK EXPERIMENTS

Tests of the electroweak theory have reached the precision that they are sensitive to the mass of the top quark. Indeed, the large top quark mass had been anticipated to some extent by these experiments. In this section we review the lowest-order theory, show how it is affected by the top quark in loop diagrams, and explore the sensitivity of precision electroweak measurements to further types of heavy particles which could shed light on the top quark's mass. We draw heavily on the discussion in Ref. [107], for which the present discussion serves in part as an update.

A. Lowest-order formalism

In order to construct a renormalizable theory of the weak interactions, it was proposed quite early [108] to regard the four-fermion interaction as the limit at zero momentum transfer of a process in which an intermediate vector boson (now called W) was exchanged, so that one identifies $G_F/\sqrt{2} = g^2/8M_W^2$. The Fermi coupling constant is well known: $G_F = 1.16639(2) \times 10^{-5}$ GeV^{-2}. However, neither the coupling constant g nor the W mass M_W were known when this proposal was made. Searches for W's as light as a couple of GeV/c^2 were undertaken in the early 1960's.

If $W^{\pm}$ exchange is to be described by a gauge interaction, the simplest group containing charged W's is SU(2), which also contains a W^0. But one cannot identify the W^0 with a photon. Aside from the fact that the symmetry must be badly broken, so that the photon remains massless while the charged W's become very heavy, there are two other difficulties. First, the charged W's couple to matter with a $V-A$ interaction [109], i.e., to left-handed particles, while the photon couples via a V interaction, i.e., to both left-handed and right-handed particles. Second, if the photon really were identified with W^0 all the charges of particles would have to be half-integer, just like values of J_3, the third component of angular momentum.

The solution [6] is to add an SU(2) singlet B to the theory, with another coupling constant g'. The photon can then be one mixture of the B and the W^0, while there will be another neutral particle which is the orthogonal mixture:

$$\begin{aligned} \text{Photon}: \; & A = B\cos\theta + W^0 \sin\theta \quad (m=0) \; ; \\ \text{New} \;\;: \; & Z = -B\sin\theta + W^0\cos\theta \quad (m \neq 0) \; . \end{aligned} \tag{49}$$

The W^0 couples to the third component I_{3L} of the SU(2). The subscript "L" stands for "left-handed" and refers to the fact that all three components of the SU(2), both the charged W's and the W^0, couple only to left-handed particles or right-handed antiparticles. The B couples to a new quantum number "weak hypercharge," invented to account for the difference between electromagnetic charge and I_{3L}:

$$Q_{\rm em} = I_{3L} + \frac{Y_W}{2} \quad . \tag{50}$$

Once the interaction Lagrangian is expressed in terms of the photon and Z, one finds that the electron charge e is related to the other parameters in the theory by $e = g\sin\theta = g'\cos\theta$, so that

$$\frac{1}{e^2} = \frac{1}{g^2} + \frac{1}{g'^2} \quad ; \quad M_W = \left(\frac{\pi\alpha}{\sqrt{2}G_F}\right)^{1/2} / \sin\theta \quad . \tag{51}$$

Moreover, the Z mass is related to the W mass:

$$M_Z = \frac{M_W}{\cos\theta} \quad ; \quad \frac{G_F}{\sqrt{2}} = \frac{g^2 + g'^2}{8M_Z^2} \quad . \tag{52}$$

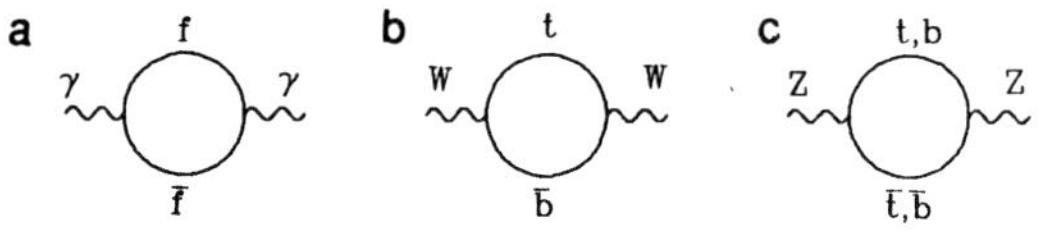

Figure 13. Loop diagrams with virtual fermions leading to important vacuum polarization effects. (a) Photons (all fermions f below the top quark are important); (b) W; (c) Z.

B. Effects in loop diagrams

The effects of fermions in loop diagrams, as shown in Fig. 13, are seen in several ways in the electroweak interactions.

The photon vacuum polarization effects shown in Fig. 13(a) lead to an effective fine structure constant $\alpha(q^2)$ at an invariant momentum scale q^2 which is related to the usual $\alpha \simeq 1/137.036$ at $q^2 = 0$ by

$$\alpha^{-1}(q) = \alpha^{-1}[1 - \Pi_{\gamma\gamma}(q^2)] \quad ; \quad \Pi_{\gamma\gamma} = \frac{\alpha}{3\pi} \sum_f Q_f^2 \left[-\frac{5}{3} + \ln \frac{q^2}{m_f^2} \right] \quad . \tag{53}$$

The quantity $\Pi_{\gamma\gamma}$ can be evaluated in part using data on $e^+e^- \to$ hadrons; several recent precise determinations [110] lead to $\alpha^{-1}(M_Z) \simeq 128.9 \pm 0.1$. This quantity is scheme-dependent; one also sees it quoted in as 127.9 ± 0.1 in the $\overline{\text{MS}}$ scheme [111].

If we had used $\alpha^{-1} = 137.036$, the expressions we have written for the W and Z masses would imply, using the latest value [7] $\sin^2\theta = 0.23165$, that $M_W = 77.5$ GeV/c^2 and $M_Z = 88.4$ GeV^2. The use of the correct value of $\alpha^{-1}(M_Z)$ changes these predictions to $M_W = 79.86$ GeV/c^2, $M_Z = 91.11$ GeV/c^2. These values are to be compared with the experimental ones $M_W = 80.34 \pm 0.10$ GeV/c^2 (see below) and $M_Z = 91.1863 \pm 0.0020$ GeV/c^2 (see [7], [112] and [113]). The use of the correct value of α considerably improves the predictions, but the experimental ratio M_W/M_Z is higher than predicted. The major part of the reason may be traced to the effect of the very heavy top quark in the loop diagrams of Figs. 13(b) and 13(c).

Both vector and axial-vector currents enter in the couplings of W and Z to fermions; the axial-vector currents are not conserved when the fermions have mass. Moreover, even the vector currents are not conserved when the fermions have unequal masses [as in Fig. 13(b)]. The result is the appearance of contributions to Π for the W and Z which are quadratic in fermion masses. No such contributions appear in $\Pi_{\gamma\gamma}$, as a result of electromagnetic gauge invariance and current conservation.

The low-energy limits of W and Z exchange are now described by:

$$\frac{G_F}{\sqrt{2}} = \frac{g^2}{8M_W^2} \quad , \quad \frac{G_F}{\sqrt{2}}\rho = \frac{g^2 + g'^2}{8M_Z^2} \quad , \tag{54}$$

where the parameter ρ, which receives contributions from quark loops to W and Z self-energies, is dominated by the top [114]:

$$\rho \simeq 1 + \frac{3G_F m_t^2}{8\pi^2\sqrt{2}} \quad , \tag{55}$$

Consequently, if we define θ by means of the precise measurement at LEP of M_Z,

$$M_Z^2 = \frac{\pi\alpha}{\sqrt{2} G_F \rho \sin^2\theta \cos^2\theta} \quad , \tag{56}$$

then θ will depend on m_t, and so will M_W in Eq. (51). The ratio $M_W/M_Z = \sqrt{\rho}\cos\theta$ then may be used to extract a value $\sqrt{\rho} = 1.0050 \pm 0.0013$, implying $m_t = 179(1 \pm 0.13)$ GeV/c^2. We shall see that the inclusion of other electroweak data does not greatly alter this simple result.

In order to display dependence of electroweak observables on such quantities as the top quark and Higgs boson masses m_t and M_H, we choose to expand the observables about "nominal" values calculated for specific m_t and M_H. We thereby bypass a discussion of "direct" radiative corrections which are independent of m_t, M_H, and new particles. We isolate the dependence on m_t, M_H, and new physics arising from "oblique" corrections [115], associated with loops in the W and Z propagators like those shown in Figs. 13(b,c).

For $m_t = 175$ GeV, $M_H = 300$ GeV, the measured value of M_Z leads to a nominal expected value of $\sin^2\theta_{\text{eff}} = 0.2315$. In what follows we shall interpret the effective value of $\sin^2\theta$ as that measured via leptonic vector and axial-vector couplings: $\sin^2\theta_{\text{eff}} \equiv (1/4)(1 - [g_V^\ell/g_A^\ell])$. We have corrected the nominal value of $\sin^2\theta_{\overline{\text{MS}}} \equiv \hat{s}^2$ as quoted by DeGrassi, Kniehl, and Sirlin [116] for the difference [117] $\sin^2\theta_{\text{eff}} - \hat{s}^2 = 0.0003$ and for the recent change in the evaluation of $\alpha(M_Z)$ [110].

Defining the parameter T by $\Delta\rho \equiv \alpha T$, we find

$$T \simeq \frac{3}{16\pi\sin^2\theta}\left[\frac{m_t^2 - (175\ \text{GeV})^2}{M_W^2}\right] - \frac{3}{8\pi\cos^2\theta}\ln\frac{M_H}{300\ \text{GeV}} \quad . \tag{57}$$

The weak mixing angle θ, the W mass, and other electroweak observables depend on m_t and M_H.

The weak charge-changing and neutral-current interactions are probed under a number of different conditions, corresponding to different values of momentum transfer. For example, muon decay occurs at momentum transfers small with respect to M_W, while the decay of a Z into fermion-antifermion pairs imparts a momentum of nearly $M_Z/2$ to each member of the pair. Small "oblique" corrections [115], logarithmic in m_t and M_H, arise from contributions of new particles to the photon, W, and Z propagators. Other (smaller) "direct" radiative corrections are important in calcuating actual values of observables.

We may then replace (54) by

$$\frac{G_F}{\sqrt{2}} = \frac{g^2}{8M_W^2}\left(1 + \frac{\alpha S_W}{4\sin^2\theta}\right) \quad , \quad \frac{G_F\rho}{\sqrt{2}} = \frac{g^2 + g'^2}{8M_Z^2}\left(1 + \frac{\alpha S_Z}{4\sin^2\theta\cos^2\theta}\right) \quad , \tag{58}$$

where S_W and S_Z are coefficients representing variation with momentum transfer. Together with T, they express a wide variety of electroweak observables in terms of quantities sensitive to new physics. (The presence of such corrections was noted quite early by Veltman [118].) The Peskin-Takeuchi [115] variable U is equal to $S_W - S_Z$, while $S \equiv S_Z$.

Expressing the "new physics" effects in terms of deviations from nominal values of top quark and Higgs boson masses, we have the expression (57) for T, while contributions of Higgs bosons and of possible new fermions U and D with electromagnetic charges Q_U and Q_D to S_W and S_Z, in a leading-logarithm approximation, are [119]

$$S_Z = \frac{1}{6\pi}\left[\ln\frac{M_H}{300\ \text{GeV}/c^2} + \sum N_C\left(1 - 4\overline{Q}\ln\frac{m_U}{m_D}\right)\right] \quad , \tag{59}$$

$$S_W = \frac{1}{6\pi}\left[\ln\frac{M_H}{300\ \mathrm{GeV}/c^2} + \sum N_C\left(1 - 4Q_D\ln\frac{m_U}{m_D}\right)\right] \quad . \tag{60}$$

The expressions for S_W and S_Z are written for doublets of fermions with N_C colors and $m_U \geq m_D \gg m_Z$, while $\overline{Q} \equiv (Q_U + Q_D)/2$. The sums are taken over all doublets of new fermions. In the limit $m_U = m_D$, one has equal contributions to S_W and S_Z. For a single Higgs boson and a single heavy top quark, Eqs. (59) and (60) become

$$S_Z = \frac{1}{6\pi}\left[\ln\frac{M_H}{300\ \mathrm{GeV}/c^2} - 2\ln\frac{m_t}{175\ \mathrm{GeV}/c^2}\right] ,$$

$$S_W = \frac{1}{6\pi}\left[\ln\frac{M_H}{300\ \mathrm{GeV}/c^2} + 4\ln\frac{m_t}{175\ \mathrm{GeV}/c^2}\right] , \tag{61}$$

where the leading-logarithm expressions are of limited validity for M_H and m_t far from their nominal values. A degenerate heavy fermion doublet with N_c colors thus contributes $\Delta S_Z = \Delta S_W = N_c/(6\pi)$. For example, in a minimal dynamical symmetry-breaking ("technicolor") scheme about which we shall have more to say presently, with a single doublet of $N_c = 4$ fermions, one will have $\Delta S = 2/(3\pi) \simeq 0.2$. This will turn out to be marginally acceptable, while many non-minimal schemes, with large numbers of doublets, will be seen to be ruled out.

C. Analysis of present data

We shall discuss a number of observables, including neutrino deep inelastic scattering, direct W mass measurements, direct measurements of a number of Z properties, and parity violation in atoms.

Among the first pieces of evidence for neutral currents was the presence of muonless events in neutrino deep inelastic scattering [120]. The ratios of neutrino and antineutrino neutral-current (NC) to charged-current (CC) cross sections, R_ν and $R_{\bar\nu}$, are predicted to be [121]

$$R_\nu \equiv \frac{\sigma_{NC}(\nu N)}{\sigma_{CC}(\nu N)} = \rho^2\left[\frac{1}{2} - x + \frac{5}{9}x^2(1+r)\right] \quad ,$$

$$R_{\bar\nu} \equiv \frac{\sigma_{NC}(\bar\nu N)}{\sigma_{CC}(\bar\nu N)} = \rho^2\left[\frac{1}{2} - x + \frac{5}{9}x^2(1+\frac{1}{r})\right] \quad , \tag{62}$$

where $x \equiv \sin^2\theta$ and $r \equiv \sigma_{CC}(\bar\nu N)/\sigma_{CC}(\nu N)$. These expressions are proportional to ρ^2 since they involve neutral current *cross sections* (squares of amplitudes).

The resulting plot of R_ν vs. $R_{\bar\nu}$ as a parametric function of $\sin^2\theta$ is shown in Fig. 14. One gets a "nose-like" curve. The experimental rate for $\bar\nu N \to \bar\nu + \ldots$ (the plotted point is based on Ref. [122]) is about as low as it can be, while the rate for $\nu N \to \nu + \ldots$ is very sensitive to $\sin^2\theta$ and provides a good measure of it. It turns out that the x and ρ dependences combine in such a way that R_ν actually depends on m_t and M_H in very much the same way as does M_W. As a result, measurements of R_ν are often quoted as effective measurements of M_W, which is what we shall do here.

Averaging a new measurement [122] with previous determinations [123, 124], one finds a result equivalent to [7, 125] $M_W = 80.306 \pm 0.218\ \mathrm{GeV}/c^2$ when one takes into account the measured value of M_Z. [A somewhat different average was quoted in [122], implying $M_W = 80.220 \pm 0.208\ \mathrm{GeV}/c^2$.]

Direct measurements of the W mass have been presented by the CDF and D0 Collaborations [126, 127]. These use kinematic fitting to the sharp Jacobian peak in

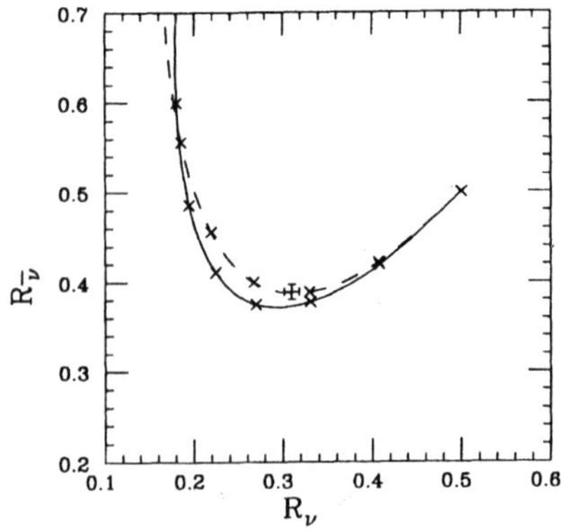

Figure 14. Ratios R_ν and $R_{\bar{\nu}}$ plotted against one another for varying values of $\sin^2\theta$ and $\rho = 1$. The $\times$ marks denote units of 0.1 in $\sin^2\theta$, starting at the point (0.5,0.5) for $\sin^2\theta = 0$. The dashed line corresponds to $r = 1/3$, which would be appropriate if the nucleon had no antiquarks but were composed only of valence quarks. The solid line corresponds to $r = 0.4$ (approximately the measured value). Plotted point illustrates measured values.

Table 4. Recent direct measurements of the W mass.

Collaboration	M_W (GeV/c^2)
CDF (Run 1A)	80.41 ± 0.18
D0 (Run 1A)	80.35 ± 0.27
D0 (Run 1B)	80.38 ± 0.17
LEP	$80.3 \pm 0.4 \pm 0.1$
UA2	80.36 ± 0.37
Average	80.356 ± 0.125

$\bar{p}p \to W + \ldots \to \ell\nu_\ell + \ldots$, where $\ell = (e, \mu)$. Now, for the first time, it is possible as a result of the increased energy at LEP to directly measure a point on the excitation curve in $e^+e^- \to W^+W^-$ [128], yielding a surprisingly accurate value for only a few dozen events. These results are compared with an earlier determination by the UA2 Collaboration [129] in Table 4. (See also Ref. [113].)

The W *width* Γ_W is an interesting quantity (see also Ref. [113]). A popular way of measuring it which is sometimes quoted as giving the total width is to measure the *ratio* of signals $\sigma(\bar{p}p \to W + \ldots \to \ell\nu_\ell + \ldots)/\sigma(\bar{p}p \to Z + \ldots \to \ell^+\ell^- + \ldots)$. In fact, however, since one needs to combine the standard model prediction for $\Gamma(W \to \ell\nu)$ with this ratio in order to extract Γ_W, what one actually ends up measuring is the *branching ratio* $B(W \to \ell\nu_\ell)$.

The standard model predicts [130]

$$B(W \to \ell\nu_\ell) \simeq \frac{1}{3 + 6[1 + (\alpha_s(M_W)/\pi)]} = 0.1084 \pm 0.0002 \tag{63}$$

if the $W \to t\bar{b}$ channel is closed and there are no unanticipated channels other than the three species of $\ell\nu_\ell$ and the lighter-quark channels. The world average [113] is

Table 5. Determinations at LEP of $\sin^2\theta_{\text{eff}}$

Observable	Name	Value	Error
A_{FB}^{ℓ}	Forward-backward asymmetry	0.23085	0.00056
A_{τ}	See notes (a,b)	0.23240	0.00085
A_e	See notes (a,b)	0.23264	0.00096
$\langle Q_{FB}\rangle$	F-B charge asymm., light quarks	0.23200	0.00100
A_{FB}^{c}	F-B charm asymmetry	0.23155	0.00112
A_{FB}^{b}	F-B bottom asymmetry	0.23246	0.00041
$\sin^2\theta_{\text{eff}}^{\text{LEP}}$	LEP Average	0.23200	0.00027

(a) $A_f \equiv 2g_V^f g_A^f/[(g_V^f)^2+(g_A^f)^2]$
(b) A_τ, A_e extracted from angular dependence of τ polarization

0.110 ± 0.003. On the other hand, a cruder but less theory-dependent measurement of the total width [131] makes use of the different dependence on Γ_W of charged lepton pair production on and off the W peak, and leads to $\Gamma_W = 2.11 \pm 0.28 \pm 0.26$ GeV in accord with the standard model prediction [113] of 2.077 ± 0.014 GeV. The "oblique" corrections mentioned above turn out not to affect the W width [130].

A number of measurements of Z properties have become possible as a result of the huge statistics amassed at the LEP e^+e^- Collider in the reaction $e^+e^- \to Z \to \ldots$. The ability of the SLC Collider at Stanford to polarize electron beams longitudinally has permitted useful information to be obtained at that machine despite a much lower collision rate. The numbers we shall quote are those contained in the latest version of the LEP Electroweak Working Group document [7], which should be consulted for further references.

The mass of the Z, now measured to be $M_Z = 91.1863 \pm 0.0020$ GeV/c^2, can serve an input to the electroweak theory when combined with G_F and $\alpha(M_Z)$ to predict all other observables as functions of m_t and M_H. Deviations of observables from "nominal" values predicted for specific choices of m_t and M_H can be described as linear functions of the parameters S, T, and U mentioned above. Every observable is a homogeneous function of degree 0, 1, or 2 of the factor $\rho = \rho^{\text{nominal}}(1+\alpha T)$ and of $\sin^2\theta(S,T)$. Thus, every observable will define a band in the $S-T$ plane (assuming for present purposes that $S_W = S_Z$, i.e., $U = 0$).

Other measured Z parameters [7] include the total width $\Gamma_Z = 2.4946 \pm 0.0027$ GeV, the hadron production cross section $\sigma_h^0 = 41.508 \pm 0.056$ nb, and $R_\ell \equiv \Gamma_{\text{hadrons}}/\Gamma_{\text{leptons}} = 20.778 \pm 0.029$, which may be combined to obtain the Z leptonic width $\Gamma_{\ell\ell}(Z) = 83.91 \pm 0.11$ MeV.

A number of determinations of $\sin^2\theta$ are obtained from asymmetries measured at LEP. These are summarized in Table 5.

The LEP average, $\sin^2\theta_{\text{eff}}^{\text{LEP}} = 0.23200 \pm 0.00027$, is to be compared with that based on the left-right asymmetry parameter A_{LR} measured with polarized electrons at SLC [132]: $\sin^2\theta_{\text{eff}}^{\text{SLC}} = 0.23061 \pm 0.00047$. The χ^2 for the average in Table 5 is 6.3 for 5 degrees of freedom, while it rises to 12.8 for 6 degrees of freedom when the SLC value is added.

Parity violation in atoms, stemming from the interference of Z and photon exchanges between the electrons and the nucleus, provides further information on electroweak couplings. The most precise constraint at present arises from the measure-

Table 6. Electroweak observables described in fit.

Quantity	Experimental value	Theoretical value
Q_W (Cs)	-71.0 ± 1.8 [a]	-73.2 [b] $- 0.80S - 0.005T$
Q_W (Tl)	-115.0 ± 4.5 [c]	-116.8 [d] $- 1.17S - 0.06T$
M_W (GeV)	80.341 ± 0.104 [e]	80.35 [f] $- 0.29S + 0.45T$
$\Gamma_{\ell\ell}(Z)$ (MeV)	83.91 ± 0.11 [g]	$83.90 - 0.18S + 0.78T$
$\sin^2\theta_{\text{eff}}$	0.23200 ± 0.00027 [h]	0.2315 [i] $+ 0.0036S - 0.0026T$
$\sin^2\theta_{\text{eff}}$	0.23061 ± 0.00047 [j]	0.2315 [i] $+ 0.0036S - 0.0026T$

[a] Weak charge in cesium [133]
[b] Calculation [135] incorporating atomic physics corrections [134]
[c] Weak charge in thallium [138, 139] (see text)
[d] Calculation [140] incorporating atomic physics corrections [141]
[e] Average of direct measurements and indirect information from neutral/charged current ratio in deep inelastic neutrino scattering [122, 123, 124]
[f] Including perturbative QCD corrections [116]
[g] LEP average [7]
[h] From leptonic asymmetries at LEP [7]
[i] As calculated [116] with correction for relation between $\sin^2\theta_{\text{eff}}$ and $\hat{s}^2$ [117]
[j] From left-right asymmetry in annihilations at SLC [132]

ment of the *weak charge* (the coherent vector coupling of the Z to the nucleus), $Q_W = \rho(Z - N - 4Z\sin^2\theta)$, in atomic cesium [133], with the result $Q_W(\text{Cs}) = -71.04 \pm 1.58 \pm 0.88$. The first error is experimental, while the second is theoretical [134]. The prediction [135] $Q_W(\text{Cs}) = -73.20 \pm 0.13$ is insensitive to standard-model parameters [135, 136, 137] once M_Z is specified; discrepancies are good indications of new physics. Recently the weak charge has also been measured in atomic thallium. Averaging determinations by the Seattle [138] and Oxford [139] in a manner described in more detail in [107], we find $Q_W(\text{Tl}) = -115.0 \pm 4.5$, to be compared with a prediction [140, 141, 142] of -116.8.

We have performed a fit to the electroweak observables listed in Table 6. The "nominal" values (including [116] $\sin^2\theta_{\text{eff}} = 0.2315$) are calculated for $m_t = 175$ GeV and $M_H = 300$ GeV. We use $\Gamma_{\ell\ell}(Z)$, even though it is a derived quantity, because it has little correlation with other variables in our fit. It is mainly sensitive to the axial-vector coupling g_A^ℓ, while asymmetries are mainly sensitive to g_V^ℓ.

In order to focus on electroweak parameters and avoid the use of highly correlated measurements, we omit several quantities from the fit which are normally included in complete analyses. (1) The total width $\Gamma_{\text{tot}}(Z)$ is omitted since it is highly correlated with $\Gamma_{\ell\ell}(Z)$ and mainly provides information on the value of the strong fine-structure constant α_s. With $\alpha_s = 0.12 \pm 0.01$, the observed total Z width is consistent with predictions. For the same reason, we omit the quantity R_ℓ. (2) The invisible width of the Z is compatible with three species of neutrinos. It would provide information on $\rho/\rho^{\text{nominal}} = 1 + \alpha T$ but is less useful as a source of that information than the leptonic width, and is highly correlated with it. (3) The experimental situation regarding the partial width $\Gamma(Z \to b\bar{b})$ has changed recently [143]. We omit it from the fit while discussing it separately below.

Each observable in Table 6 specifies a band in the $S - T$ plane with different

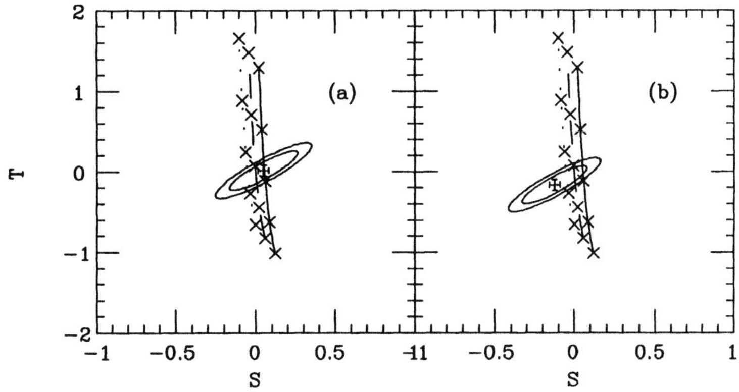

Figure 15. Allowed ranges of S and T at 68% (inner ellipses) and 90% (outer ellipses) confidence levels, corresponding to $\chi^2 = 2.3$ and 4.6 above the minimum (crosses at center of ellipses). Dotted, dashed, and solid lines correspond to standard model predictions for $M_H = 100$, 300, 1000 GeV. Symbols $\times$, from bottom to top, denote predictions for $m_t = 100$, 140, 180, 220, and 260 GeV. (a) Fit including APV experiments with present errors; (b) errors on APV experiments reduced by a factor of 4, with present central values of Q_W retained.

slope, as seen from the ratios of coefficients of S and T. Parity violation in atomic cesium and thallium is sensitive almost entirely to S [135, 137]. The impact of $\sin^2\theta_{\rm eff}$ determinations on S is considerable. The leptonic width of the Z is sensitive primarily to T. The W mass specifies a band of intermediate slope in the $S-T$ plane; here we assume $S_W = S_Z$. Strictly speaking, the ratio R_ν specifies a band with slightly more T and less S dependence than M_W [115, 135]; we have ignored this difference here.

The resulting constraints on S and T are shown in Fig. 15(a). A top quark mass of 175 ± 6 GeV/c^2 (the CDF and D0 average), corresponding as we noted in Sec. 2 to $\bar{m}_t(M_W) = 165 \pm 6$ GeV/c^2, is compatible with all Higgs boson masses between 100 and 1000 GeV/c^2, as seen by the curved lines intersecting the error ellipses. The centers of the ellipses lie at $S = 0.05$, $T = 0.01$, indicating that the fit is very comfortable with the nominal values chosen for m_t and M_H.

Independently of the standard model predictions, values of S between -0.3 and 0.3 are permitted at the 90% confidence level. The values from atomic parity violation, $S = -2.7 \pm 2.3$ [133] based on cesium and $S = -1.5 \pm 3.8$ based on thallium [138, 139], have an average $S = -2.4 \pm 2.0$. The value of S is now known much more precisely than specified by these experiments. On the other hand, since the weak charge Q_W provides unique information on S, its determination with a factor of four better accuracy than present levels could have a noticeable effect on global fits, as shown in Fig. 15(b).

In contrast to many fits in the literature (see, e.g., [144, 145, 146]), ours does not exhibit a preference for any particular Higgs boson mass in the absence of separate information about m_t. We show curves of χ^2 as a function of m_t for several values of M_H in Fig. 16.

The minima of the χ^2 curves are nearly identical (slightly above 8 for the six data points in Table 6, i.e., for 5 degrees of freedom) for all three choices of M_H. When the value $\bar{m}_t(M_W) = 165 \pm 6$ GeV/c^2 (delimited by the horizontal band) is selected, some

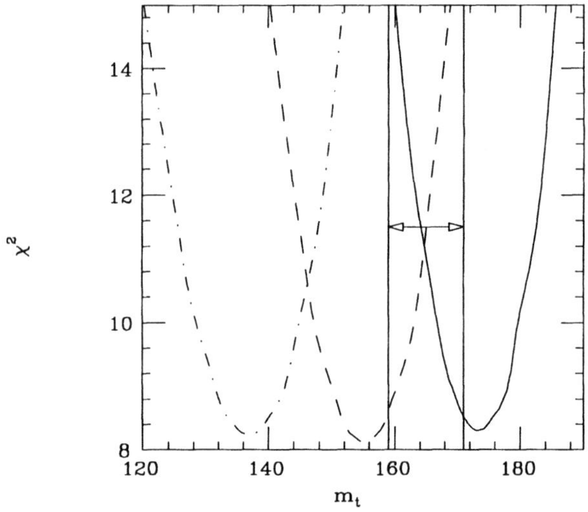

Figure 16. Values of χ^2 for fits to the data in Table 6, as function of m_t for several values of M_H. Dot-dashed, dashed, and solid curves correspond to $M_H = 100$, 300, and 1000 GeV/c^2, respectively. Horizontal bands denote the 1σ limits for $\bar{m}_t(M_W) = 165 \pm 6$ GeV/c^2.

preference arises for a Higgs boson mass in the range between 300 and 1000 GeV/c^2. To see this more clearly, we include m_t in the fit in Fig. 17.

The favored value of M_H in the fit based on the data in Table 6 and the top quark mass mentioned above is $M_H = (560 \times 1.5^{\pm 1})$ GeV/c^2. (See Fig. 17(a).) This is considerably higher than most other determinations (see, e.g., [144, 145, 146]). The experimental average of the two last rows in Table 6, $\sin^2\theta_{\text{eff}} = 0.23165 \pm 0.00024$ is slightly above our "nominal" value of $\sin^2\theta_{\text{eff}} = 0.2315$; a discrepancy in this direction tends to push the Higgs boson mass up from its nominal value of 300 GeV/c^2. To accentuate this effect we explore the effect of omitting the SLC data on A_{LR} (the last row in Table 6). The small upward change in $\sin^2\theta_{\text{eff}}$, by $+0.00035$, leads the favored M_H to increase by a factor of 1.5, to $(820 \times 1.7^{\pm 1})$ GeV/c^2. (See Fig. 17(b).) The χ^2, incidentally, drops by more than 6.

An effect which leads us to obtain a larger M_H than some other fits is that our "nominal" $\sin^2\theta_{\text{eff}} = 0.2315$ is slightly below theirs. For example [7], the fitted value $\sin^2\theta_{\text{eff}} = 0.2318 \pm 0.0002$ is quoted for $m_t = 177 \pm 7$ GeV/c^2 when $M_H = 300$ GeV/c^2.

A further possible reason for the discrepancy between our fits and others is that the others incorporate $R_\ell \equiv \Gamma(Z \to \text{hadrons})/\Gamma(Z \to \ell^+\ell^-)$, whose experimental value of 20.778 ± 0.029 is somewhat above the nominal value (20.757 in Ref. [7]). We have chosen to omit this quantity since it depends on α_s. Experimental determinations of $\alpha_s(M_Z)$ may be converging rapidly enough that one can trust them; the value [147] $\alpha_s(M_Z) = 0.118 \pm 0.003$ was quoted recently. The quantity ρ cancels in R_ℓ. Aside from its dependence on α_s (we find $\partial R_\ell/\partial\alpha_s = 6.7$), R_ℓ provides a measure of $x \equiv \sin^2\theta$; we find $\partial R/\partial x = -17.5$. Thus a high R_ℓ implies a low $\sin^2\theta$. Low experimental values of $\sin^2\theta$ push M_H down.

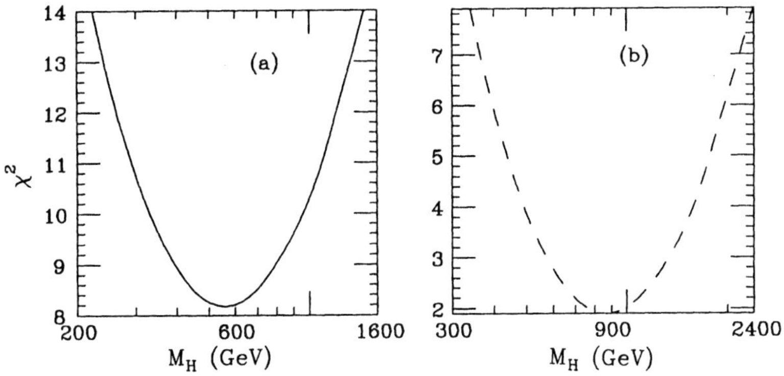

Figure 17. Values of χ^2 for fits to the data in Table 6 and the value $\bar{m}_t(M_W) = 165 \pm 6$ GeV/c^2, as function of M_H. (a) SLC data (last row of Table 6) included (6 d.o.f.); (b) SLC data omitted (5 d.o.f.).

Another possibility is that some fits allow a separate parameter to describe A_{FB}^b. This quantity, as shown by the last row of Table 5, has a small quoted error. If it is omitted, the LEP average becomes $\sin^2\theta_{\text{eff}} = 0.23163 \pm 0.00037$ with $\chi^2 = 4.0$ for 4 d.o.f., while if the SLC result is added to this sample, one finds $\sin^2\theta_{\text{eff}} = 0.23124 \pm 0.00029$ with $\chi^2 = 6.9$ for 5 d.o.f. This low value again favors low M_H.

In any event the difference between Figs. 17(a) and (b) shows the pitfalls of drawing a conclusion about M_H when its value is so sensitive to a single datum. One is really only determining $\log M_H$, whose value is affected by less than 1σ by including or dropping the SLC data.

D. The decay $Z \to b\bar{b}$

We now return to the question of the $Z \to b\bar{b}$ branching ratio, which has engaged considerable theoretical attention in the past couple of years [148]. We first discuss the calculation of $\Gamma(Z \to f\bar{f})$ for any fermion f.

The interaction Lagrangian for a Z and a fermion described by the Dirac field ψ is

$$\mathcal{L}_{\text{int}} = -\sqrt{g^2 + g'^2}\bar{\psi}\gamma^\mu Z_\mu(I_{3L}\mathcal{P}_L - Q\sin^2\theta)\psi \quad , \tag{64}$$

where $\mathcal{P}_L \equiv (1-\gamma_5)/2$. As an example, for a b quark with $I_{3L} = -(1/2)$, the axial and vector coupling constants (in a convenient normalization) are $g_A = 1/4$, $g_V = -(1/4) + (1/3)\sin^2\theta$. The Z partial width is then

$$\Gamma(Z \to f\bar{f}) = \frac{\sqrt{2}G_F M_Z^3 \rho}{3\pi} N_c \left(1 - \frac{4m_f^2}{M_Z^2}\right)^{1/2} \times \left[g_A^2\left(1 - \frac{4m_f^2}{M_Z^2}\right) + g_V^2\left(1 + \frac{2m_f^2}{M_Z^2}\right)\right] \quad . \tag{65}$$

For a massless neutrino $g_V = -g_A = 1/4$, so

$$\Gamma(Z \to \nu\bar{\nu}) = \frac{G_F M_Z^3 \rho}{12\pi\sqrt{2}} = 166 \text{ MeV} \times \rho \quad . \tag{66}$$

As mentioned, the fact that the invisible width of the Z is about 1/2 GeV leads one to conclude that there are three species of light neutrinos.

A calculation of $\Gamma(Z \to d\bar{d}) \equiv \Gamma_d$, for which $g_A^2 = 1/16 = 0.0625$ and $g_V^2 = 0.0299$, gives (368 MeV)$(\rho)(1 + [\alpha_s/\pi])$, or, with $\rho \simeq 1.01$ and $1 + (\alpha_s/\pi) = 1.04$, about 387 MeV. When compared with [7] $\Gamma_h \equiv \Gamma(Z \to \text{hadrons}) = 1743.6 \pm 2.5$ MeV, this gives $R_d^0 \equiv \Gamma(Z \to d\bar{d})/\Gamma(Z \to \text{hadrons} \simeq 0.222$. More precise calculations (see, e.g., [148]) give $R_d^0 \simeq 0.221$. We distinguish between $R_q^0 \equiv \Gamma(Z \to q\bar{q})/\Gamma(Z \to \text{hadrons})$ and what is actually measured at the Z peak, $R_q \equiv \sigma(e^+e^- \to q\bar{q})/\sigma(e^+e^- \to \text{hadrons})$. As a result of a small photon contribution, $R_b^0 = R_b + 0.0003$ [149].

The non-zero b mass must be taken into account in $Z \to b\bar{b}$. The predominance of g_A^2 over g_V^2 for the $Zb\bar{b}$ coupling (in a ratio of about 2 to 1) means that the kinematic factor $\beta^3 \simeq 0.980$ for axial couplings, where $\beta = [1 - (4m_b^2/M_Z^2)]^{1/2}$, has an appreciable effect. The corresponding kinematic factor for vector couplings, $\beta(3 - \beta^2)/2$, is nearly 1. Thus there is an overall reduction in R_b/R_d by about 1.3% just due to m_b, leading to the kinematically corrected prediction [150] $R_{b\ \text{kin}}^0 = 0.2183 \pm 0.0001$.

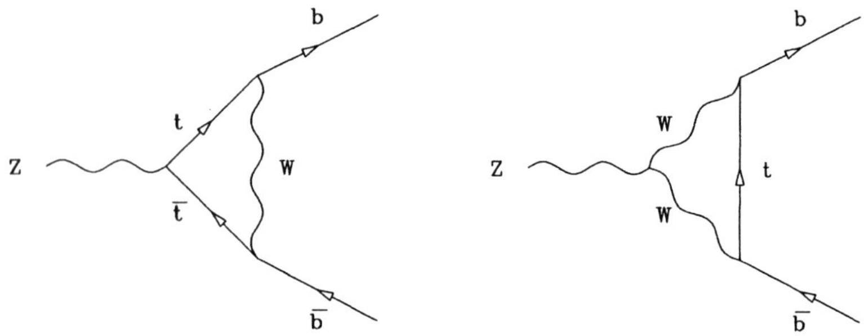

Figure 18. Examples of graphs contributing to $\Gamma(Z \to b\bar{b})$.

The top quark affects $\Gamma(Z \to b\bar{b})$ through graphs like those shown in Fig. 18. (In 't Hooft – Feynman gauge one must also include contributions of charged Higgs bosons, replacing W in all possible ways.) Contributions quadratic in m_t appear [151], as in the ρ parameter, so that $R_b^0/R_{b\ \text{kin}}^0 = 1 - aG_F m_t^2$, where a is a dimensionless constant. For the observed value of the top quark mass the vertex-corrected standard-model prediction is $R_b^0 = 0.2158$ [7], amounting to an additional correction of -1.2%.

Until this past summer, results from the LEP and SLD groups seemed to indicate a value of R_b noticeably above the standard-model prediction. More recently [143], the ALEPH Collaboration has analyzed the full LEP I data set using two different methods. Using a lifetime tag, they find $R_b = 0.2169 \pm 0.0011 \pm 0.0017$, while using five mutually exclusive hemisphere tags they find $R_b = 0.2158 \pm 0.0009 \pm 0.0011$. The current LEP average [7], incorporating some new results from other LEP groups and from SLD [152], is $R_b^0 = 0.2178 \pm 0.0011$. These results are in accord with standard model predictions. They do not yet conclusively confirm the -1.2% effect due to top-quark graphs like those in Fig. 18, but at least indicate that $R_b < R_d$.

Table 7. Final states of a $t\bar{t}$ pair as function of decays of the W bosons in $t \to W + b$. Here $\not{E}_T$ stands for "missing transverse energy," while j stands for "jet".

$W^+ \to$ $W^- \to$	1/9 $e\nu$	1/9 $\mu\nu$	1/9 $\tau\nu$	2/3 $u\bar{d}' + c\bar{s}'$
1/9 $e\nu$	$ee\not{E}_T$ $+2j$	$e\mu\not{E}_T$ $+2j$	$e\tau\not{E}_T$ $+2j$	$e\not{E}_T$ $+4j$
1/9 $\mu\nu$	$e\mu\not{E}_T$ $+2j$	$\mu\mu\not{E}_T$ $+2j$	$\mu\tau\not{E}_T$ $+2j$	$\mu\not{E}_T$ $+4j$
1/9 $\tau\nu$	$e\tau\not{E}_T$ $+2j$	$\mu\tau\not{E}_T$ $+2j$	$\tau\tau\not{E}_T$ $+2j$	$\tau\not{E}_T$ $+4j$
2/3 $\bar{u}d' + \bar{c}s'$	$e\not{E}_T$ $+4j$	$\mu\not{E}_T$ $+4j$	$\tau\not{E}_T$ $+4j$	$6j$

4. TOP QUARK OBSERVATION AT THE TEVATRON

A. Production and decay

Quite some time ago [153, 154] it was realized that the Fermilab Tevatron would be able to produce top quarks heavy enough to decay to $W + b$. An estimate [154] in 1984 predicted $\sigma(\bar{p}p \to t\bar{t} + \ldots) \sim 10\ \text{pb}(m_t/150\ \text{GeV}/c^2)^{-5}$ at c.m. energy $\sqrt{s} = 2$ TeV. The major subprocesses contributing to $t\bar{t}$ production (q and g stand for quark and gluon) are $q\bar{q} \to g^* \to t\bar{t}$ and $gg \to t\bar{t}$ (via t exchange or a gluon in the direct channel). The $q\bar{q}$ subprocess is dominant for the actual situation, so one has to know the distribution of quarks and antiquarks inside the nucleon. This information is provided, for example, by deep inelastic scattering [155]. We shall quote more precise predictions presently.

The decay of the top quark for $m_t > M_W + m_b$ is essentially 100% to $W + b$ unless other channels (like $H^+ + b$, where H^+ is a charged Higgs boson) exist. We have already mentioned that this expectation is confirmed [2], though not yet with overwhelming accuracy. Assuming that only the $W + b$ channel is significant, the decays of a $t\bar{t}$ pair then may be decomposed according to Table 7. We have used the shorthand d' and s' for the linear combinations (3) and (4) of d and s which couple to u and c via W.

Energy balance in Tevatron collisions can only be checked in the transverse direction since particles can escape undetected at small angles relative to the beam. Large missing transverse energy serves as a signature of neutrino production in W leptonic decay. One always expects at least two jets of hadrons associated with the two b quarks in $t\bar{t}$ decays. If one or two W's decays hadronically a total of 4 or 6 jets should then be produced.

All the channels in Table 7 except for those with one or two τ leptons have now been studied. These include dileptons (about 5%), leptons + jets (about 30%), and jets (about 45%). A start is even being made on channels with one τ and one e or μ. Thus a large fraction of the $t\bar{t}$ channels are seen.

B. Recent CDF and D0 analyses

We summarize in this subsection some results which were presented at the 1996 Warsaw Conference by the CDF and D0 Collaborations [26]. For CDF results, in addition to those presented at Warsaw, one may consult Refs. [156] for details. As an example of how one exploits the possibilities of Table 7, we now give some details of the CDF analyses.

Table 8. CDF top quark events in different channels based on an integrated luminosity of 110 pb^{-1}. Only the statistical error is shown for the channels with τ. Acceptances were calculated for $m_t = 175$ GeV/c^2.

Channel	N_{data}	$N_{\text{background}}$	$\sigma(t\bar{t})$ (pb)
Dilepton	10	2.1 ± 0.4	$9.3^{+4.4}_{-3.4}$
$\ell + j$ (SVX)	34	7.96 ± 1.37	$6.8^{+2.3}_{-1.8}$
$\ell + j$ (SLT)	40	24.3 ± 3.5	$8.0^{+4.4}_{-3.6}$
Hadronic	192	137 ± 11	$10.7^{+7.6}_{-4.0}$
$\tau + (e, \mu)$	4	2.0 ± 0.35	$15.6^{+18.6}_{-13.2}$
Total			$7.5^{+1.9}_{-1.6}$

1. In the *dilepton* sample, where $\ell = e$ or μ, one asks that the leptons have opposite charges, demands missing transverse energy ($\not{E}_T$) as a signature of one or more neutrinos, and requires two jets (signatures of the two b quarks). As a result, 7 events with $e\mu$, two with $\mu\mu$, and one with ee were found. Six "tagged" jets in a total of 4 events were identified as b's after the fact, but b identification was not required *a priori*.

2. In the *lepton + jets* sample, with $\ell = e$ or μ, one again asks for $\not{E}_T$, at least three observed jets, and at least one identified b quark. The b quark can be identified by its displaced vertex, since the average b lifetime is 1.5 to 1.6 ps, using the silicon vertex detector ("SVX"). Alternatively, its semileptonic decay, with a branching ratio of about 10.5% to each of $e + \ldots$ and $\mu + \ldots$, gives rise to a soft lepton tag ("SLT"). One sees 34 SVX events and 40 SLT events.

3. In the *6 jets* sample, one asks to see at least 5 of them, requiring a b to be tagged via the SVX. Here 192 events are seen.

In Table 8 we summarize the CDF measurements in different channels. The CDF cross sections in various channels average to a value $\sigma_{\text{CDF av.}}(p\bar{p} \to t\bar{t} + \ldots) = 7.5^{+1.9}_{-1.6}$ pb at $\sqrt{s} = 1.8$ TeV.

The corresponding D0 results make use of dilepton events, lepton + jets events with kinematic discrimination, and lepton + jets events with a soft muon "tag" associated with the semileptonic decay $b \to c\mu^- \bar{\nu}_\mu$. The results, based on the full 1992–5 sample with 100 pb^{-1} of integrated luminosity, are shown in Table 9. The column labeled "expected" refers to the number of events anticipated for a top quark with mass $m_t =$ 180 GeV/c^2 based on cross sections of Ref. [157]. The D0 top quark cross sections are 4.7 ± 3.2 pb in the dilepton channel, 4.6 ± 5.1 pb in the hadron channel, 4.2 ± 2.0 pb in the lepton + jets channel with kinematic discrimination, and 7.0 ± 3.3 pb in the lepton + jets/tag channel.

Combining all CDF and D0 cross section determinations, one finds

$$\sigma(p\bar{p} \to t\bar{t} + \ldots) = 6.4^{+1.3}_{-1.2} \text{ pb} \quad (\sqrt{s} = 1.8 \text{ TeV}) \quad . \tag{67}$$

This is to be compared with theoretical estimates such as

$$\sigma_{\text{Theory}}(p\bar{p} \to t\bar{t} + \ldots) = \left\{ \begin{array}{ll} 4.84^{+0.73}_{-0.69} \text{ pb} & [158] \\ 4.75^{+0.73}_{-0.62} \text{ pb} & [159] \end{array} \right\} \quad . \tag{68}$$

Table 9. D0 top quark events in different channels based on an integrated luminosity of 100 pb^{-1}.

Channel	N_{expected}	$N_{\text{background}}$	N_{data}
$e\mu$	1.7 ± 0.3	0.4 ± 0.1	3
ee	0.9 ± 0.1	0.7 ± 0.2	1
$\mu\mu$	0.5 ± 0.1	0.5 ± 0.3	1
e + jets	6.5 ± 1.4	3.8 ± 1.4	10
μ + jets	6.4 ± 1.5	5.4 ± 2.0	11
e + jets/tag	2.4 ± 0.4	1.4 ± 0.4	5
μ + jets/tag	2.8 ± 0.9	1.1 ± 0.2	6
Total	21.2 ± 3.8	13.4 ± 3.0	37

C. Mass

The most precise CDF value of the top quark mass [26] comes from a combined analysis including (a) 34 events with a lepton, jets, and at least one SVX tag (leading to $m_t = 177 \pm 7$ GeV/c^2), (b) a more refined analysis of the same sample, weighting events in terms of the quality of information they provide, leading to the value 174 ± 8 GeV/c^2, (c) the dilepton sample, leading to 160 ± 28 GeV/c^2 and 162 ± 22 GeV/c^2 in two different analyses (the second using the average $M^2_{\ell b}$ of the lepton and b), and (d) the multijet sample, leading to 187 ± 15 GeV/c^2. The resulting mass plot is shown in Fig. 19.

The CDF Collaboration has also presented distributions in top quark transverse momentum and in $m_{t\bar{t}}$, the effective mass of the $t\bar{t}$ system [156]. These data are in accord with standard model expectations.

The D0 top quark mass obtained from the lepton + jets sample is 169 ± 11 GeV/c^2, while 158 ± 26 GeV/c^2 is obtained from the dilepton sample. The mass distribution from the lepton + jets sample (before a 1 GeV/c^2 jet energy correction has been applied) is shown in Fig. 20. The world average of 175 ± 6 GeV/c^2 is based on all the CDF and D0 determinations.

In Fig. 21 we plot the expected top quark cross section [158] as a function of m_t at the c.m. energies $\sqrt{s} = 1.8$ and 2 TeV. Also shown are the individual CDF and D0 mass and cross section points. It is clear there is satisfactory agreement with the standard model, though there would be room, if one so chose, for an additional 50% extra contribution to the cross section from some non-standard process.

D. Decays

The branching ratio of the top quark to $W + b$ can be measured relative to all $W + q$ decays by comparing double b tags with single b tags [2]. The result is

$$\frac{B(t \to W + b)}{B(t \to W + q)} = 0.94 \pm 0.27 \pm 0.13 \quad , \tag{69}$$

implying $V_{tb} = 0.97 \pm 0.15 \pm 0.07$ in accord with standard-model expectations.

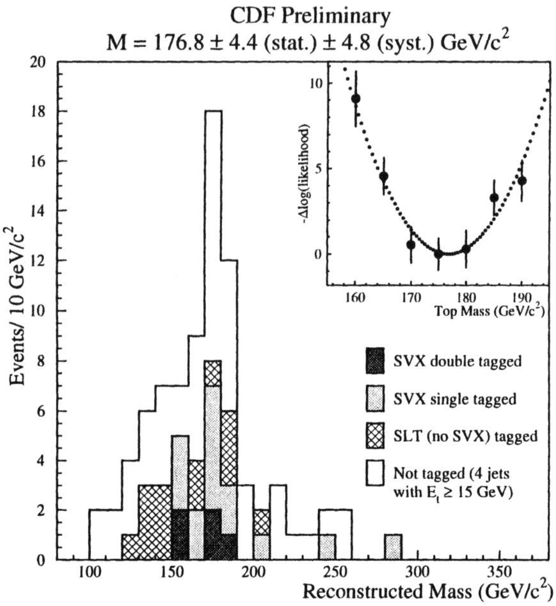

Figure 19. CDF top quark mass distributions as of July 1996.

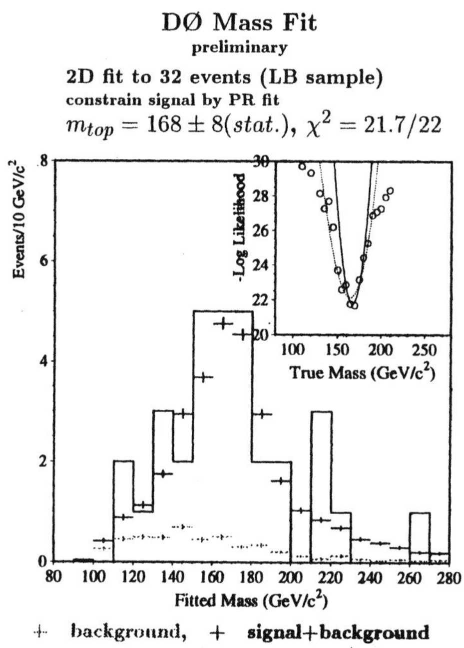

Figure 20. D0 top quark mass distribution from lepton + jets sample as of July 1996. A correction of +1 GeV/c^2 must be applied for the energy scale.

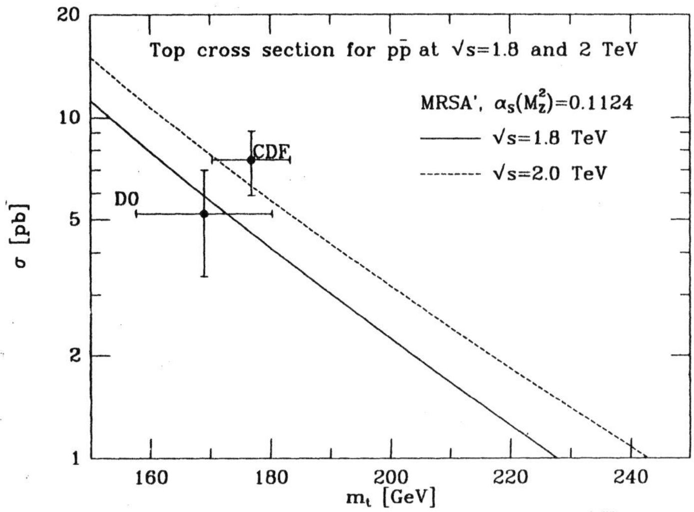

Figure 21. Top quark cross section prediction as a function of m_t.

5. TOP QUARK PRODUCTION IN ELECTRON-POSITRON COLLISIONS

A. Excitation curve

Eventually it will become possible to produce top quarks in pairs in high-energy electron-positron annihilations. Accordingly, we discuss the production of pairs of arbitrary fermions f via virtual photons (γ^*) and Z's (Z^*) in the direct channel. These results will be of use in Sec. 6 when we discuss exotic quarks and leptons.

Define $x \equiv \sin^2\theta$, $s \equiv E^2_{\text{c.m.}}$, and $r \equiv [s/(s - m_Z^2)x(1-x)]$. Then far from the Z pole, where the Z width can be neglected, the contribution of a virtual photon and Z in the direct channel to the cross section for production of a fermion with electric charge Q_f and axial and vector Z couplings $g_A = -(1/2)I^f_{3L}$ and $g_V = (1/2)I^f_{3L} - Q_f x$ is

$$\sigma(e^+e^- \to f\bar{f}) = \sigma_\gamma \left\{ Q_f^2 - 2rQ_f g_V^e g_V^f + r^2[(g_V^e)^2 + (g_A^e)^2][(g_V^f)^2 + \frac{\beta^2}{K_V}(g_A^f)^2] \right\} \ , \quad (70)$$

where

$$\sigma_\gamma \equiv \frac{4\pi\alpha^2}{3s} N_c \beta K_V \ , \quad \beta \equiv \left(1 - \frac{4m_f^2}{s}\right)^{1/2} \ , \quad K_V \equiv \frac{3-\beta^2}{2} \ , \quad (71)$$

and N_c is the number of colors of fermions. For quarks ($N_c = 3$) the cross section should be multiplied by an additional correction factor of $1 + (\alpha_s/\pi) \approx 1.04$. The values of σ/σ_0 far above pair production threshold, where $\sigma_0 \equiv \sigma(e^+e^- \to \gamma^* \to \mu^+\mu^-)$, are compared in Table 10 for various fermion species f when the energy is far below the Z pole (where only the virtual photon dominates) and when it is far above the Z (where the interference in vector contributions of the photon and Z is possible). All the cross sections far above the Z are enhanced relative to the values they would have due to the virtual photon alone.

Table 10. Cross sections σ [in units of $\sigma_0 \equiv \sigma(e^+e^- \to \gamma^* \to \mu^+\mu^-)$] for e^+e^- production of pairs of fermions $f\bar{f}$ via virtual photons and Z's in the direct channel. Here t-channel exchanges are neglected for e and ν_e. Values of g_V^f are quoted for $x = 0.2315$. QCD corrections to quark production have been neglected.

Fermion f	Q_f	g_V^f	g_A^f	σ/σ_0 far below Z	σ/σ_0 far above Z
u	2/3	0.0957	−1/4	4/3	1.80
d	−1/3	−0.1728	1/4	1/3	0.92
e^-	−1	−0.0185	1/4	1	1.13
ν_e	0	1/4	−1/4	0	0.25

The predicted top quark cross section, shown in Fig. 22, is composed of incoherent contributions corresponding to vector and axial-vector couplings to the top. The vector part involves $\gamma^* - Z^*$ interference. These contributions have rather different energy dependence as a result of different kinematic factors. The vector contribution peaks around $E_{\rm c.m.} = 410$ GeV, while the axial vector contribution peaks around 550 GeV. At 410 GeV, the axial contribution is only 5% of the total, rising to about 25% asymptotically. The sum of the two contributions peaks around 420 GeV at a value of about 0.7 pb.

B. Sensitivity to interactions

The top quark mass is too large to permit the formation of $t\bar{t}$ bound states or top mesons (t quarks bound to light antiquarks). The top quark decays to $W + b$ with a partial width of about 1.4 GeV, as noted in the Introduction. This is considerably larger than the spacings between quarkonium levels shown in Figs. 3 and 4. Nonetheless one can learn some things about the top-antitop interaction by studying the shape of the threshold function and how it is distorted in comparison with the lowest-order prediction in Fig. 22 [160, 161]. One probably sees a vestige of the $1S$ level as a slight threshold enhancement with respect to the free-quark behavior, while Higgs interactions [161] can alter the threshold shape considerably.

6. SPECULATIONS ON SOURCE OF TOP QUARK MASS

A. Mass matrix lore

In Sec. 2 A we saw that the 3×3 quark mass matrices $\mathcal{M}_{U,D}$ were diagonalized by separate unitary transformations $L_{U,D}$ and $R_{U,D}$ on left- and right-handed quarks. The CKM matrix $V = L_U^\dagger L_D$ carries no information about the transformations $R_{U,D}$. As a result, separate unitary transformations can be performed on right-handed U and D quarks, as long as corresponding redefinitions of the mass matrix are made. Moreover, the simultaneous transformations $L_U^\dagger \to L_U^\dagger A$ and $L_D \to A^\dagger L_D$, with $AA^\dagger = 1$, also leave V unchanged. Thus there is considerable freedom in the choice of basis for mass matrices, which has led to a large folklore. We shall barely scratch the surface.

An early observation [162, 163] was the relation between the Cabibbo angle and the ratio of d and s masses: $\sin\theta_c \simeq (m_d/m_s)^{1/2}$. This result can be obtained with a correction term $-(m_u/m_c)^{1/2}$ if one postulates a zero in the upper-left corner of the

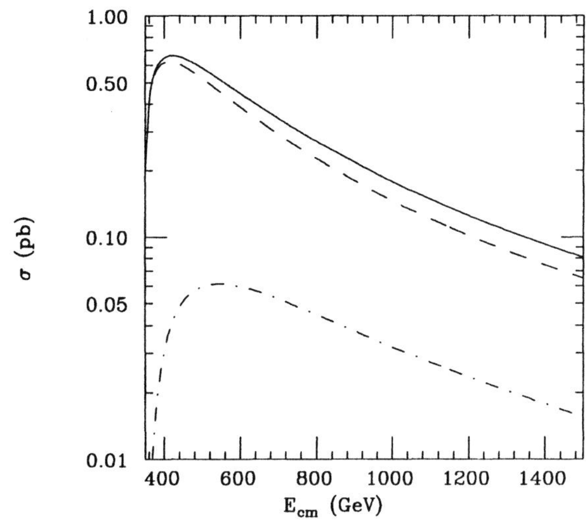

Figure 22. Cross section prediction for $e^+e^- \to (\gamma^*, Z^*) \to t\bar{t}$ as a function of $E_{\text{c.m.}}$. Solid line: full cross section; dashed line: contribution of vector couplings of t; dot-dashed line: contribution of axial coupling of t.)

mass matrices for two families of quarks:

$$\mathcal{M}_Q = \begin{bmatrix} 0 & a \\ a & b \end{bmatrix} \quad , \tag{72}$$

where $a \ll b$. Since the determinant of this matrix is $-a^2$ while its trace is b, its eigenvalues are approximately $-a^2/b$ and b, corresponding to the eigenvectors $[1, -a/b]$ and $[a/b, 1]$, respectively. We then identify

$$\begin{aligned} a_U &= \sqrt{-m_u m_c} \quad , \quad b_U = m_c \quad , \\ a_D &= \sqrt{-m_d m_s} \quad , \quad b_D = m_s \quad . \end{aligned} \tag{73}$$

The CKM matrix may be computed from the scalar product of the appropriate eigenvectors U_i and D_j via [164] $V_{ij} = U_i^\dagger \cdot D_j$. For example,

$$V_{us} = \left[1, -\sqrt{\frac{-m_u}{m_c}}\right] \cdot \left[\begin{array}{c} \sqrt{-m_d/m_s} \\ 1 \end{array}\right] = \sqrt{\frac{-m_d}{m_s}} - \sqrt{\frac{-m_u}{m_c}} \quad . \tag{74}$$

The negative sign of a quark mass is no cause for concern; it can always be changed by redefining $q \to \gamma_5 q$. The relative phases of the two terms on the right-hand side of (74) can be adjusted by an *ansatz* regarding the phases of terms in the mass matrices. For $m_s/m_d \simeq 20$ [165], the observed value $V_{us} \simeq 0.22$ is obtained for a relative phase of the two terms close to 90°.

The crucial role of zeroes in mass matrices for three-family models was first exploited by Fritzsch [166], explored in some detail by Ramond and his group [167], and has been reviewed recently by Branco *et al.* [168]. These zeroes go under the rubric of

Table 11. Quark mass ratios contributing to CKM matrix elements in one basis for quark mass matrices. Here the symbol q stands for m_q.

CKM element	Quark mass ratios	Magnitudes of ratios
V_{us}, V_{cd}	$(d/s)^{1/2}$, $(u/c)^{1/2}$	0.22, 0.06
V_{cb}, V_{ts}	s/b, c/t	0.03, 0.007
V_{ub}	$(ds)^{1/2}/b$, $(s/b)(u/c)^{1/2}$, $(uc)^{1/2}/t$	0.007, 0.002, 0.0004
V_{td}	$(ds)^{1/2}/b$, $(c/t)(d/s)^{1/2}$, $(uc)^{1/2}/t$	0.007, 0.002, 0.0004

"texture zeroes." A "democratic" basis has also been proposed [169] , which explains the hierarchies $m_u, m_c \ll m_t$ and $m_d, m_s \ll m_b$ by perturbing around matrices of the form

$$\mathcal{M} = \frac{1}{3}\begin{bmatrix} 1 & 1 & 1 \\ 1 & 1 & 1 \\ 1 & 1 & 1 \end{bmatrix} \tag{75}$$

with eigenvectors

$$\frac{1}{\sqrt{2}}\begin{bmatrix} 1 \\ -1 \\ 0 \end{bmatrix}, \quad \frac{1}{\sqrt{6}}\begin{bmatrix} 1 \\ 1 \\ -2 \end{bmatrix}, \quad \frac{1}{\sqrt{3}}\begin{bmatrix} 1 \\ 1 \\ 1 \end{bmatrix}, \tag{76}$$

corresponding to eigenvalues 0, 0, 1. By means of a unitary transformation one can end up with a matrix $\mathcal{M}$ consisting of all zeroes except for a 1 in the lower right-hand corner.

A curious basis choice seems to be encountered in several quite different approaches [164, 170]:

$$\mathcal{M} = \begin{bmatrix} 0 & \alpha p & \alpha \\ \alpha p & \beta & \beta q \\ \alpha & \beta q & \gamma \end{bmatrix}, \tag{77}$$

where p and q are numbers of order 1. For $\alpha \ll \beta \ll \gamma$ the eigenvectors and eigenvalues of this matrix are easy to compute. The CKM elements and the quark mass ratios which contribute to them (with phases which must be adjusted by hand) are summarized in Table 11. The magnitudes of the mass ratios appear to be appropriate for the sizes of the CKM elements. No insight is present in these models (nor in most others) regarding the choice of relative phases, however.

B. A two-family fable

Perhaps one of the least popular ideas these days is the notion that quarks are somehow composite objects. Nonetheless, an exercise (perhaps no more than a fable, as applied to quarks) shows the potential insight a composite model can provide into the interplay between masses and mixing [164, 171].

The exercise is based on states for which there are experimental candidates: the charmed-strange baryons known as Ξ_c illustrated in Fig. 5. These are interesting because no two quarks in them have the same mass. They consist of csq, where $q = u$ or d.

The lowest charmed-strange baryon, now called Ξ_c [47] and denoted in Fig. 5 by Ξ_c^a (originally A in Ref. [172]), has its light quarks s and q in a state primarily of zero spin and flavor SU(3) antisymmetry. The total spin of the Ξ_c^a is 1/2.

A candidate for an excited version of the Ξ_c, which we label as Ξ_c^s in Fig. 5 (originally S in Ref. [172]), has been claimed by the WA89 Collaboration at CERN [173]. It lies 95 MeV/c^2 above the Ξ_c^a and decays to it by photon emission. It, too, has total spin equal to 1/2, but its light quarks are coupled up primarily to a state of spin 1 and flavor SU(3) symmetry. This state is on shaky ground as it has only been reported in conference proceedings and is not confirmed by experiments at the CLEO detector at Cornell.

Finally, the Cornell experiments just mentioned *do* see a candidate for a Ξ_c^* state at higher mass [32], which fits well with the assignment of a state of total spin 3/2 (originally S^* in Ref. [172]). This state decays to $\Xi_c^a + \pi$. The light quarks, of course, have to be in a pure spin-1 state here.

The masses of many hadrons can be described in terms of sums of constituent-quark masses and hyperfine-interaction terms inversely proportional to quark masses [174]. Thus, for example, for a Ξ_c, one will have

$$M \simeq m_c + m_s + m_q + a\left(\frac{\sigma_q \cdot \sigma_s}{m_q m_s} + \frac{\sigma_q \cdot \sigma_c}{m_q m_c} + \frac{\sigma_s \cdot \sigma_c}{m_s m_c}\right) \tag{78}$$

We note that $m_q < m_s \ll m_c$. Now, when $m_c \to \infty$, this mass operator is diagonal in the basis labeled by S_{sq}:

$$\begin{bmatrix} S_{sq} = 0 \\ S_{sq} = 1 \end{bmatrix} \leftrightarrow \begin{bmatrix} \Xi_c^a \\ \Xi_c^s \end{bmatrix} \quad \text{for } J = 1/2 \text{ states} \quad . \tag{79}$$

When $m_c \neq \infty$, however, this is only an approximate basis: The mass eigenstates are mixed. Moreover, *changing* m_q *changes the mixing angle.* We suggest this as a paradigm for the CKM rotation. In an attempt to construct a realistic two-family model [164, 171], specific relations among parameters are necessary in order to ensure that the $J = 3/2$ state lies much higher in mass than the $J = 1/2$ ones, and in order to obtain the appropriate "texture" zero in the upper left-hand corner of the 2×2 mass matrix for the $J = 1/2$ states. The three-family model constructed in Ref. [164] from similar arguments leads to the mass matrix (77) with $p = q = \sqrt{2}$.

C. Is the top quark special?

The top quark is heavier, *so far*, than any other known fermion. We do not know if this is merely because it is near the limit of what one can produce via the strong interactions at present energies (this is certainly true), or something more fundamental. Many suggestions have been made about the crucial role played by its large mass. We have mentioned already the contribution of m_t to loop diagrams in particle-antiparticle mixing (Sec. 2) and in precise electroweak physics (Sec. 3). Here we mention some roles of a more fundamental nature which a heavy top might play.

Veltman [175] proposed that the top quark plays a crucial role in the cancellation of quadratic divergences in the effective Higgs potential, leading to the mass relation $12m_t^2 = 3M_H^2 + 3m_Z^2 + 6M_W^2$ in the limit that other quark masses are neglected. This implies that the Higgs boson lies slightly below $2m_t$ in mass, as if it were a $t\bar{t}$ bound state. In the limit in which the electroweak couplings g and g' vanish, $M_H \to 2m_t$ and the binding energy vanishes.

Nambu [176], drawing in part on his pioneering work with Jona-Lasinio [177], notes that in many dynamical symmetry breaking schemes, in which a condensate of fermion pairs $\langle f\bar{f}\rangle \neq 0$ leads to chiral symmetry breaking, there exist in the spectrum

Table 12. Analogy between states in Nambu-Jona-Lasinio (NJL) model and dynamic symmetry breaking (DSB) in electroweak interactions

State in NJL model	Role	State in DSB scheme	Role
Pions	Nambu-Goldstone bosons of spont. broken chiral symm.	Higgs bosons $\phi^{\pm}$ and $(\phi^0 - \bar{\phi}^0)/\sqrt{2}$	Provide longitudinal components of W and Z bosons
Nucleon	Constituent of pion	Top quark	Constituent of Higgs boson?
σ	Weakly bound nucleon-antinucleon state	Higgs boson $(\phi^0 + \bar{\phi}^0)/\sqrt{2}$	Preserves unitarity in scatt. of longit. W's and Z's

not only a zero-mass pseudoscalar boson, but the fermion f and a scalar excitation "σ" at $m_\sigma = 2m_f$. The analogies suggested by this observation are shown in Table 12.

Bardeen, Hill, and Lindner [178] emulate the Nambu-Jona-Lasinio model with a new interaction to generate the Higgs boson out of $t\bar{t}$. A nonzero condensate $\langle t\bar{t}\rangle \neq 0$ develops, leading to electroweak symmetry breaking. The new interactions are needed both to bind $t\bar{t}$ and to communicate the effects of the nonzero condensate to the other fermions.

In contrast to the above approach, Gérard and Weyers [179] introduce no new interaction. The Higgs field ϕ initially has $m_\phi^2 > 0$; its coupling to $t\bar{t}$ turns an effective Higgs potential with upward curvature at $\phi = 0$ into one with negative curvature, so that a non-zero value of $\langle\phi\rangle$ develops, triggering electroweak symmetry breaking. This scheme only is self-consistent for a narrow range of Higgs boson masses around 85 GeV/c^2, and thus makes a very specific prediction.

Some versions of supersymmetry, described in more detail in Ref. [180] and by Graham Ross at this Institute [1], have two Higgs doublets with m^2 values which are the same at some very large (unification) mass scale μ. One doublet (ϕ_1) interacts with down-type quarks and charged leptons, while the other (ϕ_2) interacts with up-type quarks (including the top quark) and perhaps neutrinos. As a result of the strong coupling of ϕ_2 with the top quark, $m^2(H_2)$ (where H_2 is the neutral scalar object in the complex doublet ϕ_2) evolves toward a negative value as μ is lowered, triggering electroweak symmetry breaking.

A scheme known as "reduction of coupling constants" [181] (see also [182]) or "gauge – Yukawa unification" [183] notes the presence of special relations for the renormalization-group behavior of coupling constants when gauge couplings g and Yukawa couplings g_Y are related. In contrast to supersymmetry, these schemes do not necessarily entail any extended symmetries. They posit a special role for finiteness in perturbation theory to each order.

D. Technicolor

The electroweak dynamical symmetry breaking scheme described in Table 12 makes use of the top quark to form a fermion condensate which breaks the symmetry. One can also postulate some *new* fermion which binds with its antiparticle under some new superstrong force to form the pseudo-Nambu-Goldstone bosons which are needed to make the W and Z massive. This is the essence of the scheme [184, 185] which has come to be known as "technicolor."

Suppose the u and d quark were massless. The pion would be a massless quark-antiquark bound state in QCD. Its coupling to the W via the divergence of the axial current would give the W a mass $M_W = gF_\pi/2 \simeq 30$ MeV/c^2, where we use the normalization $F_\pi = 93$ MeV here. In order to generate a W with mass $M_W = gv/2 = 80.34$ GeV/c^2, noting that $g^2/4\pi \simeq 1/30$, we need instead some fermion-antifermion pair with a decay constant $v = 246$ GeV. Thus, one needs an interaction which becomes strong at a mass scale roughly $246/0.093 = 2650$ times as high as the QCD scale. This is the postulated technicolor interaction. The fermions responding to this interaction are known as *technifermions.*

The $Z - \gamma - \gamma$ vertex described by a triangle diagram like that in Fig. 2 will be anomaly-free if the sum over technifermions F of $Q_F^2 I_{3L}^F$ vanishes. The simplest way in which this can be realized is with a single doublet with $Q_F = \pm 1/2$. The doublet may come in N_T different "technicolors." This is "minimal technicolor." It works fine for the masses of the W and Z.

An extension of technicolor to the description of quark and lepton masses immediately runs into difficulty. A new interaction ("extended technicolor" [186]) is needed to communicate the effects of $\langle \bar{F}F \rangle \neq 0$ to the ordinary quarks and leptons. This interaction has the potential for introducing flavor-changing neutral currents and steps must be taken to prevent them. The biggest mass (top) naturally poses the biggest problem for this theory.

E. Quark-lepton pattern and new physics

Returning again to Fig. 1, we see that the up-type quarks are less "dense" than the down-type quarks or charged leptons on a logarithmic plot. Moreover, their center of gravity is higher. This suggests that the down-type quarks and charged leptons are being pushed down in mass by mixing with heavier extra levels with $Q = \pm 1/3$ and $Q = \pm 1$.

The situation with the neutrinos is even more extreme. They appear to be very light if not massless. (Oscillation results, which would require non-zero masses, are not yet conclusive.) A popular way to understand such light masses is the "seesaw" mechanism [187] involving mixing with other very heavy neutrinos. The neutrino mass matrix may be written in terms of left-handed states in a basis (ν_L, ν_L^c) as

$$\mathcal{M}_\nu = \begin{bmatrix} 0 & m_D \\ m_D & M \end{bmatrix} \quad , \tag{80}$$

where m_D stands for a Dirac mass, and M is a large Majorana mass. The eigenvalues of this matrix are approximately $-m_D^2/M$ and M, corresponding to the eigenvectors $[1, -m_D/M]$ and $[m_D/M, 1]$.

Thus, the large top quark mass could be telling us that at higher masses one expects more matter which is *not* like the top quark: quarks of charge $-1/3$ or new leptons. What can serve as the source for these new forms of matter? We seek the answer in several schemes for quark-lepton unification.

F. Quark-lepton unity

The "grand unification" of strong and electroweak interactions in a larger symmetry, and the identification of quarks and leptons as objects related to one another under this symmetry, involves such groups as SU(5) [188], SO(10) [189], and E_6 [190].

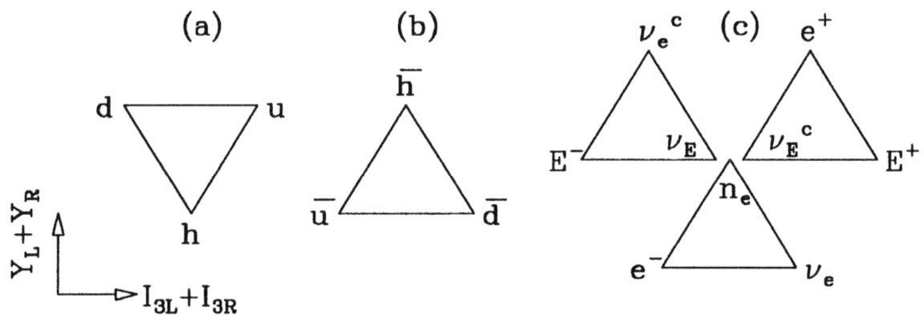

Figure 23. Left-handed fermions belonging to the **27**-plet of E_6. (a) Quarks [triangle denotes **3** of $SU(3)_L$]; (b) antiquarks [triangle denotes $\mathbf{3}^*$ of $SU(3)_R$]; (c) leptons [triangles denote $\mathbf{3}^*$'s of $SU(3)_L$; they are arranged in a **3** of $SU(3)_R$].

One forms a 5×5 matrix of SU(5) by simply putting the 3×3 matrices of color SU(3) in the upper left-hand corner and the 2×2 matrices of $SU(2)_L$ in the lower right. The miracle of SU(5) is that it also possesses a U(1) $\sim$ diag(2, 2, 2, $-3, -3$) which is just right for weak hypercharge!

Within SU(5) a specific choice of representations ($\mathbf{5}^* + \mathbf{10}$) is required for the left-handed fermions in order to accommodate the known states and to eliminate anomalies. This choice is automatic if left-handed fermions are assigned to the **16**-dimensional spinor multiplet of SO(10); the additional state is a right-handed neutrino. Thus, we have the decomposition

$$[\mathrm{SO}(10)] \quad \mathbf{16}_L = \mathbf{5}_L^* + \mathbf{10}_L + \mathbf{1}_L \quad [\mathrm{SU}(5)] \quad . \tag{81}$$

The right-handed neutrino can acquire a large Majorana mass without violating the $SU(3)_c\times SU(2)_L\times U(1)$ symmetry of the standard model. Anomalies are not present in SO(10), as long as matter belongs to complete multiplets.

The group E_6 contains SO(10). Its lowest-dimensional representation (**27**) contains the **16** of SO(10), as well as **10**- and **1**-dimensional ("exotic") representations of SO(10). These, in turn, have the following SU(5) decompositions:

$$[\mathrm{SO}(10)] \quad \mathbf{10}_L = \mathbf{5}_L + \mathbf{5}_L^* \quad [\mathrm{SU}(5)] \quad ; \quad [\mathrm{SO}(10)] \quad \mathbf{1}_L = \mathbf{1}_L \quad [\mathrm{SU}(5)] \quad . \tag{82}$$

There has been some interest in E_6 as a result of its appearance in certain versions of superstring theories [191, 192].

G. The E_6 zoo

One way to illustrate the particle content of the **27**-plet of E_6 is via the decomposition [193] $E_6 \rightarrow SU(3)_c\times SU(3)_L\times SU(3)_R$, whereby

$$\mathbf{27}_L = (\mathbf{3}_c, \mathbf{3}_L, \mathbf{1}_R) + (\mathbf{3}_c^*, \mathbf{1}_L, \mathbf{3}_R^*) + (\mathbf{1}_c, \mathbf{3}_L^*, \mathbf{3}_R) \quad . \tag{83}$$

The left-handed fermions can then be plotted in a space (Fig. 23) in which the x-axis is $I_{3L} + I_{3R}$, while the y-axis is $Y_L + Y_R$. The electric charge of particles is $Q_{\rm em} = I_{3L} + I_{3R} + (Y_L + Y_R)/2$. (I thank Graham Ross for reminding me of the utility of this picture.) The new particles [beyond those in the **16** of SO(10)] consist of h (an isosinglet $Q = -1/3$ quark), $E^\pm$ (a weak isodoublet charged lepton), ν_E and ν_E^c (a weak isodoublet Dirac neutrino, or else two Majorana neutrinos each belonging to a weak

Table 13. Cross sections σ [in units of $\sigma_0 \equiv \sigma(e^+e^- \to \gamma^* \to \mu^+\mu^-)$] for e^+e^- production of pairs of exotic fermions $f\bar{f}$ via virtual photons and Z's in the direct channel. The ν_E is assumed to be a Dirac neutrino. Values of g_V^f are quoted for $x = 0.2315$. QCD corrections to h quark production have been neglected.

Fermion f	Q_f	g_V^f	g_A^f	σ/σ_0 far below Z	σ/σ_0 far above Z
h	$-1/3$	0.0772	0	1/3	0.35
E^-	-1	-0.2685	0	1	1.20
ν_E	0	1/2	0	0	0.50

isodoublet), and an "inert" isosinglet Majorana neutrino n_e. By "weak isodoublet" or "weak isosinglet" we mean behavior under $SU(2)_L$.

Another way to visualize the additional states in E_6 is to note that the standard-model fermions belonging to $(\mathbf{16}, \mathbf{5}^*)$ of (SO(10), SU(5)) are a column vector consisting of $(\bar{d}_1, \bar{d}_2, \bar{d}_3, e^-, \nu_e)$. Then since SU(5) contains all the standard-model generators, equal SU(5) representations have equal standard-model transformation properties. (We are not considering "flipped" SU(5) models in which the electric charge lies outside SU(5) [194].) Thus, in $(\mathbf{10}, \mathbf{5}^*) = (\bar{h}_1, \bar{h}_2, \bar{h}_3, E^-, \nu_E)$, one identifies the $\bar{h}_i$ states as a color-triplet of isosinglet $Q = 1/3$ antiquarks, and (E^-, ν_E) as a weak isodoublet of leptons. In the charge-conjugate representation $(\mathbf{10}, \mathbf{5}) = (h_1, h_2, h_3, E^+, \nu_E^c)$, the $Q = -1/3$ quarks h are then weak *isosinglets*, while $E^+, \nu_E^c)$ are a weak *isodoublet.* Such fermions are not encountered in the standard model. Neither is the $(\mathbf{1}, \mathbf{1})$ consisting of the single state n_e, which is a weak isosinglet.

One can calculate the cross sections for $e^+e^- \to (\gamma^*, Z^*) \to f\bar{f}$ for the new fermions, in the manner leading to Table 10. In computing the values of g_V and g_A for E^- and a Dirac neutrino ν_E both left-handed and right-handed states have the same value of I_{3L}: $-1/2$ for E^- and $+1/2$ for ν_E. The results are shown in Table 13.

All of these states are produced exclusively via the vector current, and hence their maximum cross section is expected to occur very close to threshold. If one is sufficiently far from the Z pole, this maximum occurs at $E_{\text{c.m.}} \simeq 1.18 E_{\text{th}}$. Closer to the Z pole, the maximum occurs even lower [195].

The new fermions in E_6 are marginally consistent with the constraints on the parameters S and T mentioned in Sec. 3. The parameter T describes the effects of mass difference between $I_{3L} = +1/2$ and $I_{3L} = -1/2$ fermions. Recalling $\rho/\rho^{\text{nominal}} = 1 + \alpha T$, a more precise expression for $\Delta\rho$ than that given in terms of m_t is [196]

$$\Delta\rho = \frac{N_c G_F}{8\sqrt{2}\pi^2}\left(m_1^2 + m_2^2 - \frac{4m_1^2 m_2^2}{m_1^2 - m_2^2}\ln\frac{m_1}{m_2}\right) \quad . \tag{84}$$

The E_6 fermions thus make a significant contribution to $\Delta\rho$ only when $m(E^-) \neq m(\nu_E)$. The exotic quarks do not contribute because they are isosinglets.

The parameter S arises as a result of $W^0 - B$ mixing [115]. Consequently, it receives no contribution in E_6 from the isosinglet h quarks. There are two doublets (E^-, ν_E) and (E^+, ν_E^c) of leptons per family, leading to a contribution $\Delta S = 2(1/6\pi) \simeq 0.1$ per family. One can just barely tolerate three families.

H. Extended gauge structure

At any given mass scale, there is an interplay between the fermion spectrum and the subgroup of any grand unified theory (GUT) which remains unbroken at that mass scale. If a fermion is heavy, no GUT subgroup which relies on that fermion for anomaly cancellation should be unbroken far below that fermion's mass. Let us recapitulate how this works in several of the unified gauge groups we have mentioned.

1. In SU(5) we need a **5*** and a **10** to avoid anomalies in $I_{3L}Q^2$.

2. In SO(10) $\rightarrow$ SU(5) $\times$ U(1)$_\chi$ (the notation is that of Refs. [192] and [197]), the U(1)$_\chi$ symmetry is anomalous (the sum of $Q_\chi I_{3L}^2$ is non-zero) unless the ν_L^c is light.

3. In SO(10) $\rightarrow$ SO(6) $\times$ SO(4) (which is the same as SU(4) $\times$ SU(2)$_L\times$ SU(2)$_R$), the subsequent symmetry breakdowns are SU(4) $\rightarrow$ SU(3)$_C\times$ U(1)$_{B-L}$ and SU(2)$_R$ $\rightarrow$ U(1)$_R$. Each of these U(1)'s separately is anomalous unless the ν_L^c is light. However, one can write the electric charge as

$$Q = I_{3L} + I_{3R} + \frac{B-L}{2} = I_{3L} + \frac{Y_W}{2} \tag{85}$$

Since $I_{3L}(\nu_L^c) = -1/2$ while $(-L/2)(\nu_L^c) = +1/2$, one has $(Y_W/2)(\nu_L^c) = 0$, and the Y_W anomaly is insensitive to ν_L^c.

4. In E$_6\rightarrow$ SO(10) $\times$ U(1)$_\psi$ (see, again, [192, 197]) the cancellation of the U(1)$_\psi$ anomaly requires the entire **27**-plet to be light.

One may conjecture that the converse is also true: If an extended set of fermions is found to be light, there may exist an extended gauge structure for which this set of fermions provides the required anomaly cancellation. Thus, the search for new fermions may have bearing on extended gauge structures. One can also directly search for new gauge bosons (see, e.g., Ref. [198]) in lepton pair production, identifying them via characteristic forward-backward asymmetries [199, 200].

I. An unusual event

The CDF Collaboration [201] has reported one event with an electron-positron pair, two photons, and missing energy ($e^+e^-\gamma\gamma \not{E}_T$), produced in proton-antiproton collisions at $E_{\text{c.m.}} = 1.8$ TeV. Popular interpretations of this event have appeared within the context of supersymmetry [202], involving the production of a pair of selectrons ($\tilde{e}^+\tilde{e}^-$) or a pair of charginos ($\chi^+\chi^-$) which then decay to e^+e^- and radiatively-decaying neutral superparners. An interpretation has also appeared in one non-supersymmetric model [203].

The E$_6$ scenario mentioned above can interpret this event in the following way [195]:

$$\bar{p}p \rightarrow Z_I + \ldots \rightarrow E^+E^- + \ldots \tag{86}$$

followed by the chain

$$E^- \rightarrow e^-W_I^{(*)} \rightarrow e^-\bar{N}_e\bar{n}_e \rightarrow e^-\gamma n_e\bar{n}_e \tag{87}$$

and its charge-conjguate for E^+ decay. The n_e state is allowed to be stable as long as its mass satisfies cosmological bounds (typically less than a few tens of eV). The Z_I is a neutral gauge boson with mass greater than present limits [198] of a few hundred GeV. The W_I belongs to a subgroup SU(2)$_I$ of E$_6$ along with the Z_I [204]; it occurs in the decomposition E$_6\rightarrow$ SU(2)$_I\times$ SU(6), where the SU(6) breaks subsequently to the standard SU(5) GUT. The W_I is probably virtual, as indicated by the asterisk in parentheses. The neutral nature of all three bosons in SU(2)$_I$ is a key feature permitting the flavor of E^- to be passed on to the electron.

How does one tell the difference between these scenarios? Supersymmetry has superpartners of the gauge bosons known as "gauginos," and very specific relations between couplings involving superpartners and ordinary couplings. E$_6$ has exotic fermions

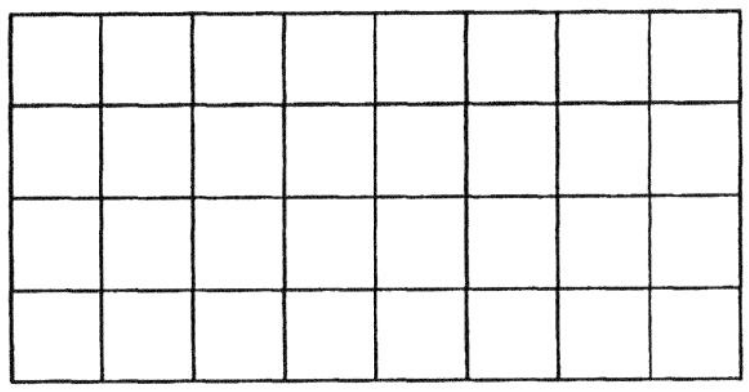

Figure 24. Part of a familiar pattern.

but they are not superpartners of Higgs bosons [despite a superficial resemblance; both belong to **10**'s of SO(10).] A typical reaction expected in E_6 is $e^+e^- \to \nu_E \bar{\nu}_E$, followed by $\nu_E \to \nu_e \bar{N}_1 N_2$, with one of the exotic neutrinos $\bar{N}_1$ decaying radiatively to $\bar{N}_2 + \gamma$.

A feature common to both scenarios is that in addition to the $e^+e^-\gamma\gamma \not{E}_T$ event, one should also see in $\bar{p}p$ collisions some diphoton events with missing transverse energy and *without* charged lepton pairs. Searches for such events have been performed in the CDF data [205]. Of all events with two photons and missing transverse energy, the event $e^+e^-\gamma\gamma \not{E}_T$ has the highest $\not{E}_T$. Thus, there are no indications yet of non-standard behavior in the diphoton events with $\not{E}_T$ but without e^+e^-.

There is still a need for extensive discussions of standard-model backgrounds to this event, such as multiple interactions, radiative production of W pairs, effects of cracks in the detector, and so on. One cannot conclude anything from a single event, but it can point the way to searches for related phenomena.

7. CONCLUSIONS

We have discussed the top quark mass more from the standpoint of what it does than where it comes from. It participates in a crucial way in loop diagrams, leading to mixing of neutral kaons and B mesons, thereby accounting for CP violation in the kaon system and predicting appreciable CP-ciolating effects for B's. It also affects electroweak observables, to the extent that its mass could be anticipated to within a few tens of GeV/c^2 before it was discovered.

The origin of the top quark mass, on the other hand, is related to the origin of *all* the quark masses, which also is linked to the curious pattern of weak charge-changing couplings expressed by the Cabibbo-Kobayashi-Maskawa (CKM) matrix. We have presented a *potpourri* of ideas about these questions while not finding any of them particularly conclusive. The best we can do is to note some overall patterns suggested by the quark and lepton masses in Fig. 1, which may indicate that we are not yet in possession of the complete experimental picture. At the very least, this is certainly the case for neutrinos, where we don't even know if they have masses. But there could be more. I leave you with the following exercise in pattern recognition: What familiar pattern do you see in Fig. 24?

One can re-express the pattern as shown in Fig. 25; perhaps it suggests something at this point.

Finally, when one adds variety to the pattern, it becomes recognizable as the periodic table of the elements (Fig. 26).

The variety of the pattern of the elementary particles laid the foundations of the quark model and our understanding of the fundamental strong interactions. Will there be a similar advance for quarks and leptons? The pattern of quarks and leptons has been quite regular up to now, just as if the periodic table of the elements consisted

Figure 25. Part of a familiar pattern, expressed differently.

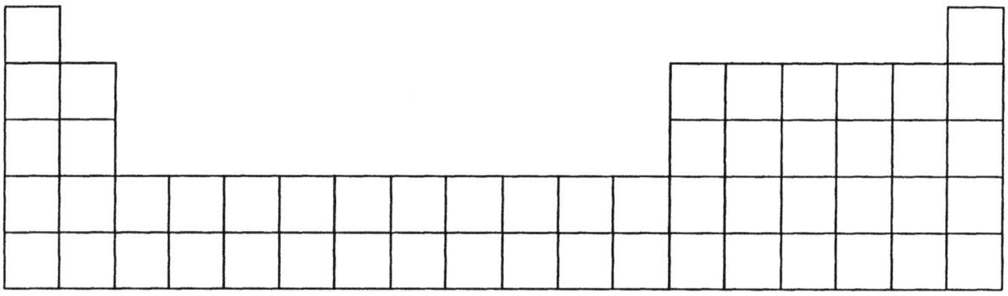

Figure 26. A larger part of a familiar pattern.

only of rows of equal length and were missing hydrogen, helium, the transition metals, the lanthanides, and the actinides. Whether one discovers superpartners of the known states, or variety such as predicted in extended gauge structures, the new states could help us to make sense of the pattern of the masses of the more familiar ones. In this sense the top quark mass may not be the end of a story, but just the beginning.

ACKNOWLEDGMENTS

I would like to thank Jim Amundson, Isi Dunietz, Aaron Grant, Michael Gronau, Oscar Hernández, Nahmin Horowitz, Mike Kelly, Harry Lipkin, David London, Bill Marciano, Alex Nippe, Sheldon Stone, Tatsu Takeuchi, and Mihir Worah for enjoyable collaborations on some of the topics mentioned in these lectures. In addition I am grateful to Gideon Alexander, Guido Altarelli, Alain Blondel, Keith Ellis, Paul Frampton, Henry Frisch, Michaelangelo Mangano, Paolo Nason, Matthias Neubert, Steve Pollock, Graham Ross, Chris Sachrajda, Bob Sachs, Pat Sandars, Michael Schmitt, Jack Steinberger, Sam Ting, Tini Veltman, Petr Vogel, Brian Webber, Larry Wilets, Bruce Winstein, and Lincoln Wolfenstein for fruitful discussions, and to the CERN Theory Group for its hospitality during the preparation of this report. Finally, I wish to thank the organizers of the Summer Institute and all of my other hosts in Cargèse for making this first visit such an enjoyable one. Parts of the research described here were performed at the Aspen Center for Physics, the Fermilab Theory Group, the Institute for Nuclear Theory at the University of Washington, and the Technion. This work was supported in part by the United States Department of Energy under Grant No. DE FG02 90ER40560.

References

[1] G. Ross, lectures at this Institute.

[2] T. J. LeCompte, Top decay physics at CDF and measurement of the CKM element V_{tb}, Fermilab report FERMILAB-CONF-96-021-E, Jan., 1996, presented at 2nd Rencontres du Vietnam, *Physics at the Frontiers of the Standard Model*, Ho Chi Minh City, Vietnam, 21 – 28 Oct., 1995.

[3] N. Cabibbo, *Phys. Rev. Lett.* 10:531 (1963).

[4] M. Kobayashi and T. Maskawa, *Prog. Theor. Phys.* 49:652 (1973).

[5] J. L. Rosner, Present and future aspects of CP violation, Enrico Fermi Institute Report No. 95-36 [hep-ph/9506364], lectures given at the VIII J. A. Swieca Summer School, *Particles and Fields*, Rio de Janeiro, February, 1995, proceedings to be published by World Scientific.

[6] S. L. Glashow, *Nucl. Phys.* 22:579 (1961); S. Weinberg, *Phys. Rev. Lett.* 19:1264 (1967); A. Salam, in *Proceedings of the Eighth Nobel Symposium*, edited by N. Svartholm (Almqvist and Wiksell, Stockholm; Wiley, New York, 1978), p. 367.

[7] LEP Electroweak Working Group, CERN report LEPEWWG/96-02, July 30, 1996, presented at XXVIII International Conference on High Energy Physics, Warsaw, July 25–31, 1996.

[8] Y. Hara, *Phys. Rev.* 134:B701 (1964); Z. Maki and Y. Ohnuki, *Prog. Theor. Phys.* 32:144 (1964); B. J. Bjorken and S. L. Glashow, *Phys. Lett.* 11:255 (1964).

[9] M. Gell-Mann and M. Lévy, *Nuovo Cim.* 19:705 (1960).

[10] S. L. Glashow, J. Iliopoulos, and L. Maiani, *Phys. Rev.* D 2:1285 (1970).

[11] C. Bouchiat, J. Iliopoulos, and Ph. Meyer, *Phys. Lett.* 38B:519 (1972). See also H. Georgi and S. L. Glashow, *Phys. Rev.* D 6:429 (1972); D. J. Gross and R. Jackiw, *ibid.* 6:477 (1972).

[12] For more details see, e.g., V. L. Fitch and J. L. Rosner, Elementary particle physics in the second half of the twentieth century, in *Twentieth Century Physics*, edited by L. M. Brown, A. Pais, and B. Pippard (AIP/IOP, New York and Bristol, 1995), ch. 9.

[13] W. Pauli, Zur älteren und neueren Geschichte des Neutrinos, in *Collected Scientific Papers*, R. Kronig and V. F. Weisskopf, eds., Interscience, New York, 1964, v. 2, p. 1313.

[14] See, e.g., L. M. Brown, in *Twentieth Century Physics*, L. M. Brown, A. Pais, and B. Pippard, eds., AIP/IOP, New York and Bristol, 1995, ch. 5.

[15] S. H. Neddermeyer and C. D. Anderson, *Phys. Rev.* 51:884 (1937). See also J. C. Street and E. C. Stevenson, *Phys. Rev.* 51:1005 (1937); Y. Nishina, M. Takeuchi, and T. Ichimaya, *Phys. Rev.* 52:1198 (1937).

[16] G. Danby *et al.*, *Phys. Rev. Lett.* 9:36 (1962).

[17] J. J. Aubert *et al.*, *Phys. Rev. Lett.* 33:1404 (1974).

[18] J.-E. Augustin *et al.*, *Phys. Rev. Lett.* 33:1406 (1974).

[19] E. G. Cazzoli *et al.*, *Phys. Rev. Lett.* 34:1125 (1975); G. Goldhaber *et al.*, *Phys. Rev. Lett.* 37:255 (1976); I. Peruzzi *et al.*, *Phys. Rev. Lett.* 37:569 (1976).

[20] M. Perl *et al.*, *Phys. Rev. Lett.* 35:1489 (1975); *Phys. Lett.* 63B:466 (1976); *Phys. Lett.* 70B:487 (1977).

[21] H. Harari, in *Proceedings of the 1975 International Symposium on Lepton and Photon Interactions at High Energies,* Stanford University, Aug. 21–27, 1975, W. T. Kirk, ed., SLAC, Stanford, CA, (1975), p. 317.

[22] S. W. Herb *et al.*, *Phys. Rev. Lett.* 39:252 (1977); W. R. Innes *et al.*, *Phys. Rev. Lett.* 39:1240, 1640(E) (1977).

[23] S. Behrends *et al.* (CLEO Collaboration), *Phys. Rev. Lett.* 50:881 (1983).

[24] F. Abe *et al.* (CDF Collaboration), *Phys. Rev. Lett.* 73:225 (1994); *Phys. Rev.* D 50:2966 (1995); *Phys. Rev. Lett.* 74:2626 (1995); S. Abachi *et al.* (D0 Collaboration), *Phys. Rev. Lett.* 74:2632 (1995).

[25] Fermilab E-872, R. Ramieka, spokesperson. See B. Lundberg *et al.*, Measurement of τ lepton production from the process $\nu_\tau + N \to \tau$, Fermilab proposal P-872, Jan., 1994.

[26] P. Tipton, Plenary talk, XXVIII International Conference on High Energy Physics, Warsaw, July 25–31, 1996; CDF Collaboration, presented by R. Hughes, reports PA-05-052, PA05-053, and PA08-108; D0 Collaboration, presented by A. Zieminski, reports PA05-027, PA05-028, and PA05-029.

[27] M. Jeżabek, J. H. Kühn, and T. Teubner, *Zeit. Phys.* C 56:653 (1992).

[28] Mark III Collaboration, R. M. Baltrusaitis *et al.*, *Phys. Rev. Lett.* 52:2126 (1984).

[29] See, e.g., W. Kwong and J. L. Rosner, *Phys. Rev.* D 38:279 (1988); W. Buchmüller and S. Cooper, Upsilon spectroscopy, in *High-Energy Electron-Positron Physics*, A. Ali and P. Söding, eds., World Scientific, Singapore (1988), p. 412.

[30] C. Quigg and J. L. Rosner, *Phys. Lett.* 71B:153 (1977); *Comments on Nucl. and Part. Phys.* 8:11 (1978); *Phys. Rep.* 56:167 (1979).

[31] A. Martin, *Phys. Lett.* 93B:338 (1980); *ibid.* 100B:511 (1981).

[32] CLEO Collaboration, P. Avery *et al.*, *Phys. Rev. Lett.* 75:4364 (1995); L. Gibbons *et al.*, *Phys. Rev. Lett.* 77:810 (1996).

[33] CLEO Collaboration, report PA01-078, presented by R. Kutschke at XXVIII International Conference on High Energy Physics, Warsaw, July 25–31, 1996. We thank J. Yelton for communicating this result to us.

[34] M. Neubert, lectures at this Institute. See also M. Neubert, CERN report CERN-TH/96-55 [hep-ph/9604412], to be published in Int. J. Mod. Phys. A.

[35] N. Gray, D. J. Broadhurst, W. Grafe, and K. Schilcher, *Zeit. Phys.* C 48:673 (1990), and references therein.

[36] A. Ali and D. London, DESY report DESY 96-140 [hep-ph/9607392], presented at High Energy Physics International Euroconference on QCD (QCD 96), Montpellier, France, July 4 – 12, 1996.

[37] L.-L. Chau and W.-Y. Keung, *Phys. Rev. Lett.* 53:1802 (1984); H. Harari and M. Leurer, *Phys. Lett.* B 181:123 (1986); J. D. Bjorken and I. Dunietz, *Phys. Rev.* D 36:2109 (1987).

[38] L. Wolfenstein, *Phys. Rev. Lett.* 51:1945 (1983).

[39] See, e.g., M. Bourquin *et al.*, *Zeit. Phys.* C 21:27 (1983); H. Leutwyler and M. Roos, *Zeit. Phys.* C 25:91 (1984); J. F. Donoghue, B. R. Holstein, and S. W. Klimt, *Phys. Rev.* D 35:934 (1987); F. J. Gilman, K. Kleinknecht, and B. Renk, in Review of particle properties, Particle Data Group, *Phys. Rev.* D 54:94 (1996).

[40] T. Browder and K. Honscheid, *Prog. Nucl. Part. Phys.* 35:81 (1995).

[41] L. Gibbons, Plenary talk, XXVIII International Conference on High Energy Physics, Warsaw, July 25–31, 1996.

[42] L.-L. Chau and W.-Y. Keung, *Phys. Rev. Lett.* 53:1802 (1984); M. Gronau and J. Schechter, *ibid.* 54:385 (1985); M. Gronau, R. Johnson, and J. Schechter, *Phys. Rev.* D 32:3062 (1985); C. Jarlskog, in *Physics at LEAR with Low Energy Antiprotons,* proceedings of the workshop, Villars-sur-Ollon, Switzerland, 1987, edited by C. Amsler *et al.* (Harwood, Chur, Switzerland, 1988), p. 571; J. D. Bjorken and I. Dunietz, *Phys. Rev.* D 36:2109 (1987).

[43] Y. Nir and H. Quinn, *Ann. Rev. Nucl. Part. Sci.* 42:211 (1992).

[44] T. P. Cheng and L. F. Li, *Gauge Theory of Elementary Particles* (Oxford University Press, 1984).

[45] J. L. Rosner, in *Testing the Standard Model* (Proceedings of the 1990 Theoretical Advanced Study Institute in Elementary Particle Physics), edited by M. Cvetič and P. Langacker (World Scientific, Singapore, 1991), p. 91.

[46] T. Inami and C. S. Lim, *Prog. Theor. Phys.* 65:297 (1981); *ibid.* 65:1772(E) (1981).

[47] Particle Data Group, *Phys. Rev.* D 54:1 (1996).

[48] A. Abada *et al.*, *Nucl. Phys.* B376:172 (1992).

[49] A. Buras, M. Jamin, and P. H. Weisz, *Nucl. Phys.* B347:491 (1990).

[50] UA1 Collaboration, C. Albajar *et al.*, *Phys. Lett.* B 186:237, 247 (1987); *ibid.* 197:565(E) (1987).

[51] ARGUS Collaboration, H. Albrecht *et al.*, *Phys. Lett.* B 192:245 (1987).

[52] ALEPH Collaboration, D. Buskulic *et al.*, *Phys. Lett.* B 313:498 (1993); DELPHI Collaboration, P. Abreu *et al.*, *Phys. Lett.* B 338:409 (1994); OPAL Collaboration, R. Akers *et al.*, *Phys. Lett.* B 327:411 (1994); *ibid.* 336:585 (1994). For a compilation of the many new LEP results presented in 1996 see Ref. [41].

[53] CDF Collaboration, F. Abe *et al.*, Fermilab report FERMILAB-CONF-95-231-E, July, 1995, contributed to *LP95: Proceedings of the International Symposium on Lepton and Photon Interactions (IHEP)*, 10–15 August 1995, Beijing, People's Republic of China, Z.-P. Zheng and H.-S. Chen, eds., World Scientific, Singapore, 1996; report PA08-032, presented by P. Sphicas at XXVIII International Conference on High Energy Physics, Warsaw, July 25–31, 1996.

[54] SLD Collaboration, reports PA08-026A, PA08-026B, and PA08-027/028, presented by D. Su at XXVIII International Conference on High Energy Physics, Warsaw, July 25–31, 1996.

[55] C. Zeitnitz, invited talk at the 4th International Workshop on B-Physics at Hadron Machines (BEAUTY 96), Rome, June 17–21, 1996, proceedings to be published in Nucl. Instr. Meth. A slightly different average was presented in Ref. [41]: $\Delta m_d = 0.464 \pm 0.017$ ps^{-1}.

[56] CERN WA75 Collaboration, S. Aoki *et al.*, *Prog. Theor. Phys.* 89:131 (1993); CLEO Collaboration, D. Acosta *et al.*, *Phys. Rev.* D 49:5690 (1994); F. Muheim and S. Stone, *Phys. Rev.* D 49:3767 (1994); BES Collaboration, J. Z. Bai *et al.*, *Phys. Rev. Lett.* 74:4599 (1995); F. Muheim, parallel session on B physics, in *The Albuquerque Meeting: DPF 94* (Division of Particles and Fields Meeting, American Physical Society, Albuquerque, NM, August 2–6, 1994), ed. by S. Seidel (World Scientific, River Edge, NJ, 1995); Fermilab E653 Collaboration, K. Kodama *et al.*, *Phys. Lett.* B 382:299 (1996).

[57] ARGUS Collaboration, H. Albrecht *et al.*, *Phys. Lett.* B 219:121 (1989); D. Bortoletto, Ph. D. Thesis, Syracuse University, 1989; D. Bortoletto and S. L. Stone, *Phys. Rev. Lett.* 65:2951 (1990); CLEO Collaboration, D. Bortoletto *et al.*, *Phys. Rev.* D 45:2212 (1992); J. L. Rosner, *Phys. Rev.* D 42:3732 (1990); in *Research Directions for the Decade* (Proceedings of the 1990 DPF Snowmass Workshop), edited by E. L. Berger (World Scientific, Singapore, 1991), p. 255; M. Paulini *et al.*, in *Proceedings of the Joint International Symposium and Europhysics Conference on High Energy Physics,* S. Hegarty, K. Potter, and E. Quercigh, eds., World Scientific, Singapore (1992), p. 592; CLEO Collaboration, A. Bean *et al.*, *Phys. Rev. Lett.* 70:2681 (1993).

[58] G. Martinelli, Plenary talk, XXVIII International Conference on High Energy Physics, Warsaw, July 25–31, 1996.

[59] Mark III Collaboration, J. Adler *et al.*, *Phys. Rev. Lett.* 60:1375 (1988); *ibid.* 63:1658(E) (1989).

[60] J. Z. Bai *et al.* (BEPC Collaboration), SLAC report SLAC-PUB-7147, 1996.

[61] J. Amundson *et al.*, *Phys. Rev.* D 47:3059 (1993); J. L. Rosner, *The Fermilab Meeting – DPF 92* (Division of Particles and Fields Meeting, American Physical Society, Fermilab, 10–14 November, 1992), ed. by C. H. Albright *et al.* (World Scientific, Singapore, 1993), p. 658.

[62] See, e.g., A. Duncan *et al.*, *Phys. Rev.* D 51:5101 (1995); UKQCD Collaboration, C. R. Allton *et al.*, *Nucl. Phys.* B437:641 (1995); MILC Collaboration, C. Bernard *et al.*, report WASH-U-HEP-96-31, June, 1996 [hep-lat/9608092], presented at Lattice 96: 14th International Symposium on Lattice Field Theory, St. Louis, MO, 4–8 June 1996; G. Martinelli, Ref. [58].

[63] B. Grinstein, *Phys. Rev. Lett.* 71:3067 (1993).

[64] J. H. Christenson, J. W. Cronin, V. L. Fitch, and R. Turlay, *Phys. Rev. Lett.* 13:138 (1964).

[65] *CP Violation,* edited by C. Jarlskog (World Scientific, Singapore, 1989).

[66] M. Gell-Mann and A. Pais, *Phys. Rev.* 97:1387 (1955).

[67] T. D. Lee, R. Oehme and C. N. Yang, *Phys. Rev.* 106:340 (1957); B. L. Ioffe, L. B. Okun' and A. P. Rudik, *Zh. Eksp. Teor. Fiz.* 32:396 (1957) [Sov. Phys. – JETP 5:328 (1957)].

[68] K. Lande, E. T. Booth, J. Impeduglia, and L. M. Lederman, *Phys. Rev.* 103:1901 (1956).

[69] L. Wolfenstein, *Phys. Rev. Lett.* 13:562 (1964).

[70] A. J. Buras, M. Jamin, and P. H. Weisz, *Nucl. Phys.* B347:491 (1990); S. Herrlich and U. Nierste, *Nucl. Phys.* B419:292 (1994); *Phys. Rev.* D 52:6505 (1995).

[71] C. Hamzaoui, J. L. Rosner, and A. I. Sanda, in *Proceedings of the Workshop on High Sensitivity Beauty Physics at Fermilab,* Fermilab, Nov. 11–14, 1987, edited by A. J. Slaughter, N. Lockyer, and M. Schmidt (Fermilab, Batavia, IL, 1988), p. 97; J. L. Rosner, A. I. Sanda, and M. P. Schmidt, *ibid.*, p. 165.

[72] See, e.g., the following sources quoted by [36]: S. Sharpe, *Nucl. Phys. B (Proc. Suppl.)* 34:403 (1994); J. Bijnens and J. Prades, *Nucl. Phys.* B444:523 (1995); A. Soni, Brookhaven National Laboratory report BNL-62284 [hep-lat/9510036]; JLQCD Collaboration, S. Aoki *et al.*, *Nucl. Phys. B (Proc. Suppl.)* 47:465 (1996); APE Collaboration, M. Crisafulli *et al.*, *Phys. Lett.* B 369:325 (1996). Our quoted error reflect a somewhat more generous spread in values obtained by different means.

[73] For discussions with references to earlier literature see, e.g., I. I. Bigi and A. I. Sanda, *Nucl. Phys.* B281:41 (1987); I. Dunietz, *Ann. Phys. (N.Y.)* 184:350 (1988); M. B. Wise, in *Particles and Fields 3* (Proceedings of the 1988 Banff Summer Institute on Particles and Fields), edited by A. N. Kamal and F. C. Khanna (World Scientific, Singapore, 1989), p. 124; I. I. Bigi, V. A. Khoze, N. G. Uraltsev, and A. I. Sanda, in Ref. [65], p. 175; J. L. Rosner, in *Testing the Standard Model* (Proceedings of the 1990 Theoretical Advanced Study Institute in Elementary Particle Physics), edited by M. Cvetič and P. Langacker (World Scientific, Singapore, 1991), p. 91; Y. Nir and H. R. Quinn, *Ann. Rev. Nucl. Part. Sci.* 42:211 (1992); B. Winstein and L. Wolfenstein, *Rev. Mod. Phys.* 65:1113 (1993).

[74] I. Dunietz and J. L. Rosner, *Phys. Rev.* D 34:1404 (1986); I. Dunietz, Ref. [73].

[75] D. London and R. Peccei, *Phys. Lett.* B 223:257 (1989); M. Gronau, *Phys. Rev. Lett.* 63:1451 (1989); B. Grinstein, *Phys. Lett.* B 229:280 (1989); M. Gronau, *Phys. Lett.* B 300:163 (1993).

[76] M. Gronau and D. London, *Phys. Rev. Lett.* 65:3381 (1990).

[77] ALEPH and DELPHI Collaborations, presented at XXVIII International Conference on High Energy Physics, Warsaw, July 25–31, 1996; L. Gibbons, Ref. [41].

[78] I. I. Bigi *et al.*, in *B Decays*, edited by S. Stone (World Scientific, Singapore, 1994), p. 132.

[79] I. Dunietz, *Phys. Rev.* D 52:3048 (1995).

[80] T. E. Browder and S. Pakvasa, *Phys. Rev.* D 52:3123 (1995).

[81] R. Aleksan, A. Le Yaouanc, L. Oliver, O. Pène, and J. C. Raynal, *Phys. Lett.* B 316:567 (1993).

[82] M. Beneke, G. Buchalla, and I. Dunietz, *Phys. Rev.* D 54:4419 (1996); M. Beneke, SLAC report SLAC-PUB-96-7258, presented at XXVIII International Conference on High Energy Physics, Warsaw, July 25–31, 1996.

[83] A. S. Dighe, I. Dunietz, H. J. Lipkin, and J. L. Rosner, *Phys. Lett.* B 369:144 (1996). See also R. Fleischer and I. Dunietz, Univ. of Karlsruhe report TTP 96-07 [hep-ph/9605220] (unpublished); A. Dighe, I. Dunietz, and R. Fleischer, in preparation.

[84] H. J. Lipkin, in *Proceedings of the SLAC Workshop on Physics and Detector Issues for a High-Luminosity Asymmetric B Factory*, edited by David Hitlin, published as SLAC, LBL and Caltech reports SLAC-373, LBL-30097 and CALT-68-1697, 1990, p. 49; I. Dunietz *et al.*, *Phys. Rev.* D 43:2193 (1991); I. Dunietz, Appendix in *B Decays*, edited by S. Stone (World Scientific, Singapore, 1994), p. 550.

[85] CLEO Collaboration, CLEO-CONF-96-22, report PA05-074, presented by R. Kutschke at XXVIII International Conference on High Energy Physics, Warsaw, July 25–31, 1996.

[86] J. L. Rosner, *Phys. Rev.* D 42:3732 (1990).

[87] L. D. Landau, *Dokl. Akad. Nauk SSSR* 60:207 (1948); C. N. Yang, *Phys. Rev.* 77:242,722 (1950). See also N. P. Chang and C. A. Nelson, Jr., *Phys. Rev. Lett.* 40:1617 (1978); *Phys. Rev.* D 20:2923 (1979).

[88] J. Ellis, M. K. Gaillard, and D. V. Nanopoulos, *Nucl. Phys.* B100:313 (1975); *ibid.* B109:213 (1976); A. I. Vaĭnshteĭn, V. I. Zakharov, and M. A. Shifman, *Pis'ma Zh. Eksp. Teor. Fiz.* 22:123 (1975) [*JETP Lett.* 22:55 (1975)]; *Zh. Eksp. Teor. Fiz.* 72:1275 (1977) [Sov. Phys. – JETP 45:670 (1977)]; J. Ellis, M. K. Gaillard, D. V. Nanopoulos, and S. Rudaz, *Nucl. Phys.* B131:285 (1977); *ibid.* B132:541(E) (1978).

[89] M. B. Wise, in *Particles and Fields 3* (Proceedings of the 1988 Banff Summer Institute on Particles and Fields), edited by A. N. Kamal and F. C. Khanna (World Scientific, Singapore, 1989), p. 124; J. M. Flynn and L. Randall, *Phys. Lett.* B 224:221 (1989); G. Buchalla, A. J. Buras, and M. Harlander, *Nucl. Phys.* B337:313 (1990); A. J. Buras, M. Jamin, and M. E. Lautenbacher, *Nucl. Phys.* B408:209 (1993); M. Ciuchini, E. Franco, G. Martinelli, and L. Reina, *Phys. Lett.* B 301:263 (1993); S. Bertolini, talk given at Workshop on *K* Physics, Orsay, France, 30 May – 4 June 1996, INFN (Trieste) report SISSA-110-96-EP [hep-ph/9607312]; A. J. Buras, Plenary talk, XXVIII International Conference on High Energy Physics, Warsaw, July 25–31, 1996.

[90] Fermilab E731 Collaboration, L. K. Gibbons *et al.*, *Phys. Rev. Lett.* 70:1203 (1993).

[91] CERN NA31 Collaboration, G. D. Barr *et al.*, *Phys. Lett.* B 317:233 (1993).

[92] A. J. Buras and R. Fleischer, *Phys. Lett.* B 341:379 (1995).

[93] CLEO Collaboration, M. Battle *et al.*, *Phys. Rev. Lett.* 71:3922 (1993).

[94] CLEO Collaboration, D. M. Asner *et al.*, *Phys. Rev.* D 53:1039 (1996).

[95] CLEO Collaboration, P. Gaidarev, *Bull. Am. Phys. Soc.* 40:923 (1995).

[96] M. Gronau, J. L. Rosner, and D. London, *Phys. Rev. Lett.* 73:21 (1994); O. F. Hernandez, D. London, M. Gronau, and J. L. Rosner, *Phys. Lett.* B 333:500 (1994); M. Gronau, O. F. Hernández, D. London, and J. L. Rosner, *Phys. Rev.* D 50:4529 (1994).

[97] M. Gronau and J. L. Rosner, *Phys. Rev.* D 53:2516 (1996).

[98] M. Gronau and J. L. Rosner, *Phys. Rev. Lett.* 76:1200 (1996); A. S. Dighe, M. Gronau, and J. L. Rosner, *Phys. Rev.* D 54:3309 (1996); A. S. Dighe and J. L. Rosner, *Phys. Rev.* D 54:4677 (1996).

[99] M. Gronau and J. L. Rosner, *Phys. Lett.* B 376:205 (1996).

[100] J. Silva and L. Wolfenstein, *Phys. Rev.* D 49:R1151 (1994); F. DeJongh and P. Sphicas, *Phys. Rev.* D 53:4930 (1996).

[101] G. Buchalla and A. J. Buras, hep-ph/9607447. See also G. Buchalla and A. J. Buras, *Nucl. Phys.* B412:106 (1994).

[102] C. O. Dib, Ph.D. Thesis, Stanford University, 1990, SLAC Report SLAC-364, April, 1990 (unpublished).

[103] B. Winstein and L. Wolfenstein, *Rev. Mod. Phys.* 65:1113 (1993).

[104] Brookhaven E787 Collaboration, S. Adler *et al.*, *Phys. Rev. Lett.* 76:1421 (1996).

[105] M. K. Gaillard and B. W. Lee, *Phys. Rev.* D 10:897 (1974).

[106] Fermilab E799 Collaboration, M. Weaver *et al.*, *Phys. Rev. Lett.* 72:3758 (1994).

[107] J. L. Rosner, *Phys. Rev.* D 53:2724 (1996).

[108] The earliest such proposals were made by H. Yukawa, *Proc. Phys. Math. Soc. Jap.* 17:48 (1935), and O. Klein, in *Les Nouvelles Théories de la Physique*, Paris, Inst. de Coöperation Intellectuelle (1939), p. 81. For others (in the 1950's) see Ref. [12].

[109] R. P. Feynman and M. Gell-Mann, *Phys. Rev.* 109:193 (1958); E. C. G. Sudarshan and R. E. Marshak, *Phys. Rev.* 109:1860 (1958).

[110] S. Eidelman and F. Jegerlehner, Paul Scherrer Institute report PSI-PR-95-1, January, 1995; H. Burkhardt and B. Pietrzyk, *Phys. Lett.* B 356:398 (1995). Both sets of authors quote $\alpha^{-1}(M_Z) = 128.89 \pm 0.09$. Values differing only slightly from this are obtained by A. D. Martin and D. Zeppenfeld, *Phys. Lett.* B 345:558 (1995) [$\alpha^{-1}(M_Z) = 128.99 \pm 0.06$] and by M. Swartz, SLAC report SLAC-PUB-95-7001, Sept. 1995, submitted to Phys. Rev. D [$\alpha^{-1}(M_Z) = 128.96 \pm 0.06$].

[111] A. Sirlin, *Phys. Rev. Lett.* 72:1786 (1994).

[112] A. Pich, lectures at this Institute.

[113] D. Treille, lectures at this Institute.

[114] M. Veltman, *Nucl. Phys.* B123:89 (1977).

[115] M. Peskin and T. Takeuchi, *Phys. Rev. Lett.* 65:964 (1990); *Phys. Rev.* D 46:381 (1992), and references therein.

[116] G. DeGrassi, B. A. Kniehl, and A. Sirlin, *Phys. Rev.* D 48:3963 (1993).

[117] P. Gambino and A. Sirlin, *Phys. Rev.* D 49:1160 (1994).

[118] M. Veltman, *Phys. Lett.* B 91:95 (1980).

[119] D. C. Kennedy and P. G. Langacker, *Phys. Rev. Lett.* 65:2967 (1990); *ibid.* 66:395(E) (1991).

[120] F. J. Hasert *et al.*, *Phys. Lett.* B 46:138 (1973); *Nucl. Phys.* B73:1 (1974); A. Benvenuti *et al.*, *Phys. Rev. Lett.* 32:800 (1974); B. Aubert *et al.*, *Phys. Rev. Lett.* 32:1454 (1974).

[121] C. H. Llewellyn Smith, *Nucl. Phys.* B228:205 (1983).

[122] CCFR Collaboration, FERMILAB-CONF-96/227-E, presented by K. S. McFarland at XXXI Rencontres de Moriond, March, 1996.

[123] CDHS Collaboration, H. Abramowicz *et al.*, *Phys. Rev. Lett.* 57:298 (1986); A. Blondel *et al.*, *Zeit. Phys.* C 45:361 (1990).

[124] CHARM Collaboration, J. V. Allaby *et al.*, *Phys. Lett.* B 177:446 (1986); *Zeit. Phys.* C 36:611 (1987).

[125] A. Blondel, Plenary talk, XXVIII International Conference on High Energy Physics, Warsaw, July 25–31, 1996.

[126] CDF Collaboration, F. Abe *et al.*, *Phys. Rev. Lett.* 75:11 (1995); *Phys. Rev.* D 52:4784 (1995).

[127] D0 Collaboration, report PA07-038, presented by A. Zieminski at XXVIII International Conference on High Energy Physics, Warsaw, July 25–31, 1996.

[128] ALEPH, DELPHI, L3, and OPAL Collaborations, presented at XXVIII International Conference on High Energy Physics, Warsaw, July 25–31, 1996. For a compilation of the results see Ref. [125].

[129] UA2 Collaboration, J. Alitti *et al.*, *Phys. Lett.* B 276:354 (1992).

[130] J. L. Rosner, M. P. Worah, and T. Takeuchi, *Phys. Rev.* D 49:1363 (1994).

[131] CDF Collaboration, F. Abe *et al.*, *Phys. Rev. Lett.* 74:341 (1995).

[132] SLD Collaboration, report PA07-063, presented by D. Su at XXVIII International Conference on High Energy Physics, Warsaw, July 25–31, 1996.

[133] M. C. Noecker, B. P. Masterson, and C. E. Wieman, *Phys. Rev. Lett.* 61:310 (1988).

[134] V. A. Dzuba, V. V. Flambaum, and O. P. Sushkov, *Phys. Lett.* A 141:147 (1989); S. A. Blundell, W. R. Johnson, and J. Sapirstein, *Phys. Rev. Lett.* 65:1411 (1990); *Phys. Rev.* D 45:1602 (1992).

[135] W. Marciano and J. L. Rosner, *Phys. Rev. Lett.* 65:2963 (1990); *ibid.* 68:898(E) (1992).

[136] J. L. Rosner, *Phys. Rev.* D 42:3107 (1990).

[137] P. G. H. Sandars, *J. Phys.* B 23:L655 (1990).

[138] P. A. Vetter, D. M. Meekhof, P. K. Majumder, S. K. Lamoreaux, and E. N. Fortson, *Phys. Rev. Lett.* 74:2658 (1995).

[139] N. H. Edwards, S. J. Phipp, P. E. G. Baird, and S. Nakayama, *Phys. Rev. Lett.* 74:2654 (1995).

[140] P. G. H. Sandars and B. W. Lynn, *J. Phys.* B 27:1469 (1994).

[141] V. A. Dzuba, V. V. Flambaum, P. G. Silvestrov, and O. P. Sushkov, *J. Phys.* B 20:3297 (1987).

[142] A. C. Hartley, E. Lindroth, and A. M. Mårtensson-Pendrill, *J. Phys.* B 23:3417 (1990); A. C. Hartley and P. G. H. Sandars, *J. Phys.* B 23:4197 (1990).

[143] ALEPH Collaboration, reports PA10-014 and PA10-015, presented by J. Carr at XXVIII International Conference on High Energy Physics, Warsaw, July 25–31, 1996.

[144] P. Langacker, M. Luo, and A. K. Mann, *Rev. Mod. Phys.* 64:87 (1992).

[145] J. Ellis, G. Fogli, and E. Lisi, CERN report CERN-TH/96-216 [hep-ph/9608239] (unpublished).

[146] W. de Boer, A. Dabelstein, W. Hollik, W. Mösle, and U. Schwickerath, Karlsruhe Univ. report IEKP-KA/96-08 [hep-ph/9609209] (unpublished).

[147] M. Schmelling, Plenary talk, XXVIII International Conference on High Energy Physics, Warsaw, July 25–31, 1996.

[148] See, e.g., R. S. Chivukula, in *The Albuquerque Meeting: DPF 94* (Division of Particles and Fields Meeting, American Physical Society, Albuquerque, NM, August 2–6, 1994), ed. by S. Seidel (World Scientific, River Edge, NJ, 1995), p. 273; T. Takeuchi, A. K. Grant, and J. L. Rosner, *ibid.*, p. 1231; A. K. Grant, *Phys. Rev.* D 51:207 (1995).

[149] ALEPH, DELPHI, L3, and OPAL Collaborations, LEP Heavy Flavor Working Group, D. Abbaneo *et al.*, CERN report CERN-PPE/96-017, Feb. 1996, submitted to *Nucl. Inst. Meth.*

[150] J. Steinberger, seminar at CERN, August 27, 1996; ALEPH Collaboration, A. O. Bazarko *et al.* [hep-ex/9609005], presented at *The Minneapolis Meeting: DPF 96* (Division of Particles and Fields Meeting, American Physical Society, Minneapolis, MN, 10–15 August, 1996), to be published.

[151] A. A. Akhundov, D. Yu. Bardin, and T. Riemann, *Nucl. Phys.* B276:1 (1986); W. Beenaker and W. Hollik, *Zeit. Phys.* C 40:141 (1988); J. Bernabeu, A. Pich, and A. Santamaria, *Phys. Lett.* B 200:569 (1988); B. A. Kniehl and J. H. Kühn, *Phys. Lett.* B 224:229 (1989); K. G. Chetyrkin and J. H. Kühn, *Phys. Lett.* B 248:359 (1990); G. Altarelli and R. Barbieri, *Phys. Lett.* B 253:161 (1991); K. G. Chetyrkin, J. H. Kühn, and A. Kwiatkowski, *Phys. Lett.* B 282:221 (1992). For a recent discussion see A. K. Grant, *Phys. Rev.* D 51:207 (1995).

[152] SLD Collaboration, Measurement of R_b at SLD, report PA10-23, presented by D. Su at XXVIII International Conference on High Energy Physics, Warsaw, July 25–31, 1996.

[153] E. Eichten, I. Hinchliffe, K. Lane, and C. Quigg, *Rev. Mod. Phys.* 56:579 (1984); *ibid.* 58:1065(E) (1986).

[154] J. L. Rosner, *Phys. Lett.* 146B:108 (1984).

[155] J. Feltesse, lectures at this Institute.

[156] CDF Collaboration, Fermilab reports FERMILAB-CONF-96-118-E, March, 1996 (presented by P. Azzi), and FERMILAB-CONF-96-146-E, June, 1996 (presented by A. Barbaro-Galtieri), at XXXI Rencontre de Moriond, *QCD and High-Energy Hadronic Interactions*, Les Arcs, France, 23–30 March 1996.

[157] E. Laenen *et al.*, *Phys. Lett.* B 321:254 (1994).

[158] R. K. Ellis, W. J. Stirling, and B. Webber, to be published by Cambridge University Press (1996). I thank R. K. Ellis for updating the plot of Fig. 21.

[159] S. Catani, M. L. Mangano, P. Nason, and L. Trentadue, *Phys. Lett.* B 378:329 (1996).

[160] W. Kwong, *Phys. Rev.* D 43:1488 (1991).

[161] M. E. Peskin and M. J. Strassler, *Phys. Rev.* D 43:1500 (1991).

[162] R. Gatto, G. Sartori, and M. Tonin, *Phys. Lett.* 28B:128 (1968); N. Cabibbo and L. Maiani, *ibid.* 28B:131 (1968); R. J. Oakes, *ibid.* 29B:683 (1969).

[163] S. Weinberg, in *A Festschrift for I. I. Rabi*, *Trans. N. Y. Acad. Sci. Ser. II* 38 (1977); F. Wilczek and A. Zee, *Phys. Lett.* 70B:418 (1977); *ibid.* 72B:504(E) (1978); H. Fritzsch, *ibid.* 70B:436 (1977).

[164] J. L. Rosner and M. P. Worah, *Phys. Rev.* D 46:1131 (1992).

[165] H. Leutwyler, lectures at this Institute.

[166] H. Fritzsch, *Phys. Lett.* 85B:81 (1979); *Nucl. Phys.* B155:189 (1979).

[167] H. Arason *et al.*, *Phys. Rev.* D 46:3945 (1992); H. Arason, D. J. Castano, E. J. Piard, and P. Ramond, *Phys. Rev.* D 47:232 (1992); P. Ramond, R. G. Roberts, and G. G. Ross, *Nucl. Phys.* B406:19 (1993); D. J. Castano, E. J. Piard, and P. Ramond, *Phys. Rev.* D 49:4882 (1994); T. Blazek, M. Carena, S. Raby, and C. Wagner, Ohio State Univ. report OHSTPY-HEP-T-96-014 [hep-ph/9608273], and references therein.

[168] G. C. Branco, W. Grimus, and L. Lavoura, *Phys. Lett.* B 380:119 (1996); G. C. Branco, D. Emmanuel-Costa, and J. I. Silva-Marcos, report [hep-ph/9608477] (unpublished).

[169] H. Harari, H. Haut and J. Weyers, *Phys. Lett.* 78B:459 (1978); Y. Nambu, in *New Theories in Physics* (23–27 May, 1988, Kazimierz, Poland), Z. Ajduk, S. Pokorski, and A. Trautman, eds., World Scientific, Singapore, 1989, p. 1; in *Strong Coupling Gauge Theories and Beyond,* Proceedings of the Second International Workshop, Nagoya, Japan, 1990, T. Muta and K. Yamawaki, eds., World Scientific, Singapore (1991), p. 3; Enrico Fermi Institute Report No. 92-03, in *Proceedings of the Workshop on Electroweak Symmetry Breaking*, Hiroshima, 1991, p. 1; Enrico Fermi Institute Report No. 92-37, Spontaneous symmetry breaking and the origin of mass, invited talk at Int. Conf. on Fluid Mech. and Theor. Phys. in honor of Professor Chou Pei-Yuan's 90th Birthday, 1992; P. Kaus and S. Meshkov, *Mod. Phys. Lett.* A 3:1251 (1988); *ibid.* 4:603(E) (1989); *Phys. Rev.* D 42:1863 (1990). See also the second of Refs. [168].

[170] K. Matumoto and T. Matsuoka, *Prog. Theor. Phys.* 83:373 (1990); *ibid.* 84:53 (1990); K. Matumoto, *ibid.* 84:185,787 (1990); *ibid.* 85:1149 (1991); second of Refs. [168].

[171] J. L. Rosner, *Prog. Theor. Phys.* 66:1421 (1981).

[172] M. K. Gaillard, B. W. Lee, and J. L. Rosner, *Rev. Mod. Phys.* 47:277 (1975).

[173] WA89 Collaboration, S. Paul *et al.*, *Nucl. Phys.* A585:183c (1995). See, however, CLEO Collaboration, report PA01-079, presented by R. Kutschke at XXVIII International Conference on High Energy Physics, Warsaw, July 25–31, 1996.

[174] A. De Rújula, H. Georgi, and S. L. Glashow, *Phys. Rev.* D 12:147 (1975); S. Gasiorowicz and J. L. Rosner, *Am. J. Phys.* 49:954 (1981).

[175] M. Veltman, in *Proc. Int. Symp. on Lepton and Photon Interactions at High Energies (Fermilab, August 23–29, 1979)*, T. B. W. Kirk and H. D. I. Abarbanel, eds., Fermilab, Batavia, IL (1979), p. 529; *Acta Physica Polonica* B12:437 (1981).

[176] Y. Nambu, Ref. [169].

[177] Y. Nambu and G. Jona-Lasinio, *Phys. Rev.* 122:345 (1961); *ibid.* 124:246 (1961).

[178] W. A. Bardeen, C. T. Hill, and M. Lindner, *Phys. Rev.* D 41:1647 (1990).

[179] J.-M. Gérard and J. Weyers, *Phys. Lett.* 146B:411 (1984); J. P. Fatelo, J.-M. Gérard, T. Hambye, and J. Weyers, *Phys. Rev. Lett.* 74:492 (1995); M. Chaichian, P. Chiappetta, J.-M. Gérard, R. Gonzalez Felipe, and J. Weyers, *Phys. Lett.* B 365:141 (1996), and references therein.

[180] L. Hall, The heavy top quark and supersymmetry, lectures given at the 1995 SLAC Summer Institute, July 10–21, 1995, report LBL-38110 [hep-ph/9605258].

[181] R. Oehme, Enrico Fermi Institute Report No. 95-47 [hep-th/9511006], and references therein.

[182] W. Marciano, *Phys. Rev. Lett.* 62:2793 (1989); *Phys. Rev.* D 41:219 (1990).

[183] J. Kubo *et al.*, Max-Planck-Institut report MPI-PHT-95-132 [hep-ph/9512400], and references therein.

[184] L. Susskind, *Phys. Rev.* D 20:2619 (1979).

[185] S. Weinberg, *Phys. Rev.* D 13:974 (1976); *ibid.* 19:1277 (1979).

[186] For a review see E. Farhi and L. Susskind, *Phys. Rep.* 74:277 (1981).

[187] M. Gell-Mann, P. Ramond, and R. Slansky, in *Supergravity*, edited by P. van Nieuwenhuizen and D. Z. Freedman (North-Holland, Amsterdam, 1979), p. 315; T. Yanagida, in *Proc. Workshop on Unified Theory and Baryon Number in the Universe*, edited by O. Sawada and A. Sugamoto (KEK Report No. 79-18, Tsukuba, Japan, 1979).

[188] H. Georgi and S. L. Glashow, *Phys. Rev. Lett.* 32:438 (1974).

[189] H. Georgi in *Proceedings of the 1974 Williamsburg DPF Meeting*, ed. by C. E. Carlson (New York, AIP, 1975) p. 575; H. Fritzsch and P. Minkowski, *Ann. Phys. (N.Y.)* 93:193 (1975).

[190] F. Gürsey, P. Ramond, and P. Sikivie, *Phys. Lett.* 60B:177 (1976).

[191] E. Witten, *Nucl. Phys.* B258:75 (1985); E. Cohen, J. Ellis, K. Enqvist, and D. V. Nanopoulos, *Phys. Lett.* 165B:76 (1985); J. L. Rosner, *Comments on Nucl. and Part. Phys.* 15:195 (1986).

[192] For a review of E_6 phenomenology in the context of superstring theories see J. L. Hewett and T. G. Rizzo, *Phys. Rep.* 183:193 (1989). More recent discussions of extended gauge structures motivated by E_6 include those by P. Langacker and M. Luo, *Phys. Rev.* D 45:278 (1992); P. Langacker, in *Precision Tests of the Standard Model*, edited by P. Langacker (World Scientific, Singapore, 1995), p. 883; M. Cvetič and S. Godfrey, in *Electro-weak Symmetry Breaking and Beyond the Standard Model*, edited by T. Barklow, S. Dawson, H. Haber, and J. Siegrist (World Scientific, Singapore, 1995), and references therein; M. Cvetič and P. Langacker, *Phys. Rev.* D 54:3570 (1996).

[193] R. Slansky, *Phys. Rep.* 79:1 (1981).

[194] See, e.g., S. M. Barr, *Phys. Lett.* 112B:219 (1982); J. Ellis, J. L. Lopez, and D. V. Nanopoulos, *Phys. Lett.* B 371:65 (1996); J. L. Lopez, D. V. Nanopoulos, and A. Zichichi, *Phys. Rev.* D 53:5253 (1996); and references therein.

[195] J. L. Rosner, CERN report CERN-TH/96-209, hep-ph/9607467, submitted to Phys. Rev. D.

[196] P. Langacker and J. Erler, in Ref. [47], p. 103.

[197] R. W. Robinett and J. L. Rosner, *Phys. Rev.* D 26:2396 (1982); P. G. Langacker, R. W. Robinett, and J. L. Rosner, *Phys. Rev.* D 30:1470 (1984).

[198] CDF Collaboration, F. Abe *et al.*, *Phys. Rev.* D 51:949 (1995); T. Kamon, Fermilab report FERMILAB-CONF-96-106-E [hep-ex/9605006], presented at XXXI Rencontre de Moriond: QCD and High-energy Hadronic Interactions, 23 – 30 March, 1996; CDF Collaboration, M. K. Pillai, E Hayashi, K. Maeshima, C. Grosso-Pilcher, P. de Barbaro, A. Bodek, B. Kim, and W. Sakumoto, report hep-ex/9608006, August, 1996, presented at *The Minneapolis Meeting: DPF 96* (Division of Particles and Fields Meeting, American Physical Society, Minneapolis, MN, 10–15 August, 1996), to be published.

[199] CDF Collaboration, F. Abe *et al.*, *Phys. Rev. Lett.* 77:2616 (1996).

[200] J. L. Rosner, *Phys. Rev.* D 54:1078 (1996).

[201] CDF Collaboration, F. Abe *et al.*, presented by S. Park, in *Proceedings of the 10th Topical Workshop on Proton-Antiproton Collider Physics,* Fermilab, May 9–13, 1995, AIP Conference Proceedings 357, edited by R. Raja and J. Yoh, (AIP, Woodbury, NY, 1996), p. 62.

[202] S. Ambrosanio, G. L. Kane, G. D. Kribs, S. P. Martin, and S. Mrenna, *Phys. Rev. Lett.* 76:3498 (1996); Univ. of Michigan reports hep-ph/9605398 and 9607414 (unpublished); S. Dimopoulos, M. Dine, S. Raby, and S. Thomas, *Phys. Rev. Lett.* 76:3494 (1996); S. Dimopoulos, S. Thomas, and J. D. Wells, *Phys. Rev.* D 54:3283 (1996); S. Dimopoulos, M. Dine, S. Raby, S. Thomas, and J. D. Wells, SLAC-PUB-7236 [hep-ph/9607450] (unpublished); J. L. Lopez and D. V. Nanopoulos, Rice University reports DOE/ER/40717-29 [hep-ph/9607220] and DOE/ER/40717-32 [hep-ph/9608275] (unpublished); K. S. Babu, C. Kolda, and F. Wilczek, Institute for Advanced Study report IASSNS-HEP-96/55 [hep-ph/9605408]; J. Hisano, K. Tobe, and T. Yanagida, Univ. of Tokyo report UT-754 [hep-ph/9607234] (unpublished).

[203] G. Bhattacharyya and R. N. Mohapatra, *Phys. Rev.* D 54:4204 (1996).

[204] D. London and J. L. Rosner, *Phys. Rev.* D 34:1530 (1986).

[205] D. Toback, Fermilab report FERMILAB-CONF-96/240-E, August 1996, presented for the CDF Collaboration at *The Minneapolis Meeting: DPF 96* (Division of Particles and Fields Meeting, American Physical Society, Minneapolis, MN, 10–15 August, 1996), to be published.

UNIFIED THEORIES OF FLAVOUR WITH U(2) AS HORIZONTAL GROUP

Andrea Romanino

Physics Department, University of Pisa
and INFN, Sez. di Pisa
I-56126 Pisa, Italy

THE U(2) SYMMETRY

The aim of this exposition is to describe a theory of flavour based on SU(5) or SO(10) as gauge group and U(2) as flavour group. Such an unified theory has been developed by R. Barbieri, L. J. Hall, S. Raby and myself [1] on the basis of previous works by R. Barbieri, G. Dvali and L.J. Hall [2] and R. Barbieri and L. J. Hall [3].

While the structure of the gauge sector of the Standard Model (SM) is well understood in terms of the unification hypothesis, a coherent and complete quantitative explanation of the flavour sector, in particular of fermion masses and mixings is still missing. Ref. [4] is an attempt in this direction. However, besides the bottom quark and τ lepton mass unification [5, 6], several relations have been noticed in the past, sometimes justified on a theoretical basis, like, e.g., $|V_{us}| \simeq (m_d/m_s)^{1/2}$ [7], $m_\mu \simeq 3m_s$ and $m_d \simeq 3m_e$ [8], or $|V_{ub}/V_{cb}| \simeq (m_u/m_c)^{1/2}$ [9], involving the masses and the CKM matrix elements renormalized at the unification scale. In a supersymmetric theory, the flavour problem extends to scalar masses and scalar-fermion mixings, that have to be consistent with experimental limits on Flavour Changing Neutral Current (FCNC) phenomena [10].

In order to explain the "horizontal" relations among masses and mixings, table 1, in terms of coupling constants of order one, we introduce as usual an horizontal symmetry acting in the same way on the family indexes of different SM representations[1], and we suppose that this symmetry is spontaneously broken by the SM-invariant vacuum expectation values of "flavon" fields.

Instead of considering the possible horizontal symmetries involving all the three families, we limit ourselves to symmetries acting on the two lightest families. Moreover

[1] As it is if the symmetry group commutes with a fully unified group.

Table 1. Fermion masses and mixing relations at the unification scale.

$U/D,E$	$m_t \gg m_b, m_\tau$	$\frac{m_c}{m_t} \ll \frac{m_s}{m_b}, \frac{m_\mu}{m_\tau}$	$\frac{m_u m_c}{m_t^2} \ll \frac{m_d m_s}{m_b^2}, \frac{m_e m_\mu}{m_\tau^2}$
D/E	$m_b \approx m_\tau$	$3\frac{m_s}{m_b} \approx \frac{m_\mu}{m_\tau}$	$\frac{m_d m_s}{m_b^2} \approx \frac{m_e m_\mu}{m_\tau^2}$
V_{CKM}	$\lvert V_{us}\rvert \sim \sqrt{\frac{m_d}{m_s}}$	$\lvert V_{cb}\rvert \sim \sqrt{\frac{m_s}{m_b}}$	$\left\lvert\frac{V_{ub}}{V_{cb}}\right\rvert \sim \sqrt{\frac{m_u}{m_c}}$

we require that these families are massless in the fermion sector and exactly degenerate in the scalar sector in the limit of unbroken symmetry, as suggested by the smallness of light fermion Yukawa couplings and by the upper limits on FCNC phenomena. The only suitable continuous non abelian unitary group is then U(2). We prefer to consider a non abelian group because the representation structure, and so the theory, is more constrained than in the abelian case, where there is a large freedom in choosing the horizontal quantum numbers. On the contrary, in the U(2) case, the transformation properties of the flavons can be guessed from the requirement of having mixing among the three fermion families.

We do not want direct couplings of the flavon fields ϕ to the fermion fields ψ through renormalizable operators. Rather, we suppose that the flavons appear only in non renormalizable effective operators generated below a "flavour scale" $M = \mathcal{O}(M_{\mathrm{G}} \div M_{\mathrm{P}})$ at which the physics that mediate the U(2) breaking is integrated out. An example of such a physics is given by the "Froggatt-Nielsen mechanism", as explained by G. Ross in his lectures [11]. The ratios $\langle\phi\rangle/M$ provide the small parameters needed to describe the fermion mass and mixing structure without using small couplings.

Let us choose the transformation properties of the two light families, ψ_a, $a = 1, 2$, under U(2):

$$U \in \mathrm{U}(2): \ \psi_a \to U_a^b \psi_b$$

(the third family, as the Higgses, is supposed to be U(2)-invariant). Then, in order to generate mixing between the third and the light families, it is necessary to have a flavon field ϕ^a that transforms in the conjugate way relative to ψ_a. This flavon gives rise to 12/3 mixing through Yukawa interactions like $((\langle\phi^a\rangle/M)\psi_a\psi_3 h$. But ϕ^a only breaks U(2) to U(1) (at a scale V with $V/M = \mathcal{O}(|V_{cb}|_{\mathrm{G}}) = \mathcal{O}(0.02)$) and the U(1) residual symmetry prevents the lightest generation to get mass and to mix with the heavier ones. In order to break U(1) and to generate the operator $\psi_a\psi_b h$, it is necessary to have a flavon field $\phi^{ab} = A^{ab} + S^{ab}$ with transformation properties under U(2) as indicated by the position of the indexes, where A^{ab} and S^{ab} are their antisymmetric and symmetric irreducible components. Since $\langle A^{ab}\rangle$ breaks U(1) and gives mass to the lightest family, the corresponding scale of breaking v has to be lower than V. Moreover, in order to generate the m_s/m_b ratio at the same scale of $|V_{cb}|$, S^{ab} should break U(2) at the V scale. If we finally suppose that all breakings of U(2) at the V scale leave unbroken the same U(1), we are left with the following breaking pattern (in a suitable basis)

$$\mathrm{U}(2) \xrightarrow{\langle\phi^a\rangle,\langle S^{ab}\rangle} \mathrm{U}(1) \xrightarrow{\langle A^{ab}\rangle} \{\mathbf{1}\} \tag{1a}$$

$$\langle\phi^a\rangle = \begin{pmatrix} 0 \\ V \end{pmatrix}, \quad \langle A^{ab}\rangle = \begin{pmatrix} 0 & v \\ -v & 0 \end{pmatrix}, \quad \langle S^{ab}\rangle = \begin{pmatrix} 0 & 0 \\ 0 & V' \end{pmatrix}, \tag{1b}$$

$$V, v > 0, \qquad v \ll V \approx |V'| \ll M. \tag{1c}$$

Also U(2) invariant fields can develop an expectation value and play a role. We denote them with Σ.

As a consequence of the previous breaking pattern, the Yukawa matrices $\boldsymbol{\lambda}$ and the scalar masses $\mathbf{m}^2$ are in the form

$$\boldsymbol{\lambda} = \begin{pmatrix} 0 & \epsilon' & 0 \\ -\epsilon' & \epsilon & x\epsilon \\ 0 & y\epsilon & 1 \end{pmatrix} \qquad \mathbf{m}^2 = \begin{pmatrix} m_1^2 & 0 & 0 \\ 0 & m_1^2(1+z\epsilon^2) & \epsilon m_2^{2*} \\ 0 & \epsilon m_2^2 & m_3^2 \end{pmatrix} \tag{2}$$

where $\epsilon \equiv V/M$, $\epsilon' \equiv v/M$, $x, y, z = \mathcal{O}(1)$ and $m_1^2, m_2^2, m_3^2 = \mathcal{O}(m_S^2)$. This leads to a CKM matrix in the form [12]

$$V_{\text{CKM}} = \begin{pmatrix} c_{12}^D & s_{12}^D - s_{12}^U e^{i\phi} & s_{12}^U s \\ s_{12}^U - s_{12}^D e^{i\phi} & c_{12}^D e^{i\phi} & -c_{12}^U s \\ -s_{12}^D s & c_{12}^D s & e^{-i\phi} \end{pmatrix} \tag{3}$$

in terms of

$$t_{12}^D = \frac{s_{12}^D}{c_{12}^D} = \sqrt{\frac{m_d}{m_s}} \qquad t_{12}^U = \frac{s_{12}^U}{c_{12}^U} = \sqrt{\frac{m_u}{m_c}} \tag{4}$$

and of the parameters $s > 0$ and $\phi \in (0, \pi)$.

The previous form of the CKM matrix is in agreement with the experimental values of $|V_{us}|$ and $|V_{ub}/V_{cb}|$. Moreover, due to the ϵ term in the λ_{22} entry, in the down quark sector we have

$$\frac{m_s}{m_b} = \mathcal{O}(\epsilon) = \mathcal{O}(|V_{cb}|) \quad \text{and} \quad \frac{\tilde{m}_d^2 - \tilde{m}_s^2}{\tilde{m}_d^2} \le \mathcal{O}\left(\epsilon^2\right) = \mathcal{O}\!\left(\frac{m_s^2}{m_b^2}\right), \tag{5}$$

in agreement with the experimental values of m_s/m_b and $|V_{cb}|$ and with the limits on $(\tilde{m}_d^2 - \tilde{m}_s^2)/\tilde{m}_d^2$ given by the value of ϵ_K, the CP violation parameter in the K^0-$\bar{K}^0$ system [10]. In models in which the 22 entry is vanishing or of order ϵ^2 it is on the contrary

$$\frac{m_s}{m_b} = \mathcal{O}\left(\epsilon^2\right) = \mathcal{O}\left(|V_{cb}|^2\right) \qquad \frac{\tilde{m}_d^2 - \tilde{m}_s^2}{\tilde{m}_d^2} \le \mathcal{O}\left(\epsilon^2\right) = \mathcal{O}\!\left(\frac{m_s}{m_b}\right). \tag{6}$$

The CKM matrix in (3) has been obtained by diagonalizing a Yukawa matrix in the form of eq. (2) in each sector of given charge U, D, E without comparing the Yukawa matrices of different sectors. By doing that, we immediately see that, to reproduce the suppression of the mass ratios in the U-sector relative to the D- and E-sector ones ("$U/D, E$" line in table 1), the 12, 21 and 22 entries of the U-sector Yukawa matrix, $\boldsymbol{\lambda}_U$, have to be suppressed relative to the corresponding entries in $\boldsymbol{\lambda}_D$ and $\boldsymbol{\lambda}_E$. This suppression can be explained in terms of couplings of order one only in presence of vertical relations among the coupling constants as those ones provided by an unified gauge group.

UNIFICATION

Let us then consider the U(2) theory in the case of the minimal unified group SU(5), where each family is represented by a tenplet T and an anti-fiveplet $\bar{F}$. The flavons must transform as **1**, **24** or **75** under SU(5), otherwise they cannot have a SM-invariant expectation value or they cannot contribute to the Yukawa matrices linearly. The only renormalizable interactions are then

$$T_3 T_3 H, \qquad \bar{F}_3 T_3 H, \tag{7}$$

while the possible non renormalizable interactions at the linear order in the flavons are

$$\bar{F}_3 T_a \bar{H}\frac{\phi^a}{M}, \qquad \bar{F}_a T_3 \bar{H}\frac{\phi^a}{M}, \qquad T_3 T_a H\frac{\phi^a}{M}, \tag{8 a}$$

$$T_a T_b H\frac{S^{ab}}{M}, \qquad T_a T_b H\frac{A^{ab}}{M}, \tag{8 b}$$

$$\bar{F}_a T_b H\frac{S^{ab}}{M}, \qquad \bar{F}_a T_b H\frac{A^{ab}}{M}. \tag{8 c}$$

The operators $T_a T_b H\langle S^{ab}\rangle/M$ and $T_a T_b H\langle A^{ab}\rangle/M$ are responsible for the contributions to m_c/m_t and $(m_u m_c/m_t^2)^{1/2}$ at linear order in ϵ, ϵ' respectively. The m_c/m_t and $(m_u m_c/m_t^2)^{1/2}$ suppressions can then be explained if this two operators are vanishing and these ratios are generated at second order in the expectation values. Can this happen in a natural way? Let us consider first the operator $T_a T_b H\langle S^{ab}\rangle/M$. Since only the symmetric part of the product $T_a T_b$ participates in this interaction, if S^{ab} transforms as **75** under SU(5) the operator $T_a T_b H S^{ab}/M$ vanishes because it cannot be constructed in a SU(5) invariant way. If S^{ab} is not a **75** of SU(5), the only case in which $T_a T_b H\langle S^{ab}\rangle/M$ vanishes is when S^{ab} has two components transforming as **1** and **24** respectively, but only for a particular value of the ratio of their expectation values. This possibility has not a natural interpretation in the SU(5) context but has to be taken in consideration when additional vertical structure is present, as it is the case in SO(10).

Analogously, the $T_a T_b H\langle A^{ab}\rangle/M$ operator vanishes only if A^{ab} is a SU(5) singlet or if A^{ab} has two components transforming as **24** and **75**, for a particular value of the ratio of their expectation values.

Thus the SU(5)×U(2) theory allows a natural explanation of the suppression of the U-sector mass ratios that requires $S^{ab} \sim$ **75** and $A^{ab} \sim$ **1**. It is very interesting that in this case the D/E relations in table 1 are *automatically* predicted as a consequence of $\lambda^D_{33} = \lambda^E_{33}$, $\lambda^D_{22} = -\frac{1}{3}\lambda^E_{22}$, $\lambda^D_{12} = \lambda^E_{12}$ that follow from the SM decomposition of the operators $\bar{F}_a T_b \bar{H}\langle S^{ab}\rangle/M$ and $\bar{F}_a T_b \bar{H}\langle S^{ab}\rangle/M$.

Let us consider now full unification of every single families through SO(10). In this case, the possible representations of the flavons are **1**, **45**, **54** and **210**. It is possible to implement the mechanism described above by using representations for A^{ab} and S^{ab} having components transforming as **1** and **75** respectively under SU(5). In this case, S^{ab} must transform as **210** under SO(10).

On the other hand, it is possible to explain the suppression of the U-sector mass ratios using only singlets and adjoints of SO(10). This is possible if S^{ab} is a SO(10) adjoint and if the particular value of the ratio of its components transforming as **1** and **24** under SU(5) is the one characteristic of the $B-L$ generator. Moreover the expectation value of A^{ab} must be SU(5) invariant.

To generate higher order contributions to the fermion masses we use the U(2)-invariant SO(10)-adjoint field Σ. If we decompose its breaking $\langle\Sigma\rangle$ along the generators X and Y, $\langle\Sigma\rangle = \Sigma_X + \Sigma_Y$, ($X$ and Y generate the SM-invariant generator subspace and are given in table 2) we see that the D- and E-sector light masses require a ratio $\langle\Sigma_X\rangle/M$ not far from 1, while $\langle\Sigma_Y\rangle/M = \mathcal{O}\,(0.02)$ is responsible for the suppression of the U-sector mass ratios. Also in this case the D/E relations in table 1 are automatically

Table 2. Values of some SM invariant generators on SM representations. X is SU(5) invariant, Y is the hypercharge, $B-L$ the barion minus lepton number and T_{3R} the third component of the right weak isospin.

	Q	u	d	L	ν	e
X	1	1	-3	-3	5	1
Y	1/6	$-2/3$	1/3	$-1/2$	0	1
$B-L$	1/3	$-1/3$	$-1/3$	-1	1	1
T_{3R}	0	1/2	$-1/2$	0	1/2	$-1/2$

predicted, but $\lambda^D_{22} = -\lambda^E_{22}/3$ becomes $\lambda^D_{22} = \lambda^E_{22}/3$. The relevant operators are then

$$\psi_3 f_1\left(\frac{\Sigma_X}{M}\right) H\psi_3 \tag{9a}$$

$$\frac{1}{M}\psi_3\langle\phi^a\rangle f_2\left(\frac{\Sigma_X}{M}\right) H\psi_a \tag{9b}$$

$$\frac{1}{M}\psi_a\left(\langle S^{ab}\rangle f_3\left(\frac{\Sigma_X}{M}\right)+\langle A^{ab}\rangle f_4\left(\frac{\Sigma_X}{M}\right)\right) H\psi_b \tag{9c}$$

$$\begin{aligned}\frac{1}{M^2}\psi_a&\left(\langle S^{ab}\rangle\Sigma_Y f_5\left(\frac{\Sigma_X}{M}\right)+\langle A^{ab}\rangle\Sigma_Y f_6\left(\frac{\Sigma_X}{M}\right)\right) H\psi_b\\ &+\frac{1}{M^2}\psi_a\langle\phi^a\rangle\langle\phi^b\rangle f_7\left(\frac{\Sigma_X}{M}\right) H\psi_b,\end{aligned} \tag{9d}$$

where the functions f take into account all orders in Σ_X/M and we only consider the leading orders in Σ_Y/M.

PREDICTIONS OF UNIFIED MODELS

In the unified models described above, the Yukawa matrices are of the form

$$\boldsymbol{\lambda}_U = \begin{pmatrix} 0 & \epsilon'\rho & 0\\ -\epsilon'\rho & \epsilon\rho' & y_U\epsilon\\ 0 & x_U\epsilon & 1\end{pmatrix}\lambda, \qquad \boldsymbol{\lambda}_{(D,E)} = \begin{pmatrix} 0 & \epsilon' & 0\\ -\epsilon' & (1,\pm 3)\epsilon & y_{(D,E)}\epsilon\\ 0 & x_{(D,E)}\epsilon & 1\end{pmatrix}\lambda\zeta, \tag{10}$$

where $x_{U,D,E}$, $y_{U,D,E}$, λ, ρ'/ρ are of order one, $\epsilon,\rho=\mathcal{O}(0.02)$, $\epsilon'=\mathcal{O}(0.004)$ and the factor $\zeta=\mathcal{O}(0.02)$ corresponds to the possibility that the two light Higgs doublets appear in the unified multiplet in the renormalizable Yukawa interaction with different weights. In this way we do not solve the $m_b/m_t \ll 1$ problem but we reexpress it in term of $\zeta \ll 1$.

From eq. (10), both qualitative and quantitative predictions follow. If we neglect the order one parameters, we can have order of magnitude expressions for the 13 Yukawa observables at the unification scale in terms of 4 small parameters, ϵ, ρ, ϵ' and ζ. This gives rise to 9 qualitative or order of magnitude predictions in the Yukawa sector[2]:

$$m_t \sim v = 174\,\text{GeV} \tag{11a}$$

$$m_b \sim m_\tau \qquad 3\frac{m_s}{m_b} \sim \frac{m_\mu}{m_\tau} \qquad \frac{m_d m_s}{m_b^2} \sim \frac{m_e m_\mu}{m_\tau^2} \tag{11b}$$

$$\frac{m_s}{m_b} \sim \left(\frac{m_c}{m_t}\right)^{1/2} \tag{11c}$$

$$\left|\frac{V_{ub}}{V_{cb}}\right| \sim \sqrt{\frac{m_u}{m_c}} \qquad \left|\frac{V_{td}}{V_{ts}}\right| \sim \sqrt{\frac{m_d}{m_s}} \tag{11d}$$

$$|V_{cb}| \sim \frac{m_s}{m_b} \qquad J_{\text{CP}} \sim \left(\frac{m_d}{m_s}\frac{m_u}{m_c}\right)^{1/2} |V_{cb}|^2, \tag{11e}$$

besides $|V_{us}| = \mathcal{O}\left(\sqrt{m_d/m_s}\right)$.

Let us come now to the quantitative predictions. Taking into account order one factors, from the diagonalization of the Yukawa matrices in eqs. (10) we obtain the 5 following relations:

$$\frac{m_\mu}{m_\tau}\left(1-\frac{m_e}{m_\mu}\right) \simeq 3\frac{m_s}{m_b}\left(1-\frac{m_d}{m_s}\right) \tag{12a}$$

$$m_b \simeq m_\tau \qquad \frac{m_e m_\mu}{m_\tau^2} \simeq \frac{m_d m_s}{m_b^2} \tag{12b}$$

$$\left|\frac{V_{ub}}{V_{cb}}\right| \simeq \sqrt{\frac{m_u}{m_c}} \qquad \left|\frac{V_{td}}{V_{ts}}\right| \simeq \sqrt{\frac{m_d}{m_s}}. \tag{12c}$$

Eq. (12a) has $\mathcal{O}(\epsilon)$ corrections from the diagonalization, while eqs. (12b,c) have only $\mathcal{O}(\epsilon^2)$ corrections. All of them, however, have corrections due to possible higher order contributions in Σ_Y/M or to weak scale radiative corrections, both of order $\mathcal{O}(\epsilon)$.

The quantitative predictions in eqs. (12) all agree with the experimental values of the quantities involved. On the other hand these errors are not so small. Nevertheless, as H. Leutwyler explained by his lectures [13], there is a combination of light quark masses that is known with good precision:

$$Q = \frac{m_s/m_d}{\sqrt{1-(m_u/m_d)^2}} = 22.7 \pm 0.08. \tag{13}$$

Taking into account $\mathcal{O}(\epsilon)$ corrections to (12a) we get our prediction

$$Q \simeq \frac{25}{\sqrt{1-(m_u/m_d)^2}} \cdot (1-2\Delta), \tag{14}$$

with

$$\Delta = \text{Re}\left[\left(x_D y_D \mp \frac{1}{3} x_E y_E\right)\epsilon\right]. \tag{15}$$

The Q value, together with $|V_{cb}| = |(x_U - x_D)\epsilon|$, is then a constraint on the $\mathcal{O}(\epsilon)$ corrections, that depend on the 23 and 32 entries of the Yukawa matrices and, in turn, on the direction of the ϕ^a expectation value in the generator space[3].

[2] The $\mathcal{O}(1)$ parameter neglected in J_{CP}, $\sin\phi$, could in principle be vanishing, making the prediction for the J_{CP} order of magnitude wrong. Actually, from the $|V_{us}|$ value it turns out that $\sin\phi = \mathcal{O}(1)$, so that also the J_{CP} relation is correct.

[3] A ϕ singlet under SO(10) is excluded.

Table 3. Results of the fit in the unified U(2) theories with ("constrained") or without ("unconstrained") inclusion of ϵ_K and Δm_{B_d} in the inputs.

	inputs	constrained	unconstrained
$m_s/$ MeV	175 ± 55	153^{+35}_{-22}	153 ± 35
$\|V_{cb}\|$	0.038 ± 0.004	$0.039^{+0.0025}_{-0.0015}$	0.038 ± 0.004
$\|V_{ub}/V_{cb}\|$	0.08 ± 0.02	0.075 ± 0.013	0.075 ± 0.016
$\epsilon_K \cdot 10^3$	2.26	2.26	$\pm(1.7^{+1.3}_{-0.1})$
B_K	0.8 ± 0.2	0.86 ± 0.16	0.8
$\Delta m_{B_d}/\text{ps}^{-1}$	0.464	0.464	$0.37^{+0.14}_{-0.05}$
$\sqrt{B} f_B/$ MeV	200 ± 40	178 ± 18	200
$\alpha_s(M_Z)$	0.117 ± 0.006	0.118 ± 0.005	0.118 ± 0.005

It is also possible to test quantitatively the class of considered models by doing a fit of the CKM matrix. Eqs. (3) and (4), that are pure U(2) consequences, allows a parameterization of V_{CKM} in terms of s and ϕ, besides m_u/m_c and m_d/m_s that, fixing Q, m_t, m_b, m_c, can be expressed in terms of m_s, $\alpha_s(M_Z)$, m_u/m_d. We can then perform a fit of the U(2) model expressing the measured $|V_{us}|$, $|V_{cb}|$, $|V_{ub}/V_{cb}|$, $\alpha_s(M_Z)$, m_s in terms of m_u/m_d, s, ϕ, $\alpha_s(M_Z)$, m_s. In the unified case, m_s can be expressed in terms of lepton masses using the relation $m_e m_\mu = m_d m_s$ valid at the unification scale. Finally, if we assume that ϵ_K and the B_d-$\bar{B}_d$ mass difference are accounted for by the usual SM diagrams, we can further constrain the fit and extend it to the SM case, in which only the parameterization in (3) is used.

The results of the fit in the unified U(2) models are summarized in table 3. Also, in figure 1 the predictions for the angles α, β of the unitarity triangle,

$$\alpha = \arg\left(-\frac{V_{tb}^* V_{td}}{V_{ub}^* V_{ud}}\right) \qquad \beta = \arg\left(-\frac{V_{cb}^* V_{cd}}{V_{tb}^* V_{td}}\right) \qquad \gamma = \arg\left(-\frac{V_{ub}^* V_{ud}}{V_{cb}^* V_{cd}}\right) \tag{16}$$

are shown. In terms of the parameterization of the CKM matrix in (3), these angles turn out to be

$$\alpha = \phi \tag{17}$$

$$\beta = \arg\left(1 - \frac{s_{12}^U}{s_{12}^D} e^{-i\phi}\right) \tag{18}$$

$$\gamma = \pi - \alpha - \beta. \tag{19}$$

In conclusion, I briefly summarize the characteristics of the unified U(2) models that we have considered.

- The models are simple and motivated.
- The order of magnitude of 13 fermion masses and mixings are qualitatively understood in terms of 4 small parameters: the ratios of the two U(2) breaking scales and the flavour scale M, the ratio of the SU(5) breaking scale and M and the ratio of the coefficients of the light Higgs doublets in the unified Higgs multiplet. The scales are then as shown in figure 2.

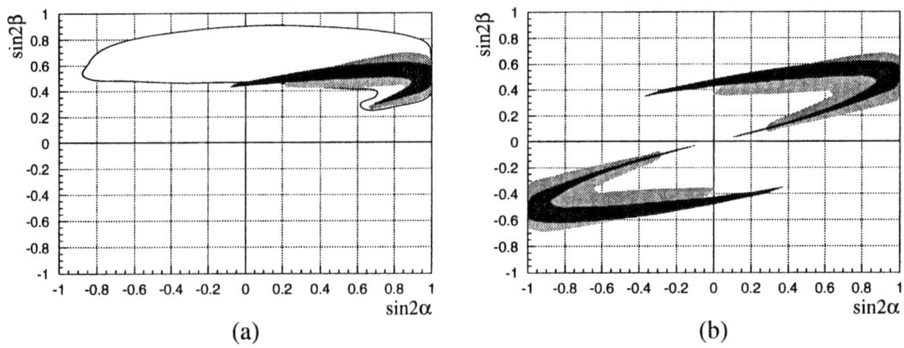

Figure 1. Regions in the $\sin 2\alpha$–$\sin 2\beta$ plane, as determined at the 90% confidence level by the fit in the SM (white area in (a)), in the "pure" U(2) model (grey area) and in the unified U(2) model (dark area), in the "constrained", (a), and "not constrained", (b), case.

- A quantitative analysis of experimental data is possible in the general model through a successful fit that predicts a strong correlation among the α and β angles of the unitarity triangle. Moreover, 5 precise relations among mass and mixings are predicted.
- The suppression of U-sector mass ratios is explained in a natural way and automatically leads to the D/E relations in table 1.
- The scalar masses and mixings fulfill the requirements following from the experimental limits on FCNC phenomena. Moreover, they are constrained by U(2) symmetry and unification. This allows an analysis of contributions from supersymmetric particles to several quantities in flavour physics [14].

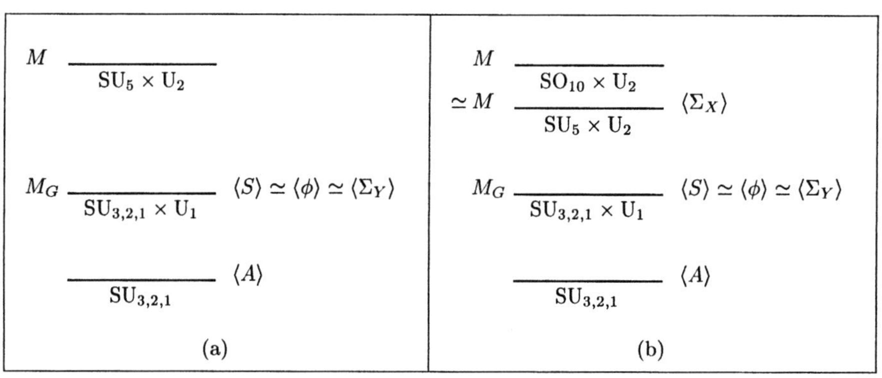

Figure 2. Scales of symmetry breaking vevs appropriate to the SU(5) (a) and SO(10) (b) cases described in the text.

REFERENCES

[1] R. Barbieri, L. J. Hall, S. Raby, and A. Romanino, Unified theories with U(2) flavor symmetry, hep-ph/9610449.

[2] R. Barbieri, G. Dvali, and L. J. Hall, Phys. Lett. **B377**, 76 (1996).

[3] R. Barbieri and L. J. Hall, A grand unified supersymmetric theory of flavor, hep-ph/9605224.

[4] G. W. Anderson, S. Raby, S. Dimopoulos, L. J. Hall, and G. D. Starkman, Phys. Rev. **D49**, 3660 (1994).

[5] M. S. Chanowitz, J. Ellis, and M. K. Gaillard, *Nucl. Phys.* **B128**, 506 (1977).

[6] L. E. Ibanez and C. Lopez, *Phys. Lett.* **126B**, 54 (1983);
H. Arason et al., *Phys. Rev. Lett.* **67**, 2933 (1991);
A. Giveon, L. J. Hall, and U. Sarid, *Phys. Lett.* **271B**, 138 (1991).

[7] R. Gatto, G. Sartori, and M. Tonin, *Phys. Lett.* **28B**, 128 (1968);
N. Cabibbo and L. Maiani, *Phys. Lett.* **28B**, 131 (1968);
S. Weinberg, in *A Festschrift for I.I. Rabi*, edited by L. Motz, New York Academy of Sciences, 1977.

[8] H. Georgi and C. Jarlskog, *Phys. Lett.* **86B**, 297 (1979).

[9] J. Harvey, P. Ramond, and D. Reiss, *Phys. Lett.* **92B**, 309 (1980).

[10] For a most recent analysis, see F. Gabbiani, E. Gabrielli, A. Masiero, and L. Silvestrini, Nucl. Phys. **B477**, 321 (1996).

[11] C. Froggatt and H. B. Nielsen, Nucl. Phys. **B147**, 277 (1979);
G. G. Ross, these proceedings.

[12] L. J. Hall and A. Rasin, Phys. Lett. **B315**, 164 (1993).

[13] H. Leutwyler, The masses of the light quarks, Talk given at the Conference on Fundamental Interactions of Elementary Particles, ITEP, Moscow, Russia, 1995;
H. Leutwyler, these proceedings.

[14] R. Barbieri, L. J. Hall and A. Romanino, work in progress.

HEAVY-QUARK MASSES

Matthias Neubert

Theory Division, CERN
CH-1211 Geneva 23, Switzerland

Abstract and Disclaimer

These lectures provide an introduction to the heavy-quark effective theory and the $1/m_Q$ expansion, which provide the modern framework for a systematic, model-independent description of the properties and decays of hadrons containing a heavy quark. I discuss the applications of these concepts to spectroscopy and to the weak decays of B mesons.

Despite the title of my lectures, I will have to say little about the numerical values of heavy-quark masses. Instead, I will discuss how the fact that these masses are large compared with the scale of non-perturbative strong-interaction effects helps to analyze much of the physics of hadrons containing a heavy quark in a model-independent way. In the course of this discussion, it will become apparent that our ability to understand heavy-flavour physics resides on the observation that heavy quarks are so heavy that their masses become irrelevant. Therefore, a better title for my lectures would have been: "Heavy-quark effective theory, and how to get rid of heavy-quark masses".

INTRODUCTION

The weak decays of hadrons containing a heavy quark are employed for tests of the Standard Model and measurements of its parameters. They offer the most direct way to determine the weak mixing angles, to test the unitarity of the Cabibbo–Kobayashi–Maskawa (CKM) matrix, and to explore the physics of CP violation. At the same time, hadronic weak decays also serve as a probe of that part of strong-interaction phenomenology which is least understood: the confinement of quarks and gluons inside hadrons.

The structure of weak interactions in the Standard Model is rather simple. Flavour-changing decays are mediated by the coupling of the charged current to the W-boson field. At low energies, the charged-current interaction gives rise to local four-fermion couplings, whose strength is governed by the Fermi constant

$$G_F = \frac{g^2}{4\sqrt{2}M_W^2} = 1.16639(2)\ \text{GeV}^{-2}\,. \tag{1}$$

According to the structure of the these interactions, the weak decays of hadrons can be divided into three classes: leptonic decays, in which the quarks of the decaying hadron annihilate each other and only leptons appear in the final state; semileptonic decays, in which both leptons and hadrons appear in the final state; and non-leptonic decays, in which the final state consists of hadrons only. Representative examples of these three types of decays are shown in Fig. 1.

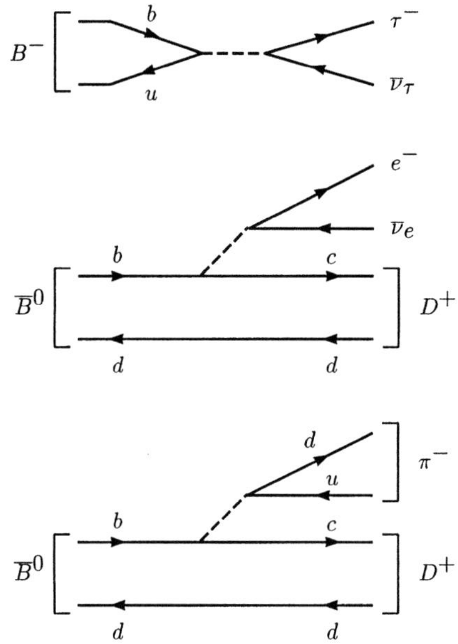

Figure 1. Examples of leptonic ($B^- \to \tau^- \bar{\nu}_\tau$), semileptonic ($\bar{B}^0 \to D^+ e^- \bar{\nu}_e$), and non-leptonic ($\bar{B}^0 \to D^+ \pi^-$) decays of B mesons.

The simple quark-line graphs shown in this figure are a gross oversimplification, however. In the real world, quarks are confined inside hadrons, bound by the exchange of soft gluons. The simplicity of the weak interactions is overshadowed by the complexity of the strong interactions. A complicated interplay between the weak and strong forces characterizes the phenomenology of hadronic weak decays. As an example, a more realistic picture of a non-leptonic decay is shown in Fig. 2. Clearly, the complexity of strong-interaction effects increases with the number of quarks appearing in the final state. Bound-state effects in leptonic decays can be lumped into a single parameter (a "decay constant"), while those in semileptonic decays are described by invariant form factors, depending on the momentum transfer q^2 between the hadrons. Approximate symmetries of the strong interactions help to constrain the properties of these form factors. For non-leptonic decays, on the other hand, we are still far from having a quantitative understanding of strong-interaction effects even in the simplest decay modes.

Over the last decade, a lot of information on heavy-quark decays has been collected in experiments at e^+e^- and hadron colliders. This has led to a rather detailed knowledge of the flavour sector of the Standard Model and many of the parameters associated with it. There have been several great discoveries in this field, such as B^0–$\bar{B}^0$ mixing[1, 2], charmless B decays[3–5], and rare decays induced by penguin operators[6, 7]. The experimental progress in heavy-flavour physics has been accompanied by a significant progress in theory, which was related to the discovery of heavy-quark symmetry and the devel-

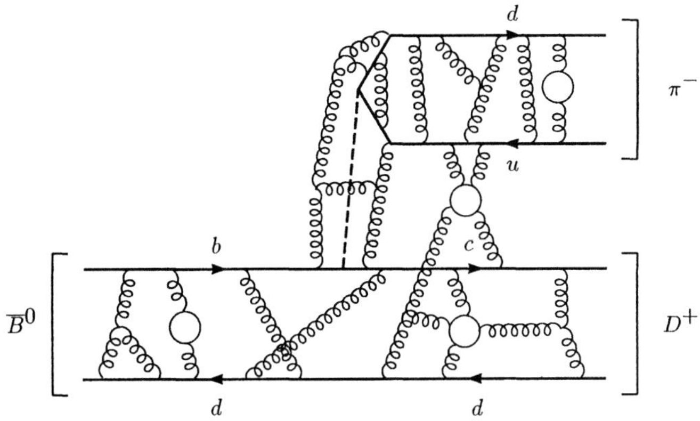

Figure 2. More realistic representation of a non-leptonic decay.

opment of the heavy-quark effective theory (HQET). The excitement about these developments is caused by the fact that they allow (some) model-independent predictions in an area in which "progress" in theory often meant nothing more than the construction of a new model, which could be used to estimate some strong-interaction hadronic matrix elements. In these notes, we explain the physical picture behind heavy-quark symmetry and discuss the construction, as well as simple applications, of the heavy-quark expansion. Because of lack of time, we will have to focus on some particularly important aspects, emphasizing the main ideas and concepts of the HQET. A more complete discussion of the applications of this formalism to heavy-flavour phenomenology can be found in some recent review articles[8, 9]. The reader is also encouraged to consult the earlier review papers in Refs. 10–14.

Hadronic bound states of a heavy quark with light constituents (quarks, antiquarks and gluons) are characterized by a large separation of mass scales: the heavy-quark mass m_Q is much larger than the mass scale $\Lambda_{\rm QCD}$ associated with the light degrees of freedom. Equivalently, the Compton wave length of the heavy quark ($\lambda_Q \sim 1/m_Q$) is much smaller than the size of the hadron containing the heavy quark ($R_{\rm had} \sim 1/\Lambda_{\rm QCD}$). Our goal will be to separate the physics associated with these two scales, in such a way that all dependence on the heavy-quark mass becomes explicit. The framework in which to perform this separation is the operator product expansion (OPE)[15, 16]. The HQET provides us with a convenient technical tool to construct the OPE. Before we start to explore in detail the details of this effective theory, however, we should mention two important reasons why it is desirable to separate short- and long-distance physics in the first place:

- A technical reason is that after the separation of short- and long-distance phenomena we can actually calculate a big portion of the relevant physics (i.e. all short-distance effects) using perturbation theory and renormalization-group techniques. In particular, in this way we will be able to control all logarithmic dependence on the heavy-quark mass.

- An important physical reason is that, after the short-distance physics has been separated, it may happen that the long-distance physics simplifies due to the realization of approximate symmetries, which imply non-trivial relations between observables.

The second point is particularly exciting, since it allows us to make statements beyond the range of applicability of perturbation theory. Notice that here we are not talking about symmetries of the full QCD Lagrangian, such as its local gauge symmetry, but approximate symmetries realized in a particular kinematic situation. In particular, we will find that an approximate spin–flavour symmetry is realized in systems in which a single heavy quark interacts with light degrees of freedom by the exchange of soft gluons.

At this point it is instructive to recall a more familiar example of how approximate symmetries relate the long-distance physics of several observables. The strong interactions of pions are severely constrained by the approximate chiral symmetry of QCD. In a certain kinematic regime, where the momenta of the pions are much less than 1 GeV (the scale of chiral-symmetry breaking), the long-distance physics of scattering amplitudes is encoded in a few "reduced matrix elements", such as the pion decay constant. An effective low-energy theory called chiral perturbation theory provides a systematic expansion of scattering amplitudes in powers of the pion momenta, and thus helps to derive the relations between different scattering amplitudes imposed by chiral symmetry[17]. We will find that a similar situation holds for the case of heavy quarks. Heavy-quark symmetry implies that, in the limit where $m_Q \gg \Lambda_{\mathrm{QCD}}$, the long-distance physics of several observables is encoded in few hadronic parameters, which can be defined in terms of operator matrix elements in the HQET.

HEAVY-QUARK SYMMETRY

The Physical Picture

There are several reasons why the strong interactions of systems containing heavy quarks are easier to understand than those of systems containing only light quarks. The first is asymptotic freedom, the fact that the effective coupling constant of QCD becomes weak in processes with large momentum transfer, corresponding to interactions at short-distance scales[18, 19]. At large distances, on the other hand, the coupling becomes strong, leading to non-perturbative phenomena such as the confinement of quarks and gluons on a length scale $R_{\mathrm{had}} \sim 1/\Lambda_{\mathrm{QCD}} \sim 1$ fm, which determines the size of hadrons[20]. Roughly speaking, $\Lambda_{\mathrm{QCD}} \sim 0.2$ GeV is the energy scale that separates the regions of large and small coupling constant. When the mass of a quark Q is much larger than this scale, it is called a heavy quark. The quarks of the Standard Model fall naturally into two classes: up, down and strange are light quarks, whereas charm, bottom and top are heavy quarks.* For heavy quarks, the effective coupling constant $\alpha_s(m_Q)$ is small, implying that on length scales comparable to the Compton wavelength $\lambda_Q \sim 1/m_Q$ the strong interactions are perturbative and similar to the electromagnetic interactions. In fact, the quarkonium systems $(\bar{Q}Q)$, whose size is of order $\lambda_Q/\alpha_s(m_Q) \ll R_{\mathrm{had}}$, are very much hydrogen-like.

Systems composed of a heavy quark and light constituents are more complicated, however. The size of such systems is determined by R_{had}, and the typical momenta exchanged between the heavy and light constituents are of order Λ_{QCD}. The heavy quark is surrounded by a most complicated, strongly interacting cloud of light quarks, antiquarks, and gluons. In this case it is the fact that $\lambda_Q \ll R_{\mathrm{had}}$, i.e. that the Compton wavelength of the heavy quark is much smaller than the size of the hadron, which leads to simplifications. To resolve the quantum numbers of the heavy quark would require a hard probe; the soft gluons exchanged between the heavy quark and

*Ironically, the top quark is of no relevance to our discussion here, since it is too heavy to form hadronic bound states before it decays.

the light constituents can only resolve distances much larger than λ_Q. Therefore, the light degrees of freedom are blind to the flavour (mass) and spin orientation of the heavy quark. They experience only its colour field, which extends over large distances because of confinement. In the rest frame of the heavy quark, it is in fact only the electric colour field that is important; relativistic effects such as colour magnetism vanish as $m_Q \to \infty$. Since the heavy-quark spin participates in interactions only through such relativistic effects, it decouples. That the heavy-quark mass becomes irrelevant can be seen as follows: As $m_Q \to \infty$, the heavy quark and the hadron that contains it have the same velocity. In the rest frame of the hadron, the heavy quark is at rest, too. The wave function of the light constituents follows from a solution of the field equations of QCD subject to the boundary condition of a static triplet source of colour at the location of the heavy quark. This boundary condition is independent of m_Q, and so is the solution for the configuration of the light constituents.

It follows that, in the limit $m_Q \to \infty$, hadronic systems which differ only in the flavour or spin quantum numbers of the heavy quark have the same configuration of their light degrees of freedom[21–26]. Although this observation still does not allow us to calculate what this configuration is, it provides relations between the properties of such particles as the heavy mesons B, D, B^* and D^*, or the heavy baryons Λ_b and Λ_c (to the extent that corrections to the infinite quark-mass limit are small in these systems). These relations result from some approximate symmetries of the effective strong interactions of heavy quarks at low energies. The configuration of light degrees of freedom in a hadron containing a single heavy quark with velocity v does not change if this quark is replaced by another heavy quark with different flavour or spin, but with the same velocity. Both heavy quarks lead to the same static colour field. For N_h heavy-quark flavours, there is thus an $\mathrm{SU}(2N_h)$ spin–flavour symmetry group, under which the effective strong interactions are invariant. These symmetries are in close correspondence to familiar properties of atoms: The flavour symmetry is analogous to the fact that different isotopes have the same chemistry, since to a good approximation the wave function of the electrons is independent of the mass of the nucleus. The electrons only see the total nuclear charge. The spin symmetry is analogous to the fact that the hyperfine levels in atoms are nearly degenerate. The nuclear spin decouples in the limit $m_e/m_N \to 0$.

Heavy-quark symmetry is an approximate symmetry, and corrections arise since the quark masses are not infinite. In many respects, it is complementary to chiral symmetry, which arises in the opposite limit of small quark masses. However, whereas chiral symmetry is a symmetry of the QCD Lagrangian in the limit of vanishing quark masses, heavy-quark symmetry is not a symmetry of the Lagrangian (not even an approximate one), but rather a symmetry of an effective theory, which is a good approximation of QCD in a certain kinematic region. It is realized only in systems in which a heavy quark interacts predominantly by the exchange of soft gluons. In such systems the heavy quark is almost on shell; its momentum fluctuates around the mass shell by an amount of order Λ_{QCD}. The corresponding fluctuations in the velocity of the heavy quark vanish as $\Lambda_{\mathrm{QCD}}/m_Q \to 0$. The velocity becomes a conserved quantity and is no longer a dynamical degree of freedom[27]. Nevertheless, results derived on the basis of heavy-quark symmetry are model-independent consequences of QCD in a well-defined limit. The symmetry-breaking corrections can, at least in principle, be studied in a systematic way. A convenient framework for analyzing these corrections is provided by the heavy-quark effective theory. Before presenting a detailed discussion of the formalism, we shall first point out some of the important implications of heavy-quark symmetry for the spectroscopy and weak decays of heavy hadrons.

Spectroscopic Implications

The spin–flavour symmetry leads to many interesting relations between the properties of hadrons containing a heavy quark. The most direct consequences concern the spectroscopy of such states[28]. In the limit $m_Q \to \infty$, the spin of the heavy quark and the total angular momentum j of the light degrees of freedom inside a hadron are separately conserved by the strong interactions. Because of heavy-quark symmetry, the dynamics is independent of the spin and mass of the heavy quark. Hadronic states can thus be classified by the quantum numbers (flavour, spin, parity, etc.) of the light degrees of freedom[29]. The spin symmetry predicts that, for fixed $j \neq 0$, there is a doublet of degenerate states with total spin $J = j \pm \frac{1}{2}$. The flavour symmetry relates the properties of states with different heavy-quark flavour.

In general, the mass of a hadron H_Q containing a heavy quark Q obeys an expansion of the form

$$m_H = m_Q + \bar{\Lambda} + \frac{\Delta m^2}{2m_Q} + O(1/m_Q^2)\,. \tag{2}$$

The parameter $\bar{\Lambda}$ represents contributions arising from all terms in the Lagrangian that are independent of the heavy-quark mass[30], whereas the quantity Δm^2 originates from the terms of order $1/m_Q$ in the effective Lagrangian of the HQET. For the moment, the detailed structure of these terms is of no relevance; it will be discussed at length in the next section. For the ground-state pseudoscalar and vector mesons, one can parametrize the contributions from the $1/m_Q$ corrections in terms of two quantities, λ_1 and λ_2, in such a way that[31]

$$\Delta m^2 = -\lambda_1 + 2\Big[J(J+1) - \tfrac{3}{2}\Big]\,\lambda_2\,. \tag{3}$$

Here J is the total spin of the meson. The first term, $-\lambda_1/2m_Q$, arises from the kinetic energy of the heavy quark inside the meson; the second term describes the interaction of the heavy-quark spin with the gluon field. The hadronic parameters $\bar{\Lambda}$, λ_1 and λ_2 are independent of m_Q. They characterize the properties of the light constituents.

Consider, as a first example, the SU(3) mass splittings for heavy mesons. The heavy-quark expansion predicts that

$$\begin{aligned} m_{B_S} - m_{B_d} &= \bar{\Lambda}_s - \bar{\Lambda}_d + O(1/m_b)\,, \\ m_{D_S} - m_{D_d} &= \bar{\Lambda}_s - \bar{\Lambda}_d + O(1/m_c)\,, \end{aligned} \tag{4}$$

where we have indicated that the value of the parameter $\bar{\Lambda}$ depends on the flavour of the light quark. Thus, to the extent that the charm and bottom quarks can both be considered sufficiently heavy, the mass splittings should be similar in the two systems. This prediction is confirmed experimentally, since[32]

$$\begin{aligned} m_{B_S} - m_{B_d} &= (90 \pm 3)\ \text{MeV}\,, \\ m_{D_S} - m_{D_d} &= (99 \pm 1)\ \text{MeV}\,. \end{aligned} \tag{5}$$

As a second example, consider the spin splittings between the ground-state pseudoscalar ($J = 0$) and vector ($J = 1$) mesons, which are the members of the spin-doublet with $j = \frac{1}{2}$. The theory predicts that

$$\begin{aligned} m_{B^*}^2 - m_B^2 &= 4\lambda_2 + O(1/m_b)\,, \\ m_{D^*}^2 - m_D^2 &= 4\lambda_2 + O(1/m_c)\,. \end{aligned} \tag{6}$$

The data are compatible with this:

$$m_{B^*}^2 - m_B^2 \simeq 0.49\ \text{GeV}^2\,,$$
$$m_{D^*}^2 - m_D^2 \simeq 0.55\ \text{GeV}^2\,. \tag{7}$$

Assuming that the B system is close to the heavy-quark limit, we obtain the value

$$\lambda_2 \simeq 0.12\ \text{GeV}^2 \tag{8}$$

for one of the hadronic parameters in (3). This quantity plays an important role in the phenomenology of inclusive decays of heavy hadrons[8].

A third example is provided by the mass splittings between the ground-state mesons and baryons containing a heavy quark. The HQET predicts that

$$m_{\Lambda_b} - m_B = \bar{\Lambda}_{\text{baryon}} - \bar{\Lambda}_{\text{meson}} + O(1/m_b)\,,$$
$$m_{\Lambda_c} - m_D = \bar{\Lambda}_{\text{baryon}} - \bar{\Lambda}_{\text{meson}} + O(1/m_c)\,. \tag{9}$$

This is again consistent with the experimental results

$$m_{\Lambda_b} - m_B = (346 \pm 6)\ \text{MeV}\,,$$
$$m_{\Lambda_c} - m_D = (416 \pm 1)\ \text{MeV}\,, \tag{10}$$

although in this case the data indicate sizeable symmetry-breaking corrections. For the mass of the Λ_b baryon, we have used the value

$$m_{\Lambda_b} = (5625 \pm 6)\ \text{MeV}\,, \tag{11}$$

which is obtained by averaging the result[32] $m_{\Lambda_b} = (5639 \pm 15)$ MeV with the value $m_{\Lambda_b} = (5623 \pm 5 \pm 4)$ MeV reported by the CDF Collaboration[33]. The dominant correction to the relations (9) comes from the contribution of the chromo-magnetic interaction to the masses of the heavy mesons,† which adds a term $3\lambda_2/2m_Q$ on the right-hand side. Including this term, we obtain the refined prediction that the values of the following two quantities should be close to each other:

$$m_{\Lambda_b} - m_B - \frac{3\lambda_2}{2m_B} = (312 \pm 6)\ \text{MeV}\,,$$
$$m_{\Lambda_c} - m_D - \frac{3\lambda_2}{2m_D} = (320 \pm 1)\ \text{MeV} \tag{12}$$

This is clearly satisfied by the data.

The mass formula (2) can also be used to derive information on the heavy-quark (pole) masses from the observed hadron masses. Introducing the "spin-averaged" meson masses $\overline{m}_B = \frac{1}{4}(m_B + 3m_{B^*}) \simeq 5.31$ GeV and $\overline{m}_D = \frac{1}{4}(m_D + 3m_{D^*}) \simeq 1.97$ GeV, we find that

$$m_b - m_c = (\overline{m}_B - \overline{m}_D)\left\{1 - \frac{\lambda_1}{2\overline{m}_B\overline{m}_D} + O(1/m_Q^3)\right\}, \tag{13}$$

where $O(1/m_Q^3)$ is used as a generic notation representing terms suppressed by three powers of the b- or c-quark masses. Using theoretical estimates for the parameter λ_1, which lie in the range[34–36]

$$\lambda_1 = -(0.3 \pm 0.2)\ \text{GeV}^2\,, \tag{14}$$

this relation leads to

$$m_b - m_c = (3.39 \pm 0.03 \pm 0.03)\ \text{GeV}\,, \tag{15}$$

where the first error reflects the uncertainty in the value of λ_1, and the second one takes into account unknown higher-order corrections.

---

†Because of the spin symmetry, there is no such contribution to the masses of the Λ_Q baryons.

Exclusive Semileptonic Decays

Semileptonic decays of B mesons have received a lot of attention in recent years. The decay channel $\bar{B} \to D^* \ell \bar{\nu}$ has the largest branching fraction of all B-meson decay modes. From a theoretical point of view, semileptonic decays are simple enough to allow for a reliable, quantitative description. The analysis of these decays provides much information about the strong forces that bind the quarks and gluons into hadrons. Heavy-quark symmetry implies relations between the weak decay form factors of heavy mesons, which are of particular interest. These relations have been derived by Isgur and Wise[26], generalizing ideas developed by Nussinov and Wetzel[23], and by Voloshin and Shifman[24, 25].

Consider the elastic scattering of a B meson, $\bar{B}(v) \to \bar{B}(v')$, induced by a vector current coupled to the b quark. Before the action of the current, the light degrees of freedom inside the B meson orbit around the heavy quark, which acts as a static source of colour. On average, the b quark and the B meson have the same velocity v. The action of the current is to replace instantaneously (at $t = t_0$) the colour source by one moving at a velocity v', as indicated in Fig. 3. If $v = v'$, nothing happens; the light degrees of freedom do not realize that there was a current acting on the heavy quark. If the velocities are different, however, the light constituents suddenly find themselves interacting with a moving colour source. Soft gluons have to be exchanged to rearrange them so as to form a B meson moving at velocity v'. This rearrangement leads to a form-factor suppression, which reflects the fact that as the velocities become more and more different, the probability for an elastic transition decreases. The important observation is that, in the limit $m_b \to \infty$, the form factor can only depend on the Lorentz boost $\gamma = v \cdot v'$ that connects the rest frames of the initial- and final-state mesons. Thus, in this limit a dimensionless probability function $\xi(v \cdot v')$ describes the transition. It is called the Isgur–Wise function[26]. In the HQET, which provides the appropriate framework for taking the limit $m_b \to \infty$, the hadronic matrix element describing the scattering process can thus be written as

$$\frac{1}{m_B} \langle \bar{B}(v') | \, \bar{b}_{v'} \gamma^\mu b_v \, | \bar{B}(v) \rangle = \xi(v \cdot v') \, (v + v')^\mu . \tag{16}$$

Here, b_v and $b_{v'}$ are the velocity-dependent heavy-quark fields of the HQET, whose precise definition will be discussed in the next lecture. It is important that the function $\xi(v \cdot v')$ does not depend on m_b. The factor $1/m_B$ on the left-hand side compensates for a trivial dependence on the heavy-meson mass caused by the relativistic normalization of meson states, which is conventionally taken to be

$$\langle \bar{B}(p') | \bar{B}(p) \rangle = 2 m_B v^0 \, (2\pi)^3 \, \delta^3(\vec{p} - \vec{p}\,') . \tag{17}$$

Note that there is no term proportional to $(v - v')^\mu$ in (16). This can be seen by contracting the matrix element with $(v - v')_\mu$, which must give zero since $\not{v} b_v = b_v$ and $\bar{b}_{v'} \not{v}' = \bar{b}_{v'}$.

It is more conventional to write the above matrix element in terms of an elastic form factor $F_{\rm el}(q^2)$ depending on the momentum transfer $q^2 = (p - p')^2$:

$$\langle \bar{B}(v') | \, \bar{b} \, \gamma^\mu b \, | \bar{B}(v) \rangle = F_{\rm el}(q^2) \, (p + p')^\mu , \tag{18}$$

where $p^{(\prime)} = m_B v^{(\prime)}$. Comparing this with (16), we find that

$$F_{\rm el}(q^2) = \xi(v \cdot v') , \qquad q^2 = -2 m_B^2 (v \cdot v' - 1) . \tag{19}$$

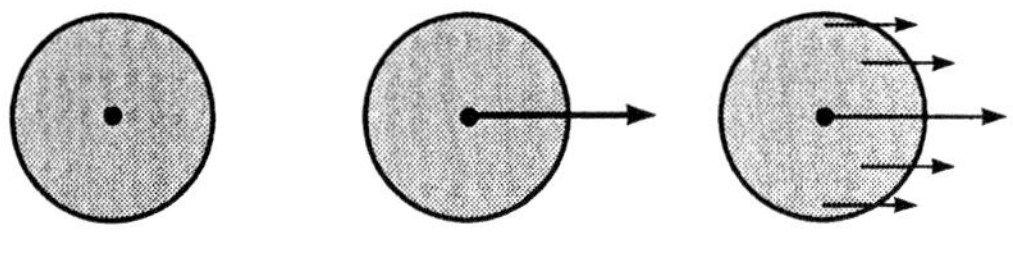

Figure 3. Elastic transition induced by an external heavy-quark current.

Because of current conservation, the elastic form factor is normalized to unity at $q^2 = 0$. This condition implies the normalization of the Isgur–Wise function at the kinematic point $v \cdot v' = 1$, i.e. for $v = v'$:

$$\xi(1) = 1 . \tag{20}$$

It is in accordance with the intuitive argument that the probability for an elastic transition is unity if there is no velocity change. Since for $v = v'$ the daughter meson is at rest in the rest frame of the parent meson, the point $v \cdot v' = 1$ is referred to as the zero-recoil limit.

We can now use the flavour symmetry to replace the b quark in the final-state meson by a c quark, thereby turning the B meson into a D meson. Then the scattering process turns into a weak decay process. In the infinite mass limit, the replacement $b_{v'} \to c_{v'}$ is a symmetry transformation, under which the effective Lagrangian is invariant. Hence, the matrix element

$$\frac{1}{\sqrt{m_B m_D}} \langle D(v')|\, \bar{c}_{v'} \gamma^\mu b_v \,|\bar{B}(v)\rangle = \xi(v \cdot v')\,(v + v')^\mu \tag{21}$$

is still determined by the same function $\xi(v \cdot v')$. This is interesting, since in general the matrix element of a flavour-changing current between two pseudoscalar mesons is described by two form factors:

$$\langle D(v')|\, \bar{c} \gamma^\mu b \,|\bar{B}(v)\rangle = f_+(q^2)\,(p + p')^\mu - f_-(q^2)\,(p - p')^\mu . \tag{22}$$

Comparing the above two equations, we find that

$$\begin{aligned} f_\pm(q^2) &= \frac{m_B \pm m_D}{2\sqrt{m_B m_D}}\, \xi(v \cdot v') , \\ q^2 &= m_B^2 + m_D^2 - 2 m_B m_D\, v \cdot v' . \end{aligned} \tag{23}$$

Thus, the heavy-quark flavour symmetry relates two a priori independent form factors to one and the same function. Moreover, the normalization of the Isgur–Wise function at $v \cdot v' = 1$ now implies a non-trivial normalization of the form factors $f_\pm(q^2)$ at the point of maximum momentum transfer, $q^2_{\rm max} = (m_B - m_D)^2$:

$$f_\pm(q^2_{\rm max}) = \frac{m_B \pm m_D}{2\sqrt{m_B m_D}} . \tag{24}$$

The heavy-quark spin symmetry leads to additional relations among weak decay form factors. It can be used to relate matrix elements involving vector mesons to those involving pseudoscalar mesons. A vector meson with longitudinal polarization is related to a pseudoscalar meson by a rotation of the heavy-quark spin. Hence, the spin-symmetry transformation $c_{v'}^{\Uparrow} \to c_{v'}^{\Downarrow}$ relates $\bar{B} \to D$ with $\bar{B} \to D^*$ transitions. The result of this transformation is[26]:

$$\frac{1}{\sqrt{m_B m_{D^*}}}\,\langle D^*(v',\varepsilon)|\,\bar c_{v'}\gamma^\mu b_v\,|\bar B(v)\rangle = i\epsilon^{\mu\nu\alpha\beta}\,\varepsilon^*_\nu\,v'_\alpha v_\beta\,\xi(v\cdot v')\,,$$
$$\frac{1}{\sqrt{m_B m_{D^*}}}\,\langle D^*(v',\varepsilon)|\,\bar c_{v'}\gamma^\mu\gamma_5\,b_v\,|\bar B(v)\rangle = \Big[\varepsilon^{*\mu}\,(v\cdot v'+1) - v'^\mu\,\varepsilon^*\cdot v\Big]\,\xi(v\cdot v')\,, \tag{25}$$

where ε denotes the polarization vector of the D^* meson. Once again, the matrix elements are completely described in terms of the Isgur–Wise function. Now this is even more remarkable, since in general four form factors, $V(q^2)$ for the vector current, and $A_i(q^2)$, $i=0,1,2$, for the axial vector current, are required to parametrize these matrix elements. In the heavy-quark limit, they obey the relations[37]

$$\begin{aligned}\frac{m_B\pm m_{D^*}}{2\sqrt{m_B m_{D^*}}}\,\xi(v\cdot v') &= V(q^2) = A_0(q^2) = A_1(q^2)\\ &= \left[1-\frac{q^2}{(m_B+m_D)^2}\right]^{-1} A_1(q^2)\,,\\ q^2 &= m_B^2+m_{D^*}^2-2m_B m_{D^*}\,v\cdot v'\,.\end{aligned} \tag{26}$$

Equations (23) and (26) summarize the relations imposed by heavy-quark symmetry on the weak decay form factors describing the semileptonic decay processes $\bar B\to D\,\ell\,\bar\nu$ and $\bar B\to D^*\ell\,\bar\nu$. These relations are model-independent consequences of QCD in the limit where $m_b, m_c \gg \Lambda_{\rm QCD}$. They play a crucial role in the determination of the CKM matrix element $|V_{cb}|$. In terms of the recoil variable $w = v\cdot v'$, the differential semileptonic decay rates in the heavy-quark limit become[38]:

$$\begin{aligned}\frac{\mathrm{d}\Gamma(\bar B\to D\,\ell\,\bar\nu)}{\mathrm{d}w} &= \frac{G_F^2}{48\pi^3}\,|V_{cb}|^2\,(m_B+m_D)^2\,m_D^3\,(w^2-1)^{3/2}\,\xi^2(w)\,,\\ \frac{\mathrm{d}\Gamma(\bar B\to D^*\ell\,\bar\nu)}{\mathrm{d}w} &= \frac{G_F^2}{48\pi^3}\,|V_{cb}|^2\,(m_B-m_{D^*})^2\,m_{D^*}^3\,\sqrt{w^2-1}\,(w+1)^2\\ &\quad\times\left[1+\frac{4w}{w+1}\,\frac{m_B^2-2w\,m_B m_{D^*}+m_{D^*}^2}{(m_B-m_{D^*})^2}\right]\xi^2(w)\,.\end{aligned} \tag{27}$$

These expressions receive symmetry-breaking corrections, since the masses of the heavy quarks are not infinitely heavy. Perturbative corrections of order $\alpha_s^n(m_Q)$ can be calculated order by order in perturbation theory. A more difficult task is to control the non-perturbative power corrections of order $(\Lambda_{\rm QCD}/m_Q)^n$. The HQET provides a systematic framework for analysing these corrections. For the case of weak-decay form factors, the analysis of the $1/m_Q$ corrections was performed by Luke[39]. Later, Falk and the present author have also analysed the structure of $1/m_Q^2$ corrections for both meson and baryon weak decay form factors[31]. We shall not discuss these rather technical issues in detail, but only mention the most important result of Luke's analysis. It concerns the zero-recoil limit, where an analogue of the Ademollo–Gatto theorem[40] can be proved. This is Luke's theorem[39], which states that the matrix elements describing the leading $1/m_Q$ corrections to weak decay amplitudes vanish at zero recoil. This theorem is valid to all orders in perturbation theory[31, 41, 42]. Most importantly, it protects the $\bar B\to D^*\ell\,\bar\nu$ decay rate from receiving first-order $1/m_Q$ corrections at zero recoil[38]. (A similar statement is not true for the decay $\bar B\to D\,\ell\,\bar\nu$, however. The reason is simple but somewhat subtle. Luke's theorem protects only those form factors not multiplied by kinematic factors that vanish for $v = v'$. By angular momentum conservation, the two pseudoscalar mesons in the decay $\bar{B} \to D\,\ell\,\bar{\nu}$ must be in a relative p wave, and hence the amplitude is proportional to the velocity $|\vec{v}_D|$ of the D meson in the B-meson rest frame. This leads to a factor $(w^2 - 1)$ in the decay rate. In such a situation, form factors that are kinematically suppressed can contribute[37].)

Model-Independent Determination of $|V_{cb}|$

We will now discuss the most important application of the HQET in the context of semileptonic decays of B mesons. A model-independent determination of the CKM matrix element $|V_{cb}|$ based on heavy-quark symmetry can be obtained by measuring the recoil spectrum of D^* mesons produced in $\bar{B} \to D^*\ell\,\bar{\nu}$ decays[38]. In the heavy-quark limit, the differential decay rate for this process has been given in (27). In order to allow for corrections to that limit, we write

$$\begin{aligned} \frac{\mathrm{d}\Gamma(\bar{B} \to D^*\ell\,\bar{\nu})}{\mathrm{d}w} &= \frac{G_F^2}{48\pi^3}\,(m_B - m_{D^*})^2\,m_{D^*}^3\sqrt{w^2-1}\,(w+1)^2 \\ &\quad\times \left[1 + \frac{4w}{w+1}\,\frac{m_B^2 - 2w\,m_B m_{D^*} + m_{D^*}^2}{(m_B - m_{D^*})^2}\right] |V_{cb}|^2\,\mathcal{F}^2(w)\,, \end{aligned} \tag{28}$$

where the hadronic form factor $\mathcal{F}(w)$ coincides with the Isgur–Wise function up to symmetry-breaking corrections of order $\alpha_s(m_Q)$ and $\Lambda_{\mathrm{QCD}}/m_Q$. The idea is to measure the product $|V_{cb}|\,\mathcal{F}(w)$ as a function of w, and to extract $|V_{cb}|$ from an extrapolation of the data to the zero-recoil point $w = 1$, where the B and the D^* mesons have a common rest frame. At this kinematic point, heavy-quark symmetry helps to calculate the normalization $\mathcal{F}(1)$ with small and controlled theoretical errors. Since the range of w values accessible in this decay is rather small ($1 < w < 1.5$), the extrapolation can be done using an expansion around $w = 1$:

$$\mathcal{F}(w) = \mathcal{F}(1)\left[1 - \widehat{\varrho}^2\,(w-1) + \dots\right]. \tag{29}$$

The slope $\widehat{\varrho}^2$ is treated as a fit parameter.

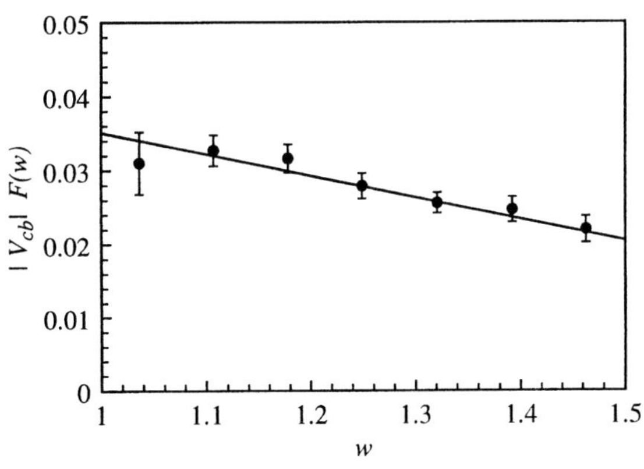

Figure 4. CLEO data for the product $|V_{cb}|\,\mathcal{F}(w)$, as extracted from the recoil spectrum in $\bar{B} \to D^*\ell\,\bar{\nu}$ decays. The line shows a linear fit to the data.

Measurements of the recoil spectrum have been performed first by the ARGUS[43] and CLEO[44] Collaborations in experiments operating at the $\Upsilon(4s)$ resonance, and more recently by the ALEPH[45] and DELPHI[46] Collaborations at LEP. As an example, Fig. 4 shows the data reported by the CLEO Collaboration. The results obtained by the various experimental groups from a linear fit to their data are summarized in Table 1. The weighted average of these results is

$$\begin{aligned} |V_{cb}|\,\mathcal{F}(1) &= (34.6 \pm 1.7) \times 10^{-3}\,, \\ \widehat{\varrho}^2 &= 0.82 \pm 0.09\,. \end{aligned} \tag{30}$$

The effect of a positive curvature of the form factor has been investigated by Stone[47], who finds that the value of $|V_{cb}|\,\mathcal{F}(1)$ may change by up to +4%. We thus increase the above value by $(2 \pm 2)\%$ and quote the final result as

$$|V_{cb}|\,\mathcal{F}(1) = (35.3 \pm 1.8) \times 10^{-3}\,. \tag{31}$$

In future analyses, the extrapolation to zero recoil should be performed including higher-order terms in the expansion (29). It can be shown in a model-independent way that the shape of the form factor is highly constrained by analyticity and unitarity requirements[48, 49]. In particular, the curvature at $w = 1$ is strongly correlated with the slope of the form factor. For the value of $\widehat{\varrho}^2$ given in (30), one obtains a small positive curvature[49], in agreement with the assumption made in Ref. 47.

Table 1. Values for $|V_{cb}|\,\mathcal{F}(1)$ (in units of 10^{-3}) and $\widehat{\varrho}^2$ extracted from measurements of the recoil spectrum in $\bar{B} \to D^*\ell\,\bar{\nu}$ decays

	$\|V_{cb}\|\,\mathcal{F}(1)$ (10^{-3})	$\widehat{\varrho}^2$
ARGUS	$38.8 \pm 4.3 \pm 2.5$	$1.17 \pm 0.22 \pm 0.06$
CLEO	$35.1 \pm 1.9 \pm 2.0$	$0.84 \pm 0.12 \pm 0.08$
ALEPH	$31.4 \pm 2.3 \pm 2.5$	$0.39 \pm 0.21 \pm 0.12$
DELPHI	$35.0 \pm 1.9 \pm 2.3$	$0.81 \pm 0.16 \pm 0.10$

Heavy-quark symmetry implies that the general structure of the symmetry-breaking corrections to the form factor at zero recoil is[38]

$$\mathcal{F}(1) = \eta_A \left(1 + 0 \times \frac{\Lambda_{\rm QCD}}{m_Q} + \mbox{const} \times \frac{\Lambda_{\rm QCD}^2}{m_Q^2} + \dots \right) \equiv \eta_A \left(1 + \delta_{1/m^2} \right), \tag{32}$$

where η_A is a short-distance correction arising from the (finite) renormalization of the flavour-changing axial current at zero recoil, and δ_{1/m^2} parametrizes second-order (and higher) power corrections. The absence of first-order power corrections at zero recoil is a consequence of Luke's theorem[39]. The one-loop expression for η_A has been known for a long time[22, 25, 50]:

$$\eta_A = 1 + \frac{\alpha_s(M)}{\pi} \left(\frac{m_b + m_c}{m_b - m_c} \ln\frac{m_b}{m_c} - \frac{8}{3} \right) \simeq 0.96\,. \tag{33}$$

The scale M in the running coupling constant can be fixed by adopting the prescription of Brodsky, Lepage and Mackenzie (BLM)[51], according to which it is identified with the average virtuality of the gluon in the one-loop diagrams that contribute to

η_A. If $\alpha_s(M)$ is defined in the modified minimal subtraction ($\overline{\rm MS}$) scheme, the result is[52] $M \simeq 0.51\sqrt{m_c m_b}$. Several estimates of higher-order corrections to η_A have been discussed. The next-to-leading order resummation of logarithms of the type $[\alpha_s \ln(m_b/m_c)]^n$ leads to[53, 54] $\eta_A \simeq 0.985$. On the other hand, the resummation of "renormalon-chain" contributions of the form $\beta_0^{n-1}\alpha_s^n$, where β_0 is the first coefficient of the QCD β-function, gives[55] $\eta_A \simeq 0.945$. Using these partial resummations to estimate the uncertainty results in $\eta_A = 0.965 \pm 0.020$. Recently, Czarnecki has improved this estimate by calculating η_A at two-loop order[56]. His result,

$$\eta_A = 0.960 \pm 0.007\,, \tag{34}$$

is in excellent agreement with the BLM-improved one-loop estimate (33). Here the error is taken to be the size of the two-loop correction.

The analysis of the power corrections δ_{1/m^2} is more difficult, since it cannot rely on perturbation theory. Three approaches have been discussed: in the "exclusive approach", all $1/m_Q^2$ operators in the HQET are classified and their matrix elements estimated, leading to[31, 57] $\delta_{1/m^2} = -(3 \pm 2)\%$; the "inclusive approach" has been used to derive the bound $\delta_{1/m^2} < -3\%$, and to estimate that[58,‡] $\delta_{1/m^2} = -(7 \pm 3)\%$; the "hybrid approach" combines the virtues of the former two to obtain a more restrictive lower bound on δ_{1/m^2}. This leads to[60]

$$\delta_{1/m^2} = -0.055 \pm 0.025\,. \tag{35}$$

Combining the above results, adding the theoretical errors linearly to be conservative, gives

$$\mathcal{F}(1) = 0.91 \pm 0.03 \tag{36}$$

for the normalization of the hadronic form factor at zero recoil. Thus, the corrections to the heavy-quark limit amount to a moderate decrease of the form factor of about 10%. This can be used to extract from the experimental result (31) the model-independent value

$$|V_{cb}| = (38.8 \pm 2.0_{\rm exp} \pm 1.2_{\rm th}) \times 10^{-3}\,. \tag{37}$$

HEAVY-QUARK EFFECTIVE THEORY

The Effective Lagrangian

The effects of a very heavy particle often become irrelevant at low energies. It is then useful to construct a low-energy effective theory, in which this heavy particle no longer appears. Eventually, this effective theory will be easier to deal with than the full theory. A familiar example is Fermi's theory of the weak interactions. For the description of weak decays of hadrons, the weak interactions can be approximated by point-like four-fermion couplings, governed by a dimensionful coupling constant G_F. Only at energies much larger than the masses of hadrons can the effects of the intermediate vector bosons, W and Z, be resolved.

The process of removing the degrees of freedom of a heavy particle involves the following steps[61–63]: one first identifies the heavy-particle fields and "integrates them out" in the generating functional of the Green functions of the theory. This is possible since at low energies the heavy particle does not appear as an external state. However, although the action of the full theory is usually a local one, what results after this

‡This bound has been criticised in Ref. 59.

first step is a non-local effective action. The non-locality is related to the fact that in the full theory the heavy particle with mass M can appear in virtual processes and propagate over a short but finite distance $\Delta x \sim 1/M$. Thus, a second step is required to obtain a local effective Lagrangian: the non-local effective action is rewritten as an infinite series of local terms in an Operator Product Expansion (OPE)[15, 16]. Roughly speaking, this corresponds to an expansion in powers of $1/M$. It is in this step that the short- and long-distance physics is disentangled. The long-distance physics corresponds to interactions at low energies and is the same in the full and the effective theory. But short-distance effects arising from quantum corrections involving large virtual momenta (of order M) are not reproduced in the effective theory, once the heavy particle has been integrated out. In a third step, they have to be added in a perturbative way using renormalization-group techniques. These short-distance effects lead to a renormalization of the coefficients of the local operators in the effective Lagrangian. An example is the effective Lagrangian for non-leptonic weak decays, in which radiative corrections from hard gluons with virtual momenta in the range between m_W and some renormalization scale $\mu \sim 1$ GeV give rise to Wilson coefficients, which renormalize the local four-fermion interactions[64–66].

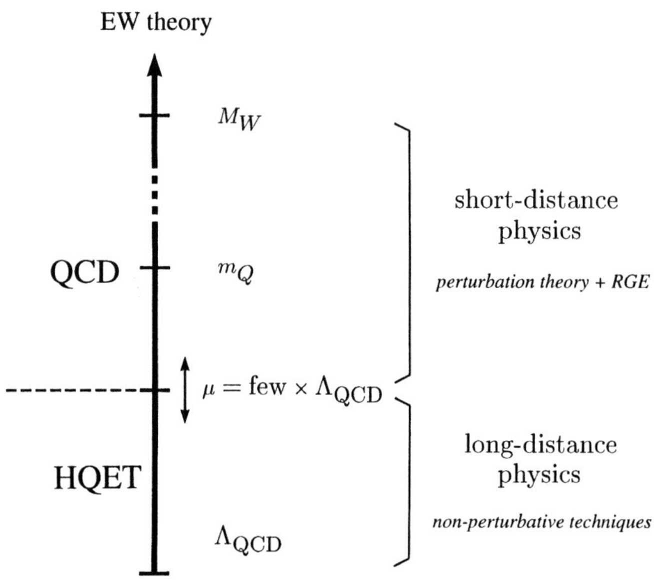

Figure 5. Philosophy of the heavy-quark effective theory.

The heavy-quark effective theory (HQET) is constructed to provide a simplified description of processes where a heavy quark interacts with light degrees of freedom predominantly by the exchange of soft gluons[67–77]. Clearly, m_Q is the high-energy scale in this case, and Λ_{QCD} is the scale of the hadronic physics we are interested in. The situation is illustrated in Fig. 5. At short distances, i.e. for energy scales larger than the heavy-quark mass, the physics is perturbative and described by ordinary QCD. For mass scales much below the heavy-quark mass, the physics is complicated and non-perturbative because of confinement. Our goal is to obtain a simplified description in this region using an effective field theory. To separate short- and long-distance effects, we introduce a separation scale μ such that $\Lambda_{QCD} \ll \mu \ll m_Q$. The HQET will be constructed in such a way that it is identical to QCD in the long-distance region, i.e. for scales below μ. In the short-distance region, the effective theory is incomplete, however, since some high-momentum modes have been integrated out from the full theory. The fact that the physics must be independent of the arbitrary scale μ allows us to derive renormalization-group equations, which we shall employ to deal with the short-distance effects in an efficient way.

Compared with most effective theories, in which the degrees of freedom of a heavy particle are removed completely from the low-energy theory, the HQET is special in that its purpose is to describe the properties and decays of hadrons which do contain a heavy quark. Hence, it is not possible to remove the heavy quark completely from the effective theory. What is possible is to integrate out the "small components" in the full heavy-quark spinor, which describe the fluctuations around the mass shell.

The starting point in the construction of the low-energy effective theory is the observation that a very heavy quark bound inside a hadron moves more or less with the hadron's velocity v, and is almost on shell. Its momentum can be written as

$$p_Q^\mu = m_Q v^\mu + k^\mu \,, \tag{38}$$

where the components of the so-called residual momentum k are much smaller than m_Q. Note that v is a four-velocity, so that $v^2 = 1$. Interactions of the heavy quark with light degrees of freedom change the residual momentum by an amount of order $\Delta k \sim \Lambda_{\rm QCD}$, but the corresponding changes in the heavy-quark velocity vanish as $\Lambda_{\rm QCD}/m_Q \to 0$. In this situation, it is appropriate to introduce large- and small-component fields, h_v and H_v, by

$$h_v(x) = e^{im_Q v\cdot x}\, P_+\, Q(x)\,, \qquad H_v(x) = e^{im_Q v\cdot x}\, P_-\, Q(x)\,, \tag{39}$$

where P_+ and P_- are projection operators defined as

$$P_\pm = \frac{1 \pm \not{v}}{2}\,. \tag{40}$$

It follows that

$$Q(x) = e^{-im_Q v\cdot x}\, [h_v(x) + H_v(x)]\,. \tag{41}$$

Because of the projection operators, the new fields satisfy $\not{v}\, h_v = h_v$ and $\not{v}\, H_v = -H_v$. In the rest frame, i.e. for $v^\mu = (1,0,0,0)$, h_v corresponds to the upper two components of Q, while H_v corresponds to the lower ones. Whereas h_v annihilates a heavy quark with velocity v, H_v creates a heavy antiquark with velocity v.

In terms of the new fields, the QCD Lagrangian for a heavy quark takes the form

$$\begin{aligned} \mathcal{L}_Q &= \bar Q\,(i\not{D} - m_Q)\,Q \\ &= \bar h_v\, iv\cdot D\, h_v - \bar H_v\,(iv\cdot D + 2m_Q)\,H_v \\ &\quad + \bar h_v\, i\not{D}_\perp H_v + \bar H_v\, i\not{D}_\perp h_v\,, \end{aligned} \tag{42}$$

where $D_\perp^\mu = D^\mu - v^\mu\, v\cdot D$ is orthogonal to the heavy-quark velocity: $v\cdot D_\perp = 0$. In the rest frame, $D_\perp^\mu = (0, \vec D\,)$ contains the spatial components of the covariant derivative. From (42), it is apparent that h_v describes massless degrees of freedom, whereas H_v corresponds to fluctuations with twice the heavy-quark mass. These are the heavy degrees of freedom that will be eliminated in the construction of the effective theory. The fields are mixed by the presence of the third and fourth terms, which describe pair creation or annihilation of heavy quarks and antiquarks. As shown in the first diagram in Fig. 6, in a virtual process a heavy quark propagating forward in time can turn into an antiquark propagating backward in time, and then turn back into a quark. The energy of the intermediate quantum state $hh\bar{H}$ is larger than the energy of the initial heavy quark by at least $2m_Q$. Because of this large energy gap, the virtual quantum fluctuation can only propagate over a short distance $\Delta x \sim 1/m_Q$. On hadronic scales set by $R_{\rm had} = 1/\Lambda_{\rm QCD}$, the process essentially looks like a local interaction of the form

$$\bar{h}_v\, i\not{D}_\perp\, \frac{1}{2m_Q}\, i\not{D}_\perp h_v\,, \tag{43}$$

where we have simply replaced the propagator for H_v by $1/2m_Q$. A more correct treatment is to integrate out the small-component field H_v, thereby deriving a non-local effective action for the large-component field h_v, which can then be expanded in terms of local operators. Before doing this, let us mention a second type of virtual corrections involving pair creation, namely heavy-quark loops. An example is shown in the second diagram in Fig. 6. Heavy-quark loops cannot be described in terms of the effective fields h_v and H_v, since the quark velocities inside a loop are not conserved and are in no way related to hadron velocities. However, such short-distance processes are proportional to the small coupling constant $\alpha_s(m_Q)$ and can be calculated in perturbation theory. They lead to corrections that are added onto the low-energy effective theory in the renormalization procedure to be discussed later.

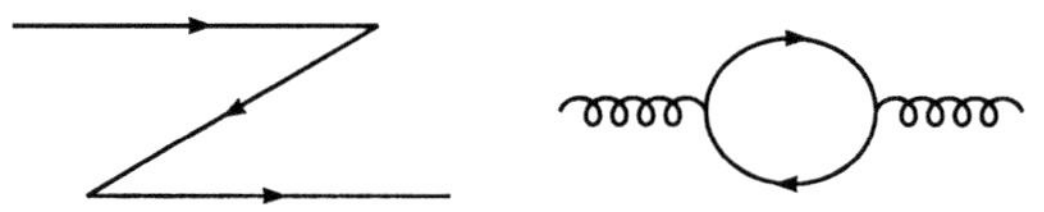

Figure 6. Virtual fluctuations involving pair creation of heavy quarks. In the first diagram, time flows to the right.

On a classical level, the heavy degrees of freedom represented by H_v can be eliminated using the equation of motion. Taking the variation of the Lagrangian with respect to the field $\bar{H}_v$, we obtain

$$(iv\cdot D + 2m_Q)\, H_v = i\not{D}_\perp h_v\,. \tag{44}$$

This equation can formally be solved to give

$$H_v = \frac{1}{2m_Q + iv\cdot D}\, i\not{D}_\perp h_v\,, \tag{45}$$

showing that the small-component field H_v is indeed of order $1/m_Q$. We can now insert this solution into (42) to obtain the "non-local effective Lagrangian"

$$\mathcal{L}_{\rm eff} = \bar{h}_v\, iv\cdot D\, h_v + \bar{h}_v\, i\not{D}_\perp\, \frac{1}{2m_Q + iv\cdot D}\, i\not{D}_\perp h_v\,. \tag{46}$$

Clearly, the second term corresponds to the first class of virtual processes shown in Fig. 6.

It is possible to derive this Lagrangian in a more elegant way by manipulating the generating functional for QCD Green's functions containing heavy-quark fields[77]. To this end, one starts from the field redefinition (41) and couples the large-component fields h_v to external sources ρ_v. Green's functions with an arbitrary number of h_v fields can be constructed by taking derivatives with respect to ρ_v. No sources are needed for the heavy degrees of freedom represented by H_v. The functional integral over these fields is Gaussian and can be performed explicitly, leading to the effective action

$$S_{\text{eff}} = \int \mathrm{d}^4x\, \mathcal{L}_{\text{eff}} - i \ln \Delta\,, \tag{47}$$

with $\mathcal{L}_{\text{eff}}$ as given in (46). The appearance of the logarithm of the determinant

$$\Delta = \exp\left(\frac{1}{2}\,\mathrm{Tr}\,\ln\left[2m_Q + iv\cdot D - i\eta\right]\right) \tag{48}$$

is a quantum effect not present in the classical derivation presented above. However, in this case the determinant can be regulated in a gauge-invariant way, and by choosing the axial gauge $v \cdot A = 0$ one shows that $\ln\Delta$ is just an irrelevant constant[77, 78].

Because of the phase factor in (41), the x dependence of the effective heavy-quark field h_v is weak. In momentum space, derivatives acting on h_v correspond to powers of the residual momentum k, which by construction is much smaller than m_Q. Hence, the non-local effective Lagrangian (46) allows for a derivative expansion in powers of iD/m_Q:

$$\mathcal{L}_{\text{eff}} = \bar{h}_v\, iv\cdot D\, h_v + \frac{1}{2m_Q} \sum_{n=0}^{\infty} \bar{h}_v\, i\not{D}_\perp \left(-\frac{iv\cdot D}{2m_Q}\right)^n i\not{D}_\perp h_v\,. \tag{49}$$

Taking into account that h_v contains a P_+ projection operator, and using the identity

$$P_+\, i\not{D}_\perp\, i\not{D}_\perp P_+ = P_+\left[(iD_\perp)^2 + \frac{g_s}{2}\,\sigma_{\mu\nu}\,G^{\mu\nu}\right] P_+\,, \tag{50}$$

where $[iD^\mu, iD^\nu] = ig_s G^{\mu\nu}$ is the gluon field-strength tensor, one finds that[75, 76]

$$\mathcal{L}_{\text{eff}} = \bar{h}_v\, iv\cdot D\, h_v + \frac{1}{2m_Q}\,\bar{h}_v\,(iD_\perp)^2\, h_v + \frac{g_s}{4m_Q}\,\bar{h}_v\,\sigma_{\mu\nu}\,G^{\mu\nu}\, h_v + O(1/m_Q^2)\,. \tag{51}$$

In the limit $m_Q \to \infty$, only the first terms remains:

$$\mathcal{L}_\infty = \bar{h}_v\, iv\cdot D\, h_v\,. \tag{52}$$

This is the effective Lagrangian of the HQET. It gives rise to the Feynman rules depicted in Fig. 7.

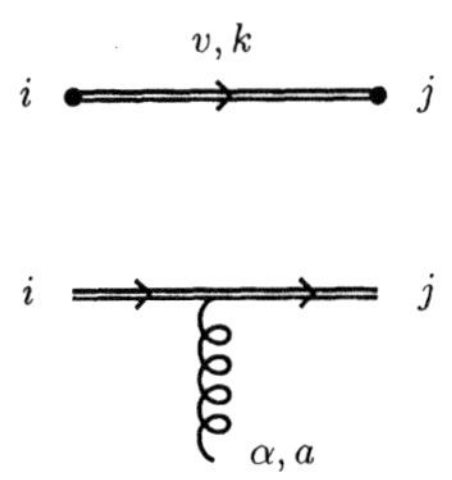

$$= \frac{i}{v\cdot k + i\eta}\,\frac{1+\not{v}}{2}\,\delta_{ji}$$

$$= i g_s\, v^\alpha\, (t_a)_{ji}$$

Figure 7. Feynman rules of the HQET (i, j and a are colour indices). A heavy quark is represented by a double line labelled by the velocity v and the residual momentum k. The velocity v is conserved by the strong interactions.

Let us take a moment to study the symmetries of this Lagrangian[27]. Since there appear no Dirac matrices, interactions of the heavy quark with gluons leave its spin unchanged. Associated with this is an SU(2) symmetry group, under which $\mathcal{L}_\infty$ is invariant. The action of this symmetry on the heavy-quark fields becomes most transparent in the rest frame, where the generators S^i of SU(2) can be chosen as

$$S^i = \frac{1}{2}\begin{pmatrix} \sigma^i & 0 \\ 0 & \sigma^i \end{pmatrix}, \qquad [S^i, S^j] = i\epsilon^{ijk} S^k . \tag{53}$$

Here σ^i are the Pauli matrices. An infinitesimal SU(2) transformation $h_v \to (1+i\vec{\epsilon}\cdot\vec{S}\,)\,h_v$ leaves the Lagrangian invariant:

$$\delta\mathcal{L}_\infty = \bar{h}_v\,[iv\cdot D, i\vec{\epsilon}\cdot\vec{S}\,]\,h_v = 0 . \tag{54}$$

Another symmetry of the HQET arises since the mass of the heavy quark does not appear in the effective Lagrangian. For N_h heavy quarks moving at the same velocity, eq. (52) can be extended by writing

$$\mathcal{L}_\infty = \sum_{i=1}^{N_h} \bar{h}_v^i\, iv\cdot D\, h_v^i . \tag{55}$$

This is invariant under rotations in flavour space. When combined with the spin symmetry, the symmetry group is promoted to SU($2N_h$). This is the heavy-quark spin–flavour symmetry[26, 27]. Its physical content is that, in the limit $m_Q \to \infty$, the strong interactions of a heavy quark become independent of its mass and spin.

Consider now the operators appearing at order $1/m_Q$ in the effective Lagrangian (51). They are easiest to identify in the rest frame. The first operator,

$$O_{\rm kin} = \frac{1}{2m_Q}\,\bar{h}_v\,(iD_\perp)^2\,h_v \to -\frac{1}{2m_Q}\,\bar{h}_v\,(i\vec{D}\,)^2\,h_v , \tag{56}$$

is the gauge-covariant extension of the kinetic energy arising from the off-shell residual motion of the heavy quark. The second operator is the non-abelian analogue of the Pauli interaction, which describes the chromo-magnetic coupling of the heavy-quark spin to the gluon field:

$$O_{\rm mag} = \frac{g_s}{4m_Q}\,\bar{h}_v\,\sigma_{\mu\nu}\,G^{\mu\nu}\,h_v \to -\frac{g_s}{m_Q}\,\bar{h}_v\,\vec{S}\cdot\vec{B}_c\,h_v . \tag{57}$$

Here $\vec{S}$ is the spin operator defined in (53), and $B_c^i = -\frac{1}{2}\epsilon^{ijk}G^{jk}$ are the components of the chromo-magnetic field. The chromo-magnetic interaction is a relativistic effect, which scales like $1/m_Q$. This is the origin of the heavy-quark spin symmetry.

Wave-Function Renormalization of the Heavy-Quark Field in the HQET

Besides being an effective theory for the strong interactions of heavy quarks with light degrees of freedom, the HQET is a consistent, renormalizable (order by order in $1/m_Q$) quantum field theory in its own right. In particular, it provides a framework for calculating radiative corrections. We shall discuss as an illustration the wave-function renormalization of the heavy-quark field h_v.

In quantum field theory, the parameters and fields of the Lagrangian have no direct physical significance. They have to be renormalized before they can be related to observable quantities. In an intermediate step the theory has to be regularized. The most convenient regularization scheme in QCD is dimensional regularization[79–81], in which the dimension of space-time is analytically continued to $D = 4 - 2\epsilon$, with ϵ being infinitesimal. Loop integrals that are logarithmically divergent in four dimensions become finite for $\epsilon > 0$. From the fact that the action $S = \int \mathrm{d}^D x\, \mathcal{L}(x)$ is dimensionless, one can derive the mass dimensions of the fields and parameters of the theory. For instance, one finds that the "bare" coupling constant $\alpha_s^{\rm bare}$ is no longer dimensionless if $D \neq 4$: $\dim[\alpha_s^{\rm bare}] = 2\epsilon$. In a renormalizable theory, it is possible to rewrite the Lagrangian in terms of renormalized quantities in such a way that Green's functions of the renormalized fields remain finite as $\epsilon \to 0$. For QCD, one introduces renormalized quantities by $Q^{\rm bare} = Z_Q^{1/2}\, Q^{\rm ren}$, $A^{\rm bare} = Z_A^{1/2} A^{\rm ren}$, $\alpha_s^{\rm bare} = \mu^{2\epsilon} Z_\alpha\, \alpha_s^{\rm ren}$, etc., where μ is an arbitrary mass scale introduced to render the renormalized coupling constant dimensionless. Similarly, in the HQET one defines the renormalized heavy-quark field by $h_v^{\rm bare} = Z_h^{1/2}\, h_v^{\rm ren}$. From now on, the superscript "ren" will be omitted.

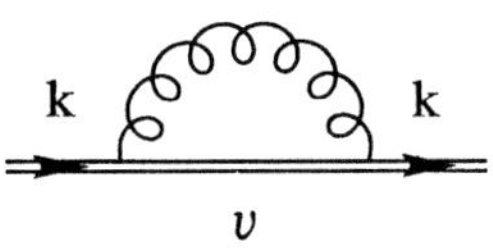

Figure 8. One-loop self-energy $-i\Sigma(v\cdot k)$ of a heavy quark in the HQET.

In the minimal subtraction (MS) scheme, Z_h can be computed from the $1/\epsilon$ pole in the heavy-quark self-energy using

$$1 - Z_h^{-1} = \frac{1}{\epsilon}\text{pole of } \frac{\partial\Sigma(v\cdot k)}{\partial v\cdot k}\,. \tag{58}$$

As long as $v\cdot k < 0$, the self-energy is infrared finite and real. The result is gauge-dependent, however. Evaluating the diagram shown in Fig. 8 in the Feynman gauge, we obtain at one-loop order

$$\begin{aligned}\Sigma(v\cdot k) &= -ig_s^2\, t_a t_a \int \frac{\mathrm{d}^D t}{(2\pi)^D}\, \frac{1}{(t^2+i\eta)[v\cdot(t+k)+i\eta]} \\ &= -2iC_F g_s^2 \int\limits_0^\infty \mathrm{d}\lambda \int \frac{\mathrm{d}^D t}{(2\pi)^D}\, \frac{1}{\left[t^2 + 2\lambda\, v\cdot(t+k) + i\eta\right]^2} \\ &= \frac{C_F\alpha_s}{2\pi}\,\Gamma(\epsilon) \int\limits_0^\infty \mathrm{d}\lambda \left(\frac{\lambda^2+\lambda\omega}{4\pi\mu^2}\right)^{-\epsilon},\end{aligned} \tag{59}$$

where $C_F = 4/3$ is a colour factor, λ is a dimensionful Feynman parameter, and $\omega = -2v\cdot k > 0$ acts as an infrared cutoff. A straightforward calculation leads to

$$\begin{aligned}\frac{\partial\Sigma(v\cdot k)}{\partial v\cdot k} &= \frac{C_F\alpha_s}{\pi}\,\Gamma(1+\epsilon)\left(\frac{\omega^2}{4\pi\mu^2}\right)^{-\epsilon} \int\limits_0^1 \mathrm{d}z\, z^{-1+2\epsilon}\,(1-z)^{-\epsilon} \\ &= \frac{C_F\alpha_s}{\pi}\,\Gamma(2\epsilon)\,\Gamma(1-\epsilon)\left(\frac{\omega^2}{4\pi\mu^2}\right)^{-\epsilon},\end{aligned} \tag{60}$$

where we have substituted $\lambda = \omega\,(1-z)/z$. From an expansion around $\epsilon = 0$, we obtain

$$Z_h = 1 + \frac{C_F\alpha_s}{2\pi\epsilon}\,. \tag{61}$$

This result was first derived by Politzer and Wise[71]. In the meantime, the calculation was also done at the two-loop order[82–85].

The Residual Mass Term and the Definition of the Heavy-Quark Mass

The choice of the expansion parameter in the HQET, i.e. the definition of the heavy-quark mass m_Q, deserves some comments. In the derivation presented earlier in this section, we chose m_Q to be the "mass in the Lagrangian", and using this parameter in the phase redefinition in (41) we obtained the effective Lagrangian (52), in which the heavy-quark mass no longer appears. However, this treatment has its subtleties. The symmetries of the HQET allow a "residual mass term" δm for the heavy quark, provided that δm is of order $\Lambda_{\rm QCD}$ and is the same for all heavy-quark flavours. Even if we arrange that such a term is not present at the tree level, it will in general be induced by quantum corrections. (This is unavoidable if the theory is regulated with a dimensionful cutoff.) Therefore, instead of (52) we should write the effective Lagrangian in the more general form[30]:

$$\begin{aligned} h_v(x) &= e^{i m_Q v\cdot x}\, P_+\, Q(x) \\ \Rightarrow \qquad \mathcal{L}_\infty &= \bar h_v\, iv\cdot D\, h_v - \delta m\, \bar h_v h_v\,. \end{aligned} \qquad (62)$$

If we redefine the expansion parameter according to $m_Q \to m_Q + \Delta m$, the residual mass changes in the opposite way: $\delta m \to \delta m - \Delta m$. This implies that there is a unique choice of the expansion parameter such that $\delta m = 0$. Requiring $\delta m = 0$, as it is usually done implicitly in the HQET, defines a heavy-quark mass, which in perturbation theory coincides with the pole mass[86]. This, in turn, defines for each heavy hadron a parameter $\bar\Lambda$ (sometimes called the "binding energy") through

$$\bar\Lambda = (m_H - m_Q)\Big|_{m_Q\to\infty}\,. \qquad (63)$$

If one prefers to work with another choice of the expansion parameter, the values of non-perturbative parameters such as $\bar\Lambda$ change, but at the same time one has to include the residual mass term in the HQET Lagrangian. It can be shown that the various parameters that depend on the definition of m_Q enter the predictions for all physical observables in such a way that the results are independent of which particular choice one adopts[30].

There is one more subtlety hidden in the above discussion. The quantities m_Q, $\bar\Lambda$ and δm are non-perturbative parameters of the HQET, which have a similar status as the vacuum condensates in QCD phenomenology[87]. These parameters cannot be defined unambiguously in perturbation theory. The reason lies in the divergent behaviour of perturbative expansions in large orders, which is associated with the existence of singularities along the real axis in the Borel plane, the so-called renormalons[88–96]. For instance, the perturbation series which relates the pole mass m_Q of a heavy quark to its bare mass,

$$m_Q = m_Q^{\rm bare}\left\{1 + c_1\,\alpha_s(m_Q) + c_2\,\alpha_s^2(m_Q) + \ldots + c_n\,\alpha_s^n(m_Q) + \ldots\right\}, \qquad (64)$$

contains numerical coefficients c_n that grow as $n!$ for large n, rendering the series divergent and not Borel summable[97, 98]. The best one can achieve is to truncate the perturbation series at the minimal term, but this leads to an unavoidable arbitrariness of order $\Delta m_Q \sim \Lambda_{\rm QCD}$ (the size of the minimal term). This observation, which at first sight seems a serious problem for QCD phenomenology, should actually not come as a surprise. We know that because of confinement quarks do not appear as physical states in nature. Hence, there is no way to define their on-shell properties such as a pole mass. In view of this, it is actually remarkable that QCD perturbation theory "knows" about its incompleteness and indicates, through the appearance of renormalon singularities, the presence of non-perturbative effects. We must first specify a scheme how to truncate the QCD perturbation series before non-perturbative statements such as $\delta m = 0$ become meaningful, and hence before non-perturbative parameters such as m_Q and $\bar{\Lambda}$ become well-defined quantities. The actual values of these parameters will depend on this scheme.

We stress that the "renormalon ambiguities" are not a conceptual problem for the heavy-quark expansion. In fact, it can be shown quite generally that these ambiguities cancel in all predictions for physical observables[99]. The way the cancellations occur is intricate, however. The generic structure of the heavy-quark expansion for an observable is of the form:

$$\text{observable} \sim C[\alpha_s(m_Q)] \left(1 + \frac{\Lambda}{m_Q} + \ldots\right). \tag{65}$$

Here $C[\alpha_s(m_Q)]$ represents a perturbative coefficient function, and Λ is a dimensionful non-perturbative parameter. The truncation of the perturbation series defining the coefficient function leads to an arbitrariness of order $\Lambda_{\rm QCD}/m_Q$, which precisely cancels against a corresponding arbitrariness of order $\Lambda_{\rm QCD}$ in the definition of the non-perturbative parameter Λ.

The renormalon problem poses itself when one imagines to apply perturbation theory in very high orders. In practise, the perturbative coefficients are known to finite order in α_s (at best to two-loop accuracy), and to be consistent one should use them in connection with the pole mass (and $\bar{\Lambda}$ etc.) defined to the same order.

MATCHING AND RUNNING

In the previous lecture, we have discussed the first two steps in the construction of the HQET. Integrating out the small components in the heavy-quark fields, a non-local effective action was derived, which was then expanded in a series of local operators. The effective Lagrangian derived that way correctly reproduces the long-distance physics of the full theory. It does not contain the short-distance physics correctly, however. The reason is obvious: A heavy quark participates in strong interactions through its coupling to gluons. These gluons can be soft or hard, i.e. their virtual momenta can be small, of the order of the confinement scale, or large, of the order of the heavy-quark mass. But hard gluons can resolve the spin and flavour quantum numbers of a heavy quark. Their effects lead to a renormalization of the coefficients of the operators in the HQET.

Consider, as an example, matrix elements of the vector current $V = \bar{q}\,\gamma^\mu Q$. In QCD this current is (partially) conserved and needs no renormalization[100]. Its matrix elements are free of ultraviolet divergences. Still, these matrix elements have a logarithmic dependence on m_Q from the exchange of hard gluons with virtual momenta of the order of the heavy-quark mass. If one goes over to the effective theory by taking the limit $m_Q \to \infty$, these logarithms diverge. Consequently, the vector current in the effective theory does require a renormalization[71]. Its matrix elements depend on an arbitrary renormalization scale μ, which separates the regions of short- and long-distance physics. If μ is chosen such that $\Lambda_{\rm QCD} \ll \mu \ll m_Q$, the effective coupling constant in the region between μ and m_Q is small, and perturbation theory can be used to compute the short-distance corrections. These corrections have to be added to the matrix elements of the effective theory, which contain the long-distance physics below the scale μ. Schematically, then, the relation between matrix elements in the full and in the effective theory is

$$\langle V(m_Q)\rangle_{\rm QCD} = C_0(m_Q,\mu)\,\langle V_0(\mu)\rangle_{\rm HQET} + \frac{C_1(m_Q,\mu)}{m_Q}\,\langle V_1(\mu)\rangle_{\rm HQET} + \dots\,, \qquad (66)$$

where we have indicated that matrix elements in the full theory depend on m_Q, whereas matrix elements in the effective theory are mass-independent, but do depend on the renormalization scale. The Wilson coefficients $C_i(m_Q,\mu)$ are defined by this relation. Order by order in perturbation theory, they can be computed from a comparison of the matrix elements in the two theories. Since the effective theory is constructed to reproduce correctly the low-energy behaviour of the full theory, this "matching" procedure is independent of any long-distance physics, such as infrared singularities, non-perturbative effects, the nature of the external states used in the matrix elements, etc.

The calculation of the coefficient functions in perturbation theory uses the powerful methods of the renormalization group. It is in principle straightforward, yet in practice rather tedious. A comprehensive discussion of most of the existing calculations of short-distance corrections in the HQET can be found in Ref. 8. Here, we shall discuss as an illustration the renormalization of the $1/m_Q$-suppressed operators in the effective Lagrangian (51). At the tree level, there appear two operators at order $1/m_Q$, which have been given in (56) and (57). Beyond the tree level, the coefficients of these operators may be modified, and other operators not present at the classical level may be induced. In general, we thus expect

$$\mathcal{L}_{1/m} = C_{\rm kin}(\mu)\,O_{\rm kin}(\mu) + C_{\rm mag}(\mu)\,O_{\rm mag}(\mu) + \text{new operators}\,, \qquad (67)$$

where μ is the renormalization scale. But how do we calculate the Wilson coefficient functions, and what are the possible new operators? To extend the classical construction of the previous lecture to include quantum corrections would be cumbersome. Fortunately, there is a systematic procedure which allows us to derive the result in the presence of quantum effects in a rather simple and straightforward way. It consists of three steps: construction of the operator basis, calculation of the "matching conditions" at $\mu = m_Q$, and renormalization-group improvement ("running"). Below, we shall first explain these steps in general and then illustrate them with the particular example of $\mathcal{L}_{1/m}$.

Construction of the Operator Basis

Similar to the fields and coupling constants, in a quantum field theory any composite operator built from quark and gluon fields may require a renormalization beyond that of its component fields. Such operators can be divided into three classes: gauge-invariant operators that do not vanish by the equations of motion (class-I), gauge-invariant operators that vanish by the equations of motion (class-II), and operators which are not gauge-invariant (class-III). In general, operators with the same dimension and quantum numbers mix under renormalization. However, things simplify if one works with the background field technique[101–104], which is an elegant method for quantizing gauge theories, preserving explicit gauge invariance. This offers the advantage that a class-I operator cannot mix with class-III operators, so that only gauge-invariant operators need to be considered[105]. Furthermore, class-II operators are irrelevant since their matrix elements vanish by the equations of motion. It it thus sufficient to consider class-I operators only.

Thus, we must find a complete set of class-I operators of the right dimension, carrying the quantum numbers allowed by the symmetries of the problem. In the case at hand, we are dealing with operators appearing at order $1/m_Q$ in a strong-interaction Lagrangian, and we thus have to find dimension-five operators containing two heavy-quark fields of the same velocity. Moreover, these operators must transform as scalars under the Lorentz group. The most general form of such operators is

$$\bar{h}_v\,\Gamma_{\mu\nu}\,iD^\mu iD^\nu\,h_v\,;\qquad \Gamma_{\mu\nu}\in\Big\{g_{\mu\nu},v_\mu v_\nu,\gamma_\mu v_\nu,\gamma_\nu v_\mu,\tfrac{1}{2}\left[\gamma_\mu,\gamma_\nu\right]\Big\}\,. \tag{68}$$

Note that the velocity is not a dynamical quantity in the HQET and thus can be used to construct the basis operators. Using that $\bar{h}_v\gamma_\mu\,h_v=\bar{h}_v\,v_\mu\,h_v$, we find that there are only three possible operators:

$$\bar{h}_v\,(iD)^2 h_v\,,\qquad \bar{h}_v\,(iv\!\cdot\! D)^2 h_v\,,\qquad \tfrac{1}{2}\,\bar{h}_v\,\sigma_{\mu\nu}g_s G^{\mu\nu}\,h_v\,. \tag{69}$$

Since the equation of motion of the HQET is $iv\!\cdot\! D\,h_v=0$, it follows that there are two class-I and one class-II operators, which we choose in the form:

$$\begin{aligned}&\text{class-I:} && \bar{h}_v\,(iD_\perp)^2 h_v\,, \qquad \tfrac{1}{2}\,\bar{h}_v\,\sigma_{\mu\nu}g_s G^{\mu\nu}\,h_v\,,\\ &\text{class-II:} && \bar{h}_v\,(iv\!\cdot\! D)^2 h_v\,.\end{aligned} \tag{70}$$

Besides a class-II operator, which has vanishing matrix elements between physical states, the kinetic and chromo-magnetic operators already present at the tree level are thus the only operators which can appear in $\mathcal{L}_{1/m}$, even in the presence of quantum corrections. Once we have found a complete basis of class-I operators, our next goal is to calculate their coefficient functions in perturbation theory.

Matching Conditions at $\mu=m_Q$

The Wilson coefficient functions $C_{\text{kin}}(\mu)$ and $C_{\text{mag}}(\mu)$ in (67) can be obtained from the comparison ("matching") of Green's functions in QCD with those in the effective theory. It is crucial that, by construction, the Wilson coefficients receive only short-distance contributions (see Fig. 5) and are thus insensitive to the properties of the external states. This ensures that once the coefficients have been determined by requiring that some particular Green's function(s) be the same in the two theories, all other Green's functions will be the same. Moreover, since the Wilson coefficients are infrared insensitive they are calculable in perturbation theory, and we can perform their calculation using quark and gluon states rather than physical hadron states.

In the example at hand, the coefficients $C_{\text{kin}}(\mu)$ and $C_{\text{mag}}(\mu)$ can be obtained from a calculation of the Green's function of two heavy quarks and a background gluon field, to one-loop order in the full and in the effective theory. The relevant vertex diagrams in QCD are shown in Fig. 9. They have to be supplemented by the wave-function renormalization of the external quark lines. The background field is not renormalized. The momentum assignments are such that p is the outgoing momentum of the background field, and k and $(k-p)$ are the residual momenta of the heavy quarks. To order $1/m_Q$, it is sufficient to keep terms linear in k or p. The quarks can be taken on shell, in which case $v\cdot k=v\cdot p=0$. A subtlety which has to be taken into account is that the QCD spinor $u_Q(P_Q,s)$ is related to the spinor $u_h(v,s)$ of the effective theory by

$$u_Q(P_Q,s)=\left(1+\frac{\not{k}}{2m_Q}+\dots\right)u_h(v,s)\,, \tag{71}$$

where $P_Q = m_Q v + k$. In the matching calculation one has to use the same spinors in both theories. We thus define a vertex function Γ^μ by writing the amplitude as $i\mu^\epsilon g_s A_{\mu,a}(p)\, \bar{u}_h \Gamma^\mu t_a u_h$, so that at the tree level in QCD

$$\begin{aligned}\Gamma^\mu_{\text{QCD},0} &= \left(1 + \frac{\not{k} - \not{p}}{2m_Q}\right)\gamma^\mu\left(1 + \frac{\not{k}}{2m_Q}\right) + \dots \\ &= v^\mu + \frac{(2k-p)^\mu}{2m_Q} + \frac{[\gamma^\mu, \not{p}]}{4m_Q} + \dots . \end{aligned} \tag{72}$$

Here the ellipses represent terms of higher order in k or p, and we have used that between the heavy-quark spinors γ^μ can be replaced by v^μ.

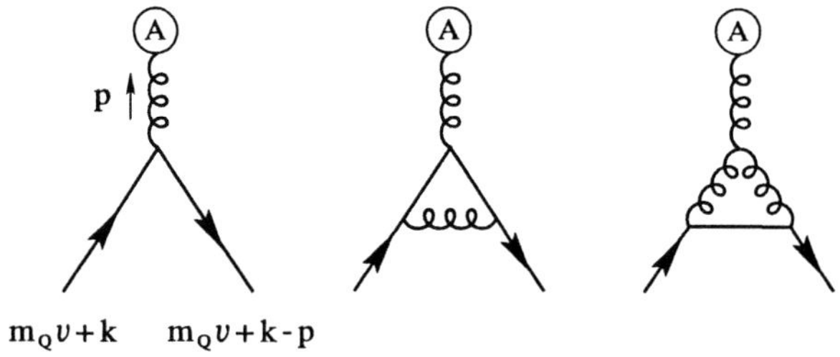

Figure 9. Diagrams for the calculation of the heavy quark-gluon vertex function in QCD. The background field is denoted by A.

The contributions to the vertex function arising at the one-loop level are also shown in Fig. 9. They contain both abelian and non-abelian vertices. Since the matching calculation is insensitive to any long-distance properties such as the nature of the infrared regulator, it is legitimate to work with any infrared regularization scheme that is convenient. Following Eichten and Hill[75, 106], we choose to regulate both ultraviolet and infrared divergences using dimensional regularization. Moreover, we expand the resulting expressions for the Feynman amplitudes to linear order in the external momenta and then set the external momenta to zero inside the loop integrals. Then the only mass scale remaining is the heavy-quark mass. In the $\overline{\text{MS}}$ scheme, the result for the one-loop contribution to the QCD vertex function is[75]

$$\Gamma^\mu_{\text{QCD},1} = \frac{[\gamma^\mu, \not{p}]}{4m_Q}\,\frac{\alpha_s}{2\pi}\left(-C_A \ln\frac{m_Q}{\mu} + C_A + C_F\right), \tag{73}$$

where $C_F = \frac{1}{2}(N_c^2 - 1)/N_c = 4/3$ and $C_A = N_c = 3$ are the eigenvalues of the quadratic Casimir operator in the fundamental and the adjoint representations.

Now comes a clue: if dimensional regularization is used to regulate both ultraviolet and infrared singularities, all loop integrals in the HQET are no-scale integrals (after a power of the external momenta has been factored out) and vanish! So only the tree-level matrix elements of the HQET operators in the effective Lagrangian multiplied by their Wilson coefficient functions remain. This is why dimensional regularization is superb for matching calculations. The result is

$$\Gamma^\mu_{\text{HQET}} = v^\mu + C_{\text{kin}}(\mu)\,\frac{(2k-p)^\mu}{2m_Q} + C_{\text{mag}}(\mu)\,\frac{[\gamma^\mu, \not{p}]}{4m_Q} + \dots . \tag{74}$$

Requiring that the vertex functions be the same in the full and in the effective theory, we find (in the $\overline{\text{MS}}$ scheme)

$$C_{\text{kin}}(\mu) = 1\,, \qquad C_{\text{mag}}(\mu) = 1 + \frac{\alpha_s}{2\pi}\left(-C_A \ln\frac{m_Q}{\mu} + C_A + C_F\right). \tag{75}$$

The fact that the kinetic operator is not renormalized is not an accident, but follows from an invariance of the HQET under small redefinitions of the velocity used in the construction of the effective Lagrangian. Clearly, the predictions of the HQET should not depend on whether v is taken to be the velocity of the hadron containing the heavy quark, the velocity of the heavy quark itself, or some other velocity differing from the hadron velocity by an amount of order Λ_{QCD}/m_Q. This so-called reparametrization invariance implies that $C_{kin}(\mu) = 1$ must hold to all orders in perturbation theory[107, 108].

In the next paragraph, we will see that the scale dependence predicted by the one-loop result quoted above cannot be trusted if $\mu \ll m_Q$; however, what can be obtained from the matching calculation are the values of the coefficient functions at the matching scale $\mu = m_Q$ as well as their logarithmic derivatives. For the coefficient of the chromo-magnetic operator, we find:

$$\begin{aligned} C_{mag}(m_Q) &= 1 + (C_A + C_F)\,\frac{\alpha_s(m_Q)}{2\pi}\,, \\ \frac{\mathrm{d}\ln C_{mag}(\mu)}{\mathrm{d}\ln\mu} &= C_A\,\frac{\alpha_s}{2\pi}\,. \end{aligned} \tag{76}$$

Renormalization-Group Evolution

The one-loop calculation presented above allows us to derive expressions for the Wilson coefficient functions provided that $m_Q/\mu = O(1)$. In practical applications of effective field theories, one is however often interested in the case where there is a large ratio of mass scales. After all, an effective theory is constructed to separate the physics on two very different energy scales. In such a situation, the coefficient functions contain large logarithms of the type $[\alpha_s \ln(m_Q/\mu)]^n$, which must be summed to all orders in perturbation theory. This is achieved by using the powerful machinery of the renormalization group.

For a set $\{O_i\}$ of n class-I operators that mix under renormalization, one defines an $n \times n$ matrix of renormalization factors Z_{ij} by $O_i^{bare} = Z_{ij}\,O_j(\mu)$, such that the matrix elements of the renormalized operators $O_j(\mu)$ remain finite as $\epsilon \to 0$. In contrast to the bare operators, the renormalized ones depend on the subtraction scale via the μ dependence of Z_{ij}:

$$\mu\frac{\mathrm{d}}{\mathrm{d}\mu}\,O_i(\mu) = \left(\mu\frac{\mathrm{d}}{\mathrm{d}\mu}Z_{ij}^{-1}\right) O_j^{bare} = -\gamma_{ik}\,O_k(\mu)\,, \tag{77}$$

where

$$\gamma_{ik} = -\left(\mu\frac{\mathrm{d}}{\mathrm{d}\mu}Z_{ij}^{-1}\right) Z_{jk} = Z_{ij}^{-1}\,\mu\frac{\mathrm{d}}{\mathrm{d}\mu}\,Z_{jk} \tag{78}$$

are called the anomalous dimensions. That under a change of the renormalization scale the operators mix among themselves follows from the fact that the basis of operators is complete. It is convenient to introduce a compact matrix notation, in which $\vec{O}(\mu)$ is the vector of renormalized operators, $\hat{Z}$ is the matrix of renormalization factors, and $\hat{\gamma}$ denotes the anomalous dimension matrix. Then the scale dependence of the renormalized operators is controlled by the renormalization-group equation (RGE)

$$\left(\mu\frac{\mathrm{d}}{\mathrm{d}\mu} + \hat{\gamma}\right)\vec{O}(\mu) = 0\,. \tag{79}$$

In the MS scheme, the matrix $\hat{Z}$ obeys an expansion of the form

$$\hat{Z} = 1 + \sum_{k=1}^{\infty}\frac{1}{\epsilon^k}\,\hat{Z}_k(\alpha_s)\,, \tag{80}$$

and by requiring that the anomalous dimensions in (78) be finite as $\epsilon \to 0$ one finds that $\hat{\gamma}$ can be computed in terms of the coefficient of the $1/\epsilon$ pole[109]:

$$\hat{\gamma} = -2\alpha_s \, \frac{\partial \hat{Z}_1(\alpha_s)}{\partial \alpha_s} \, . \tag{81}$$

The same relation holds in the $\overline{\text{MS}}$ scheme.

From (79) and the fact that the product $C_i(\mu)\, O_i(\mu)$ must be μ independent, we derive the RGE satisfied by the coefficient functions. It reads

$$\left(\mu \frac{\mathrm{d}}{\mathrm{d}\mu} - \hat{\gamma}^T \right) \vec{C}(\mu) = 0 \, , \tag{82}$$

where we have collected the coefficients into a vector $\vec{C}(\mu)$. In general, the Wilson coefficients can depend on μ both explicitly or implicitly through the running coupling. We thus have

$$\mu \frac{\mathrm{d}}{\mathrm{d}\mu} = \mu \frac{\partial}{\partial \mu} + \beta(\alpha_s) \, \frac{\partial}{\partial \alpha_s(\mu)} \, , \tag{83}$$

where the β function

$$\beta(\alpha_s) = \mu \, \frac{\partial \alpha_s(\mu)}{\partial \mu} = -2\alpha_s \left[\beta_0 \, \frac{\alpha_s}{4\pi} + \beta_1 \left(\frac{\alpha_s}{4\pi} \right)^2 + \dots \right] \tag{84}$$

describes the scale dependence of the renormalized coupling constant. The one- and two-loop coefficients are scheme independent and are given by[18, 19, 110]

$$\begin{aligned} \beta_0 &= \frac{11}{3} \, C_A - \frac{4}{3} \, T_F \, n_f \, , \\ \beta_1 &= \frac{34}{3} \, C_A^2 - \left(\frac{20}{3} \, C_A + 4C_F \right) T_F \, n_f \, , \end{aligned} \tag{85}$$

where n_f is the number of light quark flavours, and $T_F = 1/2$ is the normalization of the SU(3) generators in the fundamental representation: $\mathrm{tr}(t_a t_b) = T_F \, \delta_{ab}$. It is now straightforward to obtain a formal solution of the RGE. It reads

$$\vec{C}(\mu) = \hat{U}(\mu, m_Q) \, \vec{C}(m_Q) \, , \tag{86}$$

with the evolution matrix[111–113]

$$\hat{U}(\mu, m_Q) = T_\alpha \, \exp \int\limits_{\alpha_s(m_Q)}^{\alpha_s(\mu)} \mathrm{d}\alpha \, \frac{\hat{\gamma}^T(\alpha)}{\beta(\alpha)} \, . \tag{87}$$

Here "T_α" means an ordering in the coupling constant such that the couplings increase from right to left (for $\mu < m_Q$). This is necessary since, in general, the anomalous dimension matrices at different values of α_s do not commute. Eq. (87) can be solved perturbatively by expanding the β function (84) and the anomalous dimension matrix in powers of the renormalized coupling constant:

$$\hat{\gamma}(\alpha_s) = \hat{\gamma}_0 \, \frac{\alpha_s}{4\pi} + \hat{\gamma}_1 \left(\frac{\alpha_s}{4\pi} \right)^2 + \dots \, . \tag{88}$$

Here we shall only discuss the important case of a single coefficient function, or equivalently, when there is no operator mixing. Then the matrix $\hat{\gamma}$ reduces to a number, and the evolution is described by a function $U(\mu, m_Q)$, for which the perturbative solution of (87) at next-to-leading order yields

$$U_{\rm NLO}(\mu, m_Q) = \left(\frac{\alpha_s(m_Q)}{\alpha_s(\mu)}\right)^a \left\{1 + \frac{\alpha_s(m_Q) - \alpha_s(\mu)}{4\pi}\, S + \dots\right\}, \tag{89}$$

with

$$a = \frac{\gamma_0}{2\beta_0}, \qquad S = \frac{\gamma_1}{2\beta_0} - \frac{\gamma_0\beta_1}{2\beta_0^2}. \tag{90}$$

The theoretical framework discussed here is called "renormalization-group (RG) improved perturbation theory". In the expression for the evolution function $U(\mu, m_Q)$, there are no large logarithms of the form $\alpha_s \ln(m_Q/\mu)$ left. They are all contained in the ratio of the running couplings evaluated at the scales m_Q and μ. Thus, RG-improved perturbation theory provides the optimal method to bridge wide energy intervals. (As a side remark, we note that the same technique is used to control the evolution of gauge couplings and running mass parameters from low energies up to very high energy scales characteristic of grand unified theories.) The terms shown explicitly in (89) correspond to the so-called next-to-leading order (NLO) in RG-improved perturbation theory. In this approximation, the leading and subleading logarithms $[\alpha_s \ln(m_Q/\mu)]^n$ and $\alpha_s[\alpha_s \ln(m_Q/\mu)]^n$ are summed correctly to all orders in perturbation theory. To achieve this, it is necessary to calculate the two-loop coefficient γ_1 of the anomalous dimension. When γ_1 is not known, it is only possible to evaluate the evolution function in the so-called leading logarithmic order (LO), in which

$$U_{\rm LO}(\mu, m_Q) = \left(\frac{\alpha_s(m_Q)}{\alpha_s(\mu)}\right)^a. \tag{91}$$

This still sums the leading logarithms to all orders, but does not contain the non-logarithmic terms of order α_s.

To complete the calculation of the RG-improved coefficient function, the evolution function $U(\mu, m_Q)$ must be combined with the initial condition for the Wilson coefficient at the high energy scale $\mu = m_Q$, as shown in (86). If the operator under consideration is present at the tree level, the matching condition can be written in the form

$$C(m_Q) = 1 + c_1\, \frac{\alpha_s(m_Q)}{4\pi} + \dots, \tag{92}$$

where c_1 is obtained from a one-loop calculation. To obtain a consistent (i.e. renormalization-scheme independent) result at next-to-leading order, we have to combine the one-loop matching condition with the expression for $U(\mu, m_Q)$ given in (89). This requires the calculation of the two-loop anomalous dimension. The result is

$$C_{\rm NLO}(\mu) = \left(\frac{\alpha_s(m_Q)}{\alpha_s(\mu)}\right)^a \left\{1 + \frac{\alpha_s(m_Q)}{4\pi}\,(S + c_1) - \frac{\alpha_s(\mu)}{4\pi}\, S\right\}. \tag{93}$$

In this expression, the terms involving the coupling constant $\alpha_s(m_Q)$ are renormalization-scheme independent[111, 112]. The exponent a involves only the one-loop coefficients γ_0 and β_0 and is scheme independent by itself. For the coefficient $(S + c_1)$ of the next-to-leading term things are more complicated, however. The one-loop matching coefficient c_1, the two-loop anomalous dimension γ_1, and the QCD scale parameter $\Lambda_{\rm QCD}$ in the expression for the running coupling constant are all scheme dependent, but they conspire to give $\alpha_s(m_Q)$ a scheme-independent coefficient. On the other hand, the coefficient S of $\alpha_s(\mu)$ does depend on the renormalization procedure. This is not a surprise; only when the μ-dependent terms in the Wilson coefficient are combined with the μ-dependent matrix elements of the renormalized operator one can expect to obtain a scheme-independent result. For this reason, it is sometimes useful to factorize the solution (93) in the form $C(\mu) \equiv \widehat{C}(m_Q)\,K(\mu)$, where $\widehat{C}(m_Q)$ is RG-invariant and contains all dependence on the large mass scale m_Q. The scheme-dependent function $K(\mu)$ can be used to define a RG-invariant renormalized operator: $\widehat{\mathcal{O}} \equiv K(\mu)\,\mathcal{O}(\mu)$. At next-to-leading order, we obtain:

$$\widehat{C}(m_Q) = [\alpha_s(m_Q)]^a \left\{ 1 + \frac{\alpha_s(m_Q)}{4\pi}\,(S + c_1) \right\},$$
$$\widehat{\mathcal{O}} = [\alpha_s(\mu)]^{-a} \left\{ 1 - \frac{\alpha_s(\mu)}{4\pi}\,S \right\} \mathcal{O}(\mu)\,. \tag{94}$$

Let us finally apply this formalism to the operators appearing at order $1/m_Q$ in the effective Lagrangian of the HQET. The fact that reparametrization invariance ensures that the kinetic operator is not renormalized implies that the 2×2 anomalous dimension matrix for the operators $\mathcal{O}_{\mathrm{kin}}$ and $\mathcal{O}_{\mathrm{mag}}$ is diagonal and of the form

$$\hat{\gamma} = \begin{pmatrix} 0 & 0 \\ 0 & \gamma^{\mathrm{mag}} \end{pmatrix}. \tag{95}$$

The one-loop coefficient of the anomalous dimension of the chromo-magnetic operator, together with the one-loop matching coefficient, can be obtained from (76):

$$\gamma_0^{\mathrm{mag}} = 2C_A\,, \qquad c_1^{\mathrm{mag}} = 2(C_A + C_F)\,. \tag{96}$$

The result for γ_0^{mag} can also be obtained in a simpler way by computing only the $1/\epsilon$ poles in the matrix elements of the bare operators[76]. Unfortunately, the two-loop coefficient γ_1^{mag} is not yet known.[§] This means that the coefficient S_{mag} in the next-to-leading order solution (93) is still unknown.

Renormalization of Heavy-Quark Currents

As a final example, we discuss the renormalization of local current operators involving two heavy-quark fields. This case is of particular importance for phenomenology, as the weak current for $b \to c\,\ell\,\bar{\nu}$ transitions is of this form. In the HQET, the relevant current operators contain two heavy-quark fields at different velocity and are thus of the form $\bar{h}_{v'}\Gamma\,h_v$ (it does not matter whether the two fields have the same flavour), where Γ is some Dirac matrix, whose structure is irrelevant to our discussion. We have discussed in the first lecture that the matrix elements of such operators between meson states are proportional to the universal Isgur–Wise form factor $\xi(v\cdot v')$. We shall now derive, in leading logarithmic order, the Wilson coefficient function that relates the QCD current operators with their HQET counterparts renormalized at the scale $\mu \ll m_Q$.

To this end, we need to calculate, using dimensional regularization, the $1/\epsilon$ pole in the matrix element of the bare current operator between quark states. The relevant vertex diagram is shown in Fig. 10. Since in the effective theory the coupling of a heavy quark to a gluon does not involve a γ matrix, it is easy to see that to all

[§] This is one of the few cases of interest where an anomalous dimension in the HQET is not yet known to two-loop order. If you feel strong enough, you are invited to try the calculation of γ_1^{mag}!

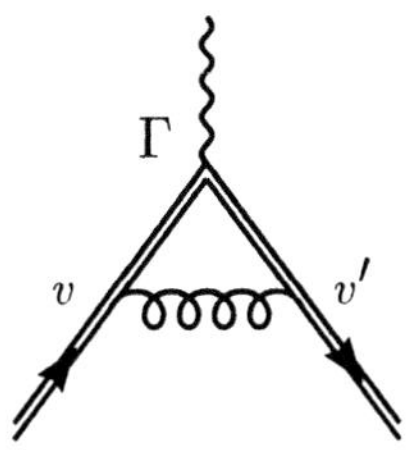

Figure 10. One-loop vertex diagram arising in the calculation of the anomalous dimension of heavy-quark currents. The external current changes the heavy-quark velocity from v to v'.

orders in perturbation theory the operator $\bar{h}_{v'}\Gamma\, h_v$ is renormalized multiplicatively and irrespective of its Dirac structure. The extraction of the one-loop ultraviolet divergence can be done in a few lines. In the Feynman gauge,¶ the value of the vertex diagram is (omitting the quark spinors):

$$\begin{aligned}
&-4ig_s^2 t_a t_a\, v\cdot v'\,\Gamma \int \frac{\mathrm{d}^D t}{(2\pi)^D}\, \frac{1}{(t^2+i\eta)(v\cdot t+i\eta)(v'\cdot t+i\eta)} \\
&= -4ig_s^2 C_F\, v\cdot v'\,\Gamma \int\limits_0^\infty \mathrm{d}\lambda \int\limits_0^\infty \mathrm{d}\rho \int \frac{\mathrm{d}^D t}{(2\pi)^D}\, \frac{1}{\left[t^2+2(\rho v+\lambda v')\cdot t+i\eta\right]^3} \\
&= -\frac{C_F\alpha_s}{\pi}\,\Gamma(1+\epsilon)\, v\cdot v'\,\Gamma\,(4\pi\mu^2)^\epsilon \int\limits_0^\infty \mathrm{d}\lambda \int\limits_0^\infty \mathrm{d}\rho\, \frac{1}{\left(\rho^2+\lambda^2+2v\cdot v'\,\rho\lambda\right)^{1+\epsilon}}\,.
\end{aligned} \tag{97}$$

Defining a new variable $z=\rho/\lambda$, and introducing an arbitrary infrared cutoff δ, we can rewrite the double integral in the form:

$$\int\limits_0^\infty \mathrm{d}\lambda\, \frac{\lambda}{(\lambda^2+\delta^2)^{1+\epsilon}} \int\limits_0^\infty \mathrm{d}z\, \frac{1}{(1+z^2+2wz)^{1+\epsilon}} = \frac{\delta^{-2\epsilon}}{2\epsilon}\, r(w) + \text{finite terms}, \tag{98}$$

where $w=v\cdot v'$, and

$$r(w) = \int\limits_0^\infty \mathrm{d}z\, \frac{1}{1+z^2+2wz} = \frac{1}{\sqrt{w^2-1}}\, \ln\left(w+\sqrt{w^2-1}\right). \tag{99}$$

Hence, the $1/\epsilon$ pole of the vertex diagram is given by

$$-\frac{C_F\alpha_s}{2\pi\epsilon}\,\Gamma\, w\, r(w)\,. \tag{100}$$

To obtain the renormalization constant Z_{hh} of the bare current operator, we have to add a contribution

$$Z_h\,\Gamma = \left(1+\frac{C_F\alpha_s}{2\pi\epsilon}\right)\Gamma \tag{101}$$

from the wave-function renormalization of the heavy-quark fields, where Z_h has been given in (61). The result is

$$Z_{hh} = 1 - \frac{C_F\alpha_s}{2\pi\epsilon}\,[w\, r(w) - 1] + \text{finite terms}. \tag{102}$$

¶The final result for the anomalous dimension is gauge independent.

By means of the relation (81), we derive from this the one-loop coefficient of the anomalous dimension of heavy-quark currents in the HQET. This is the famous velocity-dependent anomalous dimension obtained by Falk et al.[74]:

$$\gamma_0^{\rm hh}(w) = 4C_F\,[w\,r(w) - 1]\;, \tag{103}$$

In the zero-recoil limit, i.e. for $w = 1$, the heavy-quark currents are the symmetry currents of the spin–flavour symmetry, and as such they are not renormalized, since the associated charges are conserved. This implies that

$$\gamma^{\rm hh}(1) = 0 \tag{104}$$

to all orders in perturbation theory. This constraint is satisfied by the one-loop result in (103), since $r(1) = 1$.

Since in the effective theory the velocity of a heavy quark is conserved by the strong interactions, the heavy quark can be described by a Wilson line[67]. An external current can instantaneously change the velocity, resulting in a kink of that line. It is well-known that such cusps lead to singular behaviour. The renormalization of cusp singularities of Wilson lines was investigated in detail by Korchemsky and Radyushkin[114] already in 1987, prior to the development of the HQET. In particular, they calculated the one- and two-loop coefficients of the so-called cusp anomalous dimension $\gamma^{\rm cusp}(\varphi)$ as a function of the hyperbolic cusp angle φ. But this anomalous dimension is precisely that of heavy-quark currents[115], with the identification $\cosh\varphi = w$. Later, the result for $\gamma_1^{\rm hh}(w)$, which we will not present here, has been confirmed in the context of the HQET[116].

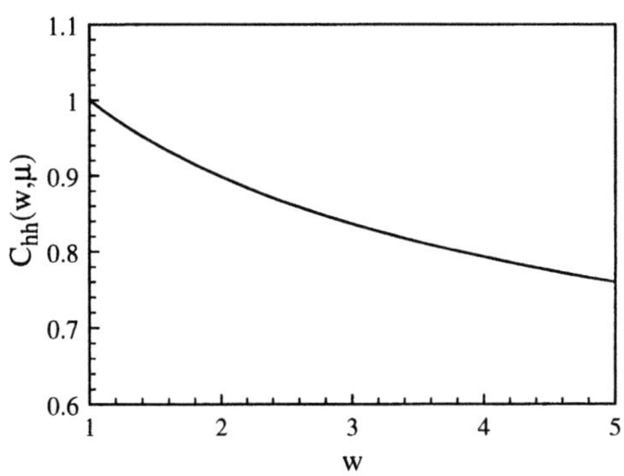

Figure 11. Velocity dependence of the Wilson coefficient $C_{\rm hh}(w,\mu)$, evaluated for $\alpha_s(\mu)/\alpha_s(m_Q) = 2$ and $n_f = 3$.

In leading logarithmic order, the expansion of heavy-quark currents in the HQET takes the form

$$\bar Q\,\Gamma\,Q \to C_{\rm hh}(w,\mu)\,\bar h_{v'}\Gamma\,h_v + O(1/m_Q)\,, \tag{105}$$

where

$$C_{\rm hh}(w,\mu) = \left(\frac{\alpha_s(m_Q)}{\alpha_s(\mu)}\right)^{a_{\rm hh}(w)}, \qquad a_{\rm hh}(w) = \frac{2C_F}{\beta_0}\,[w\,r(w) - 1]\,. \tag{106}$$

The velocity dependence of the Wilson coefficient is illustrated in Fig. 11. Physical decay or scattering amplitudes are proportional to the product

$$C_{\rm hh}(w,\mu)\,\xi(w,\mu)\,, \tag{107}$$

where $\xi(w,\mu)$ is the renormalized Isgur–Wise function, which satisfies $\xi(1,\mu)=1$. The fact that $a_{\rm hh}(1)=0$ implies that RG effects respect the normalization of form factors at zero recoil.

We like to finish this discussion with a curious remark, which illustrates the power of RG methods. For values $w=O(1)$, the leading-order expression for the coefficient $C_{\rm hh}(w,\mu)$ in (106) contains all large logarithms of the type $[\alpha_s \ln(m_Q/\mu)]^n$. For very large values of w, however, we have $w\,r(w)\to\ln(2w)-1$, so that in the RG improvement of $C_{\rm hh}(w,\mu)$ we have resummed terms of the form $[\alpha_s \ln w \ln(m_Q/\mu)]^n$. These are the well-known Sudakov double logarithms, which arise from the emission of gluon bremsstrahlungs during the scattering of the heavy quarks. This effect leads to a fractional power-like damping of the transition form factors at large recoil, which adds to the "soft" suppression contained in the Isgur–Wise form factor itself[117]. Explicitly, we obtain

$$C_{\rm hh}(w,\mu)\to\left(\frac{e}{2w}\right)^{\eta}, \qquad \eta=\frac{2C_F}{\beta_0}\ln\frac{\alpha_s(\mu)}{\alpha_s(m_Q)}. \tag{108}$$

CONCLUDING REMARKS

We have presented an introduction to heavy-quark symmetry, the heavy-quark effective theory and the $1/m_Q$ expansion, which provide the modern theoretical tools to perform quantitative calculations in heavy-flavour physics. Our hope was to convince the reader that heavy-flavour physics is a rich and diverse area of research, which is at present characterized by a fruitful interplay between theory and experiments. This has led to many significant discoveries and developments on both sides. Heavy-quark physics has the potential to determine many important parameters of the electroweak theory and to test the Standard Model at low energies. At the same time, it provides an ideal laboratory to study the nature of non-perturbative phenomena in QCD, still one of the least understood properties of the Standard Model.

Let us finish with a somewhat philosophical remark: At this school, we have heard a lot about exciting new developments related to dualities, which relate apparently very different theories to each other. So are electric, weak-coupling phenomena in one theory dual to magnetic, strong-coupling phenomena in another theory. Some people argue quite convincingly that duality seems to be everywhere in nature, and consequently there are no really difficult questions in physics; very difficult problems become trivial when approached from a different, dual point of view. There are, however, "moderately difficult" problems in physics, which are "self-dual". It is the author's opinion that real-world (i.e. non-supersymmetric) QCD at hadronic energies belongs to this category. Having said this, we conclude that heavy-quark effective theory provides a powerful tool to tackle the "moderately difficult" problems of heavy-flavour physics.

Acknowledgments

It is my pleasure to thank the organizers and the staff of the Cargèse Summer Institute on Particle Physics for the invitation to present these lectures and for making my stay in Cargèse such an enjoyable one.

REFERENCES

1. C. Albajar et al. (UA1 Collaboration), Phys. Lett. B **186**, 247 (1987).
2. H. Albrecht et al. (ARGUS Collaboration), Phys. Lett. B **192**, 245 (1987).
3. H. Albrecht et al. (ARGUS Collaboration), Phys. Lett. B **234**, 409 (1990); **255**, 297 (1991).

4. R. Fulton et al. (CLEO Collaboration), Phys. Rev. Lett. **64**, 16 (1990);
J. Bartelt et al. (CLEO Collaboration), Phys. Rev. Lett. **71**, 4111 (1993).
5. J.P. Alexander et al. (CLEO Collaboration), Phys. Rev. Lett. **77**, 5000 (1996).
6. R. Ammar et al. (CLEO Collaboration), Phys. Rev. Lett. **71**, 674 (1993).
7. M.S. Alam et al. (CLEO Collaboration), Phys. Rev. Lett. **74**, 2885 (1995).
8. M. Neubert, Phys. Rep. **245**, 259 (1994); Int. J. Mod. Phys. A **11**, 4173 (1996).
9. M. Shifman, in: QCD and Beyond, Proceedings of the Theoretical Advanced Study Institute in Elementary Particle Physics, Boulder, Colorado, June 1995, edited by D.E. Soper (World Scientific, Singapore, 1996), p. 409.
10. H. Georgi, in: Perspectives in the Standard Model, Proceedings of the Theoretical Advanced Study Institute in Elementary Particle Physics (TASI-91), Boulder, Colorado, 1991, edited by R.K. Ellis, C.T. Hill, and J.D. Lykken (World Scientific, Singapore, 1992), p. 589.
11. N. Isgur and M.B. Wise, in: Heavy Flavours, edited by A.J. Buras and M. Lindner (World Scientific, Singapore, 1992), p. 234.
12. B. Grinstein, Ann. Rev. Nucl. Part. Sc. **42**, 101 (1992).
13. T. Mannel, Chinese J. Phys. **31**, 1 (1993).
14. A.G. Grozin, preprint Budker INP 92-97 (1992), unpublished.
15. K. Wilson, Phys. Rev. **179**, 1499 (1969); D **3**, 1818 (1971).
16. W. Zimmermann, Ann. Phys. **77**, 536 and 570 (1973).
17. For a pedagocical introduction, see the lectures by H. Leutwyler in this volume.
18. D.J. Gross and F. Wilczek, Phys. Rev. Lett. **30**, 1343 (1973).
19. H.D. Politzer, Phys. Rev. Lett. **30**, 1346 (1973).
20. The increase of the running coupling constant at low energies, as predicted by QCD, has been observed by analysing the hadronic-mass distribution in decays of the τ lepton, see: M. Girone and M. Neubert, Phys. Rev. Lett. **76**, 3061 (1996).
21. E.V. Shuryak, Phys. Lett. B **93**, 134 (1980); Nucl. Phys. B **198**, 83 (1982).
22. J.E. Paschalis and G.J. Gounaris, Nucl. Phys. B **222**, 473 (1983);
F.E. Close, G.J. Gounaris, and J.E. Paschalis, Phys. Lett. B **149**, 209 (1984).
23. S. Nussinov and W. Wetzel, Phys. Rev. D **36**, 130 (1987).
24. M.B. Voloshin and M.A. Shifman, Yad. Fiz. **45**, 463 (1987) [Sov. J. Nucl. Phys. **45**, 292 (1987)].
25. M.B. Voloshin and M.A. Shifman, Yad. Fiz. **47**, 801 (1988) [Sov. J. Nucl. Phys. **47**, 511 (1988)].
26. N. Isgur and M.B. Wise, Phys. Lett. B **232**, 113 (1989); **237**, 527 (1990).
27. H. Georgi, Phys. Lett. B **240**, 447 (1990).
28. N. Isgur and M.B. Wise, Phys. Rev. Lett. **66**, 1130 (1991).
29. A.F. Falk, Nucl. Phys. B **378**, 79 (1992).
30. A.F. Falk, M. Neubert, and M.E. Luke, Nucl. Phys. B **388**, 363 (1992).
31. A.F. Falk and M. Neubert, Phys. Rev. D **47**, 2965 and 2982 (1993).
32. I.J. Kroll, in: Proceedings of the 17th International Symposium on Lepton–Photon Interactions (LP95), Beijing, P.R. China, August 1995, edited by Z. Zhi-Peng and C. He-Sheng (World Scientific, Singapore, 1996), p. 204.
33. F. Ukegawa (for the CDF Collaboration), preprint FERMILAB-Conf-96/155-E (1996), to appear in: Proceedings of the XIth Topical Workshop on $p\bar{p}$ Collider Physics, Padova, Italy, May 1996.
34. P. Ball and V.M. Braun, Phys. Rev. D **49**, 2472 (1994).
35. V. Eletsky and E. Shuryak, Phys. Lett. B **276**, 191 (1992).
36. M. Neubert, Phys. Lett. B **322**, 419 (1994); **389**, 727 (1996).
37. M. Neubert and V. Rieckert, Nucl. Phys. B **382**, 97 (1992).
38. M. Neubert, Phys. Lett. B **264**, 455 (1991).
39. M.E. Luke, Phys. Lett. B **252**, 447 (1990).
40. M. Ademollo and R. Gatto, Phys. Rev. Lett. **13**, 264 (1964).
41. M. Neubert, Nucl. Phys. B **416**, 786 (1994).
42. P. Cho and B. Grinstein, Phys. Lett. B **285**, 153 (1992).
43. H. Albrecht et al. (ARGUS Collaboration), Z. Phys. C **57**, 533 (1993).
44. B. Barish et al. (CLEO Collaboration), Phys. Rev. D **51**, 1041 (1995).
45. D. Buskulic et al. (ALEPH Collaboration), Phys. Lett. B **359**, 236 (1995).
46. P. Abreu et al. (DELPHI Collaboration), Z. Phys. C **71**, 539 (1996).
47. S. Stone, in: Proceedings of the 8th Annual Meeting of the Division of Particles and Fields of the APS, Albuquerque, New Mexico, 1994, edited by S. Seidel (World Scientific, Singapore, 1995), p. 871.
48. C.G. Boyd, B. Grinstein and R.F. Lebed, Nucl. Phys. B **461**, 493 (1996);
C.G. Boyd and R.F. Lebed, preprint UCSD/PTH 95-23 (1995) [hep-ph/9512363].

49. I. Caprini and M. Neubert, Phys. Lett. B **380**, 376 (1996).
50. M. Neubert, Nucl. Phys. B **371**, 149 (1992).
51. S.J. Brodsky, G.P. Lepage, and P.B. Mackenzie, Phys. Rev. D **28**, 228 (1983); G.P. Lepage and P.B. Mackenzie, Phys. Rev. D **48**, 2250 (1993).
52. M. Neubert, Phys. Lett. B **341**, 367 (1995).
53. A.F. Falk and B. Grinstein, Phys. Lett. B **247**, 406 (1990).
54. M. Neubert, Phys. Rev. D **46**, 2212 (1992).
55. M. Neubert, Phys. Rev. D **51**, 5924 (1995).
56. A. Czarnecki, Phys. Rev. Lett. **76**, 4124 (1996).
57. T. Mannel, Phys. Rev. D **50**, 428 (1994).
58. M. Shifman, N.G. Uraltsev, and A. Vainshtein, Phys. Rev. D **51**, 2217 (1995).
59. A. Kapustin, Z. Ligeti, M.B. Wise, and B. Grinstein, Phys. Lett. B **375**, 327 (1996).
60. M. Neubert, Phys. Lett. B **338**, 84 (1994).
61. M.A. Shifman, A.I. Vainshtein, and V.I. Zakharov, Nucl. Phys. B **120**, 316 (1977).
62. E. Witten, Nucl. Phys. B **122**, 109 (1977).
63. J. Polchinski, Nucl. Phys. B **231**, 269 (1984).
64. G. Altarelli and L. Maiani, Phys. Lett. B **52**, 351 (1974).
65. M.K. Gaillard and B.W. Lee, Phys. Rev. Lett. **33**, 108 (1974).
66. F.J. Gilman and M.B. Wise, Phys. Rev. D **27**, 1128 (1983).
67. E. Eichten and F. Feinberg, Phys. Rev. D **23**, 2724 (1981).
68. W.E. Caswell and G.P. Lepage, Phys. Lett. B **167**, 437 (1986).
69. E. Eichten, in: Field Theory on the Lattice, edited by A. Billoire et al., Nucl. Phys. B (Proc. Suppl.) **4**, 170 (1988).
70. G.P. Lepage and B.A. Thacker, in: Field Theory on the Lattice, edited by A. Billoire et al., Nucl. Phys. B (Proc. Suppl.) **4**, 199 (1988).
71. H.D. Politzer and M.B. Wise, Phys. Lett. B **206**, 681 (1988); **208**, 504 (1988).
72. E. Eichten and B. Hill, Phys. Lett. B **234**, 511 (1990).
73. B. Grinstein, Nucl. Phys. B **339**, 253 (1990).
74. A.F. Falk, H. Georgi, B. Grinstein, and M.B. Wise, Nucl. Phys. B **343**, 1 (1990).
75. E. Eichten and B. Hill, Phys. Lett. B **243**, 427 (1990).
76. A.F. Falk, B. Grinstein, and M.E. Luke, Nucl. Phys. B **357**, 185 (1991).
77. T. Mannel, W. Roberts, and Z. Ryzak, Nucl. Phys. B **368**, 204 (1992).
78. J. Soto and R. Tzani, Phys. Lett. B **297**, 358 (1992).
79. G. 't Hooft, Nucl. Phys. B **61**, 455 (1973).
80. G. 't Hooft and M. Veltman, Nucl. Phys. B **44**, 189 (1972).
81. C.G. Bollini and J.J. Giambiagi, Phys. Lett. B **40**, 566 (1972); Nuovo Cim. B **12**, 20 (1972).
82. X. Ji and M.J. Musolf, Phys. Lett. B **257**, 409 (1991).
83. D.J. Broadhurst, N. Gray, and K. Schilcher, Z. Phys. C **52**, 111 (1991).
84. D.J. Broadhurst and A.G. Grozin, Phys. Lett. B **267**, 105 (1991).
85. V. Giménez, Nucl. Phys. B **375**, 582 (1992).
86. R. Tarrach, Nucl. Phys. B **183**, 384 (1981).
87. M.A. Shifman, A.I. Vainshtein, and V.I. Zakharov, Nucl. Phys. B **147**, 385 and 448 (1979).
88. G. 't Hooft, in: The Whys of Subnuclear Physics, Proceedings of the 15th International School on Subnuclear Physics, Erice, Sicily, 1977, edited by A. Zichichi (Plenum Press, New York, 1979), p. 943.
89. B. Lautrup, Phys. Lett. B **69**, 109 (1977).
90. G. Parisi, Phys. Lett. B **76**, 65 (1978); Nucl. Phys. B **150**, 163 (1979).
91. F. David, Nucl. Phys. B **234**, 237 (1984); **263**, 637 (1986).
92. A.H. Mueller, Nucl. Phys. B **250**, 327 (1985).
93. V.I. Zakharov, Nucl. Phys. B **385**, 452 (1992); M. Beneke and V.I. Zakharov, Phys. Rev. Lett. **69**, 2472 (1992).
94. M. Beneke, Phys. Lett. B **307**, 154 (1993); Nucl. Phys. B **405**, 424 (1993).
95. D. Broadhurst, Z. Phys. C **58**, 339 (1993).
96. A.H. Mueller, in: QCD – 20 Years Later, edited by P.M. Zerwas and H.A. Kastrup (World Scientific, Singapore, 1993), p. 162; Phys. Lett. B **308**, 355 (1993).
97. M. Beneke and V.M. Braun, Nucl. Phys. B **426**, 301 (1994).
98. I.I. Bigi, M.A. Shifman, N.G. Uraltsev, and A.I. Vainshtein, Phys. Rev. D **50**, 2234 (1994).
99. M. Neubert and C.T. Sachrajda, Nucl. Phys. B **438**, 235 (1995).
100. G. Preparata and W.I. Weisberger, Phys. Rev. **175**, 1965 (1968).
101. B.S. DeWitt, Phys. Rev. **162**, 1195 (1967).

102. G. 't Hooft, in: Functional and Probabilistic Methods in Quantum Field Theory, Proceedings of the 12th Winter School of Theoretical Physics, Karpacz, Poland, Acta Univ. Wratisl. **38**, Vol. 1 (1975).
103. D. Boulware, Phys. Rev. D **23**, 389 (1981).
104. L.F. Abbott, Nucl. Phys. B **185**, 189 (1981); Acta Phys. Pol. B **13**, 33 (1982).
105. H. Kluberg-Stern and J.B. Zuber, Phys. Rev. D **12**, 3159 (1975).
106. M. Neubert, Phys. Rev. D **49**, 1542 (1994).
107. M. Luke and A.V. Manohar, Phys. Lett. B **286**, 348 (1992).
108. Y.-Q. Chen, Phys. Lett. B **317**, 421 (1993).
109. D.J. Gross, in: Methods in Field Theory, edited by C.R. Balian and J. Zinn-Justin (North-Holland, Amsterdam, 1976), p. 141.
110. A.A. Belavin and A.A. Migdal, JETP Lett. **19**, 181 (1974) [ZhETF Pis. Red. **19**, 317 (1974)].
111. W.A. Bardeen, A.J. Buras, D.W. Duke, and T. Muta, Phys. Rev. D **18**, 3998 (1978); A.J. Buras, Rev. Mod. Phys. **52**, 199 (1980).
112. E.G. Floratos, D.A. Ross, and C.T. Sachrajda, Nucl. Phys. B **129**, 66 (1977) [E: **139**, 545 (1978)].
113. A.J. Buras, M. Jamin, M.E. Lautenbacher, and P.H. Weisz, Nucl. Phys. B **370**, 69 (1992).
114. G.P. Korchemsky and A.V. Radyushkin, Nucl. Phys. B **283**, 342 (1987); Phys. Lett. B **279**, 359 (1992); G.P. Korchemsky, Mod. Phys. Lett. A **4**, 1257 (1989).
115. G.P. Korchemsky and A.V. Radyushkin, Phys. Lett. B **279**, 359 (1992).
116. W. Kilian, P. Manakos, and T. Mannel, Phys. Rev. D **48**, 1321 (1993).
117. A.G. Grozin and M. Neubert, Phys. Rev. D **55**, 272 (1997).

LIGHT-QUARK MASSES

H. Leutwyler

Institut für theoretische Physik der Universität Bern
Sidlerstr. 5, CH-3012 Berne, Switzerland and
CERN, CH-1211 Geneva, Switzerland

INTRODUCTION

The first crude estimates for the magnitude of the three lightest quark masses appeared more than 20 years ago:[1]

$$\begin{array}{lll} m_u \simeq 4\,\text{MeV} & m_d \simeq 6\,\text{MeV} & m_s \simeq 135\,\text{MeV},^2 \\ m_u \simeq 4.2\,\text{MeV} & m_d \simeq 7.5\,\text{MeV} & m_s \simeq 150\,\text{MeV}.^3 \end{array}$$

Many papers dealing with the pattern of quark masses have appeared since then, but the numbers have barely changed. In the first part of this lecture, I discuss the arguments that lead to the above pattern, emphasizing qualitative aspects concerning the breaking of isospin and eightfold way symmetries. The second part deals with the chiral perturbation theory results for the masses of the pseudoscalar octet, which lead to very stringent constraints for the *relative size* of the light quark masses, i.e. for the ratios m_u/m_d and m_s/m_d. Finally, I briefly comment on their *absolute magnitude* and then draw some conclusions.

MASS SPECTRUM OF THE PSEUDOSCALARS

The best determinations of the relative size of m_u, m_d and m_s rely on the fact that these masses happen to be small, so that the properties of the theory may be analysed by treating the quark mass term in the Hamiltonian of QCD as a perturbation. The Hamiltonian is split into two pieces:

$$H_{\text{QCD}} = H_0 + H_1 \quad ,$$

Masses of Fundamental Particles: Cargèse 1996
Edited by M. Lévy *et al.*, Plenum Press, New York, 1997

where H_0 describes the three lightest quarks as massless and H_1 is the corresponding mass term,

$$H_1 = \int d^3x \{m_u \bar{u}u + m_d \bar{d}d + m_s \bar{s}s\} \ .$$

H_0 is invariant under the group $SU(3)_R \times SU(3)_L$ of independent flavour rotations of the right- and lefthanded quark fields. The symmetry is broken spontaneously: the eigenstate of H_0 with the lowest eigenvalue, $|0\rangle$, is invariant only under the subgroup $SU(3)_V \subset SU(3)_R \times SU(3)_L$. Accordingly, the spectrum of H_0 contains eight Goldstone bosons, $\pi^\pm$, π^0, $K^\pm$, K^0, $\bar{K}^0$, η. The remaining levels form degenerate multiplets of $SU(3)_V$ of non-zero mass.

The perturbation H_1 breaks the symmetry, because it connects the right- and left-handed components: $\bar{u}u = \bar{u}_R u_L + h.c.$ In so far as the quark masses m_u, m_d, m_s are small, the entire term H_1 amounts to a small perturbation, so that the group $SU(3)_R \times SU(3)_L$ still represents an *approximate* symmetry of the full Hamiltonian. The perturbation splits the SU(3) multiplets, in particular also the Goldstone boson octet. To first order in the perturbation, the square of the pion mass is given by the expectation value of the perturbation, $M^2_{\pi^+} = \langle \pi^+ | m_u \bar{u}u + m_d \bar{d}d + m_s \bar{s}s | \pi^+ \rangle$ and is therefore linear in the quark masses. Since the eigenstates entering here are those of H_0, the matrix elements respect SU(3) symmetry. In particular, isospin conservation requires $\langle \pi^+ | \bar{u}u | \pi^+ \rangle = \langle \pi^+ | \bar{d}d | \pi^+ \rangle \equiv B$. Moreover, since the subgroup $SU(2)_R \times SU(2)_L$ becomes an exact symmetry for $m_u, m_d \to 0$, $m_s \neq 0$, the pion mass must disappear in this limit, so that the matrix element $\langle \pi^+ | \bar{s}s | \pi^+ \rangle$ vanishes. The expansion of $M^2_{\pi^+}$ in powers of m_u, m_d, m_s thus starts with

$$M^2_{\pi^+} = (m_u + m_d) B \{1 + O(m_u, m_d, m_s)\} \ . \tag{1}$$

The operation $d \to s$ takes the π^+ into the K^+, and the K^0 may be reached from there with $u \to d$. Hence the corresponding lowest order mass formulae read

$$M^2_{K^+} = (m_u + m_s) B + \dots \ , \qquad M^2_{K^0} = (m_d + m_s) B + \dots \tag{2}$$

In the ratios $M^2_{\pi^+} : M^2_{K^+} : M^2_{K^0}$, the constant B drops out, so that the quark mass ratios must approximately be given by

$$\frac{m_u}{m_d} \simeq \frac{M^2_{\pi^+} - M^2_{K^0} + M^2_{K^+}}{M^2_{\pi^+} + M^2_{K^0} - M^2_{K^+}} \simeq 0.66 \ , \qquad \frac{m_s}{m_d} \simeq \frac{M^2_{K^0} + M^2_{K^+} - M^2_{\pi^+}}{M^2_{K^0} - M^2_{K^+} + M^2_{\pi^+}} \simeq 20 \ ,$$

to be compared with $m_\mu / m_e \simeq 200$. These numbers represent rough estimates. The corrections generated by the higher order terms in the mass formulae as well as those due to the electromagnetic interaction are treated below – they do not significantly modify the above ratios.

APPROXIMATE FLAVOUR SYMMETRIES

First, however, I wish to discuss some qualitative aspects of the above relations. For this purpose, I need a crude estimate for the absolute magnitude of the light quark masses, which may be obtained with the following simple argument. The mass differences between m_u, m_d and m_s are responsible for the splittings observed within the multiplets of SU(3). The observed multiplet pattern shows that, replacing a u– or a d–quark by an s–quark, the mass of the bound state increases by about 100 or 200 MeV. Applying the rule of thumb, we infer that the mass differences $m_s - m_u$ and

$m_s - m_d$ are of this order of magnitude. Since m_u and m_d are small compared to m_s, the mass of the strange quark must be of this order, say $m_s \simeq 150$ MeV. With the above ratios, this gives $m_d \simeq \frac{1}{20}m_s \simeq 7.5$ MeV and $m_u \simeq \frac{2}{3}m_d \simeq 5$ MeV. I emphasize that these estimates only concern the order of magnitude and I will discuss our present knowledge at a quantitative level later on.

The first conclusion to draw is that m_u and m_d are surprisingly small. In particular, the mass of the proton is large compared to the sum of the masses of the quarks it consists of. Indeed, M_p does not tend to zero for $m_u, m_d \to 0$. The amount by which it decreases is known from πN scattering:[4] $\sigma = \langle N | m_u \bar{u}u + m_d \bar{d}d | N \rangle = 45 \pm 8$ MeV. This shows that the masses occurring in the Lagrangian of QCD are quite different from those used in the various bound state models, $m_{constituent} \simeq \frac{1}{3}M_p \simeq 300$ MeV.

An equally striking aspect of the above pattern is that the three masses are very different. In particular, the value for m_u/m_d shows that the masses of the u– and d–quarks are quite different. This appears to be in conflict with the oldest and best established internal symmetry of particle physics, isospin. Since u and d form an $I = \frac{1}{2}$ multiplet, isospin is a symmetry of the QCD Hamiltonian only if $m_u = m_d$.

The resolution of the paradox is that m_u, m_d are very small. Disregarding the e.m. interaction, the strength of isospin breaking is determined by the magnitude of $|m_u - m_d|$, not by the relative size m_u/m_d. The fact that m_d is larger than m_u by a few MeV implies, for instance, that the neutron is heavier than the proton by a few MeV. Compared with the mass of the proton, this amounts to a fraction of a per cent. In the case of the kaons, the relative mass splitting $(M^2_{K^0} - M^2_{K^+})/M^2_{K^+}$ is more important, because the denominator is smaller here: The effect is of order $(m_d - m_u)/(m_u + m_s) \simeq 0.02$, but this is still a small number. One might think that for the pions, where the square of the mass is proportional to $m_u + m_d$, the relative mass splitting should be large, of order $(M^2_{\pi^0} - M^2_{\pi^+})/M^2_{\pi^+} \propto (m_d - m_u)/(m_d + m_u) \simeq 0.3$, in flat contradiction with observation. It so happens, however, that the pion matrix elements of the isospin breaking part of the Hamiltonian, $\frac{1}{2}(m_u - m_d)(\bar{u}u - \bar{d}d)$, vanish because the group SU(2) does not have a d-symbol. This implies that the mass difference between π^0 and π^+ is of second order in $m_d - m_u$ and therefore tiny. The observed mass difference is almost exclusively due to the electromagnetic self energy of the π^+. So, the above quark mass pattern is perfectly consistent with the fact that isospin is an almost exact symmetry of the strong interaction: The matrix elements of the term $\frac{1}{2}(m_u - m_d)(\bar{u}u - \bar{d}d)$ are very small compared with those of H_0. In particular, the pions are protected from isospin breaking.

QCD also explains another puzzle: Apparently, the mass splittings in the pseudoscalar octet are in conflict with the claim that SU(3) represents a decent approximate symmetry. This seems to require $M^2_K \simeq M^2_\pi$, while experimentally, $M^2_K \simeq 13 M^2_\pi$. The first order mass formulae yield $M^2_K/M^2_\pi = (m_s + \hat{m})/(m_u + m_d)$, where $\hat{m} = \frac{1}{2}(m_u + m_d)$ is the mean mass of u and d. The kaons are much heavier than the pions, because it so happens that m_s is much larger than $\hat{m}$. For SU(3) to be a decent approximate symmetry, it is not necessary that the difference $m_s - \hat{m}$ is small with respect to the sum $m_s + \hat{m}$, because the latter does not represent the relevant mass scale to compare the symmetry breaking with. If the quark masses were of the same order of magnitude as the electron mass, SU(3) would be an essentially perfect symmetry of QCD; even in that world $m_s \gg \hat{m}$ implies that the ratio M^2_K/M^2_π strongly differs from 1. The strength of SU(3) breaking does not manifest itself in the mass ratios of the pseudoscalars, but in the symmetry relations between the matrix elements of the operators $\bar{u}u$, $\bar{d}d$, $\bar{s}s$, which are used in the derivation of the above mass formulae. The asymmetries in these are analogous to the one seen in the matrix elements of the axial vector currents,

$F_K/F_\pi = 1.22$, which represents an SU(3) breaking of typical size. The deviation from the lowest order mass formula,

$$\frac{M_K^2}{M_\pi^2} = \frac{m_s + \hat{m}}{m_u + m_d}\{1 + \Delta_M\} \ , \tag{3}$$

is expected to be of the same order of magnitude, $1+\Delta_M \leftrightarrow F_K/F_\pi$.

The Gell–Mann–Okubo formula yields a good check. The lowest order mass formula for the η reads

$$M_\eta^2 = \tfrac{1}{3}(m_u + m_d + 4m_s)B + \dots \ , \tag{4}$$

so that the mass relations for π, K, η lead to $3M_\eta^2 + M_\pi^2 - 4M_K^2 = 0$. The accuracy within which this consequence of SU(3) symmetry holds is best seen by working out the quark mass ratio $m_s/\hat{m}$ in two independent ways: While the mass formulae for K and π imply $m_s/\hat{m} = (2M_K^2 - M_\pi^2)/M_\pi^2 = 25.9$, those for η and π yield $m_s/\hat{m} = \frac{1}{2}(3M_\eta^2 - M_\pi^2)/M_\pi^2 = 24.2$. These numbers are nearly the same – the mass pattern of the pseudoscalar octet is a showcase for the claim that SU(3) represents a decent approximate symmetry of QCD, despite $M_K^2 \simeq 13M_\pi^2$.

MASS FORMULAE TO SECOND ORDER

The leading order mass formulae are subject to corrections arising from contributions which are of second or higher order in the perturbation H_1. A systematic method for the analysis of the higher order contributions is provided by the effective Lagrangian method.[5,6] In this approach, the quark and gluon fields of QCD are replaced by a set of pseudoscalar fields describing the degrees of freedom of the Goldstone bosons π, K, η. The effective Lagrangian only involves these fields and their derivatives, but contains an infinite string of vertices. For the calculation of the pseudoscalar masses to a given order in the perturbation H_1, however, only a finite subset contributes. The term Δ_M, which describes the SU(3) corrections in the ratio M_K^2/M_π^2 according to eq. (3), involves the two effective coupling constants L_5 and L_8, which occur in the derivative expansion of the effective Lagrangian at first non-leading order:[6]

$$\Delta_M = \frac{8(M_K^2 - M_\pi^2)}{F_\pi^2}(2L_8 - L_5) + \chi\text{logs} \ . \tag{5}$$

The term χlogs stands for the logarithms characteristic of chiral perturbation theory. They arise because the spectrum of the unperturbed Hamiltonian H_0 contains massless particles – the perturbation H_1 generates infra-red singularities. The coupling constant L_5 also determines the SU(3) asymmetry in the decay constants,

$$\frac{F_K}{F_\pi} = 1 + \frac{4(M_K^2 - M_\pi^2)}{F_\pi^2} L_5 + \chi\text{logs} \ . \tag{6}$$

The comparison of eqs. (5) and (6) confirms that the symmetry breaking effects in the decay constants and in the mass spectrum are of similar nature. The calculation[6] also reveals that the first order SU(3) correction in the mass ratio $(M_{K^0}^2 - M_{K^+}^2)/(M_K^2 - M_\pi^2)$ is the same as the one in M_K^2/M_π^2:

$$\frac{M_{K^0}^2 - M_{K^+}^2}{M_K^2 - M_\pi^2} = \frac{m_d - m_u}{m_s - \hat{m}}\{1 + \Delta_M + O(m^2)\} \ . \tag{7}$$

Hence, the first order corrections drop out in the double ratio

$$Q^2 \equiv \frac{M_K^2}{M_\pi^2} \cdot \frac{M_K^2 - M_\pi^2}{M_{K^0}^2 - M_{K^+}^2} \ . \tag{8}$$

The observed values of the meson masses thus provide a tight constraint on one particular ratio of quark masses:

$$Q^2 = \frac{m_s^2 - \hat{m}^2}{m_d^2 - m_u^2}\{1 + O(m^2)\} \ . \tag{9}$$

The constraint may be visualized by plotting the ratio m_s/m_d versus m_u/m_d.[7] Dropping the higher order contributions, the resulting curve takes the form of an ellipse:

$$\left(\frac{m_u}{m_d}\right)^2 + \frac{1}{Q^2}\left(\frac{m_s}{m_d}\right)^2 = 1 \ , \tag{10}$$

with Q as major semi–axis (the term $\hat{m}^2/m_s^2$ has been discarded, as it is numerically very small).

VALUE OF Q

The meson masses occurring in the double ratio (8) refer to pure QCD. The Dashen theorem states that in the chiral limit, the electromagnetic contributions to $M_{K^+}^2$ and to $M_{\pi^+}^2$ are the same, while the self energies of K^0 and π^0 vanish.[8] Since the contribution to the mass difference between π^0 and π^+ from $m_d - m_u$ is negligibly small, the masses in pure QCD are approximately given by

$$\begin{aligned}&(M_{\pi^+}^2)^{\rm QCD} = (M_{\pi^0}^2)^{\rm QCD} = M_{\pi^0}^2 \ ,\\ &(M_{K^+}^2)^{\rm QCD} = M_{K^+}^2 - M_{\pi^+}^2 + M_{\pi^0}^2 \ , \quad (M_{K^0}^2)^{\rm QCD} = M_{K^0}^2 \ ,\end{aligned}$$

where $M_{\pi^0}, M_{\pi^+}, M_{K^0}, M_{K^+}$ are the observed masses. Correcting for the electromagnetic self energies in this way, the lowest order formulae become[3]

$$\begin{aligned}\frac{m_u}{m_d} &\simeq \frac{M_{K^+}^2 - M_{K^0}^2 + 2M_{\pi^0}^2 - M_{\pi^+}^2}{M_{K^0}^2 - M_{K^+}^2 + M_{\pi^+}^2} = 0.55 \ ,\\ \frac{m_s}{m_d} &\simeq \frac{M_{K^0}^2 + M_{K^+}^2 - M_{\pi^+}^2}{M_{K^0}^2 - M_{K^+}^2 + M_{\pi^+}^2} = 20.1 \ .\end{aligned} \tag{11}$$

The corresponding expression for the semi–axis Q reads

$$Q_D^2 = \frac{(M_{K^0}^2 + M_{K^+}^2 - M_{\pi^+}^2 + M_{\pi^0}^2)(M_{K^0}^2 + M_{K^+}^2 - M_{\pi^+}^2 - M_{\pi^0}^2)}{4\,M_{\pi^0}^2\,(M_{K^0}^2 - M_{K^+}^2 + M_{\pi^+}^2 - M_{\pi^0}^2)} \ . \tag{12}$$

Numerically, this yields $Q_D = 24.2$. The corresponding ellipse is shown in fig. 1 as a dash–dotted line. For this value of the semi–axis, the curve passes through the point specified by Weinberg's mass ratios, eq. (11).

The Dashen theorem is subject to corrections from higher order terms in the chiral expansion. As usual, there are two categories of contributions: loop graphs of order $e^2 m$ and terms of the same order from the derivative expansion of the effective e.m. Lagrangian. The Clebsch–Gordan coefficients occurring in the loop graphs are known to be large, indicating that two–particle intermediate states generate sizeable corrections;

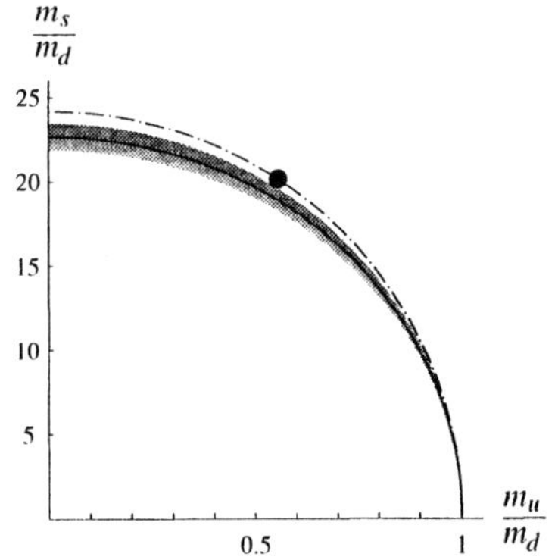

Figure 1. Elliptic constraint. The dot indicates Weinberg's mass ratios. The dash-dotted line represents the ellipse for the value $Q = 24.2$ of the semi-axis, obtained from the mass difference $K^0 - K^+$ with the Dashen theorem. The full line and the shaded region correspond to $Q = 22.7 \pm 0.8$, as required by the observed rate of the decay $\eta \rightarrow \pi^+\pi^-\pi^0$.

the corresponding chiral logarithms tend to increase the e.m. contribution to the kaon mass difference.[9] The numerical result depends on the scale used when evaluating the logarithms. In fact, taken by themselves, chiral logs are unsafe at any scale – one at the same time also needs to consider the contributions from the terms of order e^2m occurring in the effective Lagrangian. This is done in several recent papers,[10–12] but the results are controversial. Donoghue, Holstein and Wyler[10] estimate the contributions arising from vector meson exchange and conclude that these give rise to large corrections, increasing the value $(M_{K^+} - M_{K^0})_{e.m.} = 1.3$ MeV predicted by Dashen to 2.3 MeV. According to refs.[11] however, the model used is in conflict with chiral symmetry: Although the perturbations due to vector meson exchange are enhanced by a relatively small energy denominator, chiral symmetry prevents them from being large. In view of this, it is puzzling that Bijnens,[12] who evaluates the self energies within the model of Bardeen et al.,[13] finds an even larger effect, $(M_{K^+} - M_{K^0})_{e.m.} \simeq 2.6$ MeV.

Recently, the electromagnetic self energies have been analyzed within lattice QCD.[14] The result of this calculation, $(M_{K^+} - M_{K^0})_{e.m.} = 1.9$ MeV, indicates that the corrections to the Dashen theorem are indeed substantial, although not quite as large as found in refs.[10,12]. The uncertainties of the lattice result are of the same type as those occuring in direct determinations of the quark masses with this method. The mass difference between K^+ and K^0, however, is predominantly due to $m_d > m_u$, not to the e.m. interaction. An error in the self energy of 20% only affects the value of Q by about 3%. The terms neglected when evaluating Q^2 with the meson masses are of order $(M_K^2 - M_\pi^2)^2/M_0^4$, where M_0 is the mass scale relevant for the exchange of scalar or pseudoscalar states, $M_0 \simeq M_S \simeq M_{\eta'}$. The corresponding error in the result for Q is also of the order of 3% – the uncertainties in the value $Q = 22.8$ that follows from the lattice result are significantly smaller than those obtained for the quark masses with the same method. The implications of the above estimates for the value of Q are illustrated on the r.h.s. of fig. 2.

The isospin-violating decay $\eta \rightarrow 3\pi$ allows an entirely independent measurement of the semi-axis.[15] The transition amplitude is much less sensitive to the uncertainties associated with the electromagnetic interaction than the $K^0 - K^+$ mass difference: The e.m. contribution is suppressed by chiral symmetry and is negligibly small.[16] The transition amplitude thus represents a sensitive probe of the symmetry breaking generated by $m_d - m_u$. To lowest order in the chiral expansion (current algebra), the amplitude

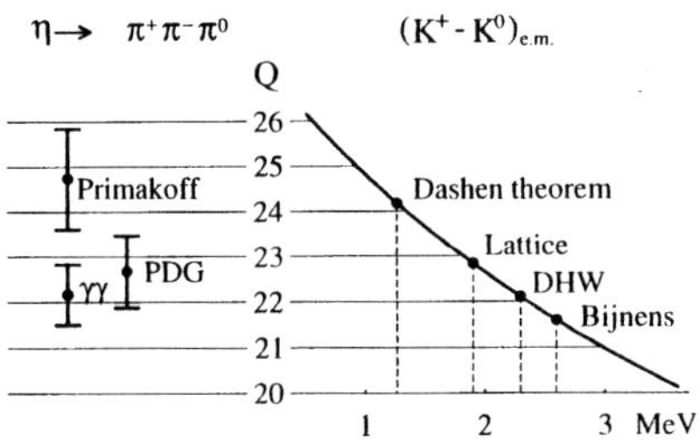

Figure 2. The l.h.s. indicates the values of Q corresponding to the various experimental results for the rate of the decay $\eta \rightarrow \pi^+\pi^-\pi^0$. The r.h.s. shows the results for Q obtained with three different theoretical estimates for the electromagnetic self energy of the kaons.

of the transition $\eta \rightarrow \pi^+\pi^-\pi^0$ is given by

$$A = -\frac{\sqrt{3}}{4}\,\frac{m_d - m_u}{m_s - \hat{m}}\,\frac{1}{F_\pi^2}\left(s - \tfrac{4}{3}M_\pi^2\right) \;,$$

where s is the square of the centre–of–mass energy of the charged pion pair. The corrections of first non–leading order (chiral perturbation theory to one loop) are also known. It is convenient to write the decay rate in the form $\Gamma_{\eta\rightarrow\pi^+\pi^-\pi^0} = \Gamma_0\,(Q_D/Q)^4$, where Q_D is specified in eq. (12). As shown in ref.[15], the one–loop calculation yields a parameter free prediction for the constant Γ_0. Updating the value of F_π, the numerical result reads $\Gamma_0 = 168 \pm 50\,\text{eV}$. Although the calculation includes all corrections of first non–leading order, the error bar is large. The problem originates in the final state interaction, which strongly amplifies the transition probability in part of the Dalitz plot. The one–loop calculation does account for this phenomenon, but only to leading order in the low energy expansion. The final state interaction is analyzed more accurately in two recent papers,[17,18] which exploit the fact that analyticity and unitarity determine the amplitude up to a few subtraction constants. For these, the corrections to the current algebra predictions are small, because they are barely affected by the final state interaction. Although the dispersive framework used in the two papers differs, the results are nearly the same: While Kambor, Wiesendanger and Wyler obtain $\Gamma_0 = 209 \pm 20\,\text{eV}$, we get $\Gamma_0 = 219 \pm 22\,\text{eV}$. This shows that the theoretical uncertainties of the dispersive calculation are small. Since the decay rate is proportional to Q^{-4}, the transition $\eta \rightarrow 3\pi$ represents an extremely sensitive probe, allowing a determination of Q to an accuracy of about $2\frac{1}{2}\%$.

Unfortunately, however, the experimental situation is not clear.[19] The value of $\Gamma_{\eta\rightarrow\pi^+\pi^-\pi^0}$ relies on the rate of the decay into two photons. The two different methods of measuring $\Gamma_{\eta\rightarrow\gamma\gamma}$ – photon–photon collisions and Primakoff effect – yield conflicting results. While the data based on the Primakoff effect are in perfect agreement with the number $Q = 24.2$ which follows from the Dashen theorem, the $\gamma\gamma$ data yield a significantly lower result (see l.h.s. of fig. 2). The statistics is dominated by the $\gamma\gamma$ data. Using the overall fit of the Particle Data Group,[19] $\Gamma_{\eta\rightarrow\pi^+\pi^-\pi^0} = 283 \pm 28\,\text{eV}$ and adding errors quadratically, we obtain $Q = 22.7 \pm 0.8$, to be compared with the value $Q = 22.4 \pm 0.9$ given in ref.[17]. The result appears to confirm the lattice calculation.[14] The above discussion makes it clear that an improvement of the experimental situation concerning $\Gamma_{\eta\rightarrow\gamma\gamma}$ is of considerable interest.

A PHENOMENOLOGICAL AMBIGUITY

Chiral perturbation theory thus fixes one of the two quark mass ratios in terms of the other, to within small uncertainties. The ratios themselves, i.e. the position on the ellipse, are a more subtle issue. Kaplan and Manohar[7] pointed out that the corrections to the lowest order result, eq. (11), cannot be determined on purely phenomenological grounds. They argued that these corrections might be large and that the u–quark might actually be massless. This possibility is widely discussed in the literature,[20] because the strong CP problem would then disappear.

The reason why phenomenology alone does not allow us to determine the two individual ratios beyond leading order is the following. The matrix

$$m' = \alpha_1 m + \alpha_2 (m^+)^{-1} \det m$$

transforms in the same manner as m. For a real, diagonal mass matrix, the transformation amounts to

$$m'_u = \alpha_1 m_u + \alpha_2 m_d m_s \quad (\text{cycl. } u \to d \to s \to u) \ . \qquad (13)$$

Symmetry does therefore not distinguish m' from m. Since the effective theory exclusively exploits the symmetry properties of QCD, the above transformation of the quark mass matrix does not change the form of the effective Lagrangian – the transformation may be absorbed in a suitable change of the effective coupling constants.[7] This implies, however, that the expressions obtained with this Lagrangian for the masses of the pseudoscalars, for the scattering amplitudes, as well as for the matrix elements of the vector and axial currents are invariant under the operation $m \to m'$. Conversely, the experimental information on these observables does not allow us to distinguish m' from m.

Within the approximations used, we may equally well write the mass ratio that characterizes the ellipse in the form $Q^2 = (m_s^2 - \frac{1}{2}m_u^2 - \frac{1}{2}m_d^2)/(m_d^2 - m_u^2)$. Up to terms of order m^4, which are beyond the accuracy of our formulae, the differences of the squares of the quark masses are invariant under the transformation (13). Hence Q^2 is invariant, i.e. the ellipse is mapped onto itself. The position on the ellipse, however, does not remain invariant and can therefore not be determined on the basis of chiral perturbation theory alone.

We are not dealing with a hidden symmetry of QCD here – this theory is not invariant under the change (13) of the quark masses. In particular, the matrix elements of the scalar and pseudoscalar operators are modified. The Ward identity for the axial current implies, for example, that the vacuum–to–pion matrix element of the pseudoscalar density is given by

$$\langle 0|\bar{d}\, i\gamma_5 u|\pi^+\rangle = \sqrt{2}\, F_\pi M_{\pi^+}^2/(m_u + m_d) \ . \qquad (14)$$

The relation is exact, except for electroweak corrections. It involves the physical quark masses and is not invariant under the above transformation. Unfortunately, however, an experimental probe sensitive to the scalar or pseudoscalar currents is not available – the electromagnetic and weak interactions happen to probe the low energy structure of the system exclusively through vector and axial currents.

ESTIMATES AND BOUNDS

I now discuss the size of the corrections to the leading order formulae (11) for the two quark mass ratios m_u/m_d and m_s/m_d. For the reasons just described, this discussion necessarily involves a theoretical input of one sort or another. To clearly identify the relevant ingredient, I explicitly formulate it as hypothesis A, B, ...

Hypothesis A: Assume that the corrections of order m^2 or higher are small and neglect these.

This is the attitude taken in early work on the problem.[1] In the notation used above, the assumption amounts to $\Delta_M \simeq 0$, so that $m_s/\hat{m} \simeq (2M_K^2 - M_\pi^2)/M_\pi^2 \simeq 26$. In the plane spanned by m_u/m_d and m_s/m_d, this represents a straight line. The intersection with the ellipse then fixes things. It is convenient to parametrize the position on the ellipse by means of the ratio R, which measures the relative size of isospin and SU(3) breaking,

$$R \equiv \frac{m_s - \hat{m}}{m_d - m_u} \quad . \tag{15}$$

With the value $Q = 24.2$ (Dashen theorem), the intersection occurs at the mass ratios given by Weinberg, which correspond to $R \simeq 43$. For the value of the semi-axis which follows from η decay, $Q = 22.7$, the intersection instead takes place at $R \simeq 39$.

The baryon octet offers a good test: Applying the hypothesis to the chiral expansion of the baryon masses, i.e. disregarding terms of order m^2, one arrives at three independent estimates for R, viz. 51 ± 10 $(N-P)$, 43 ± 4 $(\Sigma^- - \Sigma^+)$ and 42 ± 6 $(\Xi^- - \Xi^0)$.* Within the errors, these results are consistent with the values $R \simeq 43$ and 39, obtained above from $K^0 - K^+$ and from $\eta \to \pi^+\pi^-\pi^0$, respectively. A recent reanalysis of $\rho - \omega$ mixing[21] leads to $R = 41 \pm 4$ and thus corroborates the picture further.

Another source of information concerning the ratio of isospin and SU(3) breaking effects is the branching ratio $\Gamma_{\psi' \to \psi\pi^0}/\Gamma_{\psi' \to \psi\eta}$. The chiral expansion of the corresponding ratio of transition amplitudes starts with:[22]

$$\frac{\langle \psi\pi^0 | \bar{q}mq | \psi' \rangle}{\langle \psi\eta | \bar{q}mq | \psi' \rangle} = \frac{3\sqrt{3}}{4\,R}\{1 + \Delta_{\psi'} + \ldots\} \ .$$

Disregarding the correction $\Delta_{\psi'}$, which is of order $m_s - \hat{m}$, the data imply $R = 31 \pm 4$, where the error bar corresponds to the experimental accuracy of the branching ratio. The value is significantly lower than those listed above. The higher order corrections are discussed in ref.[23], but the validity of the multipole expansion used there is questionable.[24] The calculation is of interest, because it is independent of other determinations, but at the present level of theoretical understanding, it is subject to considerable uncertainties. Since the quark mass ratios given in refs.[25] rely on the value of R obtained in this way, they are subject to the same reservations. Nevertheless, the information extracted from ψ' decays is useful, because it puts an upper limit on the value of R. As an SU(3) breaking effect, the correction $\Delta_{\psi'}$ is expected to be of order 25%. The estimate $|\Delta_{\psi'}| < 0.4$ is on the conservative side. Expressed in terms of R, this implies $R < 44$.

*Note that, in this case, the expansion contains terms of order $m^{\frac{3}{2}}$, which do play a significant role numerically. The error bars represent simple rule-of-thumb estimates, indicated by the noise visible in the calculation. For details see ref.[1].

Hypothesis B: Assume that the effective coupling constants are dominated by the singularities which are closest to the origin.

This amounts to a generalization of the vector meson dominance hypothesis and yields rough estimates for the various coupling constants, for instance,[26]

$$L_5 \simeq \frac{F_\pi^2}{4\,M_{a_0}^2}\ , \quad L_7 \simeq -\frac{F_\pi^2}{48\,M_{\eta'}^2}\ , \quad L_9 \simeq \frac{F_\pi^2}{2\,M_\rho^2}\ , \ \ldots \tag{16}$$

In all cases where direct phenomenological information is available, these estimates do remarkably well. Also, this framework explains why it is justified to treat m_s as a perturbation:[27] at order p^4, the symmetry breaking part of the effective Lagrangian is determined by the constants $L_4, \ldots, L_8$. These are immune to the low energy singularities generated by spin 1 resonances, but do receive contributions from the exchange of scalar or pseudoscalar particles. Assuming that the pole terms generated by the lightest such particles dominate, the magnitude of these coupling constants is determined by the scale $M_{a_0} \simeq M_{\eta'} \simeq 1$ GeV. According to eq. (6), the asymmetry in the decay constants, for instance, is given by

$$\frac{F_K}{F_\pi} = 1 + \frac{M_K^2 - M_\pi^2}{M_{a_0}^2} + \chi\text{logs}\ . \tag{17}$$

This shows that the breaking of the chiral and eightfold way symmetries is controlled by the mass ratio of the Goldstone bosons to the non–Goldstone states of spin zero, $M_K^2/M_{a_0}^2 \simeq M_K^2/M_{\eta'}^2 \simeq \frac{1}{4}$. In chiral perturbation theory, the observation that the Goldstones are the lightest hadrons thus acquires quantitative significance.

Since the coupling constant L_5 and the combination $L_5 - 12L_7 - 6L_8$ are known experimentally (from F_K/F_π and $3M_\eta^2 + M_\pi^2 - 4M_K^2$, respectively), we may express Δ_M in terms of known quantities, except for a contribution from L_7. Inserting the estimate (16) for this constant, we obtain a small, negative number:[27] $\Delta_M \simeq -0.16$. Unfortunately, the result is rather sensitive to the uncertainties of the saturation hypothesis, because the contributions from L_7 and from the remainder are of opposite sign and thus partly cancel. This is illustrated by the following observation. The constant L_7 enters through its contribution to the mass of the η. When replacing the term with the one from η'-exchange, the corresponding energy denominator, $(M_{\eta'}^2 - p^2)^{-1}$, should be evaluated at $p^2 = M_\eta^2$. In the formula (16), the denominator is replaced by the corresponding leading order term, $(M_{\eta'}^2)^{-1}$. The neglected higher order effects are not exceedingly large, but they reduce the numerical value of the prediction for Δ_M by a factor of 2.

The breaking of SU(3) induces $\eta - \eta'$ mixing. When analyzing this effect within chiral perturbation theory,[6] we noticed that the observed value of M_η requires a mixing angle that is about twice as large as the canonical value $|\theta_{\eta'\eta}| \simeq 10°$ accepted at that time. The conclusion was confirmed experimentally soon thereafter.[28] We may now turn the argument around,[27] use the phenomenology of the mixing angle to estimate the magnitude of L_7 and then determine the size of Δ_M. For a mixing angle in the range $20° < \theta_{\eta'\eta} < 25°$, this leads to $-0.06 < \Delta_M < 0.09$. In this calculation, the energy denominator is evaluated at the proper momentum, but the uncertainties arising from the cancellation of two contributions remain.

Quite irrespective of these uncertainties, the result for Δ_M is a very small number: The hypothesis that the low energy constant L_7 is dominated by the singularity due to the η' implies that the corrections to the lowest order mass formula for M_K^2/M_π^2 are small. In view of the elliptic constraint, this amounts to the statement that A follows from B.

Hypothesis C: Assume that the large-N_c expansion makes sense for $N_c = 3$.

As noted already in ref.[29], the ambiguity discussed above disappears in the large-N_c limit, because the Kaplan–Manohar transformation violates the Zweig rule. In this limit, the structure of the effective theory is modified, because the U(1)–anomaly is then suppressed, so that the spectrum contains a ninth Goldstone boson, which is identified with the η'.[30] The implications for the effective Lagrangian are extensively discussed in the literature and the leading terms in the expansion in powers of $1/N_c$ are well-known.[33] More recently, the analysis was extended to first non-leading order, accounting for all terms which are suppressed either by one power of $1/N_c$ or by one power of the quark mass matrix.[34] This framework leads to a bound for Δ_M, which arises as follows.

At leading order of the chiral expansion, the mass of the η is given by the Gell–Mann–Okubo formula. At the next order of the expansion, there are two categories of corrections: (i) The first is of the same origin as the one which occurs in the mass formula (3) for the ratio M_K^2/M_π^2 and is also determined by Δ_M. The expression for the mass of the η, which follows from the Gell–Mann–Okubo formula, $M_\eta^2 = \frac{1}{3}(4M_K^2 - M_\pi^2)$, is replaced by

$$m_1^2 = \tfrac{1}{3}(4M_K^2 - M_\pi^2) + \tfrac{4}{3}(M_K^2 - M_\pi^2)\,\Delta_M \;.$$

(ii) In addition, there is mixing between the two states η, η'. The levels repel in proportion to the square of the transition matrix element $\sigma_1 \propto \langle \eta' | \bar{q} m q | \eta \rangle$, so that the mass formula for the η takes the form

$$M_\eta^2 = m_1^2 - \frac{\sigma_1^2}{M_{\eta'}^2 - m_1^2} \;. \tag{18}$$

This immediately implies the inequality $M_\eta^2 < m_1^2$, i.e.

$$\Delta_M > -\frac{4M_K^2 - 3M_\eta^2 - M_\pi^2}{4(M_K^2 - M_\pi^2)} = -0.07 \;.$$

At leading order of the expansion, the transition matrix element σ_1 is given by $\sigma_0 = \frac{2}{3}\sqrt{2}\,(M_K^2 - M_\pi^2)$. There are again two corrections of first non-leading order: $\sigma_1 = \sigma_0(1 + \Delta_M - \Delta_N)$. The first is an SU(3) breaking effect of order $m_s - \hat{m}$, determined by Δ_M, while Δ_N represents a correction of order $1/N_c$ of unknown size – the mass formula (18) merely fixes Δ_N as a function of Δ_M or vice versa: As Δ_M grows, Δ_N decreases. A coherent picture, however, only results if both $|\Delta_M|$ and $|\Delta_N|$ are small compared with unity. If the above inequality were saturated, σ_1 would have to vanish, i.e. $1 + \Delta_N - \Delta_M = 0$. In other words, the corrections would have to cancel the leading term. It is clear that, in such a situation, the expansion is out of control. Accordingly, Δ_M must be somewhat larger than -0.07. Even $\Delta_M = 0$ calls for large Zweig rule violations, $\Delta_N \simeq \frac{1}{2}$. The condition

$$\Delta_M > 0 \tag{19}$$

thus represents a generous lower bound for the region where a truncated $1/N_c$ expansion leads to meaningful results. It states that the current algebra formula, which relates the quark mass ratio $m_s/\hat{m}$ to the meson mass ratio M_K^2/M_π^2, represents an upper limit, $m_s/\hat{m} < 2M_K^2/M_\pi^2 - 1 = 25.9$.

This shows that A, B and C are mutually consistent, provided Δ_M is small and positive. The bound (19) is shown in fig. 3: Mass ratios in the hatched region are in conflict with the hypothesis that the first two terms of the $1/N_c$ expansion yield meaningful results for $N_c = 3$. Since the Weinberg ratios correspond to $\Delta_M = 0$, they are located at the boundary of this region. In view of the elliptic constraint, the bound in particular implies $m_u/m_d \gtrsim \frac{1}{2}$.

Hypothesis D: Assume that m_u vanishes.

It is clear that this assumption violates the large-N_c bound just discussed. D is also inconsistent with A and B. In fact, as pointed out in refs.[35], this hypothesis leads to a very queer picture, for the following reason.

The lowest order mass formulae (1) and (2) imply that the ratio m_u/m_d determines the K^0/K^+ mass difference, the scale being set by M_π:

$$M_{K^0}^2 - M_{K^+}^2 = \frac{m_d - m_u}{m_u + m_d} M_\pi^2 + \ldots$$

The formula holds up to corrections from higher order terms in the chiral expansion and up to e.m. contributions. Setting $m_u = 0$, the relation predicts $M_{K^0} - M_{K^+} \simeq 16\,\mathrm{MeV}$, four times larger than the observed mass difference. The disaster can only be blamed on the higher order terms, because the electromagnetic self energies are much too small. Under such circumstances, it does not make sense to truncate the expansion at first non-leading order. The conclusion to be drawn from the assumption $m_u = 0$ is that chiral perturbation theory is unable to account for the masses of the Goldstone bosons. It is difficult to understand how a framework with a basic flaw like this can be so successful.

The assumption $m_u = 0$ also implies that the matrix elements of the scalar and pseudoscalar currents must exhibit very strong SU(3) breaking effects.[35] Consider e.g. the pion and kaon matrix elements of the scalar operators $\bar{u}u, \bar{d}d, \bar{s}s$. In the limit $m_d = m_s$, the ratio

$$r = \frac{\langle \pi^+ | \bar{u}u - \bar{s}s | \pi^+ \rangle}{\langle K^+ | \bar{u}u - \bar{d}d | K^+ \rangle}$$

is equal to 1. The SU(3) breaking effects are readily calculated by working out the derivatives of $M_{\pi^+}^2, M_{K^+}^2$ with respect to m_u, m_d, m_s. Neglecting the chiral logarithms which turn out to be small in this case, the first order corrections may be expressed in terms of the masses,

$$r = \left(\frac{m_s - m_u}{m_d - m_u} \cdot \frac{M_{K^0}^2 - M_{\pi^+}^2}{M_{K^0}^2 - M_{K^+}^2} \right)^2 \left\{ 1 + O(m^2) \right\} \; .$$

The relation is of the same character as the one that leads to the elliptic constraint: The corrections are of second order in the quark masses. For $m_u = 0$, this constraint reduces to $m_s/m_d = Q + \frac{1}{2}$, so that the relation predicts $r \simeq 3$, the precise value depending on the number used for the electromagnetic contribution to $M_{K^+} - M_{K^0}$. So, $m_u = 0$ leads to the prediction that the evaluation of the above matrix elements with sum rule or lattice techniques will reveal extraordinarily strong flavour symmetry breaking effects – a bizarre picture. For me this is enough to stop talking about $m_u = 0$ here.

MAGNITUDE OF m_s

Finally, I briefly comment on the absolute magnitudes of the quark masses. The effective low energy theory does not allow us to determine these phenomenologically,

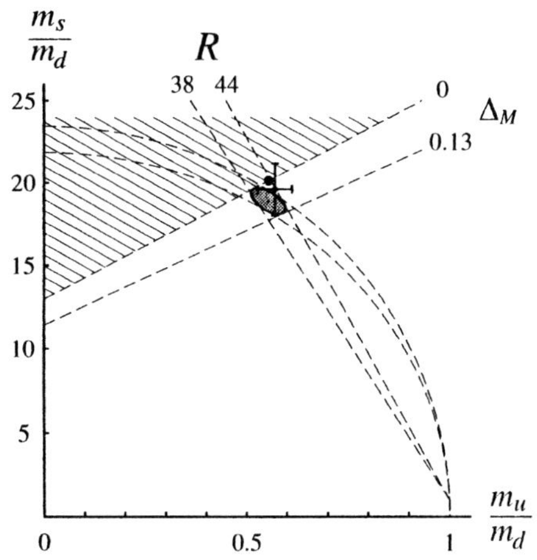

Figure 3. Quark mass ratios. The dot corresponds to Weinberg's values, while the cross represents the estimates given in ref.[1]. The hatched region is excluded by the bound $\Delta_M > 0$. The error ellipse shown is characterized by the constraints $Q = 22.7 \pm 0.8$, $\Delta_M > 0$, $R < 44$, which are indicated by dashed lines.

because the low energy constant B cannot be measured directly. Since the ratios $m_u : m_d : m_s$ are known to within small uncertainties, it suffices to discuss the available information in terms of one of the three masses – I consider the value of m_s.

The best determinations of the absolute magnitude of m_u, m_d and m_s rely on QCD sum rules.[36] A detailed discussion of the method in application to the mass spectrum of the quarks was given in 1982.[1] Fixing the running scale at $\mu = 1\,\text{GeV}$, the result quoted for the $\overline{\text{MS}}$ running mass of the strange quark is $m_s = 175 \pm 55\,\text{MeV}$. The issue has been investigated in considerable detail since then.[37] The value given in the most recent paper,[38]

$$m_s = 175 \pm 25\,\text{MeV} \quad , \tag{20}$$

summarizes the state of the art: The central value is confirmed and the error bar is reduced by about a factor of two. The residual uncertainty mainly reflects the systematic errors of the method, which it is difficult to narrow down further.

There is considerable progress in the numerical simulation of QCD on a lattice.[39] For gluodynamics and bound states of heavy quarks, this approach already yields significant results. The values obtained for m_s are somewhat smaller than the one given above. The APE collaboration,[40] for instance, reports $m_s = 128 \pm 18\,\text{MeV}$ for the $\overline{\text{MS}}$ running mass at $\mu = 2\,\text{GeV}$. It is difficult, however, to properly account for the vacuum fluctuations generated by quarks with small masses. Further progress with light dynamical fermions is required before the numbers obtained for m_u, m_d or m_s can be taken at face value. In the long run, however, this method will allow an accurate determination of all of the quark masses.

CONCLUSION

Light quark effective theory represents a coherent theoretical framework for the analysis of the low energy structure of QCD. The method has been applied to quite a few matrix elements of physical interest and the predictions so obtained have survived the experimental tests performed until now. In these lectures, I focussed on the implications for the ratios of the light quark masses, where the method leads to the following results:

1. The ratios m_u/m_d and m_s/m_d are constrained to an ellipse, whose small semi-axis is equal to 1,

$$\left(\frac{m_u}{m_d}\right)^2 + \frac{1}{Q^2}\left(\frac{m_s}{m_d}\right)^2 = 1 \ .$$

η decay yields a remarkably precise measurement of the large semi-axis,

$$Q = 22.7 \pm 0.8 \ .$$

Unfortunately, however, the experimental situation concerning the lifetime of the η is not satisfactory – the given error bar relies on the averaging procedure used by the Particle Data Group.

2. The position on the ellipse cannot accurately be determined from phenomenology alone. The theoretical arguments given imply that the corrections to Weinberg's leading order mass formulae are small. In particular, there is a new bound based on the $1/N_c$ expansion, which requires $m_u/m_d \gtrsim \frac{1}{2}$ and thereby eliminates the possibility that the u-quark is massless.

3. The final result for the quark mass ratios is indicated by the shaded error ellipse in fig. 3, which is defined by the following three constraints: (i) On the upper and lower sides, the ellipse is bounded by the two dashed lines that correspond to $Q = 22.7 \pm 0.8$. (ii) To the left, it touches the hatched region, excluded by the large-N_c bound. (iii) On the right, I use the upper limit $R < 44$, which follows from the observed value of the branching ratio $\Gamma_{\psi' \to \psi\pi^0}/\Gamma_{\psi' \to \psi\eta}$. The corresponding range of the various parameters of interest is[41]

$$\frac{m_u}{m_d} = 0.553 \pm 0.043 \ , \quad \frac{m_s}{m_d} = 18.9 \pm 0.8 \ , \quad \frac{m_s}{m_u} = 34.4 \pm 3.7 \ ,$$
$$\frac{m_s - \hat{m}}{m_d - m_u} = 40.8 \pm 3.2 \ , \quad \frac{m_s}{\hat{m}} = 24.4 \pm 1.5 \ , \quad \Delta_M = 0.065 \pm 0.065 \ .$$

While the central value for m_u/m_d happens to coincide with the leading order formula, the one for m_s/m_d turns out to be slightly smaller. The difference, which amounts to 6%, originates in the fact that the available data on the η lifetime imply a somewhat smaller value of Q than what is predicted by the Dashen theorem.

4. The theoretical arguments discussed as hypotheses A, B and C are perfectly consistent with these numbers. In particular, the early determinations of R, based on the baryon mass splittings and on $\rho - \omega$ mixing,[1] are confirmed. The rough estimate $m_s/\hat{m} = 29 \pm 7$, obtained by Bijnens, Prades and de Rafael from QCD sum rules,[38] provides an independent check: The lower end of this interval corresponds to $\Delta_M < 0.17$. Fig. 3 shows that this constraint restricts the allowed region to the right and is only slightly weaker than the condition $R < 44$ used above.

5. The mass of the strange quark is known quite accurately from QCD sum rules:

$$m_s = 175 \pm 25\,\text{MeV} \quad (\overline{\text{MS}} \text{ scheme at } \mu = 1\,\text{GeV}) \ .$$

Using this value, the above ratios then determine the size of m_u and m_d:

$$m_u = 5.1 \pm 0.9\,\text{MeV} \ , \quad m_d = 9.3 \pm 1.4\,\text{MeV} \ .$$

REFERENCES

1. The literature contains even earlier estimates, which however were based on the assumption that the strong interaction conserves isospin and thus took $m_u = m_d$ for granted. The prehistory is reviewed in
   J. Gasser and H. Leutwyler, *Phys. Rep.* 87 (1982) 77.
2. J. Gasser and H. Leutwyler, *Nucl. Phys.* B94 (1975) 269.
3. S. Weinberg, in *A Festschrift for I.I. Rabi*, ed. L. Motz, *Trans. New York Acad. Sci.* Ser. II 38 (1977) 185.
4. J. Gasser, H. Leutwyler and M. E. Sainio, *Phys. Lett.* B253 (1991) 252.
5. S. Weinberg, *Physica* A96 (1979) 327.
6. J. Gasser and H. Leutwyler, *Nucl. Phys.* B250 (1985) 465.
7. D. B. Kaplan and A. V. Manohar, *Phys. Rev. Lett.* 56 (1986) 2004.
8. R. Dashen, *Phys. Rev.* 183 (1969) 1245.
9. P. Langacker and H. Pagels, *Phys. Rev.* D8 (1973) 4620;
   K. Maltman and D. Kotchan, *Mod. Phys. Lett.* A5 (1990) 2457;
   G. Stephenson, K. Maltman and T. Goldman, *Phys. Rev.* D43 (1991) 860.
10. J. Donoghue, B. Holstein and D. Wyler, *Phys. Rev.* D47 (1993) 2089.
11. R. Urech, *Nucl. Phys.* B433 (1995) 234;
    R. Baur and R. Urech, *Phys. Rev.* D53 (1996) 6552.
12. J. Bijnens, *Phys. Lett.* B306 (1993) 343.
13. W. A. Bardeen, A. Buras and J.-M. Gérard, *Phys. Lett.* B211 (1988) 343;
    W. A. Bardeen, J. Bijnens and J.-M. Gérard, *Phys. Rev. Lett.* 62 (1989) 1343.
14. A. Duncan, E. Eichten and H. Thacker, *Phys. Rev. Lett.* 76 (1996) 3894.
15. J. Gasser and H. Leutwyler, *Nucl. Phys.* B250 (1985) 539.
16. D. G. Sutherland, *Phys. Lett.* 23 (1966) 384;
    J. S. Bell and D. G Sutherland, *Nucl. Phys.* B4 (1968) 315;
    For a recent analysis of the electromagnetic contributions, see
    R. Baur, J. Kambor and D. Wyler, *Nucl. Phys.* B460 (1996) 127.
17. J. Kambor, C. Wiesendanger and D. Wyler, *Nucl. Phys.* B465 (1996) 215.
18. A. V. Anisovich and H. Leutwyler, *Phys. Lett.* B375 (1996) 335;
    H. Leutwyler, *Phys. Lett.* B374 (1996) 181.
19. Review of Particle Properties, *Phys. Rev.* D45 (1992).
20. For a recent review, see
    T. Banks, Y. Nir and N. Seiberg, in *Yukawa couplings and the origin of mass, Proc. 2nd IFT Workshop, University of Florida, Gainesville, Feb. 1994*, ed. P. Ramond, to be published by International Press.
21. R. Urech, *Phys. Lett.* B355 (1995) 308.
22. B. L. Ioffe and M. A. Shifman, *Phys. Lett.* B95 (1980) 99 and the references therein.
23. J. Donoghue and D. Wyler, *Phys. Rev.* D45 (1992) 892.
24. M. Luty and R. Sundrum, *Phys. Lett.* B312 (1993) 205.
25. J. Donoghue, B. Holstein and D. Wyler, *Phys. Rev. Lett.* 69 (1992) 3444;
    J. Donoghue, *Lectures given at the Theoretical Advanced Study Institute*, Boulder, Colorado (1993);
    D. Wyler, *Proc. XVI Kazimierz Meeting on Elementary Particle Physics*, eds. Z. Ajduk et al. (World Scientific, Singapore, 1994).
26. G. Ecker et al., *Nucl. Phys.* B321 (1989) 311; *Phys. Lett.* B223 (1989) 425.
27. H. Leutwyler, *Nucl. Phys.* B337 (1990) 108.
28. J. F. Donoghue, B. R. Holstein and Y. C. R. Lin, *Phys. Rev. Lett.* 55 (1985) 2766;
    F. Gilman and R. Kaufmann, *Phys. Rev.* D36 (1987) 2761;
    Riazuddin and Fayazuddin, *Phys. Rev.* D37 (1988) 149;
    J. Bijnens, A. Bramon and F. Cornet, *Phys. Rev. Lett.* 61 (1988) 1453;
    ASP Collaboration, N. A. Roe et al., *Phys. Rev.* D41 (1990) 17.
29. J.-M. Gerard, *Mod. Phys. Lett.* A5 (1990) 391.
30. G. 't Hooft, *Nucl. Phys.* B72 (1974) and B75 (1974) 461;
    G. Veneziano, *Nucl. Phys.* B117 (1974) 519 and B159 (1979) 213;
    R. J. Crewther, *Phys. Lett.* 70B (1977) 349 and in *Field Theoretical Methods in Particle Physics*, ed. W. Rühl (Plenum, New York, 1980);
    E. Witten, *Nucl. Phys.* B156 (1979) 213 and B117 (1980) 57;

P. Di Vecchia, *Phys. Lett.* 85B (1979) 357;
P. Nath and R. Arnowitt, *Phys. Rev.* D23 (1981) 473.
31. S. Coleman and E. Witten, *Phys. Rev. Lett.* 45 (1980) 100.
32. P. Minkowski, *Phys. Lett.* B237 (1990) 531, discusses an alternative picture, where the topological susceptibility diverges in the large-N_c limit.
33. P. Di Vecchia and G. Veneziano, *Nucl. Phys.* B171 (1980) 253;
C. Rosenzweig, J. Schechter and T. Trahern, *Phys. Rev.* D21 (1980) 3388;
E. Witten, *Ann. Phys. (N.Y.)* 128 (1980) 363.
34. H. Leutwyler, *Phys. Lett.* B374 (1996) 163.
35. H. Leutwyler, *Proc. XXVI Int. Conf. on High Energy Physics, Dallas, Aug. 1992*, ed. J. R. Sanford (AIP Conf. Proc. No. 272, New York 1993) and in *Yukawa couplings and the origin of mass, Proc. 2nd IFT Workshop, University of Florida, Gainesville, Feb. 1994*, ed. P. Ramond, to be published by International Press.
36. A. I. Vainshtein et al., *Sov. J. Nucl. Phys.* 27 (1978) 274;
B. L. Ioffe, *Nucl. Phys.* B188 (1981) 317; B191 (1981) 591(E).
37. I only quote the papers published in the last two years – earlier work may be found in the references therein:
V. L. Eletsky and B. L. Ioffe, *Phys. Rev.* D48 (1993) 1441;
C. Adami, E. G. Drukarev and B. L. Ioffe, *Phys. Rev.* D48 (1993) 2304;
X. Jin, M. Nielsen and J. Pasupathy, *Phys. Rev.* D51 (1995) 3688;
M. Jamin and M. Münz, *Z. Phys.* C66 (1995) 633;
K. G. Chetyrkin, C. A. Dominguez, D. Pirjol and K. Schilcher, *Phys. Rev.* D51 (1995) 5090.
38. J. Bijnens, J. Prades and E. de Rafael, *Phys. Lett.* B348 (1995) 226.
39. P. B. Mackenzie, *Nucl. Phys.* B (Proc. Suppl.) 34 (1994) 400;
B. Gough, T. Onogi and J. Simone, in *Proc. Int. Symp. on Lattice Field Theory, Melbourne, Australia (1995)*, preprint hep-lat/9510009.
40. C. R. Allton, M. Ciuchini, M. Crisafully, E. Franco, V. Lubicz and G. Martinelli, *Nucl. Phys.* B431 (1994) 667.
41. H. Leutwyler, *Phys. Lett.* B378 (1996) 313.

WEAK MATRIX ELEMENTS ON THE LATTICE: RECENT DEVELOPMENTS IN K-PHYSICS

M. Talevi

Dip. di Fisica, Univ. di Roma "La Sapienza" and INFN, Sezione di Roma

P.le A. Moro 2, I-00185 Roma, Italy

INTRODUCTION

In this talk we present some recent developments in the calculation of weak matrix elements on the lattice. Lattice QCD is one of the few systematically improvable methods for computing them from first principles, and has proven a powerful and appealing approach. In spite of the successes, progress has been slow due to the presence of systematic effects, such as discretization and non-perturbative renormalization effects. In the following, we concentrate on the applications in K-physics of a recently introduced method for non-perturbative (NP) renormalization [1].

Renormalization of lattice operators is a crucial ingredient in the calculation of physical weak matrix elements on the lattice. A physical amplitude $A_{\alpha\to\beta}$ of a weak transition $\alpha \to \beta$ is calculated via the Operator Product Expansion (OPE) by

$$A_{\alpha\to\beta} = C_W(\mu/M_W)\langle\alpha|\hat{O}(\mu)|\beta\rangle \tag{1}$$

where C_W is the Wilson coefficient of the OPE, M_W is the mass of the W boson, μ is the renormalization scale and $\langle\alpha|\hat{O}(\mu)|\beta\rangle$ is the matrix element of the renormalized operator (at the scale μ) relevant to the physical process. The Wilson coefficient $C_W(\mu/M_W)$ contains the short-distance information and can be calculated in Perturbation Theory (PT) in the continuum at the renormalization scale μ. The matrix element contains the long-distance dynamics and thus must be calculated non-perturbatively on the lattice. Renormalization relates the regularized lattice matrix elements to its continuum counterpart.

On the lattice, chiral symmetry is explicitly broken with Wilson-like fermions. The possibility of recovering the chiral symmetry in the continuum limit was shown in [2]. The general prescription is to subtract from the bare operator $O(a)$ all the operators of dimension less or

Masses of Fundamental Particles: Cargèse 1996
Edited by M. Lévy *et al.*, Plenum Press, New York, 1997

equal than $O(a)$, which have the same quantum numbers left unbroken by the regularization:

$$\langle\alpha|\hat{O}(\mu)|\beta\rangle = \lim_{a\to 0}\langle\alpha|Z_O(\mu a)[O(a) + \sum_i Z_i O_i(a)]|\beta\rangle, \tag{2}$$

If the subtracted operators O_i have lower dimension than O, the mixing constants are power-divergent in the cutoff, $Z_i \sim 1/a^d$ with $d > 0$. These divergent factors can pick up exponentially small contributions in the strong coupling, yielding a finite contribution as $a \to 0$, i.e.

$$\frac{1}{a}e^{-1/\alpha_s(a)} \sim \Lambda_{\rm QCD}. \tag{3}$$

These divergences must be subtracted in a completely non-perturbative way. Recently, non-perturbatively renormalization has witnessed a great progress [1, 3, 4]. Here, we want to discuss the applications of the method of ref. [1], which in the following will be refered to as the NP method (NPM).

We will concentrate here on two physical processes of phenomenological interest:

- $K^0 - \bar{K}^0$ mixing, in which the bare operator $O^{\Delta S=2} = \bar{s}\gamma^L_\mu d\bar{s}\gamma^L_\mu d$ mixes only with operators of the same dimension [5, 6].

  From the matrix element of the $\Delta S = 2$ operator the kaon B-parameter B_K is obtained, which is a quantity of great phenomenological interest, being related to the ϵ-parameter which measures CP-violation in the K^0–$\bar{K}^0$ system. With the measured value of the top quark mass, an accurate prediction of B_K enables us to limit the range of values of the CP-violation phase δ.

- $K \to \pi\pi$ decay, in which the bare operators $O^{\pm}_{LL}$ also mix with operators of lower dimensionality [7].

  These decays are relevant to the study of $\Delta I = 1/2$ rule, the enhancement of the $\Delta I = 1/2$ with respect to the $\Delta I = 3/2$ rate, a long-standing question which lattice QCD should be able to understand in a quantitative fashion. Yet, this has proven a formidable task due to the power-divergent mixing. The NPM allows us to make the subtraction without loosing predictive power, as one would do imposing renormalization conditions directly on hadronic states [8].

NON-PERTURBATIVE METHOD

In the NPM, the renormalization conditions are applied directly to the Green functions of quarks and gluons, in a fixed gauge, with given off-shell external states of large virtualities [1]. The method mimicks what is usually done in the perturbative calculation, but the Green functions are evaluated in a non-perturbative fashion from Monte Carlo simulations.

To give the flavour of the method, let us consider the simplified case of a multiplicatively renormalizable operator, e.g. a two-quark operator $O = \bar{q}\Gamma q$. Given the bare lattice operator $O^{\rm latt}(a)$, the renormalization condition we impose is [1]

$$Z^{\rm latt}_{\rm RI}(\mu a)\langle p|O^{\rm latt}(a)|p\rangle|_{p^2=\mu^2} = \langle p|O^{\rm latt}(a)|p\rangle|^{\rm tree}_{p^2=\mu^2}, \tag{4}$$

where $\langle p|\cdots|p\rangle$ denotes the matrix element of external quarks of momenta p which can be calculated to all orders in the QCD coupling via Monte Carlo simulations. The renormalized operator obtained with the NPM is then

$$\hat{O}_{\rm RI}(\mu) = Z^{\rm latt}_{\rm RI}(\mu a)O^{\rm latt}(a), \tag{5}$$

which depends on the external states and the gauge, but not on method used to regulate the ultra-violet divergences. To stress this point, we call the NP renormalization scheme Regularization Independent (RI) [9]. The physical operator

$$O^{\rm phys}(M_W) = C_{\rm RI}(M_W/\mu)\hat{O}_{\rm RI}(\mu) \qquad (6)$$

is independent of external momenta and gauge (up to higher orders in continuum PT and lattice systematic effects) if the Wilson coefficient function $C_{\rm RI}(M_W/\mu)$ in the RI scheme is calculated with the same external momenta and gauge of $\hat{O}_{\rm RI}(\mu)$. The advantage of the RI scheme is that it completely avoids the use of lattice PT, which is expected to have a worse convergence than the continuum expansion [10]. The coefficient function $C_{\rm RI}(M_W/\mu)$ are instead calculated in continuum PT, which cannot be avoided since the Wilson OPE is defined perturbatively.

The NPM is valid for any composite operator, as long as we can can find a window in the range of μ such that $\Lambda_{\rm QCD} \ll \mu \ll O(1/a)$, in order to keep under control both the higher-order effects in the (continuum) perturbative calculation of $C_{\rm RI}$ and discretization errors [1]. We stress that this requirement is common to all NP methods on the lattice.

$\Delta S = 2$

We consider the renormalization of the four-fermion operator

$$O^{\Delta S=2} = (\bar{s}\gamma_\mu^L d)(\bar{s}\gamma_\mu^L d)\,,\ \gamma_\mu^L = \frac{1}{2}\gamma_\mu(1-\gamma_5)\,, \qquad (7)$$

which appears in the weak effective Hamiltonian relevant for K^0–$\bar{K}^0$ mixing.

The Wilson term in the quark action induces the mixing of the operator (7) with dimension-six operators of different chirality, determined by CPS symmetry of the action (the S stands for the $s \leftrightarrow d$ flavour exchange symmetry) [14, 5, 6]

$$\begin{aligned}
O_1 &\equiv -\frac{1}{16N_c}[O_{SS} - O_{PP}] \\
O_2 &\equiv -\frac{(N_c^2+N_c-1)}{32N_c}[O_{VV} - O_{AA}] \\
O_3 &\equiv \frac{(N_c-1)}{16N_c}[O_{SS} + O_{PP} + O_{TT}] \\
O_4 &\equiv \frac{(N_c-1)}{16N_c}[O_{SS} + O_{PP} - \tfrac{1}{3}O_{TT}]
\end{aligned} \qquad (8)$$

where N_c denotes the number of colours and $O_{\Gamma\Gamma} = (\bar{s}\Gamma d)(\bar{s}\Gamma d)$, with Γ one of the Dirac matrices. Note that O_4 is not present in 1-loop PT [11], so the mixing is at most of $\mathcal{O}(g_0^4(a))$, where $g_0(a)$ is the bare lattice QCD coupling. We define the renormalized operator as

$$\hat{O}_{\rm RI}^{\Delta S=2}(\mu) = Z^{\Delta S=2}(\mu a)O_{\rm sub}^{\Delta S=2}(\mu a) \equiv Z^{\Delta S=2}(O^{\Delta S=2} + \sum_{i=1}^{4} Z_i O_i) \qquad (9)$$

Since there are no $\Delta S = 2$ operators of dimension lower than six, the mixing constants Z_i are finite, while the logarithmically divergent $Z^{\Delta S=2}$ renormalizes multiplicatively the subtracted operator $O_{\rm sub}^{\Delta S=2}$.

We determine the mixing constants Z_i with a projection method on the four-point amputated Green functions [5, 6]. Denoting by O_0 the operator $O^{\Delta S=2}$, we define a set of mutually

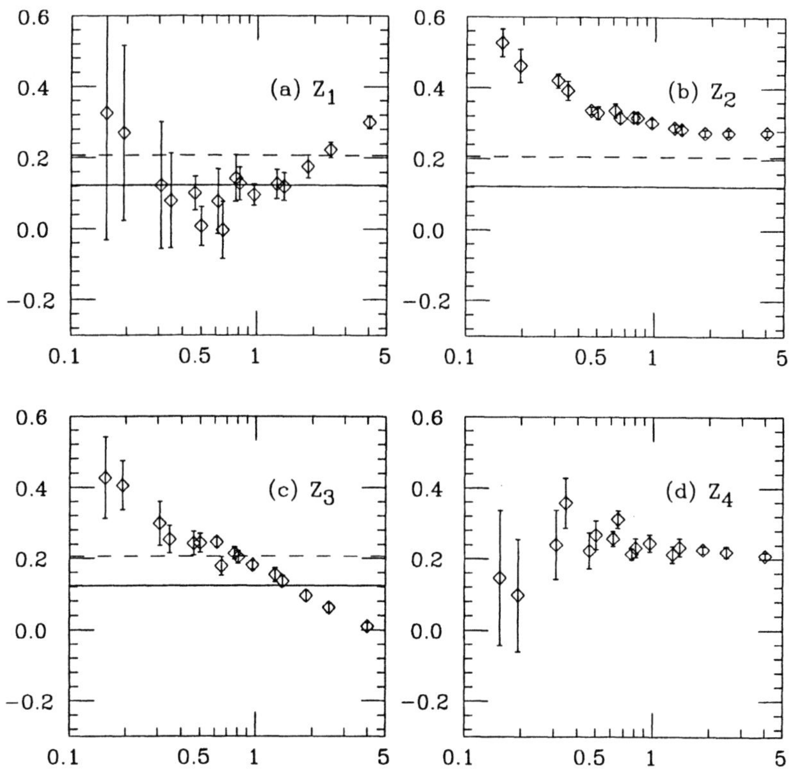

Figure 1. Mixing constants $Z_i (i = 1, \ldots, 4)$ at $\kappa = 0.1432$, for several renormalization scales $\mu^2 a^2$. The solid line is the result from "standard" PT, while the dashed line comes from "boosted" PT with $\alpha_V \simeq 1.68\ \alpha_s^{\rm latt}$.

orthogonal projectors $\mathbf{P}_i\ i = 0, \ldots, 4$ on the amputated tree-level four-quark Green functions $\Lambda_i^{(0)}$ of the operators O_i, by the relation Tr $\mathbf{P}_i \Lambda_j^{(0)} = \delta_{ij}$, where the trace over colour and spin is understood. The mixing constants Z_i are fixed by the condition that the subtracted operator $O_{\rm sub}^{\Delta S=2}$ be proportional to the bare free operator, i.e.

$$\mathrm{Tr}\ \mathrm{P}_k \Lambda_{\rm sub}^{\Delta S=2}(pa) = 0,\ k = 1, \ldots, 4, \tag{10}$$

where $\Lambda_{\rm sub}^{\Delta S=2}(pa)$ is the amputated Green function of $O_{\rm sub}^{\Delta S=2}$, which is calculated at equal external momenta p and in the Landau gauge in a completely NP fashion from numerical simulations. Eq. (10) yields a linear non-homogeneous system, from which we determine the mixing constants Z_i. Once we have determined the mixing constants, the overall renormalization constant $Z^{\Delta S=2}$ is determined by [5, 6]

$$Z^{\Delta S=2}(\mu a) Z_q^{-2}(\mu a) \Gamma_{\rm sub}^{\Delta S=2}(pa)|_{p^2=\mu^2} = 1, \tag{11}$$

where $\Gamma_{\rm sub}^{\Delta S=2}(pa) = \mathrm{Tr}\ \mathrm{P}_0 \Lambda_{\rm sub}^{\Delta S=2}(pa)$, and Z_q is the light-quark renormalization constant, determined from the conserved vector current in a NP way [1].

Table 1. Values of α, β and γ for several renormalization scales $\mu^2 a^2$. In the last row we present results obtained with "boosted" PT renormalization constants.

$\mu^2 a^2$	α	β	γ
0.31	0.030(18)	0.27(21)	0.90(15)
0.62	−0.027(16)	0.36(18)	0.75(13)
0.96	−0.012(14)	0.24(17)	0.69(12)
1.27	0.005(13)	0.14(16)	0.68(12)
1.39	−0.009(13)	0.24(16)	0.67(12)
1.85	−0.003(13)	0.18(16)	0.66(11)
2.46	−0.001(12)	0.24(15)	0.65(11)
4.01	−0.002(12)	0.44(15)	0.67(11)
BPT	−0.052(12)	0.16(15)	0.62(11)

The quark Green functions have been calculated using Monte Carlo simulations. We have used an ensemble of 100 configurations, on a $16^3 \times 32$ lattice, at $\beta = 6.0$, and with three values of the hopping parameter (which is related to the quark mass) $\kappa = 0.1425, 0.1432, 0.1440$ for the quark propagator, in the lattice Landau gauge. To reduce lattice artefacts, we have used an $\mathcal{O}(a)$ improved lattice quark action [12]. We have performed the calculation for a wide range of scales $\mu^2 a^2$. In fig. 1, we show the NP mixing constants at $\kappa = 0.1432$ as a function of $\mu^2 a^2$, and compare them with the PT result in the same gauge and external momenta [6]. We also report the PT result using a "boosted" coupling $\alpha_V \simeq 1.68\, \alpha_s^{\rm latt}$ [10]. We have checked that the dependence of the Z's on κ is very mild. As expected the higher-order contributions differentiate the mixing constants with respect to the PT value, which at one-loop is the same for all the constants [11]. We note that Z_2 and Z_4 are very well defined and almost scale independent in a large "window" of $\mu^2 a^2$, whereas Z_1 and Z_3 are more scale-dependent. Moreover, Z_4 which is absent in 1-loop PT is not neglegible.

The effects of the NP corrections can be most clearly seen in the study of the chiral behaviour of the matrix element $\langle \bar{K}^0 | \hat{O}^{\Delta S=2} | K^0 \rangle_{\rm latt}$. The bare matrix elements have been computed from an ensemble of 460 configurations with the same $\mathcal{O}(a)$ improved action, on a $18^3 \times 64$ lattice, at $\beta = 6.0$ [13]. Parametrizing the lattice matrix element as

$$\langle \bar{K}^0(p) | \hat{O}_{\rm RI}^{\Delta S=2} | K^0(q) \rangle = \alpha + \beta m_K^2 + \gamma (p \cdot q) + \ldots \tag{12}$$

one finds the values of α, β and γ at several renormalization scales $\mu^2 a^2$, cf. tab. 1. The B_K parameter in the RI scheme is given by $B_K^{\rm RI} = \gamma(\mu)/Z_A^2$, where $Z_A = 1.06(3)$ is the axial current renormalization constant. α is a lattice artefact and should vanish in the continuum limit.

As can be seen, in the window of $\mu^2 a^2$ values in which the Z's were found to be reliable, α is compatible with zero within one standard deviation, whereas β is compatible with zero within 1.5 standard deviations. On the other hand, in the same window of $\mu^2 a^2$, γ stabilizes to a non-zero value. The fluctuations of Z_1 and Z_3 do not influence the stability of the results since the relative matrix elements $\langle \bar{K}^0(p) | O_{1,3} | K^0(q) \rangle$ are quite small compared to $\langle \bar{K}^0(p) | O_2 | K^0(q) \rangle$, whose mixing constant Z_2 is extremely well determined. The vanishing of α allows us to conclude that the use of the NP Z's improves the chiral behaviour for a large range of values of $\mu^2 a^2$ [6].

One we are confident that the correct chiral behaviour is recovered, we can reliably calculate the physical value of B_K, obtained by calculating the Wilson coefficient in the same

RI scheme [13, 6].

$\Delta I = 1/2$

In the continuum, with an active charm quark and the GIM mechanism at work, the operator basis is given by [8]

$$O^{\pm}_{LL} = \frac{1}{2}[(\bar{s}d)_L(\bar{u}u)_L \pm (\bar{s}u)_L(\bar{u}d)_L] - (u \to c). \tag{13}$$

Again, the renormalization strategy is complicated by chiral symmetry breaking. In fact, the Wilson term induces the mixing of $O^{\pm}_{LL}$ with lower-dimensional operators, with power-divergent coefficients, which need to be subtracted non-perturbatively. In order to renormalize the operators (13) on the lattice we find it convenient to separate the $\{8,1\}$ and the $\{27,1\}$ components under the $SU(3)_L \otimes SU(3)_R$ chiral group. In the following, we shall concentrate only on the octet component of $O^{\pm}_{LL}$, which we will denote with $O^{\pm}_0$ [7][1].

There are different ways of calculating the $K \to \pi\pi$ matrix elements, which correspond to different renormalization structures. We considering here a general structure of the form

$$\hat{O}^{\pm} = Z^{\pm}\left[O^{\pm}_0 + \sum_{i=1}^{4} Z^{\pm}_i O^{\pm}_i + Z^{\pm}_5 O_5 + Z^{\pm}_3 O_3\right], \tag{14}$$

refering to [7] for a more detailed analysis. In eq. (14) $O^{\pm}_0$ are the bare operators, $O^{\pm}_i$, $i = 1,\ldots,4$ are dimension-six operators of wrong chirality (similarly to the $\Delta S = 2$ case), O_5 is a dimension-five of the form $\bar{s}\Sigma_{\mu\nu}F_{\mu\nu}d$ ($\Sigma_{\mu\nu} = \sigma_{\mu\nu}$ or $\tilde{\sigma}_{\mu\nu}$) and O_3 is a dimension-three operator of the form $\bar{s}\Gamma d$ ($\Gamma = \mathbf{1}$ or γ_5).

According to the NPM, the mixing Z's are determined by finding a set of projectors on the tree-level amputated Green functions (GF), with off-shell quark and gluon external states, the choice of which depends on the nature of the operators at hand. For the $\Delta I = 1/2$ operators we choose the following set of external states: $q\bar{q}$, $q\bar{q}g$, $q\bar{q}q\bar{q}$, with the momenta given below in eq. (15). For each choice of external states, i.e. for each different set of GF, we need different type of projectors. Let us denote with $\mathbf{P}_3$ the projector on the $q\bar{q}$ GF of the operator O_3, with $\mathbf{P}_5$ the projector on the $q\bar{q}g$ GF of the operator O_5, and with $\mathbf{P}_j$, $j = 1,\ldots,4$ the set of mutually orthogonal projectors on the operators O_i, $i = 1,\ldots,4$. We refer the reader to refs. [7, 6] for the explicit expressions of the projectors. Applying the projectors to the corresponding NP GF of the renormalized operators $\hat{O}^{\pm}$, with an appropriate choice of the external states, we require that the renormalized operators be proportional to the bare operators, $\hat{O}^{\pm}(\mu) \propto O^{\pm}_0(a)$ (up to terms of $\mathcal{O}(a)$), i.e. we impose the following renormalization conditions (trace over colour and spin is understood in the projection operation):

$$\begin{aligned} &\mathbf{P}_3\langle q(p)|\hat{O}^{\pm}|\bar{q}(p)\rangle = 0 \\ &\mathbf{P}_5\langle q(p-k)g(k)|\hat{O}^{\pm}|\bar{q}(p)\rangle = 0 \\ &\mathbf{P}^{\pm}_j\langle q(p)\bar{q}(p)|\hat{O}^{\pm}|q(p)\bar{q}(p)\rangle = 0,\ j = 1,\ldots,4 \end{aligned} \tag{15}$$

where p and k denote the momentum of the external quark and gluon legs. The system of equations (15) completely determines in a NP way the renormalization constants, as we have six conditions (non-homogeneous due to the matrix elements of $O^{\pm}_0$, cf. eq. (14)) in six unknown mixing constants, $Z^{\pm}_i$, $i = 1,\ldots,4$, $Z^{\pm}_5$, $Z^{\pm}_3$.

[1] The lattice penguin operators, being proportional to $(m_c^2 - m_u^2)a^2 \ll 1$, will be neglected in the following.

Unfortunately, since solving eq. (15) involves delicate cancellations between large contributions, it may very likely result in a very noisy determination, even with large statistics. An equivalent strategy we can adopt is:
1. We introduce an intermediate subtraction for the dimension-five and -six operators

$$\begin{aligned} \bar{O}_i^{\pm} &= O_i^{\pm} + C_3^{(\pm,i)} O_3, \quad i = 0, \dots, 4, \\ \bar{O}_5 &= O_5 + C_3^{(5)} O_3, \end{aligned} \tag{16}$$

and determine the power-divergent mixing constants $C_3^{(\pm,i)}$ and $C_3^{(5)}$ by imposing

$$\begin{aligned} \mathrm{P}_3 \langle q(p) | \bar{O}_i^{\pm} | \bar{q}(p) \rangle &= 0, \quad i = 0, \dots, 4, \\ \mathrm{P}_3 \langle q(p) | \bar{O}_5 | \bar{q}(p) \rangle &= 0. \end{aligned} \tag{17}$$

2. The finite mixing constants $Z_i^{\pm}$ and $Z_5^{\pm}$, which in principle can be calculated in perturbation theory (PT), are then determined from the system

$$\begin{aligned} \mathrm{P}_5 \langle q(p-k) g(k) | \textstyle\sum_i Z_i^{\pm} \bar{O}_i^{\pm} + Z_5^{\pm} \bar{O}_5 | \bar{q}(p) \rangle &= \mathrm{P}_5 \langle q(p-k) g(k) | \bar{O}_0^{\pm} | \bar{q}(p) \rangle \\ \mathrm{P}_j \langle q(p) \bar{q}(p) | \textstyle\sum_i Z_i^{\pm} \bar{O}_i^{\pm} + Z_5^{\pm} \bar{O}_5 | q(p) \bar{q}(p) \rangle &= \mathrm{P}_j \langle q(p) \bar{q}(p) | \bar{O}_0^{\pm} | q(p) \bar{q}(p) \rangle, \ j = 1, \dots, 4. \end{aligned} \tag{18}$$

The numerical implementation of this program is in progress [7].

CONCLUSIONS

Over the last few years there has been considerable progress in the application of lattice QCD to calculate weak matrix elements. Much control has been gained over the two major systematic errors one had to face, discretization effects and higher orders effects in the lattice perturbative expansion. Discretization effects have been greatly reduced by the "improvement" program [12, 15], while the results obtained with the non-perturbative renormalization program make us confident that we can realiably calculate weak matrix elements in the low-energy regime.

ACKNOWLEDGEMENTS

The results presented here have been obtained in a most enjoyable and fruitful collaboration with A. Donini, V. Gimènez, G. Martinelli, G.C. Rossi, C.T. Sachrajda, M. Testa and A. Vladikas, to whom I am much indebted for all that I've learned.

References

[1] G. Martinelli, C. Pittori, C.T. Sachrajda, M. Testa and A. Vladikas, Nucl. Phys. B445 (1995) 81.

[2] M. Bochicchio, L. Maiani, G. Martinelli, G.C. Rossi and M. Testa, Nucl. Phys. B262 (1985) 331.

[3] K. Jansen *et al.*, Phys. Lett. B372 (1996) 275.

[4] S. Aoki *et al.* (JLQCD Collaboration), talk given at Lattice 96, 14th International Symposym on Lattice Field Theory, 4-8 June 1996, St. Louis, MO (USA), to appear in the proceedings.

[5] A. Donini, G. Martinelli, C.T. Sachrajda, M. Talevi and A. Vladikas, Phys. Lett. B360 (1996) 83.

[6] A. Donini, V. Giménez, G. Martinelli, M. Talevi and A. Vladikas, ROME1-1154/96.

[7] G. Martinelli, G.C. Rossi, M. Talevi, M. Testa and A. Vladikas, in preparation.

[8] L. Maiani, G. Martinelli, G.C. Rossi and M. Testa, Phys. Lett. B176 (1986) 445; Nucl. Phys. B289 (1987) 505.

[9] M. Ciuchini, E. Franco, G. Martinelli, L. Reina and L. Silvestrini, Z. Phys. C68 (1995) 239.

[10] G.P. Lepage and P.B. Mackenzie, Phys. Rev. D48 (1993) 2250.

[11] G. Martinelli, Phys. Lett. B141 (1984) 395.
C. Bernard, A. Soni and T. Draper, Phys. Rev. D36 (1987) 3224.

[12] G. Heatlie, G. Martinelli, C. Pittori, G.C. Rossi and C.T. Sachrajda, Nucl. Phys. B352 (1991) 266.

[13] M. Crisafulli *et al.*, Phys. Lett. B369 (1996) 325.

[14] C. Bernard, T. Draper, G. Hockney and A. Soni, Nucl. Phys. B(Proc. Suppl.)4 (1988) 483.

[15] M. Lüscher *et al.*, talk given at Lattice 96, 14th International Symposyum on Lattice Field Theory, 4-8 June 1996, St. Louis, MO (USA), to appear in the proceedings.

LEPTON UNIVERSALITY

A. Pich

Departament de Física Teòrica, IFIC
CSIC — Universitat de València
Dr. Moliner 50, E–46100 Burjassot, València, Spain

Abstract

The Standard Model requires the three known leptonic families to have identical couplings to the gauge bosons. The present experimental tests on lepton universality are reviewed, both for the charged and neutral current sectors. Our knowledge about the Lorentz structure of the $l^- \to \nu_l l'^- \bar{\nu}_{l'}$ transition amplitudes is also discussed.

INTRODUCTION

The Standard Model (SM) is a gauge theory, based on the group $SU(3)_C \otimes SU(2)_L \otimes U(1)_Y$, which describes strong, weak and electromagnetic interactions, via the exchange of the corresponding spin–1 gauge fields: 8 massless gluons and 1 massless photon for the strong and electromagnetic interactions, respectively, and 3 massive bosons, $W^\pm$ and Z, for the weak interaction. The fermionic matter content is given by the known leptons and quarks, which are organized in a 3–fold family structure:

$$\begin{bmatrix} \nu_e & u \\ e^- & d \end{bmatrix}, \qquad \begin{bmatrix} \nu_\mu & c \\ \mu^- & s \end{bmatrix}, \qquad \begin{bmatrix} \nu_\tau & t \\ \tau^- & b \end{bmatrix}, \tag{1}$$

where (each quark appears in 3 different *colours*)

$$\begin{bmatrix} \nu_l & q_u \\ l^- & q_d \end{bmatrix} \equiv \begin{pmatrix} \nu_l \\ l^- \end{pmatrix}_L, \quad \begin{pmatrix} q_u \\ q_d \end{pmatrix}_L, \quad l^-_R, \quad (q_u)_R, \quad (q_d)_R, \tag{2}$$

plus the corresponding antiparticles. Thus, the left-handed fields are $SU(2)_L$ doublets, while their right-handed partners transform as $SU(2)_L$ singlets. The 3 fermionic families in (1)

Masses of Fundamental Particles: Cargèse 1996
Edited by M. Lévy *et al.*, Plenum Press, New York, 1997

appear to have identical properties (gauge interactions); they only differ by their mass and their flavour quantum number.

The gauge symmetry is broken by the vacuum, which triggers the Spontaneous Symmetry Breaking (SSB) of the electroweak group to the electromagnetic subgroup:

$$SU(3)_C \otimes SU(2)_L \otimes U(1)_Y \stackrel{\text{SSB}}{\longrightarrow} SU(3)_C \otimes U(1)_{QED} . \qquad (3)$$

The SSB mechanism generates the masses of the weak gauge bosons, and gives rise to the appearance of a physical scalar particle in the model, the so-called *Higgs*. The fermion masses and mixings are also generated through the SSB mechanism.

The SM constitutes one of the most successful achievements in modern physics. It provides a very elegant theoretical framework, which is able to describe all known experimental facts in particle physics. A detailed description of the SM and its present phenomenological status can be found in Refs. 1 and 2, which discuss the electroweak and strong sectors, respectively.

In spite of its enormous phenomenological success, the SM leaves too many unanswered questions to be considered as a complete description of the fundamental forces. We do not understand yet why fermions are replicated in three (and only three) nearly identical copies? Why the pattern of masses and mixings is what it is? Are the masses the only difference among the three families? What is the origin of the SM flavour structure? Which dynamics is responsible for the observed CP violation?

The fermionic flavour is the main source of arbitrary free parameters in the SM: 9 fermion masses, 3 mixing angles and 1 complex phase (assuming the neutrinos to be massless). The problem of fermion–mass generation is deeply related with the mechanism responsible for the SSB. Thus, the origin of these parameters lies in the most obscure part of the SM Lagrangian: the scalar sector. Clearly, the dynamics of flavour appears to be "terra incognita" which deserves a careful investigation.

The flavour structure looks richer in the quark sector, where mixing phenomena among the different families occur (leptons would also mix if neutrino masses were non-vanishing). Since quarks are confined within hadrons, an accurate determination of their mixing parameters requires first a good understanding of hadronization effects in flavour–changing transitions. A rather exhaustive description of our present knowledge on the different quark couplings has been given in Ref. 3.

The leptonic sector is easier to analyze. The absence of a direct lepton–gluon vertex provides a much cleaner environment to study the structure of the weak currents and the universality of their couplings to the gauge bosons. In the pure leptonic transitions, strong interactions are only present through small higher–order corrections (vacuum polarization, ...). Thus, it is possible to obtain precise theoretical predictions which can be compared with the available data. Although hadronization is of course present in semileptonic decays, such as $\tau^- \to \nu_\tau\pi^-$, $\pi^- \to \mu^-\bar{\nu}_\mu, \ldots$, it only involves gluonic exchanges between the quarks of a single hadronic current. Taking appropriate ratios of different semileptonic transitions with identical hadronic components, the QCD effects cancel to a very good approximation. Therefore, semileptonic decays also provide accurate tests of the leptonic couplings.

The measured masses and lifetimes of the known leptons, shown in table 1, are very different. The mass spectrum indicates a hierarchy of the original Yukawa couplings, which increase from one generation to the other. A similar pattern occurs in the quark sector. The huge lifetime differences can be simply understood as a kinematic reflection of the different masses [see Eq. (18)]. How precisely we know that the underlying interactions are actually identical for the three lepton generations is the main question we want to address in the following.

Table 1. Masses and lifetimes of the known leptons.[4,5]

	Mass	Lifetime
e	$0.51099907 \pm 0.00000015$ MeV	$> 4.3 \times 10^{23}$ yr
μ	105.658389 ± 0.000034 MeV	$(2.19703 \pm 0.00004) \times 10^{-6}$ s
τ	$1777.00\,^{+0.30}_{-0.27}$ MeV	$(290.21 \pm 1.15) \times 10^{-15}$ s
ν_e	< 10–15 eV	> 300 s $\times (m_{\nu_e}/\text{eV})$ (90% CL)
ν_μ	< 0.17 MeV (90% CL)	> 15.4 s $\times (m_{\nu_\mu}/\text{eV})$ (90% CL)
ν_τ	< 18.2 MeV (95% CL)	Model dependent

QED COUPLINGS

A general description of the electromagnetic coupling of a spin–$\frac{1}{2}$ charged lepton to the virtual photon involves three different form factors:

$$T[l\bar{l}\gamma^*] = e\,\varepsilon_\mu(q)\,\bar{l}\left[F_1(q^2)\gamma^\mu + i\frac{F_2(q^2)}{2m_l}\sigma^{\mu\nu}q_\nu + \frac{F_3(q^2)}{2m_l}\sigma^{\mu\nu}\gamma_5 q_\nu\right] l\,, \tag{4}$$

where q^μ is the photon momentum. Owing to the conservation of the electric charge, $F_1(0) = 1$. At $q^2 = 0$, the other two form factors reduce to the lepton magnetic dipole moment, $\mu_l \equiv (e/2m_l)\,(g_l/2) = e(1 + F_2(0))/2m_l$, and electric dipole moment $d_l = eF_3(0)/2m_l$.

The $F_i(q^2)$ form factors are sensitive quantities to a possible lepton substructure. Moreover, $F_3(q^2)$ violates T and P invariance; thus, the electric dipole moments, which vanish in the SM, constitute a good probe of CP violation. Owing to their chiral changing structure, the magnetic and electric dipole moments may provide important insights on the mechanism responsible for mass generation. In general, one expects[6] that a fermion of mass m_f (generated by physics at some scale $M \gg m_f$) will have induced dipole moments proportional to some power of m_f/M.

The measurement of the $e^+e^- \to l^+l^-$ cross-section has been used to test the universality of the leptonic QED couplings. At low energies, where the Z contribution is small, the deviations from the QED prediction are usually parameterized through*

$$\sigma(e^+e^- \to l^+l^-) = \sigma_{\text{QED}}\left(1 \mp \frac{s}{s - \Lambda_\pm^2}\right)^2 . \tag{5}$$

The cut-off parameters $\Lambda_\pm$ characterize the validity of QED and measure the point-like nature of the leptons. From PEP and PETRA data, one finds[7]: $\Lambda_+(e) > 435$ GeV, $\Lambda_-(e) > 590$ GeV, $\Lambda_+(\mu) > 355$ GeV, $\Lambda_-(\mu) > 265$ GeV, $\Lambda_+(\tau) > 285$ GeV and $\Lambda_-(\tau) > 246$ GeV (95% CL), which correspond to upper limits on the lepton charge radii of about 10^{-3} fm.

The most stringent QED test comes of course from the high–precision measurements of the e and μ anomalous magnetic moments[8–13] $a_l \equiv (g_l - 2)/2$:

$$a_e = \begin{cases} (115\,965\,214.0 \pm 2.8) \times 10^{-11} & \text{(Theory)} \\ (115\,965\,219.3 \pm 1.0) \times 10^{-11} & \text{(Experiment)} \end{cases}, \tag{6}$$

$$a_\mu = \begin{cases} (1\,165\,917.1 \pm 1.0) \times 10^{-9} & \text{(Theory)} \\ (1\,165\,923.0 \pm 8.4) \times 10^{-9} & \text{(Experiment)} \end{cases}. \tag{7}$$

*A slightly different parameterization is adopted for $e^+e^- \to e^+e^-$, to account for the t–channel contribution.[7]

Experimentally, very little is known about a_τ since the spin precession method used for the lighter leptons cannot be applied due to the very short lifetime of the τ. The effect is however visible in the $e^+e^- \to \tau^+\tau^-$ cross-section. The limit $|a_\tau| < 0.023$ (95% CL) has been derived[14,15] from PEP and PETRA data. This limit actually probes the corresponding form factor $F_2(s)$ at $s \sim 35$ GeV. A more direct bound at $q^2 = 0$ has been extracted[5] from the decay $Z \to \tau^+\tau^-\gamma$:

$$|a_\tau| < 0.0104 \qquad (95\%\,\mathrm{CL})\,. \tag{8}$$

A slightly better, but more model–dependent, limit has been derived[16] from the $Z \to \tau^+\tau^-$ decay width: $-0.004 < a_\tau < 0.006$.

In the SM the overall value of a_τ is dominated by the second order QED contribution,[17] $a_\tau \approx \alpha/2\pi$. Including QED corrections up to $O(\alpha^3)$, hadronic vacuum polarization contributions and the corrections due to the weak interactions (which are a factor 380 larger than for the muon), the tau anomalous magnetic moment has been estimated to be[18,19]

$$a_\tau\big|_{\mathrm{th}} = (1.1773 \pm 0.0003) \times 10^{-3}\,. \tag{9}$$

So far, no evidence has been found for any CP–violation signature in the lepton sector. The present limits on the leptonic electric dipole moments are[4,5]:

$$\begin{aligned} d_e &= (-0.3 \pm 0.8) \times 10^{-26}\, e\,\mathrm{cm}, \\ d_\mu &= (3.7 \pm 3.4) \times 10^{-19}\, e\,\mathrm{cm}, \\ |d_\tau| &< 5.8 \times 10^{-17}\, e\,\mathrm{cm}. \end{aligned} \tag{10}$$

CHARGED CURRENT UNIVERSALITY

In the SM, the charged–current interactions are governed by an universal coupling g:

$$\mathcal{L}_{CC} = \frac{g}{2\sqrt{2}} \left\{ W_\mu^\dagger \left[\sum_{ij} \bar{u}_i \gamma^\mu (1-\gamma_5) V_{ij} d_j + \sum_l \bar{\nu}_l \gamma^\mu (1-\gamma_5) l \right] + \mathrm{h.c.} \right\}. \tag{11}$$

In the original basis of weak eigenstates quarks and leptons have identical interactions. The diagonalization of the fermion masses gives rise to the unitary quark mixing matrix V_{ij}, which couples any *up–type* quark with all *down–type* quarks. For massless neutrinos, the analogous leptonic mixing matrix can be eliminated by a redefinition of the neutrino fields. The lepton flavour is then conserved in the minimal SM without right–handed neutrinos.

$\mu^- \to e^-\bar{\nu}_e\nu_\mu$

The simplest flavour–changing process is the leptonic decay of the muon, which proceeds through the W–exchange diagram shown in Fig. 1. The momentum transfer carried by the intermediate W is very small compared to M_W. Therefore, the vector–boson propagator reduces to a contact interaction,

$$\frac{-g_{\mu\nu} + q_\mu q_\nu / M_W^2}{q^2 - M_W^2} \quad \xrightarrow{q^2 \ll M_W^2} \quad \frac{g_{\mu\nu}}{M_W^2}\,. \tag{12}$$

The decay can then be described through an effective local 4–fermion Hamiltonian,

$$\mathcal{H}_{\mathrm{eff}} = \frac{G_F}{\sqrt{2}} \left[\bar{e}\gamma^\alpha(1-\gamma_5)\nu_e\right] \left[\bar{\nu}_\mu \gamma_\alpha (1-\gamma_5)\mu\right], \tag{13}$$

where

$$\frac{G_F}{\sqrt{2}} = \frac{g^2}{8M_W^2} \tag{14}$$

is called the Fermi coupling constant. G_F is fixed by the total decay width,

$$\frac{1}{\tau_\mu} = \Gamma(\mu^- \to e^-\bar{\nu}_e\nu_\mu) = \frac{G_F^2 m_\mu^5}{192\pi^3}\,(1+\delta_{\rm RC})\, f\left(m_e^2/m_\mu^2\right), \tag{15}$$

where $f(x) = 1 - 8x + 8x^3 - x^4 - 12x^2 \ln x$, and

$$(1+\delta_{\rm RC}) = \left[1 + \frac{\alpha(m_\mu)}{2\pi}\left(\frac{25}{4} - \pi^2\right)\right]\left[1 + \frac{3}{5}\frac{m_\mu^2}{M_W^2} - 2\frac{m_e^2}{M_W^2}\right] = 0.9958 \tag{16}$$

takes into account the leading higher-order corrections.[20,21] The measured lifetime,[4] $\tau_\mu = (2.19703 \pm 0.00004) \times 10^{-6}$ s, implies the value

$$G_F = (1.16639 \pm 0.00002) \times 10^{-5}\ \text{GeV}^{-2} \approx \frac{1}{(293\ \text{GeV})^2}\,. \tag{17}$$

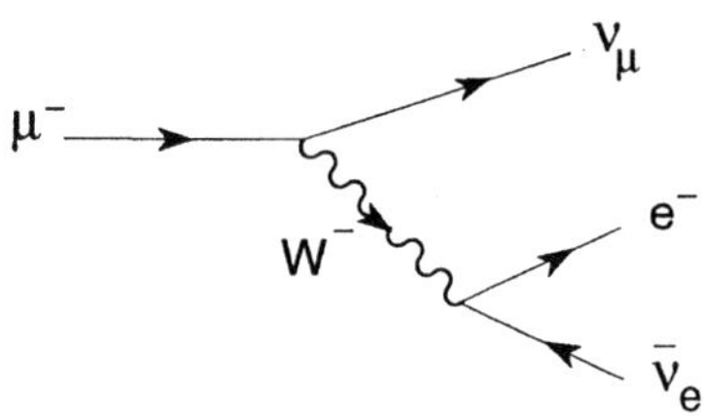

Figure 1. μ-decay diagram.

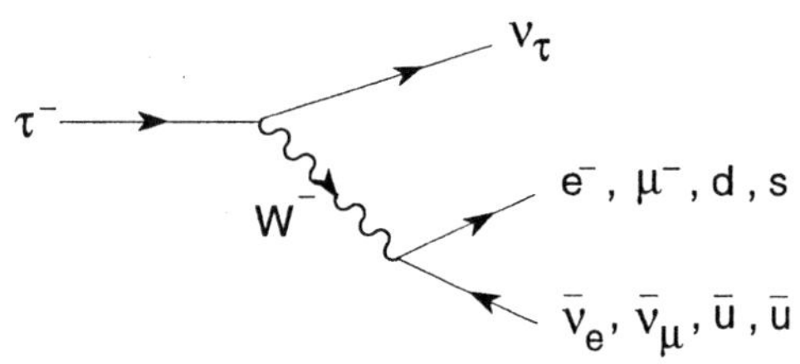

Figure 2. τ-decay diagram.

τ Decay

The decays of the τ lepton proceed through the same W–exchange mechanism as the leptonic μ decay. The only difference is that several final states are kinematically allowed: $\tau^- \to \nu_\tau e^-\bar{\nu}_e$, $\tau^- \to \nu_\tau\mu^-\bar{\nu}_\mu$, $\tau^- \to \nu_\tau d\bar{u}$ and $\tau^- \to \nu_\tau s\bar{u}$. Owing to the universality of the W–couplings, all these decay modes have equal amplitudes (if final fermion masses and QCD interactions are neglected), except for an additional $N_C|V_{ui}|^2$ factor ($i = d, s$) in the semileptonic channels, where $N_C = 3$ is the number of quark colours. Making trivial kinematical changes in Eq. (15), one easily gets the lowest–order prediction for the total τ decay width:

$$\frac{1}{\tau_\tau} \equiv \Gamma(\tau) \approx \Gamma(\mu)\left(\frac{m_\tau}{m_\mu}\right)^5 \left\{2 + N_C\left(|V_{ud}|^2 + |V_{us}|^2\right)\right\} \approx \frac{5}{\tau_\mu}\left(\frac{m_\tau}{m_\mu}\right)^5, \tag{18}$$

where we have used the unitarity relation $|V_{ud}|^2 + |V_{us}|^2 = 1 - |V_{ub}|^2 \approx 1$. From the measured muon lifetime, one has then $\tau_\tau \approx 3.3 \times 10^{-13}$ s, to be compared with the experimental value[5] $\tau_\tau^{\rm exp} = (2.9021 \pm 0.0115) \times 10^{-13}$ s.

The branching ratios into the different decay modes are predicted to be:

$$\text{Br}(\tau^- \to \nu_\tau l^-\bar{\nu}_l) \approx \frac{1}{5} = 20\%\,,$$
$$R_\tau \equiv \frac{\Gamma(\tau \to \nu_\tau + \text{Hadrons})}{\Gamma(\tau^- \to \nu_\tau e^-\bar{\nu}_e)} \approx N_C\,, \tag{19}$$

Table 2. Experimental values[5] of some basic τ decay branching fractions.

B_e	$(17.786 \pm 0.072)\%$
B_μ	$(17.317 \pm 0.078)\%$
$R_\tau^B \equiv (1 - B_e - B_\mu)/B_e$	3.649 ± 0.019
$\mathrm{Br}(\tau^- \to \nu_\tau \pi^-)$	$(11.01 \pm 0.11)\%$
$\mathrm{Br}(\tau^- \to \nu_\tau K^-)$	$(0.692 \pm 0.028)\%$

in good agreement with the measured numbers,[5] given in table 2. Our naive predictions only deviate from the experimental results by about 20%. This is the expected size of the corrections induced by the strong interactions between the final quarks, that we have neglected. Notice that the measured τ hadronic width provides strong evidence for the colour degree of freedom.

The pure leptonic decays $\tau^- \to e^-\bar{\nu}_e\nu_\tau, \mu^-\bar{\nu}_\mu\nu_\tau$ are theoretically understood at the level of the electroweak radiative corrections.[21] The corresponding decay widths are given by Eqs. (15) and (16), making the appropriate changes for the masses of the initial and final leptons.

Using the value of G_F measured in μ decay, Eq. (15) provides a relation between the τ lifetime and the leptonic branching ratios $B_l \equiv \mathrm{Br}(\tau^- \to \nu_\tau l^- \bar{\nu}_l)$:

$$B_e = \frac{B_\mu}{0.972564 \pm 0.000010} = \frac{\tau_\tau}{(1.6321 \pm 0.0014) \times 10^{-12}\,\mathrm{s}}\,. \tag{20}$$

The errors reflect the present uncertainty of 0.3 MeV in the value of m_τ.

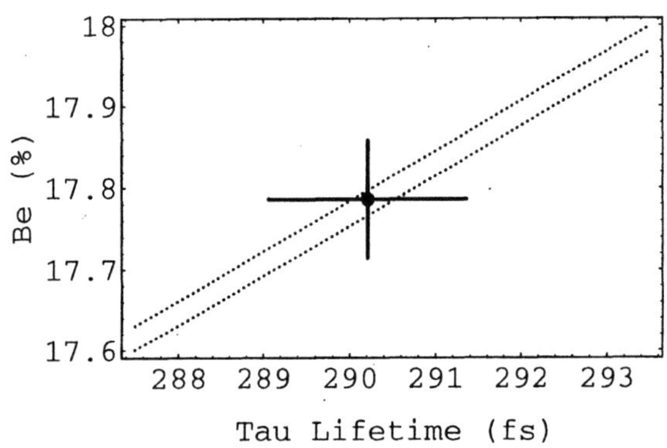

Figure 3. Relation between B_e and τ_τ. The dotted band corresponds to the prediction in Eq. (20).

The predicted B_μ/B_e ratio is in perfect agreement with the measured value $B_\mu/B_e = 0.974 \pm 0.006$. As shown in Fig. 3, the relation between B_e and τ_τ is also well satisfied by the present data. Notice, that this relation is very sensitive to the value of the τ mass $[\Gamma(\tau^- \to l^-\bar{\nu}_l\nu_\tau) \propto m_\tau^5]$. The most recent measurements of τ_τ, B_e and m_τ have consistently moved the world averages in the correct direction, eliminating the previous ($\sim 2\sigma$) disagreement.[22] The experimental precision (0.4%) is already approaching the level where a possible non-zero ν_τ mass could become relevant; the present bound[5] $m_{\nu_\tau} < 18.2$ MeV (95% CL) only guarantees that such effect is below 0.08%.

Semileptonic Decays

Semileptonic decays such as $\tau^- \to \nu_\tau P^-$ or $P^- \to l^-\bar{\nu}_l$ [$P = \pi, K$] can be predicted in a similar way. The effects of the strong interactions are contained in the so–called decay constants f_P, which parameterize the hadronic matrix element of the corresponding weak current:

$$\begin{aligned} \langle \pi^-(p)|\bar{d}\gamma^\mu\gamma_5 u|0\rangle &\equiv -i\sqrt{2}f_\pi p^\mu , \\ \langle K^-(p)|\bar{s}\gamma^\mu\gamma_5 u|0\rangle &\equiv -i\sqrt{2}f_K p^\mu . \end{aligned} \tag{21}$$

Taking appropriate ratios of different semileptonic decay widths involving the same meson P, the dependence on these decay constants factors out. Therefore, those ratios can be predicted rather accurately:

$$\begin{aligned} R_{e/\mu} &\equiv \frac{\Gamma(\pi^- \to e^-\bar{\nu}_e)}{\Gamma(\pi^- \to \mu^-\bar{\nu}_\mu)} = \frac{m_e^2(1-m_e^2/m_\pi^2)^2}{m_\mu^2(1-m_\mu^2/m_\pi^2)^2}(1+\delta R_{e/\mu}) = (1.2351 \pm 0.0005)\times 10^{-4}, \\ R_{\tau/\pi} &\equiv \frac{\Gamma(\tau^- \to \nu_\tau\pi^-)}{\Gamma(\pi^- \to \mu^-\bar{\nu}_\mu)} = \frac{m_\tau^3}{2m_\pi m_\mu^2}\frac{(1-m_\pi^2/m_\tau^2)^2}{(1-m_\mu^2/m_\pi^2)^2}\left(1+\delta R_{\tau/\pi}\right) = 9774 \pm 15 , \\ R_{\tau/K} &\equiv \frac{\Gamma(\tau^- \to \nu_\tau K^-)}{\Gamma(K^- \to \mu^-\bar{\nu}_\mu)} = \frac{m_\tau^3}{2m_K m_\mu^2}\frac{(1-m_K^2/m_\tau^2)^2}{(1-m_\mu^2/m_K^2)^2}\left(1+\delta R_{\tau/K}\right) = 480.4 \pm 1.1 , \end{aligned} \tag{22}$$

where $\delta R_{e/\mu} = -(3.76 \pm 0.04)\%$, $\delta R_{\tau/\pi} = (0.16 \pm 0.14)\%$ and $\delta R_{\tau/K} = (0.90 \pm 0.22)\%$ are the computed[23,24] radiative corrections. These predictions are in excellent agreement with the measured ratios[5,25,26]: $R_{e/\mu} = (1.2310 \pm 0.0037)\times 10^{-4}$, $R_{\tau/\pi} = 9878 \pm 106$ and $R_{\tau/K} = 465 \pm 19$.

Universality Tests

All these measurements can be used to test the universality of the W couplings to the leptonic charged currents. Allowing the coupling g in Eq. (11) to depend on the considered lepton flavour (i.e., g_e, g_μ, g_τ), the ratios B_μ/B_e and $R_{e/\mu}$ constrain $|g_\mu/g_e|$, while B_e/τ_τ and $R_{\tau/P}$ provide information on $|g_\tau/g_\mu|$. The present results are shown in tables 3 and 4, together with the values obtained from the comparison of the $\sigma \cdot B$ partial production cross-sections for the various $W^- \to l^-\bar{\nu}_l$ decay modes at the p-$\bar{p}$ colliders.[27–29]

The present data verify the universality of the leptonic charged–current couplings to the 0.15% (μ/e) and 0.30% (τ/μ) level. The precision of the most recent τ–decay measurements is becoming competitive with the more accurate π–decay determination. It is important to realize the complementarity of the different universality tests. The pure leptonic decay modes probe the charged–current couplings of a transverse W. In contrast, the decays $\pi/K \to l\bar{\nu}$ and $\tau \to \nu_\tau\pi/K$ are only sensitive to the spin–0 piece of the charged current; thus, they could unveil the presence of possible scalar–exchange contributions with Yukawa–like couplings proportional to some power of the charged–lepton mass. One can easily imagine new physics scenarios which would modify differently the two types of leptonic couplings.[6] For instance, in the usual two Higgs doublet model, charged–scalar exchange generates a correction to the ratio B_μ/B_e, but $R_{\pi\to e/\mu}$ remains unaffected. Similarly, lepton mixing between the ν_τ and an hypothetical heavy neutrino would not modify the ratios B_μ/B_e and $R_{\pi\to e/\mu}$, but would certainly correct the relation between B_l and the τ lifetime.

Table 3. Present constraints on $|g_\mu/g_e|$.

	$|g_\mu/g_e|$
B_μ/B_e	1.0005 ± 0.0030
$R_{\pi\to e/\mu}$	1.0017 ± 0.0015
$\sigma \cdot B_{W\to\mu/e}$	1.01 ± 0.04

Table 4. Present constraints on $|g_\tau/g_\mu|$.

	$|g_\tau/g_\mu|$
$B_e\tau_\mu/\tau_\tau$	1.0001 ± 0.0029
$R_{\tau/\pi}$	1.005 ± 0.005
$R_{\tau/K}$	0.984 ± 0.020
$\sigma \cdot B_{W\to\tau/\mu}$	0.99 ± 0.05

NEUTRAL CURRENT UNIVERSALITY

In the SM, all leptons with equal electric charge have identical couplings to the Z boson:

$$\mathcal{L}_{\rm NC}^{Z} = \frac{g}{2\cos\theta_W} Z_\mu \sum_l \bar{l}\gamma^\mu(v_l - a_l\gamma_5)l\,, \tag{23}$$

where

$$v_l = T_3^l(1 - 4|Q_l|\sin^2\theta_W)\,, \qquad a_l = T_3^l\,. \tag{24}$$

This has been tested at LEP and SLC,[30] where the *effective* vector and axial–vector couplings of the three charged leptons have been determined.

For unpolarized e^+ and e^- beams, the differential $e^+e^- \to \gamma, Z \to l^+l^-$ cross-section can be written, at lowest order, as

$$\frac{d\sigma}{d\Omega} = \frac{\alpha^2}{8s}\left\{A\,(1+\cos^2\theta) + B\,\cos\theta - h_l\left[C\,(1+\cos^2\theta) + D\cos\theta\right]\right\}, \tag{25}$$

where h_l $(= \pm 1)$ is the l^- helicity and θ is the scattering angle between e^- and l^-. Here,

$$\begin{aligned} A &= 1 + 2v_e v_l\,\mathrm{Re}(\chi) + (v_e^2 + a_e^2)\,(v_l^2 + a_l^2)\,|\chi|^2, \\ B &= 4a_e a_l\,\mathrm{Re}(\chi) + 8v_e a_e v_l a_l|\chi|^2, \\ C &= 2v_e a_l\,\mathrm{Re}(\chi) + 2\,(v_e^2 + a_e^2)\,v_l a_l|\chi|^2, \\ D &= 4a_e v_l\,\mathrm{Re}(\chi) + 4v_e a_e\,(v_l^2 + a_l^2)\,|\chi|^2, \end{aligned} \tag{26}$$

and χ contains the Z propagator

$$\chi = \frac{G_F M_Z^2}{2\sqrt{2}\pi\alpha}\,\frac{s}{s - M_Z^2 + is\Gamma_Z/M_Z}\,. \tag{27}$$

The coefficients A, B, C and D can be experimentally determined, by measuring the total cross-section, the forward–backward asymmetry, the polarization asymmetry and the forward–backward polarization asymmetry, respectively:

$$\sigma(s) = \frac{4\pi\alpha^2}{3s} A,$$
$$\mathcal{A}_{\rm FB}(s) \equiv \frac{N_F - N_B}{N_F + N_B} = \frac{3}{8}\frac{B}{A},$$
$$\mathcal{A}_{\rm Pol}(s) \equiv \frac{\sigma^{(h_l=+1)} - \sigma^{(h_l=-1)}}{\sigma^{(h_l=+1)} + \sigma^{(h_l=-1)}} = -\frac{C}{A}, \tag{28}$$
$$\mathcal{A}_{\rm FB,Pol}(s) \equiv \frac{N_F^{(h_l=+1)} - N_F^{(h_l=-1)} - N_B^{(h_l=+1)} + N_B^{(h_l=-1)}}{N_F^{(h_l=+1)} + N_F^{(h_l=-1)} + N_B^{(h_l=+1)} + N_B^{(h_l=-1)}} = -\frac{3}{8}\frac{D}{A}.$$

Here, N_F and N_B denote the number of l^-'s emerging in the forward and backward hemispheres, respectively, with respect to the electron direction.

For $s = M_Z^2$, the real part of the Z propagator vanishes and the photon exchange terms can be neglected in comparison with the Z–exchange contributions ($\Gamma_Z^2/M_Z^2 \ll 1$). Eqs. (28) become then,

$$\sigma^{0,l} \equiv \sigma(M_Z^2) = \frac{12\pi}{M_Z^2}\frac{\Gamma_e\Gamma_l}{\Gamma_Z^2}, \qquad \mathcal{A}_{\rm FB}^{0,l} \equiv \mathcal{A}_{FB}(M_Z^2) = \frac{3}{4}\mathcal{P}_e\mathcal{P}_l,$$
$$\mathcal{A}_{\rm Pol}^{0,l} \equiv \mathcal{A}_{\rm Pol}(M_Z^2) = \mathcal{P}_l, \qquad \mathcal{A}_{\rm FB,Pol}^{0,l} \equiv \mathcal{A}_{\rm FB,Pol}(M_Z^2) = \frac{3}{4}\mathcal{P}_e, \tag{29}$$

where Γ_l is the Z partial decay width to the l^+l^- final state, and

$$\mathcal{P}_l \equiv \frac{-2v_l a_l}{v_l^2 + a_l^2} \tag{30}$$

is the average longitudinal polarization of the lepton l^-, which only depends on the ratio of the vector and axial–vector couplings. $\mathcal{P}_l$ is a sensitive function of $\sin^2\theta_W$.

The Z partial decay width to the l^+l^- final state,

$$\Gamma_l \equiv \Gamma(Z \to l^+l^-) = \frac{G_F M_Z^3}{6\pi\sqrt{2}}\left(v_l^2 + a_l^2\right)\left(1 + \frac{3\alpha}{4\pi}\right), \tag{31}$$

determines the sum $(v_l^2 + a_l^2)$, while the ratio v_l/a_l is derived from the asymmetries†. The signs of v_l and a_l are fixed by requiring $a_e < 0$.

The measurement of the final polarization asymmetries can (only) be done for $l = \tau$, because the spin polarization of the τ's is reflected in the distorted distribution of their decay products. Therefore, $\mathcal{P}_\tau$ and $\mathcal{P}_e$ can be determined from a measurement of the spectrum of the final charged particles in the decay of one τ, or by studying the correlated distributions between the final products of both $\tau's$.[33]

With polarized e^+e^- beams, one can also study the left–right asymmetry between the cross-sections for initial left- and right-handed electrons. At the Z peak, this asymmetry directly measures the average initial lepton polarization, $\mathcal{P}_e$, without any need for final particle identification:

$$\mathcal{A}_{\rm LR}^0 \equiv \mathcal{A}_{\rm LR}(M_Z^2) = \frac{\sigma_L(M_Z^2) - \sigma_R(M_Z^2)}{\sigma_L(M_Z^2) + \sigma_R(M_Z^2)} = -\mathcal{P}_e. \tag{32}$$

Tables 5 and 6 show the present experimental results for the leptonic Z–decay widths and asymmetries. The data are in excellent agreement with the SM predictions and confirm the

†The asymmetries determine two possible solutions for $|v_l/a_l|$. This ambiguity can be solved with lower–energy data or through the measurement of the transverse spin–spin correlation[31] of the two τ's in $Z \to \tau^+\tau^-$, which requires[32] $|v_\tau/a_\tau| << 1$.

Table 5. Measured values[30] of $\Gamma_l \equiv \Gamma(Z \to l^+l^-)$ and the leptonic forward–backward asymmetries. The last column shows the combined result (for a massless lepton) assuming lepton universality.

	e	μ	τ	l
Γ_l (MeV)	83.96 ± 0.15	83.79 ± 0.22	83.72 ± 0.26	83.91 ± 0.11
$\mathcal{A}_{\rm FB}^{0,l}$ (%)	1.60 ± 0.24	1.62 ± 0.13	2.01 ± 0.18	1.74 ± 0.10

Table 6. Measured values[30] of the different polarization asymmetries.

$\mathcal{A}_{\rm Pol}^{0,\tau} = \mathcal{P}_\tau$	$\frac{4}{3}\mathcal{A}_{\rm FB,Pol}^{0,\tau} = \mathcal{P}_e$	$-\mathcal{A}_{\rm LR}^{0} = \mathcal{P}_e$	$-\{\frac{4}{3}\mathcal{A}_{\rm FB}^{0,l}\}^{1/2} = \mathcal{P}_l$
-0.1401 ± 0.0067	-0.1382 ± 0.0076	-0.1542 ± 0.0037	-0.1523 ± 0.0044

universality of the leptonic neutral couplings‡. There is however a small ($\sim 2\sigma$) discrepancy between the $\mathcal{P}_e$ values obtained[30] from $\mathcal{A}_{\rm FB,Pol}^{0,\tau}$ and $\mathcal{A}_{\rm LR}^{0}$. Assuming lepton universality, the combined result from all leptonic asymmetries gives

$$\mathcal{P}_l = -0.1500 \pm 0.0025\,. \tag{33}$$

The measurement of $\mathcal{A}_{\rm Pol}^{0,\tau}$ and $\mathcal{A}_{\rm FB,Pol}^{0,\tau}$ assumes that the τ decay proceeds through the SM charged–current interaction. A more general analysis should take into account the fact that the τ–decay width depends on the product $\xi\mathcal{P}_\tau$ (see the next section), where ξ is the corresponding Michel parameter in leptonic decays, or the equivalent quantity ξ_h ($= h_{\nu_\tau}$) in the semileptonic modes. A separate measurement of ξ and $\mathcal{P}_\tau$ has been performed by ALEPH[34] ($\mathcal{P}_\tau = -0.139 \pm 0.040$) and L3[35] ($\mathcal{P}_\tau = -0.154 \pm 0.022$), using the correlated distribution of the $\tau^+\tau^-$ decays.

The combined analysis of all leptonic observables from LEP and SLD ($\mathcal{A}_{\rm LR}^{0}$) results in the effective vector and axial–vector couplings given in table 7.[30] The corresponding 68% probability contours in the a_l–v_l plane are shown in Fig. 4. The measured ratios of the e, μ and τ couplings provide a test of charged–lepton universality in the neutral–current sector.

The neutrino couplings can be determined from the invisible Z–decay width, by assuming three identical neutrino generations with left–handed couplings (i.e., $v_\nu = a_\nu$), and fixing the sign from neutrino scattering data.[36] The resulting experimental value,[30] given in table 7, is in perfect agreement with the SM. Alternatively, one can use the SM prediction for $\Gamma_{\rm inv}/\Gamma_l$ to get a determination of the number of (light) neutrino flavours[30]:

$$N_\nu = 2.989 \pm 0.012\,. \tag{34}$$

The universality of the neutrino couplings has been tested with $\nu_\mu e$ scattering data, which fixes[37] the ν_μ coupling to the Z: $v_{\nu_\mu} = a_{\nu_\mu} = 0.502 \pm 0.017$.

The measured leptonic asymmetries can be used to obtain the effective electroweak mixing angle in the charged–lepton sector:[30]

$$\sin^2\theta_{\rm eff}^{\rm lept} \equiv \frac{1}{4}\left(1 - \frac{v_l}{a_l}\right) = 0.23114 \pm 0.00031\,. \tag{35}$$

Including also the hadronic asymmetries, one gets[30] $\sin^2\theta_{\rm eff}^{\rm lept} = 0.23165 \pm 0.00024$ with a $\chi^2/\text{d.o.f.} = 12.8/6$.

‡A small 0.2% difference between Γ_τ and $\Gamma_{e,\mu}$ is generated by the m_τ corrections.

Table 7. Effective vector and axial–vector lepton couplings derived from LEP and SLD data.[30]

	Without Lepton Universality	
	LEP	LEP + SLD
v_e	-0.0368 ± 0.0015	-0.03828 ± 0.00079
v_μ	-0.0372 ± 0.0034	-0.0358 ± 0.0030
v_τ	-0.0369 ± 0.0016	-0.0367 ± 0.0016
a_e	-0.50130 ± 0.00046	-0.50119 ± 0.00045
a_μ	-0.50076 ± 0.00069	-0.50086 ± 0.00068
a_τ	-0.50116 ± 0.00079	-0.50117 ± 0.00079
v_μ/v_e	1.01 ± 0.11	0.935 ± 0.085
v_τ/v_e	1.001 ± 0.062	0.959 ± 0.046
a_μ/a_e	0.9989 ± 0.0018	0.9993 ± 0.0017
a_τ/a_e	0.9997 ± 0.0019	1.0000 ± 0.0019
	With Lepton Universality	
	LEP	LEP + SLD
v_l	-0.03688 ± 0.00085	-0.03776 ± 0.00062
a_l	-0.50115 ± 0.00034	-0.50108 ± 0.00034
$a_\nu = v_\nu$	$+0.5009 \pm 0.0010$	$+0.5009 \pm 0.0010$

LORENTZ STRUCTURE OF THE CHARGED CURRENTS

Let us consider the decay $l^- \to \nu_l l'^- \bar{\nu}_{l'}$, where the lepton pair (l, l') may be (μ, e), (τ, e), or (τ, μ). The most general, local, derivative–free, lepton–number conserving, four–lepton interaction Hamiltonian, consistent with locality and Lorentz invariance,[38–43]

$$\mathcal{H} = 4\frac{G_{l'l}}{\sqrt{2}} \sum_{n,\epsilon,\omega} g^n_{\epsilon\omega} \left[\overline{l'_\epsilon}\Gamma^n (\nu_{l'})_\sigma\right] \left[\overline{(\nu_l)_\lambda}\Gamma_n l_\omega\right] , \tag{36}$$

contains ten complex coupling constants or, since a common phase is arbitrary, nineteen independent real parameters which could be different for each leptonic decay. The subindices $\epsilon, \omega, \sigma, \lambda$ label the chiralities (left–handed, right–handed) of the corresponding fermions, and n the type of interaction: scalar (I), vector (γ^μ), tensor ($\sigma^{\mu\nu}/\sqrt{2}$). For given n, ϵ, ω, the neutrino chiralities σ and λ are uniquely determined.

Taking out a common factor $G_{l'l}$, which is determined by the total decay rate, the coupling constants $g^n_{\epsilon\omega}$ are normalized to[41]

$$\begin{aligned} 1 = \frac{1}{4} &\left(|g^S_{RR}|^2 + |g^S_{RL}|^2 + |g^S_{LR}|^2 + |g^S_{LL}|^2\right) + 3\left(|g^T_{RL}|^2 + |g^T_{LR}|^2\right) \\ &+ \left(|g^V_{RR}|^2 + |g^V_{RL}|^2 + |g^V_{LR}|^2 + |g^V_{LL}|^2\right) . \end{aligned} \tag{37}$$

In the SM, $g^V_{LL} = 1$ and all other $g^n_{\epsilon\omega} = 0$.

For an initial lepton polarization $\mathcal{P}_l$, the final charged–lepton distribution in the decaying–lepton rest frame is usually parameterized[39] in the form

$$\frac{d^2\Gamma}{dx\, d\cos\theta} = \frac{m_l \omega^4}{2\pi^3} G^2_{l'l} \sqrt{x^2 - x_0^2} \left\{ F(x) - \frac{\xi}{3} \mathcal{P}_l \sqrt{x^2 - x_0^2} \cos\theta\, A(x) \right\} , \tag{38}$$

where θ is the angle between the l^- spin and the final charged–lepton momentum, $\omega \equiv (m_l^2 + m_{l'}^2)/2m_l$ is the maximum l'^- energy for massless neutrinos, $x \equiv E_{l'^-}/\omega$ is the reduced energy, $x_0 \equiv m_{l'}/\omega$ and

$$\begin{aligned} F(x) &= x(1-x) + \frac{2}{9}\rho\left(4x^2 - 3x - x_0^2\right) + \eta\, x_0(1-x)\,, \\ A(x) &= 1 - x + \frac{2}{3}\delta\left(4x - 4 + \sqrt{1-x_0^2}\right)\,. \end{aligned} \tag{39}$$

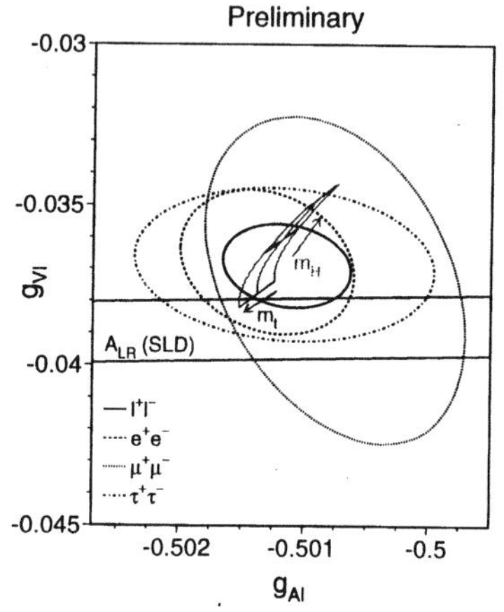

Figure 4. 68% probability contours in the a_l-v_l plane from LEP measurements.[30] The solid contour assumes lepton universality. Also shown is the 1σ band resulting from the $\mathcal{A}_{\mathrm{LR}}^0$ measurement at SLD. The grid corresponds to the SM prediction.

For unpolarized $l's$, the distribution is characterized by the so-called Michel[38] parameter ρ and the low–energy parameter η. Two more parameters, ξ and δ, can be determined when the initial lepton polarization is known. If the polarization of the final charged lepton is also measured, 5 additional independent parameters[4] (ξ', ξ'', η'', α', β') appear.

For massless neutrinos, the total decay rate is given by[43]

$$\Gamma = \frac{m_l^5 \widehat{G}_{l'l}^2}{192\pi^3}\, f\left(\frac{m_{l'}^2}{m_l^2}\right)\,(1+\delta_{\mathrm{RC}})\,, \tag{40}$$

where

$$\widehat{G}_{l'l} \equiv G_{l'l}\sqrt{1 + 4\,\eta\,\frac{m_{l'}}{m_l}\,\frac{g(m_{l'}^2/m_l^2)}{f(m_{l'}^2/m_l^2)}}\,, \tag{41}$$

and $g(z) = 1 + 9z - 9z^2 - z^3 + 6z(1+z)\ln z$. Thus, $\widehat{G}_{e\mu}$ corresponds to the Fermi coupling G_F, measured in μ decay. The B_μ/B_e and $B_e\tau_\mu/\tau_\tau$ universality tests, discussed in the previous section, actually prove the ratios $|\widehat{G}_{\mu\tau}/\widehat{G}_{e\tau}|$ and $|\widehat{G}_{e\tau}/\widehat{G}_{e\mu}|$, respectively. An important point, emphatically stressed by Fetscher and Gerber,[42] concerns the extraction of $G_{e\mu}$, whose uncertainty is dominated by the uncertainty in $\eta_{\mu\to e}$.

In terms of the $g^n_{\epsilon\omega}$ couplings, the shape parameters in Eqs. (38) and (39) are:

$$\begin{aligned}
\rho &= \frac{3}{4}(\beta^+ + \beta^-) + (\gamma^+ + \gamma^-), \\
\xi &= 3(\alpha^- - \alpha^+) + (\beta^- - \beta^+) + \frac{7}{3}(\gamma^+ - \gamma^-), \\
\xi\delta &= \frac{3}{4}(\beta^- - \beta^+) + (\gamma^+ - \gamma^-), \\
\eta &= \frac{1}{2}\mathrm{Re}\left[g^V_{LL} g^{S*}_{RR} + g^V_{RR} g^{S*}_{LL} + g^V_{LR}\left(g^{S*}_{RL} + 6 g^{T*}_{RL}\right) + g^V_{RL}\left(g^{S*}_{LR} + 6 g^{T*}_{LR}\right)\right],
\end{aligned} \tag{42}$$

where[44]

$$\begin{aligned}
\alpha^+ &\equiv |g^V_{RL}|^2 + \frac{1}{16}|g^S_{RL} + 6 g^T_{RL}|^2, & \alpha^- &\equiv |g^V_{LR}|^2 + \frac{1}{16}|g^S_{LR} + 6 g^T_{LR}|^2, \\
\beta^+ &\equiv |g^V_{RR}|^2 + \frac{1}{4}|g^S_{RR}|^2, & \beta^- &\equiv |g^V_{LL}|^2 + \frac{1}{4}|g^S_{LL}|^2, \\
\gamma^+ &\equiv \frac{3}{16}|g^S_{RL} - 2 g^T_{RL}|^2, & \gamma^- &\equiv \frac{3}{16}|g^S_{LR} - 2 g^T_{LR}|^2,
\end{aligned} \tag{43}$$

are positive–definite combinations of decay constants, corresponding to a final right– (α^+, β^+, γ^+) or left– ($\alpha^-, \beta^-, \gamma^-$) handed lepton. In the SM, $\rho = \delta = 3/4$, $\eta = \eta'' = \alpha' = \beta' = 0$ and $\xi = \xi' = \xi'' = 1$.

The normalization constraint (37) is equivalent to $\alpha^+ + \alpha^- + \beta^+ + \beta^- + \gamma^+ + \gamma^- = 1$. It is convenient to introduce[41] the probabilities $Q_{\epsilon\omega}$ for the decay of an ω–handed l^- into an ϵ–handed daughter lepton,

$$\begin{aligned}
Q_{LL} &= \frac{1}{4}|g^S_{LL}|^2 + |g^V_{LL}|^2 &&= \frac{1}{4}\left(-3 + \frac{16}{3}\rho - \frac{1}{3}\xi + \frac{16}{9}\xi\delta + \xi' + \xi''\right), \\
Q_{RR} &= \frac{1}{4}|g^S_{RR}|^2 + |g^V_{RR}|^2 &&= \frac{1}{4}\left(-3 + \frac{16}{3}\rho + \frac{1}{3}\xi - \frac{16}{9}\xi\delta - \xi' + \xi''\right), \\
Q_{LR} &= \frac{1}{4}|g^S_{LR}|^2 + |g^V_{LR}|^2 + 3|g^T_{LR}|^2 &&= \frac{1}{4}\left(5 - \frac{16}{3}\rho + \frac{1}{3}\xi - \frac{16}{9}\xi\delta + \xi' - \xi''\right), \\
Q_{RL} &= \frac{1}{4}|g^S_{RL}|^2 + |g^V_{RL}|^2 + 3|g^T_{RL}|^2 &&= \frac{1}{4}\left(5 - \frac{16}{3}\rho - \frac{1}{3}\xi + \frac{16}{9}\xi\delta - \xi' - \xi''\right).
\end{aligned} \tag{44}$$

Upper bounds on any of these (positive–semidefinite) probabilities translate into corresponding limits for all couplings with the given chiralities.

For μ decay, where precise measurements of the polarizations of both μ and e have been performed, there exist[41] upper bounds on Q_{RR}, Q_{LR} and Q_{RL}, and a lower bound on Q_{LL}. They imply corresponding upper bounds on the 8 couplings $|g^n_{RR}|$, $|g^n_{LR}|$ and $|g^n_{RL}|$. The measurements of the μ^- and the e^- do not allow to determine $|g^S_{LL}|$ and $|g^V_{LL}|$ separately.[41,45] Nevertheless, since the helicity of the ν_μ in pion decay is experimentally known[46] to be -1, a lower limit on $|g^V_{LL}|$ is obtained[41] from the inverse muon decay $\nu_\mu e^- \to \mu^- \nu_e$. The present (90% CL) bounds[4] on the μ–decay couplings are shown in Fig. 5. These limits show nicely that the bulk of the μ–decay transition amplitude is indeed of the predicted V−A type.

The experimental analysis of the τ–decay parameters is necessarily different from the one applied to the muon, because of the much shorter τ lifetime. The measurement of the τ polarization and the parameters ξ and δ is possible due to the fact that the spins of the $\tau^+\tau^-$ pair produced in e^+e^- annihilation are strongly correlated.[31,33,47–52] Another possibility is to use the beam polarization, as done by SLD. However, the polarization of the charged lepton emitted in the τ decay has never been measured. In principle, this could be done for the decay $\tau^- \to \mu^- \bar{\nu}_\mu \nu_\tau$ by stopping the muons and detecting their decay products.[51] An alternative method would be[53] to use the radiative decays $\tau \to l^- \bar{\nu}_l \nu_\tau \gamma$ ($l = e, \mu$), since the distribution

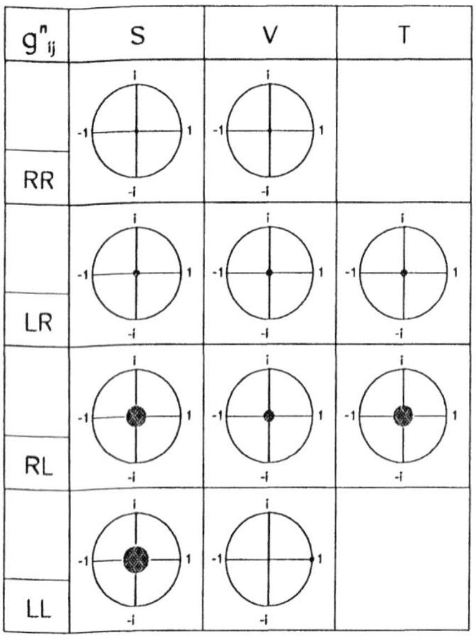

Figure 5. 90% CL experimental limits[4] for the normalized μ–decay couplings $g'^n_{\epsilon\omega} \equiv g^n_{\epsilon\omega}/N^n$, where $N^n \equiv \max(|g^n_{\epsilon\omega}|) = 2,\ 1,\ 1/\sqrt{3}$ for $n =$ S, V, T. (Taken from Ref. 54).

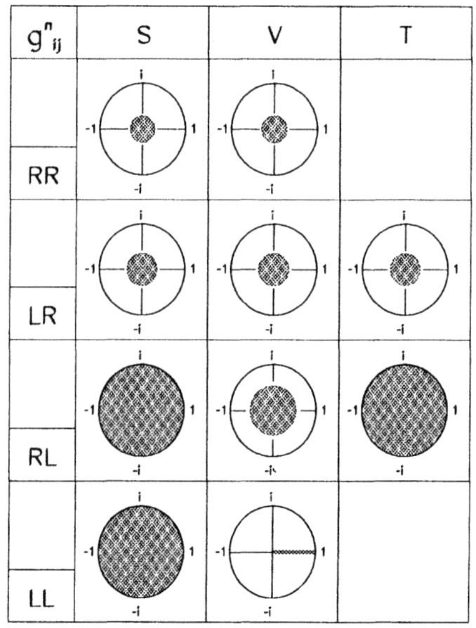

Figure 6. 90% CL experimental limits for the normalized τ–decay couplings $g'^n_{\epsilon\omega} \equiv g^n_{\epsilon\omega}/N^n$, assuming e/μ universality.

of the photons emitted by the daughter lepton is sensitive to the lepton spin. The measurement of the inverse decay $\nu_\tau l^- \to \tau^- \nu_l$ looks far out of reach.

The present experimental status[5] on the τ–decay Michel parameters is shown in table 8. For comparison, the values measured in μ decay[4] are also given. The improved accuracy of the most recent experimental analyses has brought an enhanced sensitivity to the different shape parameters, allowing the first measurements[5] of $\eta_{\tau\to\mu}$, $\xi_{\tau\to e}$, $\xi_{\tau\to\mu}$, $(\xi\delta)_{\tau\to e}$ and $(\xi\delta)_{\tau\to\mu}$ without any e/μ universality assumption.

Table 8. World average[4,5] Michel parameters. The last column ($\tau \to l$) assumes identical couplings for $l = e, \mu$. $\xi_{\mu\to e}$ refers to the product $\xi_{\mu\to e}\mathcal{P}_\mu$, where $\mathcal{P}_\mu \approx 1$ is the longitudinal polarization of the μ from π decay.

	$\mu \to e$	$\tau \to \mu$	$\tau \to e$	$\tau \to l$
ρ	0.7518 ± 0.0026	0.733 ± 0.031	0.734 ± 0.016	0.741 ± 0.014
η	-0.007 ± 0.013	-0.04 ± 0.20	—	0.047 ± 0.076
ξ	1.0027 ± 0.0085	1.19 ± 0.18	1.09 ± 0.16	1.04 ± 0.09
$\xi\delta$	0.7506 ± 0.0074	0.73 ± 0.11	0.80 ± 0.18	0.73 ± 0.07

The determination of the τ polarization parameters allows us to bound the total probability for the decay of a right–handed τ,[51]

$$Q_{\tau_R} \equiv Q_{RR} + Q_{LR} = \frac{1}{2}\left[1 + \frac{\xi}{3} - \frac{16}{9}(\xi\delta)\right] . \tag{45}$$

One finds (ignoring possible correlations among the measurements):

$$\begin{aligned} Q_{\tau_R}^{\tau\to\mu} &= 0.05 \pm 0.10 \;<\; 0.20 \quad (90\%\ \mathrm{CL}), \\ Q_{\tau_R}^{\tau\to e} &= -0.03 \pm 0.16 \;<\; 0.25 \quad (90\%\ \mathrm{CL}), \\ Q_{\tau_R}^{\tau\to l} &= 0.02 \pm 0.06 \;<\; 0.12 \quad (90\%\ \mathrm{CL}), \end{aligned} \tag{46}$$

where the last value refers to the τ decay into either $l = e$ or μ, assuming identical e/μ couplings. Since these probabilities are positive semidefinite quantities, they imply corresponding limits on all $|g^n_{RR}|$ and $|g^n_{LR}|$ couplings.

A measurement of the final lepton polarization could be even more efficient, since the total probability for the decay into a right–handed lepton depends on a single Michel parameter:

$$Q_{l'_R} \equiv Q_{RR} + Q_{RL} = \frac{1}{2}(1 - \xi') \,. \tag{47}$$

Thus, a single polarization measurement could bound the five RR and RL complex couplings.

Another useful positive–definite quantity is[54]

$$\rho - \xi\delta = \frac{3}{2}\beta^+ + 2\gamma^- \,, \tag{48}$$

which provides direct bounds on $|g^V_{RR}|$ and $|g^S_{RR}|$. A rather weak upper limit on γ^+ is obtained from the parameter ρ. More stringent is the bound on α^+ obtained from $(1 - \rho)$, which is also positive–definite; it implies a corresponding limit on $|g^V_{RL}|$.

Table 9 gives the resulting (90% CL) bounds on the τ–decay couplings. The relevance of these limits can be better appreciated in Fig. 6, where e/μ universality has been assumed.

Table 9. 90% CL limits for the $g^n_{\epsilon\omega}$ couplings.

	$\mu \to e$	$\tau \to \mu$	$\tau \to e$	$\tau \to l$
$\lvert g^S_{RR}\rvert$	< 0.066	< 0.71	< 0.83	< 0.57
$\lvert g^S_{LR}\rvert$	< 0.125	< 0.90	< 1.00	< 0.70
$\lvert g^S_{RL}\rvert$	< 0.424	≤ 2	≤ 2	≤ 2
$\lvert g^S_{LL}\rvert$	< 0.55	≤ 2	≤ 2	≤ 2
$\lvert g^V_{RR}\rvert$	< 0.033	< 0.36	< 0.42	< 0.29
$\lvert g^V_{LR}\rvert$	< 0.060	< 0.45	< 0.50	< 0.35
$\lvert g^V_{RL}\rvert$	< 0.110	< 0.56	< 0.54	< 0.53
$\lvert g^V_{LL}\rvert$	> 0.96	≤ 1	≤ 1	≤ 1
$\lvert g^T_{LR}\rvert$	< 0.036	< 0.26	< 0.29	< 0.20
$\lvert g^T_{RL}\rvert$	< 0.122	$\leq 1/\sqrt{3}$	$\leq 1/\sqrt{3}$	$\leq 1/\sqrt{3}$

If lepton universality is assumed, the leptonic decay ratios B_μ/B_e and $B_e\tau_\mu/\tau_\tau$ provide limits on the low–energy parameter η. The best sensitivity[55] comes from $\widehat{G}_{\mu\tau}$, where the term proportional to η is not suppressed by the small m_e/m_l factor. The measured B_μ/B_e ratio implies then:

$$\eta_{\tau\to l} \;=\; 0.005 \pm 0.027 \,. \tag{49}$$

This determination is more accurate that the one in table 8, obtained from the shape of the energy distribution, and is comparable to the value measured in μ decay.

A non-zero value of η would show that there are at least two different couplings with opposite chiralities for the charged leptons. Assuming the V$-$A coupling g^V_{LL} to be dominant, the second one would be[51] a Higgs–type coupling g^S_{RR}. To first order in new physics contributions, $\eta \approx \mathrm{Re}(g^S_{RR})/2$; Eq. (49) puts then the (90% CL) bound: $-0.08 < \mathrm{Re}(g^S_{RR}) < 0.10$.

High–precision measurements of the τ decay parameters have the potential to find signals for new phenomena. The accuracy of the present data is still not good enough to provide strong constraints; nevertheless, it shows that the SM gives indeed the dominant contribution to the decay amplitude. Future experiments should then look for small deviations of the SM predictions and find out the possible source of any detected discrepancy.

In a first analysis, it seems natural to assume[43] that new physics effects would be dominated by the exchange of a single intermediate boson, coupling to two leptonic currents. Table 10 summarizes the expected changes on the measurable shape parameters,[43] in different new physics scenarios. The four general cases studied correspond to adding a single intermediate boson exchange, V^+, S^+, V^0, S^0 (charged/neutral, vector/scalar), to the SM contribution (a non-standard W would be a particular case of the SM + V^+ scenario).

Table 10. Changes in the Michel parameters induced by the addition of a single intermediate boson exchange (V^+, S^+, V^0, S^0) to the SM contribution[43]

	V^+	S^+	V^0	S^0
$\rho - 3/4$	< 0	0	0	< 0
$\xi - 1$	$\pm$	< 0	< 0	$\pm$
$\delta\xi - 3/4$	< 0	< 0	< 0	< 0
η	0	$\pm$	$\pm$	$\pm$

SUMMARY

The flavour structure of the SM is one of the main pending questions in our understanding of weak interactions. Although we do not know the reason of the observed family replication, we have learned experimentally that the number of SM fermion generations is just three (and no more). Therefore, we must study as precisely as possible the few existing flavours to get some hints on the dynamics responsible for their observed structure.

The lepton sector provides a clean environment to test the universality and Lorentz structure of the electroweak couplings. We want to investigate whether the mass is the only difference among the three fermion families. Naïvely, one would expect the τ to be much more sensitive than the e or the μ to new physics related to the flavour and mass–generation problems. While many precision measurements of the electron and muon properties have been done in the past, it is only recently that τ experiments have achieved a comparable accuracy.[5]

Lepton universality has been tested quite precisely, both in the charged and neutral current sectors. The leptonic couplings to the charged W have been verified to be universal at the 0.15% (g_μ/g_e) and 0.30% (g_τ/g_μ) level. The axial couplings of the Z boson to the charged leptons have been measured with a comparable accuracy; universality is satisfied to the 0.17% (a_μ/a_e) and 0.19% (a_τ/a_e) level. The experimental precision is worse for the Z vector couplings, which are known to be the same for the three charged leptons to 9% (v_μ/v_e) and 5% (v_τ/v_e) accuracy.

The Lorentz structure of the $l^- \to \nu_l l'^- \bar{\nu}_{l'}$ decay amplitudes has been investigated by many experiments. The present data nicely show that the bulk of the μ–decay transition amplitude in indeed of the predicted V$-$A type. The available information on the leptonic τ decays, is still not good enough to determine the underlying dynamics; nevertheless, useful limits on possible new physics contributions start to emerge.

At present, all experimental results are consistent with the SM. There is, however, large room for improvements. Future experiments will probe the SM to a much deeper level of sensitivity and will explore the frontier of its possible extensions.

ACKNOWLEDGEMENTS

I would like to thank the organizers for the charming atmosphere of this meeting. I am indebted to Manel Martinez for keeping me informed about the most recent LEP averages, and to Wolfgang Lohmann for providing the PAW files to generate figures 5 and 6. This work has been supported in part by CICYT (Spain) under grant No. AEN-96-1718.

REFERENCES

1. A. Pich, *The Standard Model of Electroweak Interactions*, Proc. XXII International Winter Meeting on Fundamental Physics: *The Standard Model and Beyond* (Jaca, 7–11 February 1994), eds. J.A. Villar and A. Morales (Editions Frontières, Gif-sur-Yvette, 1995), p. 1 [hep-ph/9412274].
2. A. Pich, *Quantum Chromodynamics*, Proc. 1994 European School of High Energy Physics (Sorrento, 29 August – 11 September 1994), eds. N. Ellis and M.B. Gavela, Report CERN 95-04 (Geneva, 1995), p. 157 [hep-ph/9505231].
3. A. Pich, *Flavourdynamics*, Proc. XXIII International Winter Meeting on Fundamental Physics: *The Top Quark, Heavy Flavour Physics and Symmetry Breaking* (Comillas, 22–26 May 1995), eds. T. Rodrigo and A. Ruiz (World Scientific, Singapore, 1996), p. 1 [hep-ph/9601202].
4. Particle Data Group, *Review of Particle Properties, Phys. Rev.* D54 (1996) 1.
5. A. Pich, *Tau Lepton Physics: Theory Overview*, Proc. Fourth Workshop on Tau Lepton Physics –TAU96– (Colorado, 16–19 September 1996), ed. J. Smith, *Nucl. Phys. B (Proc. Suppl.)* in press [IFIC/96-96; hep-ph/9612308].
6. W.J. Marciano, *Nucl. Phys. B (Proc. Suppl.)* 40 (1995) 3.
7. H.-U. Martyn, *Test of QED by High Energy Electron–Positron Collisions*, in Ref. 8.
8. T. Kinoshita (editor), *Quantum Electrodynamics*, Advanced Series on Directions in High Energy Physics, Vol. 7 (World Scientific, Singapore, 1990).
9. M. Hayakawa, T. Kinoshita and A.I. Sanda, *Phys. Rev.* D54 (1996) 3137.
10. T. Kinoshita and A.I. Sanda, *Phys. Rev. Lett.* 75 (1995) 790.
11. J. Bijnens, E. Pallante and J. Prades, *Nucl. Phys.* B474 (1996) 379.
12. B. Krause, *Higher–Order Hadronic Contributions to the Anomalous Magnetic Moment of Leptons*, Karlsruhe preprint TTP96-26 (1996).
13. R. Alemany, *Tau Vector Spectral Functions and the Hadronic Contributions to* $(g-2)_\mu$ *and* $\alpha(M_Z^2)$, Proc. Fourth Workshop on Tau Lepton Physics –TAU96– (Colorado, 16–19 September 1996), ed. J. Smith, *Nucl. Phys. B (Proc. Suppl.)* in press.
14. D.J. Silverman and G.L. Shaw, *Phys. Rev.* D27 (1983) 1196.
15. R. Marshall, *Rep. Prog. Phys.* 52 (1989) 1329.
16. R. Escribano and E. Masso, *Phys. Lett.* B301 (1993) 419; *Nucl. Phys.* B429 (1994) 19; hep-ph/9609423.
17. J.S. Schwinger, *Phys. Rev.* 73 (1948) 416.
18. S. Narison, *J. Phys. G: Nucl. Phys.* 4 (1978) 1849.
19. M.A. Samuel, G. Li and R. Mendel, *Phys. Rev. Lett.* 67 (1991) 668.
20. T. Kinoshita and A. Sirlin, *Phys. Rev.* 113 (1959) 1652.
21. W.J. Marciano and A. Sirlin, *Phys. Rev. Lett.* 61 (1988) 1815.

22. A. Pich, *Tau Physics*, in *Heavy Flavours*, eds. A.J. Buras and M. Lindner, Advanced Series on Directions in High Energy Physics – Vol. 10 (World Scientific, Singapore, 1992), p. 375.
23. W.J. Marciano and A. Sirlin, *Phys. Rev. Lett.* 71 (1993) 3629.
24. R. Decker and M. Finkemeier, *Nucl. Phys.* B438 (1995) 17; *Nucl. Phys. B (Proc. Suppl.)* 40 (1995) 453.
25. D.I. Britton *et al*, *Phys. Rev. Lett.* 68 (1992) 3000.
26. G. Czapek *et al*, *Phys. Rev. Lett.* 70 (1993) 17.
27. C. Albajar *et al* (UA1), *Z. Phys.* C44 (1989) 15.
28. J. Alitti *et al* (UA2), *Phys. Lett.* B280 (1992) 137.
29. F. Abe *et al* (CDF), *Phys. Rev. Lett.* 68 (1992) 3398; 69 (1992) 28.
30. The LEP Electroweak Working Group and the SLD Heavy Flavour Group, *A Combination of Preliminary LEP and SLD Electroweak Measurements and Constraints on the Standard Model*, CERN preprint LEPEWWG/96-02 (30 July 1996).
31. J. Bernabéu, A. Pich and N. Rius, *Phys. Lett.* B257 (1991) 219.
32. F. Sánchez, *Measurement of the Transverse Spin Correlations in the Decay $Z \to \tau^+\tau^-$*, Proc. Fourth Workshop on Tau Lepton Physics –TAU96– (Colorado, 16–19 September 1996), ed. J. Smith, *Nucl. Phys. B (Proc. Suppl.)* in press.
33. R. Alemany *et al*, *Nucl. Phys.* B379 (1992) 3.
34. D. Buskulic *et al* (ALEPH), *Phys. Lett.* B321 (1994) 168.
35. M. Acciarri *et al* (L3), *Phys. Lett.* B377 (1996) 313.
36. P. Vilain *et al* (CHARM II), *Phys. Lett.* B335 (1994) 246.
37. P. Vilain *et al* (CHARM II), *Phys. Lett.* B320 (1994) 203.
38. L. Michel, *Proc. Phys. Soc.* A63 (1950) 514; 1371.
39. C. Bouchiat and L. Michel, *Phys. Rev.* 106 (1957) 170;
T. Kinoshita and A. Sirlin, *Phys. Rev.* 107 (1957) 593; 108 (1957) 844.
40. F. Scheck, *Leptons, Hadrons and Nuclei* (North-Holland, Amsterdam, 1983); *Phys. Rep.* 44 (1978) 187.
41. W. Fetscher, H.-J. Gerber and K.F. Johnson, *Phys. Lett.* B173 (1986) 102.
42. W. Fetscher and H.-J. Gerber, *Precision Measurements in Muon and Tau Decays*, in *Precision Tests of the Standard Electroweak Model*, ed. P. Langacker, Advanced Series on Directions in High Energy Physics – Vol. 14 (World Scientific, Singapore, 1995), p. 657.
43. A. Pich and J.P. Silva, *Phys. Rev.* D52 (1995) 4006.
44. A. Rougé, *Results on τ Neutrino from Colliders*, Proc. XXXth Rencontre de Moriond, *Dark Matter in Cosmology, Clocks and Tests of Fundamental Laws*, eds. B. Guiderdoni *et al* (Editions Frontières, Gif-sur-Yvette, 1995) p. 247.
45. C. Jarlskog, *Nucl. Phys.* 75 (1966) 659.
46. W. Fetscher, *Phys. Lett.* 140B (1984) 117.
47. Y.S. Tsai, *Phys. Rev.* D4 (1971) 2821;
S. Kawasaki, T. Shirafuji and S.Y. Tsai, *Progr. Theor. Phys.* 49 (1973) 1656.
48. S.-Y. Pi and A.I. Sanda, *Ann. Phys., NY* 106 (1977) 171.
49. C.A. Nelson, *Phys. Rev.* D43 (1991) 1465; *Phys. Rev. Lett.* 62 (1989) 1347; *Phys. Rev.* D40 (1989) 123 [*Err:* D41 (1990) 2327].
50. S. Goozovat and C.A. Nelson, *Phys. Rev.* D44 (1991) 2818; *Phys. Lett.* B267 (1991) 128 [*Err:* B271 (1991) 468].
51. W. Fetscher, *Phys. Rev.* D42 (1990) 1544.
52. M. Davier *et al*, *Phys. Lett.* B306 (1993) 411.
53. A. Stahl and H. Voss, BONN-HE-96-02.
54. W. Lohmann and J. Raab, *Charged Current Couplings in τ Decay*, DESY 95-188.
55. A. Stahl, *Phys. Lett.* B324 (1994) 121.

NEUTRINO MASS

Rudolf L. Mössbauer

Department of physics
Technical University of Munich
D-85747 Garching, Germany

HISTORICAL REMARKS

The neutrino was hypothetically introduced into physics in 1930 by W.Pauli in order to explain the radioactive ß-decay with its continuous ß-spectrum. The name neutrino, the Italian word for little neutron, was coined in 1933 after the discovery of the neutron by E. Fermi. A first direct experimental observation of neutrinos was achieved only around 1956 by F.Reines and C.Cowan,[1] based on a reaction which today is written as $\bar{\nu}_e + p \rightarrow n + e^+$. In 1962, Lederman, Schwartz, Steinberger and coworkers,[2] employing the in-flight decay of pions and kaons, discovered a second type of neutrino, the muon neutrino ν_μ . In 1975, M. Perl discovered a third charged lepton, the tauon.[3] Theoretical arguments and recoil observations suggest its connection with still another kind of neutrino, the tauon neutrino ν_τ This third type of neutrino, due to lack of intensity, has not yet been directly observed.

NEUTRINO PROPERTIES

In 1989, the LEP accelerator (large electron-positron collider) of CERN has demonstrated, that three and only three types of lepton families do exist.[4] Experiments lead to a V-A weak interaction, which permits only left-handed neutrinos and right-handed antineutrinos. Experimental observation of neutrinos have played a major role in verifying the Glashow, Salam Weinberg $SU(2)_L \times U(1)_Y$ model of electro-weak interactions, in particular through the discovery of the weak neutral current interaction[5] in 1973. Neutrinos via deep inelastic scattering have also played a significant role in establishing the quark structure of nuclei. Yet in spite of all these spectacular applications, hardly anything is known about the neutrino itself. Table 1 gives a survey of some of the unknown properties of the neutrino. In particular, in spite of many experiments we have up to this date no knowledge about neutrino masses. There is no a priori reason for neutrino masses to be equal to zero. In particular, we know of no symmetry property for such a value. The following arguments hint instead upon finite values for neutrino masses:

Masses of Fundamental Particles: Cargèse 1996
Edited by M. Lévy *et al.*, Plenum Press, New York, 1997

Table 1. Unknown properties of neutrinos

- What are the masses of the neutrinos and why presumably are they so small ?
- Is there flavor mixing in vacuum ?
- Is there flavor mixing in matter ?
- Is there particle $\leftrightarrow$ antiparticle mixing ?
- Is there total lepton number and family lepton number conservation ?
- Are neutrinos Dirac or Majorana particles ?
- Are neutrinos stable particles ?
- Are neutrinos providing dark matter ?
- Are there relic neutrinos ?
- Is there coherent elastic scattering of neutrinos ?
- Is the measured solar neutrino flux indicative of physics beyond the Standard Solar Model
- Are there right-handed neutrinos ?
- Is CP violated in the lepton sector ?
- why are there just three neutrino flavors ?
- is there a static or a dynamic magnetic moment ?

a) there might be an analogy in the lepton sector to the mixing which is experimentally well known in the quark sector (Cabbibo-Kobayashi-Maskawa mixing matrix).
b) A mass term with $SU(2)_L$ symmetry may readily be introduced into the Dirac equation via the standard Higgs particle (Yukawa coupling), see below.
c) for Majorana neutrinos there exists even a bilinear Lorentz-covariant.
d) Grand unified theories (GUT's) favor the existence of massive Majorana neutrinos, which would permit the transfer of quarks into leptons. The resulting decay of the proton, however, up to this date has experimentally not been observed.
e) the deficit observed in the flux of solar neutrinos might be due to neutrinos with masses.
f) The noted presence of mass related to Dark Matter might be due to neutrinos.

MASS TERMS IN THE LAGRANGIAN

For free Dirac neutrinos, we have the eigenstates

$$\Psi = \nu = \nu_L + \nu_R = \tfrac{1}{2}(1 - \gamma_5)\nu + \tfrac{1}{2}(1 + \gamma_5)$$

and the mass term in the Lagrangian

$$(\mathcal{L}_m)^0_D = -\, m^0_D (\overline{\nu}_R \nu_L + \overline{\nu}_L \nu_R)$$

This mass term is not invariant under $SU(2)_L$ transformations, because there is no term ν_R in $SU(2)_L$.

For free Majorana neutrinos, which are feasible due to the neutral character of neutrinos, we have the eigenstates.

$$\chi = \chi^C = \nu_L + \nu^C_R = (\nu_L)^C + (\nu^C_R)^C = (\nu^C)_R + (\nu^{CC})_L = \nu^C_R + \nu_L \qquad \text{and}$$

$$\omega = \omega^C = \nu_R + \nu^C_L$$

leading to the following mass terms in the Lagrangian:

$$(\mathcal{L}_m)^0_{M(A)} = M_A \bar{\chi}\chi = M_A (\bar{\nu}^C_R \nu_L + \bar{\nu}_L \nu^C_R)$$

$$(\mathcal{L}_m)^0_{M(B)} = M_B \bar{\omega}\omega = M_B (\bar{\nu}^C_L \nu_R + \bar{\nu}_R \nu^C_L)$$

Note, that the mass term $(\mathcal{L}_m)^0_{M(B)}$ is not invariant under $SU(2)_L$ transformations, because there exists no term ν_R in $SU(2)_L$; by contrast, there might very well exist a term $(\mathcal{L}_m)^0_{M(A)}$.

For fermions subject to weak interactions, invariance under $SU(2)_L$ transformations can be established by introducing additional particles, i.e. the same scalar Higgs doublets, which attribute masses to the $W^{\pm}$ and the Z^0 bosons. For fermions such as quarks or leptons, we must choose ϕ for the lower and ϕ^C for the upper members of weak interaction isospin doublets, requiring Yukawa couplings for neutrinos to involve ϕ^C. We thus obtain for e.g. Dirac neutrinos the mass term.

$$(\mathcal{L}_m)^0_D = -g_\nu \phi^C \bar{\Psi}\Psi = -g_\nu \begin{pmatrix} -\eta \\ 0 \end{pmatrix} (\bar{\nu}_e, \bar{e}) \begin{pmatrix} \nu_e \\ e^- \end{pmatrix} = -g_\nu (-\eta \bar{\nu}_e, 0) \begin{pmatrix} \nu_e \\ e^- \end{pmatrix} = g_\nu \eta \bar{\nu}_e \nu_e = m_D \bar{\nu}_e \nu_e$$

SEESAW MECHANISM

One here assumes the simultaneous existence of Dirac (m_D) and Majorana (M) masses. For one family one then obtains the following mass terms:

$$\mathcal{L}_m = m_D \bar{\nu}_L \nu_R + M \bar{\nu}^C_L \nu_R + \text{h.c.} = \left(\overline{\nu_L}, \overline{\nu^C_L}\right) \begin{pmatrix} 0 & \frac{1}{2} m_D \\ \frac{1}{2} m_D & M \end{pmatrix} \begin{pmatrix} \nu^C_R \\ \nu_R \end{pmatrix}$$

Diagonalization of the mass matrix yields

the mixing angle $\quad \text{tg}\,\theta = m_D/2M \quad$ and

the mass eigenvalues $m_1 \approx m_D^2/4M \ll m_D$ and

$$m_2 \approx M \gg m_D$$

This description is called seesaw mechanism, because larger values of M lead to smaller mass eigenvalues. One might choose, e.g., $M \approx M_{GUT} \approx 10^{16}$ GeV, while m_D might (for the first family) be of order m_e or m_u. This seesaw mechanism thus leads to an „explanation“ of the small values of neutrino masses as well as to a quadratic mass hierarchy such as, e.g., $m(\nu_e) : m(\nu_\mu) : m(\nu_\tau) = m_e^2 : m_\mu^2 : m_\tau^2$.

NEUTRINOS OF COSMIC ORIGIN

The critical density ρ_c of the Universe, corresponding to $\Omega = \rho/\rho_c = 1$, which is equivalent to a curvature $k = 1$ and to a deceleration parameter $q = ½$, is given by $\rho_c = 3H_0^2/8\pi G = 1.9 \cdot 10^{-29}\, h_0^2\, \text{gm} \cdot \text{cm}^{-3}$, where the dimensionless quantity h_0 is related to the present Hubble parameter H_0 via $H_0 = h_0 H_0'$ with $H_0' = 100\ \text{km} \cdot \text{s}^{-1} \cdot \text{Mpc}^{-1}$.

The density of stable relic neutrinos must not exceed the upper limit of the density of the Universe. This requirement yields an upper limit on the masses for stable light neutrinos, the number of which according to LEP equals three.[6] Specifying the number density of neutrinos of one type by $n_{\nu+\bar{\nu}}$ we have $n_{\nu+\bar{\nu}} \approx \sum_{i=1}^{3}(m_\nu)_i c^2 < 2\rho_e$. Using $n_{\nu+\bar{\nu}} = \frac{3}{11} n_\gamma$, $n_\gamma \approx \approx 400/\text{cm}^3$ and $h_0 \approx \frac{1}{2}$, where the factor $\frac{3}{11}$ arises from the differences of Fermi- and Bose-distributions and from the annihilation of electrons and positrons and adding a factor of 2 uncertainty, we obtain a cosmological mass limit $\sum_{i=1}^{3}(m_\nu)_i c^2 < 40\,\text{eV}$ for three types of stable neutrinos.

Besides light neutrinos, there might be heavy neutrinos with masses larger than the neutrino decoupling temperature T_f ($T_f \approx 1$ MeV). Stable heavy neutrinos would drop out of thermal equilibrium very early during the expansion of the Universe. The evolution equations reveal that the present mass density of such objects is given by[6]

$$\rho = \frac{4.5 \cdot 10^{-29}\ [\text{g} \cdot \text{cm}^{-3}]}{m^2\ [\text{GeV}/c^2]^2} \tag{1}$$

To avoid overclosing the Universe, masses of light neutrinos must not exceed some 40 eV, while masses of heavy leptons according to (1) should not be lower as 2 GeV. The LEP accelerator at CERN has already excluded neutrino masses < 45 GeV, but other neutral weakly interacting particles (WIMP's) might still exist with masses as low as 2 GeV and contribute, for instance, to the Dark Matter.

DIRECT EXPERIMENTAL SEARCH FOR NEUTRINO MASSES

Table 2 gives limits obtained in direct experimental searches for neutrino masses.

Table 2. Experimental limits for neutrino masses

Neutrino mass	decay mode	confidence level	reference
$m(\nu_e) < 4.35$ eV	$^3H \rightarrow {}^3He + e^- + \bar{\nu}_e$	95 %	7
$m(\nu_\mu) < 0.16$ MeV	$\pi^+ \rightarrow \mu^+ + \nu_\mu$	90 %	8
$m(\nu_\tau) < 23.8$ MeV	$\tau^- \rightarrow 3\pi^- + 2\pi^+ + \nu_\tau$	95 %	9

In the case of electron neutrinos these limits have been obtained with an isotope with a particularly small total energy. Most of the pertinent experiments obtain negative errors for the neutino masses entering in a quadratic fashion, for reasons which are not understood. The theoretical electron spectrum for allowed ß-decays, given by

$$\frac{dN}{dE_\nu} = G_F^2 \frac{m_e^5 c^4}{2\pi^3 \hbar^7} \cos^2\theta_c |M|^2 F(Z, E_e) p_e E_e (E_0 - E_e)\left[(E_0 - E_e)^2 - m_\nu^2\right]^{\frac{1}{2}}$$

where M is the nuclear matrix element, θ_c is the Cabbibo angle, $F(Z,E_e)$ is a Coulomb correction factor for the energy loss (e^-) or gain (e^+) and E_0 is the total decay energy, can be exactly calculated. This electron spectrum is usually linearized, leading to a Kurie-plot:

$$K(E_e) = \left[\frac{dN/dE_e}{p_e E_e F(Z,E_e)}\right]^{\frac{1}{2}} \propto \left\{(E_0 - E_e)\left[(E_0 - E_e)^2 - m_\nu^2\right]^{\frac{1}{2}}\right\}^{\frac{1}{2}}$$

For $m_\nu = 0$ a straight line arrives, while for $m\nu \neq 0$ one obtains a vertical drop at the energy $E_0 - m_\nu$, as shown in figure 1.

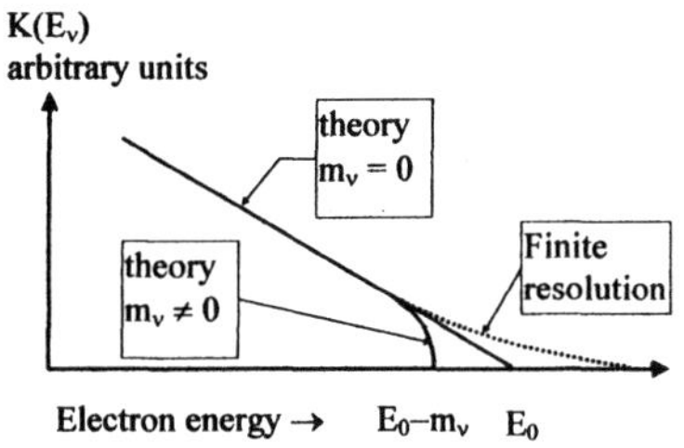

Figure 1. Shape of allowed beta-spectrum with and without a finite neutrino mass. The effect of the spectrometer resolution is likewise illustrated.

DOUBLE BETA -DECAY

Double beta decay (ßß-decay) may be observed, if the ordinary beta decay to states differing by only one unit in charge number, is energetically forbidden. There are in principal three modes of ßß-decay:

1) 2ν-ßß-decay: $(A,Z) \rightarrow (A,Z+2) + 2e^- + 2\bar{\nu}_e$
2) 0ν-ßß-decay with only Majoron (χ) emission: $(A,Z) \rightarrow (A,Z+2) + 2e^- + \chi$
3) 0ν-ßß-decay (neutrinoless ßß-decay) $(A,Z) \rightarrow (A,Z+2) + 2e^-$

The only practical way to distinguish between Dirac and Majorana neutrinos seems to be the observation of the 0ν-ßß-decay, which would result in a single narrow line at maximum electron energy E_0. Such a 0ν-ßß-decay may be visualized by the successive virtual emission of an electron anti-neutrino and the reabsorption of a neutrino. This process thus is subject to two conditions:

1) The neutrino must be identical with ist antiparticle, i.e. it must be a Majorana particle.
2) The decay is associated with either the emission of a massless Goldstone boson (the Ma joron) or it might be interpreted in terms of a massive Majorana neutrino or of a right-handed current in the weak interaction.

Thus far, only the ordinary 2ν–ßß-decay has been measured, while neither a Majoron emission, which anyhow would hardly be compatible with the Z^0 decay-width measured at LEP, nor the 0ν-ßß-decay have been observed. Many searches for 0ν-ßß-decay have been performed, thus far only yielding limits to (Majorana) neutrino masses.[11] Calorimetric detectors, involving the measurement of very small temperature changes, are increasingly involved in such searches for 0ν-ßß-decays. These devices, which are operated at very low temperatures in low background underground laboratories, exhibit extremely high resolution and very low thresholds. Temperature measurements are performed by either thermistors (highly doped semiconductors) or by means of transition edge thermometers, which employ the transition between normal and superconducting states. Other detectors are based on superconducting tunnel diodes, generating likewise many secondary particles (phonons) with a correspondingly high resolution in energy. Mass limits observed in direct experimental searches for 0ν-ßß-decay seem to be ultimately confined in the case of Dirac electron neutrinos to a few eV, in the case of Majorana electron neutrinos to a few 0.1 eV,[10] with limits for other types of neutrinos according to table 2 being much worse. Actual masses of neutrinos presumably are very much lower than the limits stated.

NEUTRINO OSCILLATIONS

The observation of neutrino oscillations should offer a much higher sensitivity to mass, because such oscillations would involve a coherent superposition of eigenstates instead of an incoherent superposition as typical for direct measurements of weak interactions. Neutrino oscillations would arise, if there is a distinction between eigenstates ν_l ($l = e,\mu,\tau$) produced in weak interactions and mass eigenstates ν_i ($i = 1,2,3$). Both eigenstates are connected by a unitary transformation. The connection is equivalent to the experimentally well-known relation which exists in the quark sector (Cabbibo-Kobayashi-Maskawa mixing). In the lepton sector we would have instead

$$\underbrace{\begin{pmatrix} \nu_e \\ \nu_\mu \\ \nu_\tau \end{pmatrix}}_{\substack{\text{weak}\\ \text{interaction}\\ \text{eigenstates}}} = \underbrace{\begin{pmatrix} U_{e1} & U_{e2} & U_{e3} \\ U_{\mu 1} & U_{\mu 2} & U_{\mu 3} \\ U_{\tau 1} & U_{\tau 2} & U_{\tau 3} \end{pmatrix}}_{\text{mixing matrix}} \underbrace{\begin{pmatrix} \nu_1 \\ \nu_2 \\ \nu_3 \end{pmatrix}}_{\substack{\text{mass}\\ \text{(propagation)}\\ \text{eigenstates}}}$$

Due to redundancies, including the unitary character of the matrix, we are dealing besides three mass parameters $\Delta m^2_{ki} = |m^2_k - m^2_i|$ ($k \neq i$) with only four mixing parameters U_{lk}, consisting of three mixing angles and one phase. One often uses a two-neutrino approximation, which then involves only one mass parameter Δm^2 and one mixing angle θ. We have in this case, for example

$$\underbrace{\begin{pmatrix} \nu_e \\ \nu_\mu \end{pmatrix}}_{\substack{\text{flavor}\\ \text{eigenstates}}} = \underbrace{\begin{pmatrix} \cos\theta & \sin\theta \\ -\sin\theta & \cos\theta \end{pmatrix}}_{\text{mixing matrix}} \underbrace{\begin{pmatrix} \nu_1 \\ \nu_2 \end{pmatrix}}_{\substack{\text{mass}\\ \text{eigenstates}}}$$

It should be noted, that due to redundancies it is sufficient to introduce mixing between either the upper or the lower states of isospin doublets. It thereby is customary to choose for quarks mixing between the lower states (d,s,b) and for leptons mixing between the upper states (i.e. between the neutrino states ν_e, ν_μ, ν_τ).

If neutrino oscillations exist, we must distinguish eigenstates ν_l associated with weak interactions, for instance weak decays; such states enter weak V-A currents and are called states with definite flavor.

eigenstates ν_i associated with definite masses; such states propagate independently in vacuum; transitions $\nu_i \leftrightarrow \nu_k$ are absent.

A weak interaction eigenstate of momentum $\vec{p}$, which in the presence of oscillations is composed of e.g. two different mass eigenstates, will propagate in space or time with different phase velocities $(v_{Phase})_i = E_i / p = \sqrt{p^2 + m_i^2} / p \approx 1 + m_i^2 / 2p^2$. The probability of observing at time t or distance L (for relativistic particles $t \approx L$) a new eigenstate $|\nu_\mu(L)\rangle$ at position L, if one has started with an eigenstate $|\nu_e(0)\rangle$ at the origin, is given by

$$P(\nu_e \to \nu_\mu) = P(\bar{\nu}_e \to \bar{\nu}_\mu) = |\langle \nu_e(0)|\nu_\mu(L)\rangle|^2 =$$

$$= \tfrac{1}{2}\sin^2 2\theta \left[1 - \cos\frac{m_2^2 - m_1^2}{2E_\nu}\right] = \sin^2 2\theta \left[\sin^2 \frac{1.27(\Delta m^2 / eV^2)(L/m)}{(E_\nu / MeV)}\right] \qquad (2)$$

We also have

$$P(\nu_e \to \nu_e) = P(\bar{\nu}_e \to \bar{\nu}_e) = |\langle \nu_e(0)|\nu_e(L)\rangle|^2 = 1 - P(\nu_e \to \nu_e) \qquad (3)$$

It should be noted, that smaller values of Δm^2 require larger distances L. It should also be noted, that if $m_1 \ll m_2$, which usually seems to be the case, we obtain directly $m_2 \approx \sqrt{\Delta m^2}$.

Figure 2 gives a survey of the range of mass parameters available in some experiments.

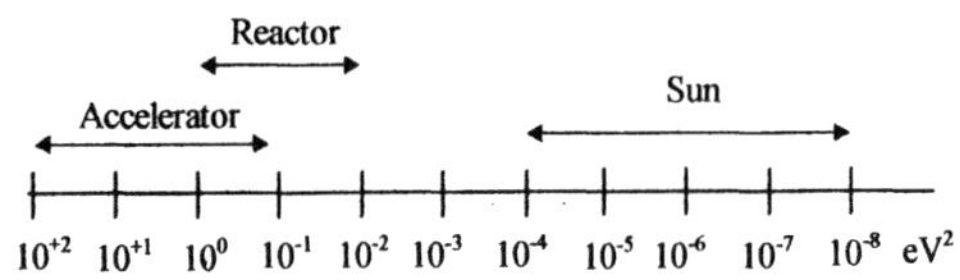

Figure 2. Mass parameter range Δm^2 in [ev^2] of some neutrino oscillation experiments

Apparently there exists a gap between terrestrial experiments and experiments involving solar neutrinos. Disappearance experiments at Chooz and at Palos Verde try to narrow this gap, albeit with only limited sensitivity to mixing angles.

It should be noted, that CC-interactions are flavor dependent: $\nu_f + N_1(Z) \to f + N_2(Z-1)$, while NC interactions are flavor independent:

Elastic processes $\nu_f + N(Z) \to \nu_f + N(Z)$,

Inelastic processes $\nu_1 + N(Z) \to \nu_2 + N^*(Z)$,

where CC and NC represent charged current and neutral current interactions, respectively while for inelastic processes $E_2 < E_1$ and $E(N^*) > E(N)$.

Concerning neutrino oscillation experiments, there are four cases of particular interest:

1) Disappearance (inclusive) experiments, for example experiments of the type $\bar{\nu}_e \to \bar{\nu}_e$. Such experiments are essentially performed with reactor neutrinos. The energy of the neutrino generated in an oscillation, i.e. $\bar{\nu}_\mu$ in the example given, is here often too low to generate charged leptons by means of charged current reactions., e.g. reactions of the type $\nu_\mu \to \mu$ .One thus may observe mere changes in the intensity of one and the same neu

trino, which in the example given may be observed via the reaction $\bar{\nu}_e + p \rightarrow n + e^+$. One thereby chooses as signature a coincidence between neutron and positron. By consequence, disappearance experiments are associated with only a low sensitivity on mixing angle ($\sin^2 2\theta > 0.05$), but due to the low energies involved exhibit a relatively high sensitivity on the mass parameter ($\Delta m^2 > 0.02$ eV^2).

2) Appearance (exclusive) experiments, which are largely performed at high energy accelerators. The high energies involved permit in oscillations the appearance of new types of charged leptons, leading to a high sensitivity on mixing angle ($\sin^2 2\theta > 10^{-4}$). At the same time, the high energies allow only, if at all, a measurement of low values of mass parameters ($\Delta m^2 > 10^{-1}$ eV^2).
3) An averaging of the oscillatory term in (2) due to large distances L leaves a sensitivity only on the mixing angle, e.g. $P(\nu_e \leftrightarrow \nu_\mu) = \frac{1}{2}\sin^2 2\theta$.
4) Of particular interest are very long baseline experiments, i.e. experiments performed at high energy and at long distance. Such experiments according to (2) would permit a high sensitivity on both mixing angle and mass parameter, because they would on the one hand permit appearance experiments with their high sensitivity on mixing parameters, while on the other hand the long distances applied would compensate the high energies and thus facilitate a high sensitivity also on mass parameters. The actual situation is somewhat more complicated.

It is crucial to arrange for proper initial beam divergencies, for instance by the application of special horns. For the MINUS detector at FERMI Lab, one anticipates a beam width of some 500 m at the distant detector some 740 km away in the Soudan Mine.

SOLAR NEUTRINO OSCILLATIONS

The word neutrino oscillations is synonymously used for flavor oscillations, with particle $\leftrightarrow$ antiparticle oscillations requiring in addition a change in helicity.The observation of solar neutrino oscillations would offer two possibilities, the observation of vacuum oscillations in the space between Sun and Earth and the observation of resonance transitions within the solar matter, the so-called Mikhaeyev-Smirnov-Wolfenstein (MSW)-effect.[11,12] In passing it should be noted in this context, that the coherence of neutrinos emitted by the Sun is always assured in terrestrial experiments. The most simple observation of solar neutrino oscillations would arise in the case of complete mixing, with electron neutrinos being produced in the solar core, while complete mixing would produce one third of ν_e, ν_μ and ν_τ each in the adjacent vaccum between Sun and Earth. Because terrestrial detectors can only observe neutrinos of the type ν_e, one would indeed observe a solar neutrino deficit by a factor of three. However, such a scenario requires complete mixing, which presumably would already have been observed in previous experiments.

The Mikhaeyev-Smirnov-Wolfenstein (MSW) Effect

The MSW-effect may arise due to an asymmetry in the weak interaction of neutrinos: Electron neutrinos are subject to both charged (CC) and neutral current (NC) interactions, while ν_μ and ν_τ are subject to only neutral currents. By consequence, the following picture arises: Electron neutrinos generated in the solar core and travelling to the solar surface may pass through a resonant range, where they convert efficiently into neutrinos of another flavor, thus escaping detection in a terrestrial detector. Such a resonant conversion of neutrinos appears, if the relation holds

$$\cos 2\theta = G_F \rho_e 2\sqrt{2} E_\nu / \Delta m^2$$

In the two-neutrino approximation, vacuum oscillations according to equation (2) are characterized by the oscillation parameters Δm^2 and $\sin^2 2\theta$, while the MSW–effect, i.e. the actual appearance of a resonance, depends in addition on the electron density ρ_e in the Sun, reflecting the actual position of the resonance along the solar radius, as well as on the energy E_ν of the emitted solar neutrinos. It should be noted, that the MSW-effect gives rise to a connection between the mixing angle θ and the mass parameter Δm^2. The eigenvalues present with the MSW-effect are plotted in figure 3 as function of the electron density ρ_e within the Sun. The effects of the common NC interaction are dropped in the figure. The CC interaction of ν_e grows linearly with the electronic density ρ_e within the Sun, while neutrinos of the type ν_μ are not influenced by CC interactions. A ν_e emitted in the core of the Sun, where ρ_e is highest, travels to the surface and may at a certain density ρ_e experience a resonance conversion into ν_μ or ν_t. Whether or not the resonance does occur, depends on the actual and presently unknown vacuum values of Δm^2 or θ. Neutrinos of the type ν_μ leaving the Sun are not registered in a terrestrial neutrino detector, which thus notes a solar neutrino deficit. In addition to this phenomenon, there may be non-adiabatic transitions between the two branches of the curves, causing the reappearance of electron neutrinos and reducing a possible neutrino deficit.

Solar Neutrino Spectra

Measurements of solar neutrinos involve neutrinos from various solar fusion reactions, as listed in table 3 and with spectra shown in figure 4. Detection of solar neutrinos is performed by means of the inverse ß-decay, where solar neutrinos generate transitions from a lower state (A,Z) to upper states (A,Z+1). Knowing the Gamow-Teller matrix elements of the ß-transitions, the cross sections for the inverse transitions to the lowest states of neighboring nuclei can readily be worked out. Matrix elements for higher final states may be deduced from e.g. p,n forward scattering experiments, but the evaluation of such matrix elementsis much less certain. The inverse ß-processes require in any case an excitation energy, whence such processes give rise to threshold energies.

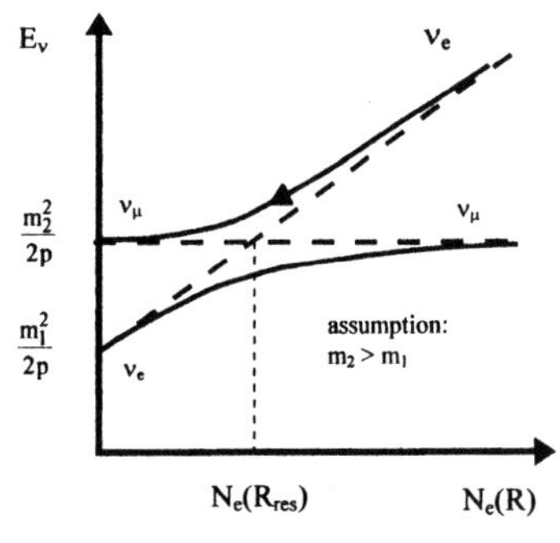

Figure 3. Energy eigenvalues for neutrinos subject to weak interactions in solar matter as function of the electron density ρ_e within the Sun. For pure adiabatic conditions neutrinos ν_e generated in the solar core will appear as ν_μ or ν_τ at the solar surface, in case these ν_e passed a conversion resonance on their way to the solar surface.

Table 3. Fusion reactions within the Sun. The CNO-cycle contributing only 1.5 % to the solar luminosity has been left out, together with the hep production cycle, which are of even minor interest as concerns their luminosity contribution.

	Overall fusion reaction: $4p \rightarrow {}^4He + 2e^+ + 2\nu_e + 26.7$ MeV	
	Detailed fusion reactions:	
	$p + p \rightarrow d + e^+ + \nu_e$	(ν_e-spectrum; $E_\nu \leq 422$ keV)
	$d + p \rightarrow {}^3He$	
a) 85%	${}^3He + {}^3He \rightarrow {}^4He + 2p$	
b) 15%	${}^3He + {}^4He \rightarrow {}^7Be$	
either	${}^7Be + e^- \rightarrow {}^7Li + \nu_e$	(ν_e monochromatic; $E_\nu = 861$ keV (90%))
	${}^7Li + p \rightarrow {}^8Be \rightarrow {}^4He + {}^4He$	
or	${}^7Be + p \rightarrow {}^8B$	
	${}^8B \rightarrow {}^4He + {}^4He + e^+ + \nu_e$	(ν_e-spectrum; $E_\nu \leq 14$ MeV)

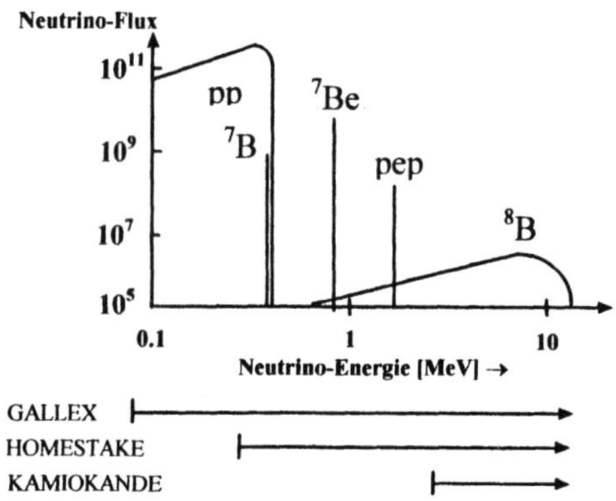

Figure 4. Pertinent components of the solar pp-cycle neutrino spectrum according to the Standard Solar Model. Shown are line spectra in [$cm^{-2}s^{-1}$] and continuous spectra in [$cm^{-2}s^{-1}MeV^{-1}$]. Different thresholds of various experiments are also indicated.

The GALLEX Solar Neutrino Experiment

Two experiments are presently measuring those neutrinos which essentially determine the well-known solar luminosity, GALLEX (Gallium experiment) and SAGE (Soviet-American Gallium experiment). The other solar neutrino experiments (Homestake and Kamiokande) measure only higher neutrino branches, which can more readily be detected, because the cross section in the nuclear energy range varies in proportion to the square of the energy in the laboratory. GALLEX is performed in the Gran Sasso Neutrino Underground Laboratory some 150 km North-East of Rome, where a highway crosses the Appenino mountain range (Abruzzi mountains). Underground laboratories have been drilled in the center of the tunnel, yielding a shielding of detectors by some 3300 m.w.e. against cosmic radiation. Background arises from cosmic muons, which relative to the surface are depressed by a factor of 10^{-6}, as well as from such internal sources as radon and fast neutrons. The access of the laboratory by a big highway offers advantages over the usual small mine shafts, permitting to employ one big absorption tank rather than a number of smaller ones. Table 4 gives a list of the contributing laboratories.

Table 4. Laboratories participating in GALLEX

MPI	Heidelberg/Germany (T.Kirsten, spokesman of the collaboration)
KFK	Karlsruhe/Germany
TU	München/Germany
INFN	Milano/Italy
INFN	Rome/Italy
DAPNIA	Saclay/France
CEN	Grenoble/France
U	Nice/France
WIS	Rehovoth/Israel
BNL	Brookhaven/USA

The Gallex Solar Neutrino Detection procedure

GALLEX employs 30 t of Gallium in the form of 100 t of $GaCl_3$, corresponding to about 10^{29} nuclei of the pertinent isotope ^{71}Ga, which capture according to the Standard Solar Model $\approx$ 1.1 neutrinos/d; relative abundances are 60.5% (^{71}Ga) and 39.5% (^{69}Ga). As capture reaction serves $^{71}GaCl_3$ + solar $\nu_e \rightarrow {}^{71}GeCl_4 + e^-$. For comparison we give the half-lifes of some relevant isotopes: $T_{1/2}(^{69}Ge)$ = 1.6 d; $T_{1/2}(^{71}Ge)$ = 11.4 d; $T_{1/2}(^{68}Ge)$ = 271 d. Extraction of the few atoms of $^{71}GeCl_4$, excited mainly by solar neutrinos within one experimental cycle of about 4 weeks, proceeds by adding 1 mg of isotopic Ge-carrier to the solution and by flushing the tank with some 2000 m^3 of nitrogen gas. The concentration of the extracted material is achieved by alternately acidifying and de-acidifying the solution, giving rise to alternate desorption and absorption processes, whereby the solution is reduced in three successive steps from 50 l to 50 ml. Because the electronegative $GeCl_4$ cannot be used as counter gas, it is chemically converted to GeH_4 and then gaschromatographically purified. A carrier weight determination serves to determine the extraction efficiency. The final measurement of the decay from ^{71}Ge back to ^{71}Ga, from which follows the solar neutrino flux, is performed by means of proportional counters of 0.5 cm^3 volume, which are filled with the GeH_4 plus 30% added Xe-gas. Counting is performed in both the L-peak (1.2 keV) and the K-peak (10.4 keV) of the counters, employing fast transient pulse shape recorders. Appropriate signals are selected by either choosing pulses with proper rise times in a 10-70 % interval (old method) or by making use of the entire pulse shape in the analysis (present method). Both methods agree within less than 2 SNU's out of a total of some 70 SNU's.

The entire procedure has been checked by employing an artificial neutrino source of ^{51}Cr with a source strength of about 70 PBq ($\approx$ 2 MCi), which at the beginning yielded a neutrino flux about 20 times the flux arriving from the Sun and with a similar spectrum. Table 5 shows experimental results and theoretical predictions of present solar neutrino experiments.

All experiments measure solar neutrinos with energies from their thresholds up to a maximum energy of about 14 MeV. Predictions based on the Standard Solar Model differ somewhat for ^7Be and ^8B due to uncertainties in cross sections of excited states and in opacities, but nearly coincide for pp solar neutrinos.

There are two possible reasons for the apparent deficit of solar neutrinos:

Table 5. Measurements and predictions for solar neutrino experiments.

EXPERIMENT			Prediction [SNU]*	
collaboration	measurement [SNU]*	threshold energy	reference	Bahcall [3σ]
HOMESTAKE $^{8}B + (^{7}Be)$	$2.55 \pm 0.17 \pm 0.18$	0.814 MeV	13	8.0 ± 3.0
KAMIOKA ^{8}B	$(0.51 \pm 0.07)\times$pred.	7.5 MeV	14	$(1.0 \pm 0.14)\times$pred.
SAGE $E_\nu > 233$ keV	$69 \pm 10^{+5}_{-7}$	233 keV	15	132^{+21}_{-17}
GALLEX $E_\nu > 233$ keV	$69.7 \pm 6.7^{+3.9}_{-4.5}$	233 keV	16	132^{+21}_{-17}

errors are usually stated as sequence of statistical and systematic errors

*1 [SNU] = 1 [Solar Neutrino Unit] = 10^{-36} [captures/target atom/s]

except for Kamioka, where only relative numbers apply

Predictions based on the Standard Solar Model are taken from Bahcall and Pinsonne ault.[17]

1) Particle physics, such as the presence of neutrino oscillations, in particular of the MSW-effect.
2) Astrophysics, in particular a failure of the Standard Solar Model.

A possible experimental distinction would require either a measurement of the neutrino flux leaving the Sun by measuring the flavor independent NC interaction or by determining in a suitable detector such as a TPC chamber the flavor contents within the detector.

The future of Solar Neutrino Observations

There are at present four installations in the preparatory stage (with the exception of the already operative SUPERKAMIOKANDE), which aim at improving our knowledge on solar neutrinos:

1) SUPERKAMIOKANDE (Japan)

   This experiment is a blow-up of the original KAMIOKANDE facility. Like its predecessor it is a water Cerenkov detector employing the scattering of neutrinos off electrons. A fiducial volume of 50 000 m^3 is viewed by 11 000 Photomultiplieres each of 50 cm diameter. The center detector is surrounded by a shielding layer of 3 m water viewed by another 1800 PM's.The on-line nature of the experiment allows for a directional sensitivity; with an anticipated threshold of 5 MeV, permitting only measurements of solar ^{8}B neutrinos. The expected rate is 10 000 neutrinos/a, which corresponds to roughly 80 times the rate of KAMIOKANDE. The experiment should be able to verify the presence of the MSW effect, which would give rise to a rather distorted ß-spectrum. SUPERKAMIOKANDE has already started in April 1996

2) SNO (Sudbury Neutrino Observatory; Canada)

   The experiment employs a heavy water Cerenkov detector with its compared to water larger cross section.

   CC interactions: $d + \nu_e \rightarrow p + p + e^-$

   NC interactions: $d + \nu \rightarrow p + n + \nu$ with direct or indirect detection of neutrons.

   Anticipated thresholds: CC interactions 5 MeV NC interactions 2.2 MeV

   Anticipated rates: CC interactions 9 000 ν/a NC interactions 3 000 ν/a

   Expected start: 1997 (?)

3) BOREXINO (Gran Sasso/Italy)
The experiment uses the elastic scattering off electrons in an organic scintillator, producing recoil electrons with a continuous Compton-like spectrum. The scintillator (which originally was a Boron-derivative, whence the name) yields a light-output some 100 times higher than in a Cerenkov detector, yet at the expense of no directional sensitivity, requiring a particularly low background. The scintillation photons are viewed by 2 000 PM's in the 100 m^3 liquid scintillator. The large light output should yield a threshold of only 250 keV, permitting the observation of the neutrino flux from the ^7Be reaction, which according to the other experiments should be anomalously low. The expected rate is 18 000 recoil-electrons/a in an energy range 250-665 keV; the experimental start is anticipated in 1999.

4) HELLAZ (proposed helium liquid nitrogen (= azote) detector)
This detector, under development at Saclay/France, is supposed to measure the elastic scattering of solar pp-neutrinos off electrons in a He-methane gas mixture (2000 m^3; 5 bar; 77°C). This time projection chamber detector, which would have directional capacity, is supposed to distinguish the tracks of different neutrino flavors, thus enabling the terrestrial observation of neutrino oscillations. A 40 m^3 prototype is presently being installed in the Gran Sasso laboratory in Italy.

LITERATURE REFERENCES

1. C.L.Cowan, Jr., F.Reines et al., Science 124:103 (1956).
2. G.Danby et al., Phys.Rev.Lett. 9:36 (1962).
3. M.Perl et al. Phys. Rev. Lett. 35:1489 (1975).
4. see, e.g., J.R.Carter, Joint Intl. Lepton-Photon Symposium & Europhysics Conf. on High Energy Physics, World Scientific, Singapore, 2:3 (1991).
5. F.J.Hasert et al., Phys.Lett. 46B:121 (1973).
6. B.W.Lee and S.Weinberg, Phys.Rev.Lett. 39:165 (1977).
7. E.Belesev et al., Phys.Lett. B350:263 (1995)
8. K.Assamagan B335:231 (1994)
9. ALEPH Collaboration, D.Buskulic et al., Phys.Lett. B 349:585 (1995)
10. see e.g. A.Balysh et al., Proc. XXVII Intl.Conf. on High Energy Phys., Glasgow (1994)
11. L.Wolfenstein, Phys.Rev. D 17:2369 (1978)
12. S.P.Mikhaeyev and A.Yu Smirnov, Nuovo Cimento 9C:17 (1986)
13. B.Cleveland et al., Nucl.Phys.B Proc. Suppl. 38: 47 (1995)
14. Y.Suzuki et al., Nucl.Phys.B Proc. Suppl. 38:54 (1995)
15. W.Hampel et al., submitted June 1996 to Physics Letters B
16. J.N.Abdurashitor et al., Nucl.Phys.B Proc. Suppl. 48:299 (1996)
17. J.N. Bahcall and M.H. Pinsonneault, Rev. Mod. Phys. 64:885 (1992)

GALLEX NEUTRINO OBSERVATIONS AND ITS IMPLICATIONS ON SOLAR NEUTRINO SPECTROSCOPY

Michael Altmann

Technische Universität München
Physik Department E15
D - 85747 Garching

1 INTRODUCTION: SOLAR NEUTRINOS

Solar neutrino measurements provide a unique experimental tool to test both, intrinsic neutrino properties, and stellar models. Whereas the sun is highly opaque for photons, the neutrinos produced in nuclear fusion reactions deep inside the solar core can unhinderedly escape, thus carrying real-time information about the solar interior.

Several experiments have been set up to detect these neutrinos, utilizing different measurement techniques. The radiochemical Chlorine experiment at Homestake, USA, pioneering this field of research for more than one decade, exploits the neutrino capture on ^{37}Cl, i.e. $^{37}\mathrm{Cl} + \nu_e \rightarrow {}^{37}\mathrm{Ar} + e^-$. The energy threshold for this reaction is 814 keV. The KAMIOKANDE collaboration operated a water Cherenkov detector, initially designed to search for proton decay, to perform real-time measurements of high energy neutrinos ($E_{\mathrm{threshold}} \simeq 7.5\,\mathrm{MeV}$) via the scattering reaction $\nu_e + e^- \rightarrow \nu_e + e^-$.

Both experiments, however, have too high energy thresholds to be sensitive to the by far most abundant solar neutrino species, the so-called pp-neutrinos from the primary fusion reaction $p + p \rightarrow {}^2\mathrm{D} + e^+ + \nu_e$, whose maximal energy is only $E^{pp}_{max} = 422\,\mathrm{keV}$. In fact, HOMESTAKE and KAMIOKANDE can only detect higher energetic neutrinos originating from side branches of the solar proton-proton fusion cycle, which is sketched in Figure 1 together with the neutrino spectrum.

Compared to the pp-flux, however, the solar model predictions of the high energy neutrino fluxes are still associated with rather large uncertainties. Thus it is desirable to detect the bulk of the neutrinos emitted from the sun, i.e. the low energy primary pp-neutrinos. To achieve this goal, at present and most probably also in the next decade(s), only a radiochemical method utilizing the inverse beta decay reaction $^{71}\mathrm{Ga} + \nu_e \rightarrow {}^{71}\mathrm{Ge} + e^-$ is feasible. The energy threshold for this process is 233 keV, well below the endpoint energy of the pp-neutrinos. At present, two gallium experiments are in operation, the Russian SAGE experiment, installed at the Baksan Neutrino Laboratory in the Kaukasian Mountains, Russia, and GALLEX, operated in the Gran Sasso Underground Laboratory in Italy.

Masses of Fundamental Particles: Cargèse 1996
Edited by M. Lévy *et al.*, Plenum Press, New York, 1997

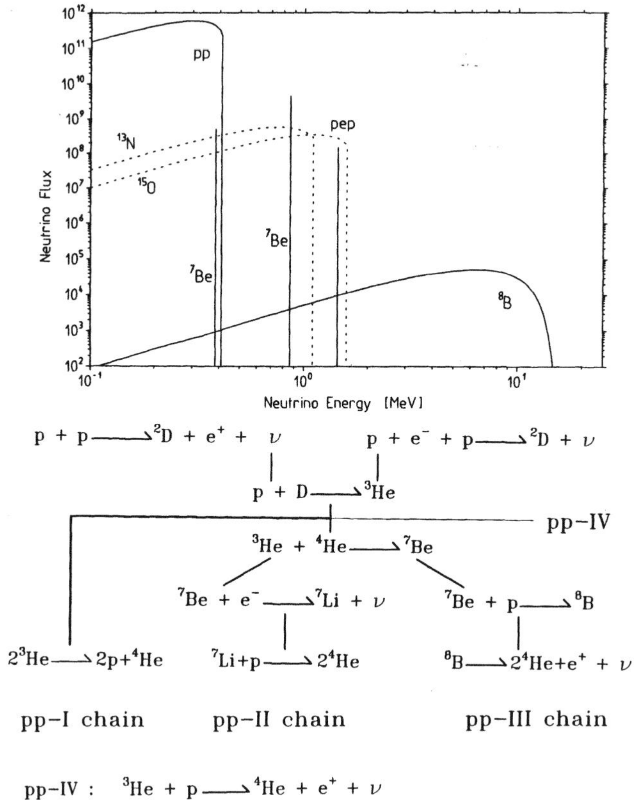

Figure 1. The top figure shows the pp-cycle, being dominant for solar energy generation, with its various terminations. From solar model calculations one expects about 85% of the terminations to occur via the pp-I chain and about 15% via pp-II, whereas pp-III and in particular pp-IV terminations are comparably rare. On bottom the resulting neutrino spectrum is shown. Units for the line spectra of pep and ^{7}Be neutrinos are $cm^{-2}s^{-1}$, for continuous spectra $MeV^{-1}cm^{-2}s^{-1}$. Spectra originating from the CNO-cycle, which plays a minor role for solar energy generation, are plotted with dashed lines.

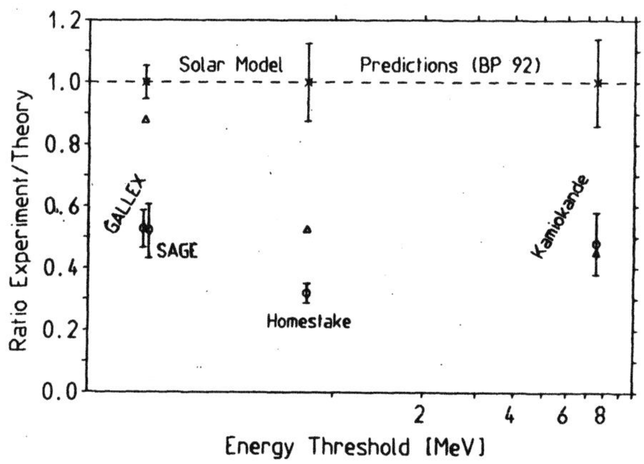

Figure 2. Present status of the solar neutrino puzzle. The experimental results normalised to the respective theoretical predictions are plotted as a function of the threshold energy of the respective detector. For normalisation the SSM [Bah92] is used, with the theoretical uncertainties of its predictions indicated. For comparison, the predictions of the low-flux model [Dar94] are also shown (open triangles).

All the solar neutrino experiments have reported fluxes significantly lower than the standard solar model (SSM) predictions [Bah92,Bah95,Tur93]. This intriguing result commonly is referred to as the *Solar Neutrino Puzzle*. So-called low flux solar models, e.g. [Dar94], were designed to tackle this problem by using other input data and different modelling of the processes taking place in the solar interior (e.g. electron screening, plasma effects, element diffusion). Although they succeeded in reproducing the KAMIOKANDE result for the high energy ^8B neutrino flux, a persisting discrepancy for the chlorine and gallium experiments remained. The present experimental situation is depicted schematically in figure 2.

Apart from the unjustified "ad hoc claim" of large unknown systematic errors in the experiments, several serious attempts have been proposed to explain the apparent neutrino deficit. Firstly, the description of some physical processes in the standard solar models and/or some of the input parameters of these models (e.g. opacities, heavy element abundances, etc.) might be wrong (so-called *astrophysical solution*). Secondly, the nuclear cross sections used to calculate the solar fusion and the terrestrial detection rates might be significantly inaccurate (*nuclear physics solution*), and, thirdly, neutrino oscillations, either in vacuum or matter-enhanced, might occur (cf. e.g. [Obe92]) (*particle physics solution*). The latter explanation, however, requires not only the neutrino to be massive but also flavour mixing in the lepton sector, i.e. phenomena beyond the electroweak standard model.

Which of these proposals is the correct one is a long-standing question which has been controversly discussed for many years. However, meanwhile it became possible to severely constrain these explanations *alone from the increasingly more accurate* GALLEX *solar neutrino measurements* in conjunction with the results of the integral efficiency tests of the GALLEX detector by means of a man-made high intensity neutrino source.

2 THE GALLEX EXPERIMENT

2.1 Detector Setup and Experimental Procedure

The GALLEX target consists of 30 tons of natural gallium in the form of 101 tons of aqueous $GaCl_3$ solution, acidified in HCl. It is stored in one of two $70\,m^3$ glass fibre reinforced vinylester resin tanks, lined with PVDF on the inside.

In an usual solar neutrino run, the target is, after the addition of ≈ 1 mg of a stable germanium carrier which allows for chemical yield determination, exposed to the solar neutrino flux for 4 weeks. Then both, the neutrino produced ^{71}Ge and the stable carrier isotope, present in the form of $GeCl_4$ which is highly volatile in acidic environment, is extracted by purging the tank with nitrogen.

The $GeCl_4$ then is dissolved in water and after some concentration steps, organic extraction and aqueous back-extraction, finally germane GeH_4 is synthesized, which is a suitable counting gas [Scha93]. Together with old xenon, it is filled into a miniaturized low-level proportional counter [Win93] (Xe/GeH_4 : 70/30), inserted in an air-tight, low background shielding tank, calibrated [Urb89,Scha95] and counted for six months.

The shielding tank itself and the data acquisition electronics are installed inside a Faraday cage [Heu94]. With an optical fibre cable the electronics is linked to a μVAX outside the Faraday cage, which allows for continuous online monitoring of the entire counting system. Up to 23 counters can be operated simultaneously. In the proportional counter, the decay $^{71}Ge + e^- \rightarrow {}^{71}Ga + \nu_e$ is observed. ^{71}Ge decays via electron capture with a half-life of 11.4 days. Thereby Auger electrons and/or X-rays are emitted leading to energy depositions mainly in two peaks at 10.4 keV (K peak) and 1.1 keV (L peak).

The overall goal of the offline data analysis is to separate those events which fulfill a number of acceptance criteria characteristic for ^{71}Ge decays from the bulk of recorded events. This is achieved by performing an energy and in particular a pulse shape analysis. Then, in a maximum likelihood analysis a time constant background plus a component exponentially decreasing with the known life-time of ^{71}Ge is fitted to the time sequence of these accepted events, yielding the initial ^{71}Ge activity (and therefrom the solar neutrino flux).

In GALLEX several different and largely independent analysis methods are employed, cf. e.g. [Alt96a,GAL92,Scha93a]. Their results are in remarkable good agreement [Alt96,GAL96], which underlines the quality of the GALLEX data set and the reliability and trustworthiness of the published results.

2.2 The Chromium Experiment

However, what this cross-check cannot do is to exclude effects of hot-atom chemistry: It may be possible, that the neutrino-produced germanium behaves chemically different than the added stable germanium carrier (recovery yield: on average about 96%).

Although extensive search for such effects has shown no evidence at all, to unquestionably exclude these and other unknown systematic effects and experimental artefacts, the best possible integral test of the whole experimental procedure has been carried out: We have irradiated our target with a man-made source of known activity, emitting neutrinos with energies typical for the sun.

As a source isotope showing the desired properties, ^{51}Cr was selected. It decays via electron capture with a Q-value of 751 keV and a half-life of 27.7 d: $^{51}\mathrm{Cr} + e^- \rightarrow {}^{51}\mathrm{V} + \nu_e$. Four monoenergetic neutrino lines with energies 746 keV (81%), 751 keV (9%), 426 keV (9%) and 431 keV (1%) are emitted.

The source test has been performed twice, using the same neutrino source, consisting of $\approx$ 36 kg of Chromium enriched to 38.6% in ^{50}Cr. It was prepared in Moscow and Saclay, for both experiments irradiated for about one month in the Siloe reactor in Grenoble and afterwards transported to Gran Sasso and immediately inserted into the reentrant tube of the target tank. The initial source activity at end of the irradiation (EoB: end of bombardment) was $A_0^{\mathrm{EoB}} = (62.4 \pm 0.4)$ PBq for the first and $A_0^{\mathrm{EoB}} = (68.7 \pm 0.7)$ PBq for the second experiment. It was determined with several independent methods, as described in [GAL95]: calorimetry, neutronics and gamma scanning, ionization chamber technique and spectroscopy of small samples as well as measurement of the Vanadium content of the chromium after several ^{51}Cr half-lives.

For the first source test altogether 11 germanium extractions were performed during the period of more than 3 months the source was installed, whereas the second experiment added another 7 measurements[1]. The individual run results of these 18 measurement are depicted in figure 3.

The "neutrino results" for the initial source strengths, deduced from the first and second experiment, are $A_\nu^{\mathrm{EoB}} = (62.5 \pm 6.5)$ PBq and $A_\nu^{\mathrm{EoB}} = (57.1 \pm 6.8)$ PBq, respectively, to be compared with the known activities listed above. Performing a combined analysis of all 18 runs, we arrive at $R = A_\nu^{\mathrm{EoB}}/A_0^{\mathrm{EoB}} = 0.92 \pm 0.08$ [GAL96a] for the ratio of the initial activity deduced from the neutrino measurements and the directly measured source strength.

This result provides the desired overall check of the entire GALLEX experiment and moreover demonstrates the reliability of the radiochemical method for rare event

[1] Counting is not yet completed for all of the latter runs. Thus, the result for the second source experiment is still to be considered preliminary.

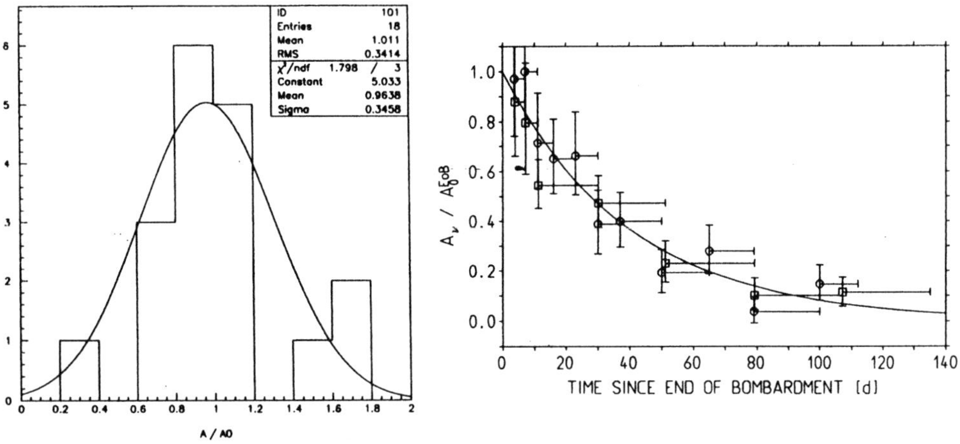

Figure 3. Results of the GALLEX neutrino source experiments. The left figure shows the individual run results of the first (circles) and second experiment (boxes) as a function of time after the end of irradiation, normalized to A_0^{EoB}, the known source activity at EoB. The horizontal bars correspond to the length of exposure. The solid line is not a fit to the data but depicts the expectation, showing the exponential decay of the source. The picture on the right shows the distribution of the individual run results, re-traced to $t = 0$ and normalized to A_0^{EoB}, together with a gaussian fit.

detection in general. A similar test has been performed by the SAGE collaboration, also showing an R-value close to 1.

2.3 Solar Neutrino Results

The result of the source experiment does reassure, that there are no major unidentified systematic errors and experimental artefacts. This has important consequences for the interpretation of the GALLEX result from the solar neutrino measurements.

Up to now[2] 53 solar neutrino runs have been successfully conducted and counting completed, corresponding to about 4 years of target exposure. The result of a combined maximum likelihood analysis of these runs is $(69.7 \pm 6.7\,(\mathrm{stat.})\,^{+3.9}_{-4.5}\,(\mathrm{syst.}))$ SNU, or, with errors added in quadrature, $(69.7\,^{+7.8}_{-8.1})$ SNU [GAL96]. This result is to be compared to the SSM-predictions, ranging between 137 SNU [Bah95] and 116 SNU for the low-flux model [Dar94]. Hence, the GALLEX result is substantially below all the predictions, as has already been shown in figure 2.

3 IMPLICATIONS

3.1 A Detection Solution for the Gallium Experiments?

Obviously, the ^{51}Cr neutrino source experiment strongly disfavours attempts to ascribe the almost 50% solar neutrino deficit observed in GALLEX to inappropriate cross sections for neutrino capture on ^{71}Ga. This conclusion holds, even if one allows for a hypothetical cross section suppression near threshold due to final state interaction, which recently has been speculated about [Dar95]. Hence, a "detection solution" to the solar neutrino puzzle seems, at least for the Gallium experiments, to be highly unlikely.

---

[2] August 1996

Table 1. Predictions of the solar model [Bah92]. Mean neutrino energies $\langle E_\nu \rangle$, fluxes Φ_ν at 1 AU from the sun, flux uncertainties $\Delta\Phi_\nu$ and predicted contributions to the signals of the ^{37}Cl- and ^{71}Ga-experiments.

	pp	pep	^{7}Be	^{8}B	^{13}N	^{15}O
$\langle E_\nu \rangle$ [MeV]	0.26	1.44	0.86 (89.7%) 0.38 (10.3%)	6.7	0.71	1.00
Φ_ν [cm^{-2}s^{-1}]	6.0×10^{10}	1.4×10^8	4.9×10^9	5.7×10^6	4.9×10^8	4.3×10^8
$\Delta\Phi_\nu (1\sigma)$ [%]	0.7	1.3	6	14.3	17	19.3
$\langle \sigma_\nu \Phi_\nu \rangle$, Cl [SNU]	0	0.2	1.2	6.2	0.1	0.3
$\langle \sigma_\nu \Phi_\nu \rangle$, Ga [SNU]	70.8	3.1	35.8	13.8	3.0	4.9

3.2 The Crucial Role of the Beryllium Neutrinos

To value the remaining explanation attempts, a closer look to both the solar fusion cycle and the spectral detector sensitivities is inevitable. In the pp-chain (cf. fig. 1), to which more than 98% of the solar luminosity is attributed to, neutrino emission takes place in several reactions, where the relative fluxes depend on the branching ratios at the vertices. For the SSM [Bah92] the partial fluxes and flux uncertainties, mean neutrino energies and spectral contributions to the predicted detection rates of the chlorine and gallium experiments are listed in table 1.

From table 1 it is obvious, that the chlorine experiment is mainly sensitive to boron neutrinos, with only a small additional contribution attributed to the 862 keV beryllium line. As KAMIOKANDE, due to its high energy threshold of 7.5 MeV, is exclusively sensitive to ^{8}B neutrinos[3] and the measured suppression for KAMIOKANDE is weaker than for the chlorine experiment, there is essentially left no room for beryllium neutrinos. Moreover, it appears puzzling, that the high energy part of the boron spectrum observed by KAMIOKANDE seems to indicate a weaker suppression than the more extended measurement of the same spectrum by the chlorine experiment. This has been speculatively interpreted as a possible shape distortion of the beta-decay spectrum of the ^{8}B neutrinos ("Homestake-Kamiokande anomaly").

In any case, however, the scenario of boron neutrinos suppressed to about one half (or essentially not suppressed at all, if the predictions of the low-flux models are correct), but beryllium neutrinos, in contrast, suppressed almost completely is complementary to what is obtained by changing the SSM input parameters. Therefrom, it has been concluded that astrophysical solutions to the solar neutrino puzzle are incompatible with the experimental data, e.g. [Ber93,Hat94].

However, by some persons it repeatedly is claimed, (though there are no convincing arguments supporting this statement) that the chlorine experiment might not be trustworthy since it never has been calibrated. In any case, even if one completely discards the chlorine result, the above conclusion is not altered.

In this context it is crucial to realize that the neutrino lines emitted from the ^{51}Cr source nicely accommodate the beryllium neutrino energies. In particular, chromium neutrinos not only give rise to the ^{71}Ga$-^{71}$Ge ground state to ground state transition, but also initiate the population of the two lowest excited nuclear levels at 175 keV and 500 keV, respectively. It is exactly these three levels which also can be populated from pp and beryllium neutrino interactions. With GALLEX in the chromium experiment having measured about the expected signal, it seems conclusive that the experiment is also sensitive to beryllium neutrinos. In other words, if there really are beryllium neutrinos, GALLEX should see them.

[3] And also to higher energetic neutrinos from the pp-IV chain (cf. fig. 1), the flux of which, however, is negligibly tiny.

Relying on the reasonable assumptions that GALLEX is fully sensitive to pp- and ^{7}Be neutrinos and the sun is in an equilibrium state, i.e. producing energy at the same rate as radiating it away, one can impose a severe upper constraint on the ^{7}Be neutrino flux[4] *alone* from the GALLEX result, which is independent of solar model details. This is accomplished in figure 4, making the most conservative assumption that no pep-, ^{8}B , hep and CNO-neutrinos are present, but the solar luminosity being exclusively due to the pp-I and pp-II chains, cf. figure 1.

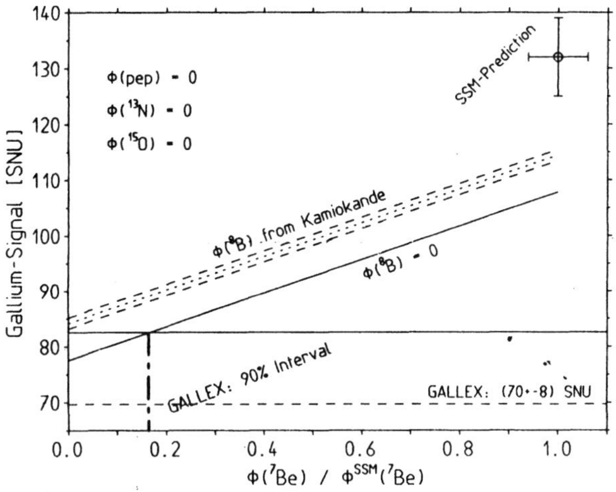

Figure 4. Solar model independent upper constraint for the ^{7}Be neutrino flux. The solid line depicts the signal caused in the gallium detectors as a function of the beryllium neutrino flux, conservatively assuming that pep, CNO and ^{8}B neutrinos do not contribute at all. To at least marginally accommodate the GALLEX 90% confidence region, the beryllium flux must not be higher than at most 17% of the [Bah92] prediction. If one takes into account boron neutrinos with a flux as measured by Kamiokande, essentially no room is left for ^{7}Be neutrinos (dashed band).

In figure 4 the signal to be seen in the gallium detectors is calculated as a function of the beryllium neutrino flux for the very conservative cases that (a) no pep, CNO and boron neutrinos contribute and (b) that pep and CNO-neutrinos do not contribute, but ^{8}B neutrinos do with a flux as measured by KAMIOKANDE. A comparison with the actual GALLEX result of (70 ± 8) SNU imposes severe upper constraints on the beryllium flux — without any regard to the chlorine experiment. To be at least marginally compatible with the GALLEX 90% confidence interval, the beryllium flux is – even under assumptions that seem unreasonably conservative – required to be less than about $\Phi_{\rm Be}/\Phi_{\rm Be}^{\rm SSM} \lesssim 0.17$, i.e. $\Phi_{\rm Be} \lesssim 8 \times 10^8\,{\rm cm}^{-2}{\rm s}^{-1}$. Hence, again the suppression for beryllium neutrinos is required to be significantly higher than for boron neutrinos.

It should be strengthened again, that this conclusion is independent of any solar model details but relies only on the solar luminosity constraint, i.e. the sun being in an equilibrium state on timescales compatible with the time required for photons to escape the solar interior. From this constraint, any astrophysical solution to the solar

---

[4] The ratio between the two lines is assumed to stick to the value listed in table 1. If the 384 keV line is hypothetically enhanced with respect to the high energy line, e.g. caused by polarization of plasma electrons, the bound on the ^{7}Be neutrino flux is weaker. On the other hand, including the SAGE result strengthens the constraint.

neutrino puzzle seems to be excluded, provided not at least both gallium detectors report considerably wrong results.

3.3 A Nuclear Physics Solution at the Production Stage?

It has been pointed out [Alt95,Ber95], however, that in principle there still exists a possibility to (at least marginally) accommodate both the GALLEX and either the KAMIOKANDE or the Chlorine results by suitably adjusting the astrophysical cross section parameters (S-factors) which determine the branching at the vertices of the pp-chain[5]. However, such an adjustment requires drastic modifications far beyond the uncertainties quoted for the respective S-factors.

To elucidate this point, we shall consider an example: In a first order approximation the S-factor dependence of the relative branching towards the $^3He - ^4He$-initiated subchains (cf. fig.1) is given as $\lambda_{34} \propto n_3 \cdot n_4 \cdot S_{34} \propto S_{34}/\sqrt{S_{33}}$, where S_{33} enters, since the equilibrium density n_3 of ^3He depends on the $^3He - {}^3He$ cross section. Then, to achieve the above mentioned beryllium flux suppression to 17% of the SSM prediction – which still ends up with a signal prediction being about 2 sigma off from the GALLEX result if the ^8B flux is taken from KAMIOKANDE – a narrow low energy resonance in the $^3He - {}^3He$ cross section is required: $S_{33} \gtrsim 30\ S_{33}^{SSM}$ (with $S_{34} = S_{34}^{SSM}$). This factor 30 is horribly far outside the 6% uncertainty [Bah92] quoted for the (flat) extrapolation of S_{33} from the laboratory measurements down to the energies relevant in the sun. To experimentally clarify the situation, the LUNA experiment at Gran Sasso has started to measure this cross section at lower energies than achieved in previous experiments. Meanwhile they have performed measurements down to an energy of 21 keV, corresponding to the center of the Gamow-peak for the solar $^3He - {}^3He$ reaction, without any indication for a resonant enhancement of the cross section [Lun96].

However, though with this modified S_{33}-parameter the branching towards the pp-I chain is enhanced and thus the desired strong beryllium flux suppression can be achieved, we immediately run into problems with the boron neutrino flux: In order to get the weaker suppression of about $\Phi(^8B) \approx 0.5\,\Phi^{SSM}(^8B)$ as measured by Kamiokande, it is necessary to fine-tune also the S_{17} factor[6]. S_{17} has to be about twice as large as assumed in [Bah92], again a value far outside the estimated uncertainty for S_{17}. It is interesting to note that, firstly, the required change is in the opposite direction than what has been discussed all the years to explain the chlorine result (i.e. a small S_{17}), and, secondly, a new measurement [Mot94] seems to indicate that S_{17} indeed may be smaller than assumed in the [Bah92] SSM.

In summary, a nuclear physics solution to the solar neutrino puzzle at the production stage doesn't look appealing. Moreover, it principally cannot resolve the HOMESTAKE– KAMIOKANDE anomaly. To get a reasonably good fit to both experiments, it is required, apart from a significant flux suppression for beryllium neutrinos, that also the spectral shape of the boron neutrinos is distorted in a sense that the

[5] Part of the outcome of these S-factor modifications might perhaps also be achieved by allowing element concentrations (in particular ^3He) far from local equilibrium, as this can help to modify the branching ratio at the ppII–ppIII vertex. However, in any case this would require rapid diffusion or mixing of ^3He deep in the solar core, for which there is no indication at all from helioseismological or other measurements. We thus do not consider this hypothetical possibility any further.

[6] An increased (not decreased!) solar core temperature does work in the same direction, but may cause problems with the CNO-cycle which becomes more important as the core temperature rises.

suppression increases towards lower energies. Obviously, this cannot be achieved by simply adjusting the reaction rates in the sun: the spectral shapes remain unaltered.

3.4 Nonstandard Neutrinos

Reviewing all the arguments, what remains is nonstandard neutrino properties being responsible for the observed deficit. Although a vacuum oscillation solution is not excluded, matter-enhanced oscillations via the celebrated MSW-mechanism look much more promising, indicating a neutrino mass difference of $|m^2_{\nu_x} - m^2_{\nu_e}| \approx 10^{-5}\,\mathrm{eV}^2$, where ν_x is a neutrino state different from ν_e. The mixing angle $\sin^2 2\theta$ of the best-fit parameter set depends somewhat on the solar model one uses. It lies in the region $10^{-3} \lesssim \sin^2 2\theta \lesssim 10^{-2}$ [Ber94] for the models mentioned here.

In the best-fit MSW-case (the so-called "small angle solution"), the solar ν_e-spectrum at earth significantly deviates from the shape shown in figure 1. Depending on what solar model one bases the calculation on, the shape of the ^{8}B spectrum is more or less distorted, showing an increasing suppression towards lower energies. In any case, both the pep-line and the 862 keV ^{7}Be line should be strongly suppressed, while the pp-spectrum remains essentially unaffected. If neutrino oscillations really are the reason for the observed deficit, this is for the first time an experimental manifestation of physics beyond the standard model.

3.5 Next Generation Experiments

Obviously, to further clarify the situation, apart from the high statistics measurements of the boron neutrino spectrum envisaged by the SNO and Superkamiokande experiments[7], a high statistics measurement of the beryllium flux is crucial. BOREXINO, a 100 ton liquid scintillator detector for observation of neutrino-electron scattering events, is currently built at Gran Sasso to accomplish this task.

The direct measurement of the ^{7}Be flux is in particular important with respect to the low-flux solar models. If they really were the correct explanation for the KAMIOKANDE measurement, and the MSW-mechanism were at work accommodating the other experimental results, the shape distortion of the ^{8}B neutrino spectrum above 5 MeV might be too tiny to be detected. Moreover, also the CC/NC ratio measurement of the interactions of ^{8}B neutrinos with deuterium in SNO may not deliver further indication for neutrino conversion, as the high energy part of the ^{8}B spectrum in that case is almost unaffected by the conversion mechanism. The 862 keV ^{7}Be line, however, shall be strongly suppressed with respect to the model prediction.

GALLEX itself will stop solar neutrino recording in autumn 1996, with counting ongoing till in 1997. The statistical error then will have almost reached the systematic uncertainty. Whether or not there will be a continuation of neutrino measurements with gallium at Gran Sasso beyond 1996 (GNO: Gallium Neutrino Observatory [GNO96]) shall be decided until fall 1996. Clearly, for obvious reasons it is highly desirable to have the low energy pp-neutrinos observed in parallel to the higher energy parts of the solar neutrino spectrum also in the near future.

Acknowledgements

I would like to thank F.v.Feilitzsch, T.Goldbrunner, R.Mößbauer, and L.Oberauer for numerous discussions. It is a pleasure to thank the organizers of the school for

[7] Superkamiokande is an upgrade of Kamiokande by a factor of about 30. It started operation in spring 1996, did, however, due to problems with radon not yet reach the energy threshold of 5 MeV. SNO is a 1000 ton heavy water detector, currently under construction.

providing me the opportunity to spend two weeks in such a stimulating atmosphere and beautiful environment. GALLEX is supported by the German Bundesministerium für Bildung, Wissenschaft, Forschung und Technologie (BMBF).

References

Alt95 M.Altmann: GALLEX after the Source Experiment: Status and Implications, Proc. "Ringberg Workshop on Astro-Particle-Physics", Tegernsee; ed.: A.Weiss et al., 1995

Alt96 M.Altmann: Doctoral Dissertation, Technische Universität München, 1996

Alt96a M.Altmann, F.v.Feilitzsch, U.Schanda: A Parameterized Pulse Shape Analysis Method for the GALLEX Solar Neutrino Experiment, accepted for publication in Nucl. Inst. Meth. A, 1996

Bah92 J.Bahcall, M.Pinsonneault: Rev. Mod. Phys. 64 (1992) 885

Bah95 J.Bahcall, M.Pinsonneault: Rev. Mod. Phys. 67 (1995) 781

Ber93 V.Berezinsky: On the Exclusion of an Astrophysical Solution to the Solar Neutrino Problem, Preprint LNGS 93/86

Ber94 V.Berezinsky et al.: Phys. Lett. B 341 (1994) 38

Ber95 V.Berezinsky et al.: Last Hope for an Astrophysical Solution to the Solar Neutrino Problem, Preprint LNGS-95/56 and Phys. Lett. B 365 (1996) 185

Dar94 A.Dar, G.Shaviv: A Standard Model Solution to the Solar Neutrino Problem?, Preprint Technion-Ph-94-5

Dar95 A.Dar: Nucl. Phys. B (Proc. Suppl.) 48 (1996) 335

GAL92 GALLEX Collaboration: Phys. Lett. B 285 (1992) 376

GAL95 GALLEX Collaboration: Phys. Lett. B 342 (1995) 440

GAL96 GALLEX Collaboration: GALLEX Solar Neutrino Observations: Results for GALLEX III, accepted for publication in Phys. Lett. B, 1996

GAL96a GALLEX Collaboration: Preliminary Results from the Second ^{51}Cr Neutrino Source Experiment in GALLEX, GALLEX internal note GX-90, 1996

GNO96 GNO Collaboration: Proposal for a permanent Gallium Solar Neutrino Observatory at LNGS, 1996

Hat94 N.Hata et al.: Phys. Rev. D 49 (1994) 3622

Heu94 G.Heusser: Characteristics of the GALLEX Spectrometer, Proc. 2nd Int. Conf. Trends in Astropart. Phys., Aachen; ed. P.Bosetti, Teubner, Leipzig, 1994

Lun96 Luna Collaboration: First measurements of the ^{3}He(^{3}He, 2p)^{4}He cross section within the solar Gamow peak, submitted to Phys. Lett. B, 1996

Mot94 T.Motobayashi et al.: Phys. Rev. Lett. 73 (1994) 2680

Obe92 L.Oberauer, F.v.Feilitzsch: Rep. Prog. Phys. 55 (1992) 1093

Scha93 U.Schanda: Chemical Aspects of GALLEX, Proc. Int. School "Particles and Cosmology 1993", Baksan; ed.: E.Alexeev et al., World Scientific, Singapore, 1994

Scha93a U.Schanda: Doctoral Dissertation, Technische Universität München, 1993

Scha95 U.Schanda, A.Urban: Energy Calibration of Xe-filled Proportional Counters in the Range from 1 to 10 keV and Determination of their total active Volume, accepted for publication in Nucl. Inst. Meth. A, 1995

Tur93 S.Turck-Chieze et al.: Phys. Rep. 230 (1993) 59

Urb89 A.Urban: Doctoral Dissertation, Technische Universität München, 1989

Win93 R.Wink et al.: Nucl. Inst. Meth. A 329 (1993) 541

DOUBLE BETA DECAY RESEARCH - PRESENT AND FUTURE

Yuri G. Zdesenko

Institute for Nuclear Research of the National Academy of Sciences
252028 Kiev, Ukraine

INTRODUCTION

During the last fifteen years remarkable results have been obtained in double beta decay investigation, level of which looked like unattainable during the whole history of this research beginning from the theoretical prediction of 2β decay by M. Goeppert-Mayer [1] in 1935 and the first experimental work by E. Fireman [2] in 1948: half-life limit of greater than 10^{24} y for neutrinoless (0ν) 2β decay of ^{76}Ge [3-6]; large scale experiments on ^{76}Ge with enriched HP Ge detectors [5, 6]; half-life limits of greater than 10^{23} y for $0\nu2\beta$ decay of ^{136}Xe [7] and 10^{22} y for ^{82}Se [8], ^{100}Mo [9], ^{116}Cd [10] and ^{130}Te [11]; observation of the two neutrino (2ν) 2β decay of ^{76}Ge [4-6], ^{82}Se [8], ^{100}Mo [12-14], ^{116}Cd [15-17] and ^{150}Nd [13, 18] in direct counting experiments; progress in the theoretical interpretation of double beta decay [19-21].

These achievements are related with key role of neutrino for the modern particle physics theories, astrophysics and cosmology and with particular place of 2β decay investigations among many approaches to study the neutrino properties.

In the pioneer work by M. Goeppert-Mayer [1] the possible double beta decay was pointed out as a process in which "... a metastable isobar can change into a more stable one by simultaneous emission of two electrons". Usually it is considered two main modes of $2\beta^-$ decay, two-neutrino (2ν) and neutrinoless (0ν):

$$(A, Z) \dashrightarrow (A, Z+2) + 2e^- + 2\nu_e, \quad (1)$$

$$(A, Z) \dashrightarrow (A, Z+2) + 2e^- . \quad (2)$$

Analogous equations can be written also for $2\beta^+$ decay as well as for electron capture with β^+ decay ($\varepsilon\beta^+$) or for double electron capture (2ε).

Masses of Fundamental Particles: Cargèse 1996
Edited by M. Lévy *et al.*, Plenum Press, New York, 1997

The two-neutrino decay (1) is a conventional (but very rare) decay described in the Standard Model (SM) as a second order weak transition between two even-even nuclei. The second process (2) is forbidden in the SM and requires a violation of lepton number conservation and a non-vanishing Majorana neutrino mass or/and an admixture of a right-handed weak interaction to the left-handed one. In gauge theories, which suggest that neutrinos are Majorana particles, there are three ways to produce a Majorana mass: explicit B-L breaking and spontaneous breaking of a local or global B-L symmetry. In last case 2β decay could occur with emission of light pseudo-scalar Goldstone boson called Majoron:

$$(A, Z) \dashrightarrow (A, Z+2) + 2e^- + M. \qquad (3)$$

Different models for Majoron as extensions of the SM were proposed [22-25]. Recent LEP measurement of Z^0 width [26] ruled out possible triplet [22] and doublet [24, 25] Majoron but existence of singlet Majoron [23] or so called new Majorons are still possible [27].

Also supersymmetric (SUSY) mechanism of 0ν2β decay was introduced [28] based on SUSY theories with R-parity nonconservation.

At present there are several indications of a physics beyond the SM - the solar and atmospheric neutrinos deficit and mixed dark matter models which can be accorded with the degenerate neutrino mass scenario with $m_\nu \approx 1$ eV. Also the possible neutrino oscillation claimed by the LSND Collaboration [29] can be explained with a preferred δm^2 of either 1 eV^2 or 6 eV^2. The extended SO(10) grand unified model with S4 horizontal symmetry was recently proposed [30], which leads naturally to such a scenario. The decisive test of these models can be fulfilled in 2β decay research.

Therefore existence of neutrinoless 2β decay requires a new physics beyond the Standard Model and experiments to study this process even with negative results (lower limits of half-lives) can provide very stringent constrains on neutrino mass, right-handed admixtures in the weak interaction, neutrino-Majoron coupling constants, SUSY R-parity nonconservation parameter space etc. This can explain why double beta decay investigation continue to be very interesting and important field and now is even more popular than 20 years ago.

The main goal of present article is not to cover the material contained in many capital reviews [19-21, 31-52] but pay attention only to recent developments in 2β decay research.

THEORETICAL SITUATION

Trustworthy restrictions for neutrino mass, right-handed admixtures and other parameters from measured limits on 0ν2β decay rates are possible only in case of reliable calculations of nuclear matrix elements which enter the half-life formulae. Since the 0ν2β decay has not been observed up to now, it is impossible to check the theoretical 0ν2β rates directly and only way is to use the positive 2ν2β decay data to test the assumptions made in nuclear structure calculations.

Three main approaches have been used most intensively to calculate nuclear matrix elements: the nuclear shell model, the quasiparticle random phase approximation (QRPA) and, last time, the operator expansion method (OEM). Full scale computations in the shell model [38,53-56] were unattainable because the number of basic states in model space increased explosively for nuclei heavier than ^{48}Ca and various simplifications had to be introduced. From this point of view the QRPA [19,44,57-66] is more attractive so far as the new vacuum is determined for quasiparticles and nuclear wave function can be described in terms of small number of quasiparticles' degrees of freedom.

Calculations in the framework of QRPA revealed that the 2ν2β decay rates could be strongly suppressed by taking into account the ground state correlations in nuclei which are enhanced by the particle-particle interaction. Predicted 2ν2β decay rates are found to be very sensitive to the value of the strength of the particle-particle interaction g_{pp}, and calculated half-lives can be tuned to all experimental data in region of $g_{pp}\approx1$. In order to remove or at least decrease these difficulties several attempts were undertaken. In [67] the 2ν2β nuclear matrix elements were calculated taking into account the self-consistent self-energy corrections to the single particle spectra. In this way the QRPA instability around $g_{pp}\approx1$ was avoided and a reasonably good agreement of calculated matrix elements with experimental values was achieved. Paper [68] reports the results of the investigation of high order corrections upon the QRPA nuclear matrix elements - it was found that HQRPA corrections display a weak dependence on a particle-particle strength g_{pp} and become important in the region of $g_{pp}\approx1$ where QRPA values vanish.

As regards the 0ν2β decay, QRPA calculations of 0ν rates are comparably insensitive to various types of ground state correlations (and to particle-particle correlations in particular). This is because in neutrinoless mode the transitions are possible not through intermediate states with $J^{\pi}=1^{+}$ only but mainly through another multipoles J^{π} whose corresponding matrix elements are much less affected by the change of g_{pp}. Most of the authors agree that comparably certain computation of 0ν2β decay rates are possible.

The tedious calculations of matrix elements in shell model and drastic sensitivity of 2ν2β decay rates on the value of g_{pp} in QRPA stimulated the development of alternative approaches. In operator expansion method [69-74] the intermediate 1^{+} energy spectrum is not explicitly used in calculation of 2ν2β decay amplitude. In this way the dependence on g_{pp} is highly reduced and 2ν nuclear matrix elements become comparatively constant in physical region of g_{pp}. OEM assumptions are criticized in [75] as generating too small values of nuclear matrix elements. Instead, another approach eliminating the explicit summation over intermediate states is proposed [75] for exact evaluation of Green's function based on a Lanczos algorithm for inverting linear operators. This method removes the need of Green's function approximation used in OEM.

Calculations of the 2ν2β half-lives for several deformed nuclei were performed in [76,77], where the pseudo-SU(3) shell model scheme and a summation procedure without closure approximation were used.

The approach to calculate $2\nu2\beta$ decay rates in the framework of a relativistic quark confinement model was developed in [78-81].

References [19-21, 44, 47, 48, 82-82] may be consulted for a discussion and comparison between calculations of matrix elements in shell model, QRPA and OEM.

EXPERIMENTAL STATUS

There are two different classes of the direct double β decay experiments: a) with active source (source = detector); b) with passive source. Both of them could be divided into two kinds - with detector for measurement of the energy of the electrons and more complicated with tracking and energy detectors. In the first class of the experiments the required effect can be distinguished on the basis of only one property of 2β decay - the distribution of the total energy of the electrons. In the experiments with passive source and energy detectors it is possible to use time coincidence between two electrons and to measure the single and sum energy distributions. Full information about all properties of 2β decay events can be obtained in the most complete class of the experiments - time coincidence, tracks and vertex of the electrons, energy and angular distributions.

Let us consider briefly some recent results of the direct counting 2β decay experiments. In order to obtain the restrictions on the neutrino mass from experimental results in the same scale we will use the calculations [64] with the most extensive list of nuclei (because of a difference up to a few times between $T_{1/2}$ calculated in different approaches).

^{76}Ge

During 8 years Max-Planck Institute (Heidelberg) in collaboration with Kurchatov Institute (Moscow) have performed the most sensitive experiment to study ^{76}Ge by using HP Ge semiconductor detectors enriched in ^{76}Ge (86%). They used 16.9 kg of enriched ^{76}Ge to produce HP Ge detectors for ultra low-background measurements. At first three HP Ge detectors (0.927, 2.758 and 2.6 kg) enriched in ^{76}Ge to 86% with energy resolution 2.5 keV at 1.3 MeV were running in the Gran Sasso Underground Laboratory (Italy). The background rate around 2 MeV was 0.25 counts/y/keV/kg. Now the experiment is running with two more HP Ge enriched detectors so full mass of five detectors is equal to 11.5 kg. From 13.60 kg·y of measuring time the following half-life limit for the $0\nu2\beta$ decay of the ^{76}Ge has been obtained: $T_{1/2}^{0\nu} > 7.4\cdot10^{24}$ y (90% CL). This result corresponds to a neutrino mass less than 0.56 eV. Therefore during the next 4 years of measurements one can expect a limit for neutrino mass less than 0.4 eV. Also $2\nu2\beta$ decay of ^{76}Ge has been confirmed with high statistics: $T_{1/2}^{2\nu} = (1.42\pm0.03\pm0.13)\cdot10^{21}$ y [86].

There is another big international project IGEX [6] to study 2β decay of ^{76}Ge. In the first stage three detectors each composed of ≈0.7 kg of germanium enriched to 87.5% of ^{76}Ge were fabricated and tested. In the second phase of IGEX, three more detectors were produced with fiducial masses of about 2 kg each. All detectors (full mass about 8 kg) are in operating now in different underground laboratories: Homestake gold mine in USA (4000 mwe), Canfranc laboratory in Spain (2450 mwe) and Baksan Neutrino Observatory in Russia (660 mwe). The background rate (around 2 MeV) of less than 0.2 counts/y/keV/kg has been achieved. From 18.5 mole·y of data 2ν2β decay of ^{76}Ge half-life $T^{2\nu}_{1/2} = (1.1\pm0.2)\cdot10^{21}$ y [87] was deduced, which is differ slightly from Heidelberg-Moscow Collaboration result [86]. A sum of 23.7 mole·y of data were used to obtain the limit on 0ν2β decay: $T^{0\nu}_{1/2} > 4.3\cdot10^{24}$ y (90% CL) [87]. The corresponding bound on neutrino mass is less than 0.74 eV [87]. With further background improvement (due to pulse shape discrimination technique) the IGEX Collaboration could also provide a sensitive measurement of 0ν2β decay of ^{76}Ge in the sub-eV range for the neutrino mass within several years of data taking.

^{82}Se

The two-neutrino 2β decay of ^{82}Se has been measured by Irvine group from the University of California with help of the time projection chamber (TPC) placed in magnetic field of ≈700 G [8]. The TPC records the electron tracks, and the energies and opening angle are determined from each track. The discovery of two-neutrino double beta decay of ^{82}Se with $T^{2\nu}_{1/2} = 1.08^{+0.26}_{-0.06}\cdot10^{20}$ y (68% CL) has been claimed after ≈21000 h measuring time [8]. It was the first observation of the 2β decay in direct counting experiment. Also the limit $T^{0\nu}_{1/2} > 2.7\cdot10^{22}$ y (68% CL) has been set for 0ν2β decay of ^{82}Se [8].

^{100}Mo

Five groups have made the measurements with ^{100}Mo. In 1982 the Institute for Nuclear Research (Kiev) group has used [88] the plastic scintillator-wafer stacks with sheets of the ^{100}Mo. The limit $T^{0\nu}_{1/2} > 2.6\cdot10^{21}$ y (68% CL) has been set [88]. Recently about the same technique (but with semiconductor-wafer stacks) was used by LBL+MHC+UNM+INEL Collaboration [9]. The limit $T^{0\nu}_{1/2} > 4.4\cdot10^{22}$ y (68% CL) has been reached in this work. Three another groups have used the apparatus with tracking and energy detectors and all groups have claimed the discovery of two-neutrino 2β decay of the ^{100}Mo: Moscow group - $T^{2\nu}_{1/2} = 3.3\cdot10^{18}$ y [89]; Osaka - $T^{2\nu}_{1/2} = 1.15\cdot10^{19}$ y [11]; Irvine - $T^{2\nu}_{1/2} = 1.16\cdot10^{19}$ y [13]. Recently Osaka and Irvine results were confirmed by the NEMO Collaboration with help of NEMO 2 apparatus with tracking device (source ≈1 m^2) and 2x64 plastic scintillators for energy measurements: $T^{2\nu}_{1/2} = (0.95\pm0.10)\cdot10^{19}$ y [14]. The NEMO collaboration is going to build the NEMO 3 tracking detector scale up in comparison with previous one by factor of 20 in size (source square ≈ 20 m^2) and by factor of 1000 in sensitivity. It should be the biggest detector worldwide for double beta decay research. With source's mass of ≈10 kg of ^{100}Mo they plan to reach the limit $T^{0\nu}_{1/2} \approx 10^{25}$ y.

^{116}Cd

To study double beta decay of ^{116}Cd the $CdWO_4$ crystal scintillators enriched in ^{116}Cd to 83% have been developed by Kiev group (Institute for Nuclear Research) [10, 90, 91]. Experiments have been carried out in the Solotvina Underground Laboratory of INR built in the salt mine on the depth more than 1000 mwe, where cosmic muon flux is suppressed by a factor of 10^4. Three enriched $^{116}CdWO_4$ crystal scintillators (19, 14 and 13 cm^3) as well as natural detectors $CdWO_4$ (8, 9 and 57 cm^3) were used for the measurements. In the low background installation the cadmium tungsten crystal ($\approx$15 cm^3) is viewed by a PMT (type FEU-110) through a light-guide of 51 cm long. The energy resolution of the $^{116}CdWO_4$ crystal is equal to 14.1 and 7.1% at the energy 662 and 2615 keV, respectively. The passive shielding of OFHC copper (5 cm) and lead (23 cm) surrounds the plastic scintillator which was used as an active shielding of the main $^{116}CdWO_4$ detector. The background rate of the $^{116}CdWO_4$ detector in the region of 2.7-2.9 MeV was reduced to 0.55 counts/(y·kg·keV) [92]. This value is similar to that of the best low background HP Ge detectors. Moreover the off-line analysis of the time distribution of the measured events was developed and fulfilled, which allowed to measure the extremely low level of the intrinsic contamination of the crystal by the ^{232}Th ($1.8(2)\cdot10^{-5}$ Bk/kg) and to set the limits of $\leq 7\cdot10^{-6}$ and $\leq 2\cdot10^{-5}$ Bk/kg for ^{235}U and ^{226}Ra, respectively.

After 11888 h of measuring time the half-life limit for neutrinoless 2β decay of ^{116}Cd has been achieved: $T_{1/2}^{0\nu} \geq 3.2(5.4)\cdot10^{22}$ y with 90% (68%) CL [92]. This limit corresponds to the following restrictions on an effective Majorana neutrino mass and right-handed admixtures in a weak interactions: $\langle m_\nu \rangle \leq 3.9$ eV, $\langle \eta \rangle \leq 5.6\cdot10^{-8}$, $\langle \lambda \rangle \leq 5.1\cdot10^{-6}$ (90% CL). At the near future it is possible to promote this result in the range of about 2 eV for the neutrino mass limit. INR (Kiev) and MPI (Heidelberg) have recently proposed [15] a project of the big scale experiment with $^{116}CdW0_4$ crystal scintillators ($\approx$20 kg of the ^{116}Cd) in which the limit $T_{1/2}^{0\nu} > 10^{25}$ y can be reached ($m_\nu < 0.3$ eV). In the second phase of this experiment $^{116}CdW0_4$ crystals could be used as the cryogenic thermal detectors (the energy resolution of about 5 keV at the energy of 2.6 MeV has recently been obtained in the first test of $CdW0_4$ crystal with mass of 58 g [93]).

Two-neutrino 2β decay of the ^{116}Cd was discovered by three groups. Osaka University in collaboration with INR (Kiev) have performed the experiment in the Kamioka Underground Laboratory, where the half-life of $T_{1/2}^{2\nu}$ =(2.6\S\DO3+0.9,−0.5)·10^{21} was measured [16]. The INR (Kiev) group in their own experiment [15] has obtained for 2ν2β decay of ^{116}Cd the value of $T_{1/2}^{2\nu} = \{2.7^{+0.5}_{-0.4}(\text{stat.})^{+0.9}_{-0.6}(\text{syst.})\}\cdot10^{19}$ y, by subtracting the spectra of natural $CdWO_4$ scintillators from the spectra of enriched $^{116}CdWO_4$ detectors [15]. The Osaka and Kiev groups' results were confirmed [17] by NEMO Collaboration with the better statistics and ratio signal to background ($\approx$4:1): $T_{1/2}^{2\nu} = \{3.6^{+0.3}_{-0.3}(\text{stat.})^{+0.3}_{-0.3}(\text{syst.})\}\cdot10^{19}$ y .

^{130}Te

Milano group have used the crystal TeO_2 (334 g) as a cryogenic thermal detector for 2β decay research of ^{130}Te [11]. At present array of 4 crystals, 340 g each, has been installed [94] in the Gran Sasso Underground Laboratory. Energy resolution of the crystals is ≈12 keV at 2.6 MeV and background rate in the vicinity of 2.5 MeV is about 4 counts/y/keV/kg. With total measuring time of 0.73 kg·y the limit $T^{0\nu}_{1/2} > 3.25 \cdot 10^{22}$ y (90% CL) has been established for $0\nu2\beta$ decay of the ^{130}Te ($m_\nu < 3.9$ eV) [94]. For the future with 20 natural TeO_2 crystals (full mass ≈7 kg) the limit $T^{0\nu}_{1/2} > 10^{23}$ y can be set ($m_\nu < 2.2$ eV) and with enriched $^{130}TeO_2$ crystals level of 10^{24} y can be reached during several years of measurements ($m_\nu < 0.7$ eV). It should be mentioned that development of this technique is an important step in 2β decay research because it can be applied to different crystals which contain the possible 2β active nuclei.

^{136}Xe

Caltech (Pasadena)+Institut de Physique (Neuchâtel)+Paul Scherrer Institute (Villigen) Collaboration [7] have constructed the time projection chamber with an active volume of 207 l, which has operated with 5 atm of xenon enriched in ^{136}Xe to 62.5%. The energy resolution of the TPC is 6.6% at 1.6 MeV. The track reconstruction capability of the TPC provides a powerful means of background rejection. As a result a background rate around 2.48 MeV (within FWHM energy interval) is 0.005 counts/y/keV/kg [95]. From 10480 h of data taking in the Gotthard Underground Laboratory the limit $T^{0\nu}_{1/2} > 4.2 \cdot 10^{23}$ y (90% CL) has been set for $0\nu2\beta$ decay of the ^{136}Xe [95]. Upper limit of the neutrino mass is equal to 2.3 eV. At the near future and after some improvements of the apparatus this collaboration plan to reach the limit $T^{0\nu}_{1/2} > 10^{24}$ y ($m_\nu < 1.5$ eV). One can note that equipment developed in this work joints the advantages of both classes of direct 2β decay experiments: the reasonable high efficiency and tracking information. As the natural next step in this direction the large-volume liquid Xe TPC [96] and self-triggered drift chamber filled by liquid enriched ^{136}Xe [97] have been proposed allowing to reach the sensitivity of $\approx 10^{25}$ y ($m_\nu < 0.5$ eV). In order to overcome even this level a new approach was considered, which used the coincident detection of $^{136}Ba^{2+}$ ions (result of 2β decay of ^{136}Xe on the atomic level) and $0\nu2\beta$ signal of ^{136}Xe with energy 2.5 MeV in a TPC filled by liquid Xe [98, 99]. Such apparatus with background-free 1000 kg of ^{136}Xe can investigate the Majorana neutrino mass in 0.01 eV region [98].

^{150}Nd

In early experiment of Moscow group the plastic scintillators with sheets of ^{150}Nd were used and the limit $T^{0\nu}_{1/2} > 1.7 \cdot 10^{21}$ y (95% CL) was established for $0\nu2\beta$ decay of ^{150}Nd ($m_\nu < 4.5$ eV) [100]. Irvine group using 6342 h TPC run with 11.2 g of ^{150}Nd has set the limit

$T^{0\nu}_{1/2} > 2.1\cdot10^{21}$ y (90% CL), corresponding to neutrino mass $m_\nu < 4.0$ eV [101]. Due to the high sensitivity of ^{150}Nd to neutrino mass (comparing with another candidate nuclei) the new and reasonable scale up TPC with ≈400 g of ^{150}Nd can provide the investigation of the Majorana neutrino mass in 0.5 eV domain [101]. Two-neutrino 2β decay of ^{150}Nd was recently observed by ITEP (Moscow) and Irvine groups (preliminary results are: $T^{2\nu}_{1/2} = 1.7\cdot10^{19}$ y [18] and $T^{2\nu}_{1/2} = 1.0\cdot10^{19}$ y [13]). The final ITEP value [102] is:

$$T^{2\nu}_{1/2}=\{1.88^{+0.69}_{-0.39}(\text{stat.})^{+0.19}_{-0.19}(\text{syst.})\}\cdot10^{19}\ \text{y}\ .$$

^{160}Gd

Despite of the low 2β decay energy (1729.7 keV), ^{160}Gd is interesting candidate for research because the high natural abundance of ^{160}Gd (21.86%) and due to background problem from 2ν2β decay in the region of 0ν2β signal, which is very serious limitation for detectors with poor energy resolution. It was shown [76] that 2ν2β decay of ^{160}Gd is strongly forbidden due to heavy deformation of ^{160}Gd. If so the background from 2ν2β mode will be absent. The calculated value of $T^{0\nu}_{1/2}\cdot\langle m_\nu\rangle^2 = 8.6\cdot10^{23}$ y·eV^2 is nearly three times lower than for ^{76}Ge and ^{136}Xe [64].

In light of these reasons the experiment to study 0ν2β decay of ^{160}Gd [103] was performed by INR (Kiev) group in the Solotvina Underground Laboratory with help of cerium-doped gadolinium silicate Gd_2SiO_5:Ce (GSO) crystal grown by Czochralski method with a mass of 698 g (volume of 104 cm^3). The GSO scintillator has a large density (6.71 g/cm^3), fast response (decay constant is equal to 40-60 ns) and quite large light output (20% of NaI(Tl); peak emission - 440 nm). The number of ^{160}Gd nuclei in this crystal is $4.346\cdot10^{23}$. In the low background installation the GSO crystal is viewed by PMT through a plastic light-guide 18 cm long. The energy resolution is 15.5% and 7.9% at energy 662 and 2615 keV, respectively. The passive shielding of OFHC copper (5 cm), mercury (7 cm) and lead (15 cm) surrounds the GSO scintillator to reduce the external background. The total measuring time is equal to 5421 h. To reduce the background from intrinsic contamination of the crystal the off-line time analysis of the measured events was performed also. For example, the sequence of two α decays belonging to ^{232}Th family was searched for: ^{220}Rn(E_α=6.29 MeV, $T_{1/2}$=55.6 s) –> ^{216}Po(E_α=6.78 MeV, $T_{1/2}$=146 ms) –> ^{212}Pb. The background events with the energy in the interval of 1.2 - 2.2 MeV were used as a trigger (α particle energy of ^{220}Rn corresponds to 1.66 MeV in γ scale). Then any events following trigger in the time interval 10 ms - 500 ms (it contains ~86% of all decays from ^{216}Po) were selected. The obtained spectra are in excellent agreement with what expected from α particles on ^{220}Rn and ^{216}Po decay. The distribution of the time intervals between the first and second α particles is also in a clear accordance with expected half-life of 146 ms. Using these results, the ^{228}Th content was found to be equal to 2.79(5) mBq/kg. The same technique was applied for the ^{235}U and ^{226}Ra family, which leads to the intrinsic activity of the crystal equal to 0.995(23) and 0.53(3) mBq/kg for ^{227}Ac and ^{226}Ra, respectively. All selected events were excluded from the measured spectrum of the GSO crystal. The background rate in the region of $0\nu2\beta$ decay of ^{160}Gd (1650-1810 keV) was reduced [103] to 0.045 counts/(h·kg·keV).

The half-life limits have been set [103] for neutrinoless double beta decay of ^{160}Gd to the ground and first excited levels of ^{160}Dy: $T^{0\nu}_{1/2} \geq 8.2(13) \cdot 10^{20}$ y with 90% (68%) CL.

The main conclusion of this study is the possibility to perform more sensitive experiment on $\beta\beta$ decay of ^{160}Gd using the GSO crystals. The INR (Kiev) and MPI (Heidelberg) Collaboration has recently proposed such a project with approximately one thousand of pure GSO crystals with full mass about one ton. Due to high abundance of ^{160}Gd the GSO crystals could be grown up from natural Gd and as a result the cost of such a multidetector array should be well less expensive than for experiments and projects based on the enriched isotopes and detectors. The reduction of the internal impurities of the crystals by more than order of magnitude is under progress now. With further improved background the limit of $T^{0\nu}_{1/2} \approx 10^{25}$ y could be reached that corresponds the restriction of the Majorana neutrino mass less than 0.3 eV, which is comparable with the most sensitive experiments with ^{76}Ge [85, 87].

The results of the recent and most advanced 2β decay experiments are summarized in the table, where theoretical estimations [64] are presented also. It is clear that investigations of ^{76}Ge and ^{136}Xe give us the most stringent restrictions on the Majorana neutrino mass in the range of 0.6 - 2 eV and in other five experiments the obtained limits are ranged within the interval of 4-6 eV.

Table 1. Limits on neutrino mass from the most advanced experiments

Isotope	Experiment lim $T^{0\nu}_{1/2}$, y	Theory [64] $T^{0\nu}_{1/2} \cdot \langle m_\nu \rangle^2$, $y \cdot eV^2$	Limit $\langle m_\nu \rangle$, eV 68% CL	90% CL
^{76}Ge	$7.4 \cdot 10^{24}$ (90% CL) [85]	$2.33 \cdot 10^{24}$		0.56
	$1.3 \cdot 10^{25}$ (68% CL) [85]		0.43	
^{82}Se	$2.7 \cdot 10^{22}$ (68% CL) [8]	$6.03 \cdot 10^{23}$	4.7	-
^{100}Mo	$4.4 \cdot 10^{22}$ (68% CL) [9]	$1.27 \cdot 10^{24}$	5.4	-
^{116}Cd	$3.2 \cdot 10^{22}$ (90% CL) [92]	$4.87 \cdot 10^{23}$		3.9
	$5.4 \cdot 10^{22}$ (68% CL) [92]		3.0	
^{130}Te	$3.2 \cdot 10^{22}$ (90% CL) [94]	$4.89 \cdot 10^{23}$		3.9
	$5.5 \cdot 10^{22}$ (68% CL) [94]		3.0	
^{136}Xe	$4.2 \cdot 10^{23}$ (90% CL) [95]	$2.21 \cdot 10^{24}$		2.3
^{150}Nd	$2.1 \cdot 10^{21}$ (90% CL) [101]	$3.37 \cdot 10^{22}$	-	4.0

CONCLUSIONS

Up to now only 18 nuclei from all 35 potential $2\beta^-$ active candidates were studied in direct counting experiments. The highest limits of half-lives for neutrinoless 2β decay have been set for ^{76}Ge - $7\cdot10^{24}$ y, ^{136}Xe - $4\cdot10^{23}$ y, ^{82}Se, ^{100}Mo, ^{116}Cd, ^{130}Te - $(1\text{-}3)\cdot10^{22}$ y, ^{48}Ca, ^{150}Nd - $\approx10^{21}$ y. The $0\nu2\beta$ results for other nuclei are placed in the region of 10^{16} - 10^{19} y. The two-neutrino 2β decay of ^{76}Ge ($T_{1/2} \approx 10^{21}$ y), ^{82}Se ($\approx10^{20}$ y), ^{100}Mo and ^{150}Nd ($\approx10^{19}$ y) were observed surely.

The most sensitive results for $0\nu2\beta$ decay (^{76}Ge, ^{136}Xe , ^{116}Cd and ^{130}Te) as well as for 2ν decay mode (^{76}Ge) were obtained by using the technique with active source. The devices with passive source and tracking detectors are more convenient for detail investigation of $2\nu2\beta$ decay (the present day level of sensitivity is 10^{19} - 10^{20} y). In not so far future one can expect the limit $T^{0\nu}_{1/2}$ for ^{76}Ge - 10^{25} y ($m_\nu \leq 0.5$ eV), ^{116}Cd and ^{130}Te - $\approx10^{23}$ y ($m_\nu \leq 2$ eV), ^{136}Xe - $\approx10^{24}$ y ($m_\nu \leq 1.5$ eV).

There are several projects to increase the sensitivity of $0\nu2\beta$ decay experiments to the level of 10^{24}-10^{25} years. The most promising proposals among them are the following: the development of large volume liquid Xe TPC [96] and self-triggered drift chamber filled by liquid enriched ^{136}Xe [97]; the building of the biggest tracking detector for 2β investigations of ^{100}Mo by NEMO collaboration [14]; using in the experiment of Milano group the array from natural and enriched TeO_2 crystals with total mass about 10 kg [94]; the Kiev-Heidelberg project with approximately one thousand of pure GSO crystals with full mass about one ton and with enriched $^{116}CdWO_4$ crystal scintillators [103]. However for so sensitive experiments the serious problem exists connected with background occasioned by the two-neutrino 2β decay [97, 103]. It should be stressed that these background events in principle could not be distinguished from $0\nu2\beta$ signal because all features are the similar: the same particles simultaneously emitted from one point, the same energy region and the right angular distribution. Only difference in the electrons' sum energy distributions can be used. In this case the energy resolution of device will play a decisive role, like for calorimeters which measure only the released energy. It was shown [49, 96, 104] that reaching the sensitivity of more than 10^{24} y with poor resolution (worse than 10%) looks like very questionable especially if we take into account the problems of low statistics. For instance, in 10 kg of ^{100}Mo only 42 (4.2) $0\nu2\beta$ decays will occur for a period of one year if $T^{0\nu}_{1/2} = 10^{24}$ (10^{25}) y).

The usual question is "why we need to perform the research with several nuclides, but not with ^{76}Ge or ^{136}Xe only?". There are many reasons to spread the efforts to more candidate nuclei. First of all, a neutrinoless 2β decay is unobserved phenomenon still now. Despite the big progress in understanding and interpretation of this process, new development in calculations of the nuclear matrix elements and half-lives, the present theoretical situation does not look completely definite. Thus it is rather difficult to suppose that on the base of theory anybody can point out the best nucleus for discovery of neutrinoless 2β decay. On the other hand, a success in these research could be achieved on the border of modern technology as an alloy of science, experimental art and even luck, hence nobody can know *a priori* where the highest sensitivity will be reached. The new and sometimes unexpected advancements in experimental technique could give the advantages to particular 2β decay candidates. The good example is $CdWO_4$ crystal scintillators which were developed and used for study of 2β decay of ^{116}Cd [15, 90, 91]. Recent discovery that $CdW0_4$ crystals can work as the cryogenic thermal detectors with high energy resolution (≈5 keV) [93] gives the excellent chance to increase the sensitivity of experiment with ^{116}Cd. So the present experimental and theoretical situation in 2β decay problem makes necessary to extend the number of nuclides studied with a sensitivity comparable or better than for ^{76}Ge. In case of a possible observation of 0ν2β peak of ^{76}Ge or ^{136}Xe, existence of such a signal has to be confirmed on other nuclides and also by another technique.

Therefore we can conclude that in course of about 60 years the 2β decay investigations have been and continue to be popular due to their fundamental importance. Last experiments with enriched ^{76}Ge and ^{136}Xe, the development of $^{116}CdWO_4$ crystals and TeO_2 cryogenic detectors, discovery of two-neutrino 2β decay of ^{76}Ge, ^{82}Se, ^{100}Mo, ^{116}Cd and ^{150}Nd give us the new time scale for research in this important field of modern physics. In not so far future we can expect the new distinguished results which will provide a test of current theory as well as an unique and a very sensitive probe of new physics beyond the Standard Model.

REFERENCES

1. M.Goeppert-Mayer, Phys. Rev. **48**, 512 (1935).
2. E.L.Fireman. Phys. Rev. **74**, 1238 (1948).
3. D.O. Caldwell et al., Nucl. Phys. B (Proc. Suppl.) **13**, 547 (1990).
4. I.V. Kirpichnikov, Nucl. Phys. B (Proc. Suppl.) **28A**, 210 (1992).
5. H.V. Klapdor-Kleingrothaus (Heidelberg-Moscow Collabor.), Nucl. Phys. B (Proc. Suppl.) **31**, 72 (1993).
6. F.T. Avignone et al. (IGEX Collabor.), Nucl. Phys. B (Proc. Suppl.) **35**, 354 (1994).
7. J.-C.Vuilleumieret al., Phys. Rev. D **48**, 1009 (1993).
8. S.R. Elliot et al., Phys. Rev. C **46**, 1535 (1992).
9. M. Alston-Garnjost et al., Phys. Rev. Lett. **71**, 831 (1993).
10. F.A. Danevich et al., Proc. 3-rd Int. Symp. on Weak and Electromagn. Inter. in Nuclei WEIN-92 (Dubna, June 1992), ed. Ts. Vylov (World Sci. Publ. Co., 1993) p.575.
11. A. Alessandrello et al., Phys. Lett. B **335**, 519 (1994).
12. H. Ejiri et al., Nucl. Phys. B (Proc. Suppl.) **28A**, 219 (1992).
13. S.R. Elliot et al., Nucl. Phys. B (Proc. Suppl.) **31**, 68 (1993).
14. D. Dassie et al. (NEMO Collaboration), Phys. Rew. D **51**, 2090 (1995).
15. F.A. Danevich et al., Phys. Lett. B **344**, 72 (1995).
16. H. Ejiri et al., Journal of the Phys. Society of Japan **64**, 339 (1995).

17. R.Arnold et al., JETP Lett. **61**, 170 (1995).
18. V.A. Artemiev et al., JETP Lett. **58**, 262 (1993).
19. T. Tomoda, Rep. Prog. Phys. **54**, 53 (1991).
20. H.V. Klapdor-Kleingrothaus, Proc. 3-rd Int. Symp. on Weak and Electromagn. Inter.in Nuclei WEIN-92, Dubna, June 1992, ed. Vylov (World Sci. Publ. Co., 1993) p.201.
21. M. Moe and P. Vogel, Ann. Rev. Nucl. Part. Sci. **44**, 247 (1994).
22. G.B.Gelmini and M.Roncadelli, Phys. Lett. B **99**, 411 (1981).
23. Y. Chikashige, R.N. Mohapatra, R.D. Peccei, Phys. Lett. B **98**, 265 (1981).
24. C. Aulakh and R.N. Mohapatra, Phys. Lett. B **119**, 136 (1982).
25. R.N. Mohapatra and P.B. Pal, Massive Neutrinos in Phys. and Astrophys. (World Sci. Publ. Co., 1991).
26. J. Steinberger, Phys. Rep. **203**, 345 (1991).
27. J. Helmig et al., Proc.Int. Workshop on Double Beta Decay and Related Topics, Trento, Italy, April 24 - May 5, 1995 (World Sci. Publ. Co., 1996) p.130.
28. M. Hirsch et al., Phys. Rev. D. **53**, 1329 (1996).
29. Athanassopoulos et al., Phys.Rev. Lett. **75**, 2650 (1995).
30. D.G. Lee and R.N. Mohapatra, Phys. Lett. B **329**, 463 (1994).
31. Ya.B. Zeldovich, S.Yu. Lukyanov and Ya.A. Smorodinski, Sov. Phys.Usp. **54**, 361 (1954).
32. G.F.Dell'Antonio and E.Fiorini, Suppl. Nuovo Cim. **17**, 132 (1960).
33. V.R.Lazarenko, Sov. Phys. Usp. **9**, 860 (1967).
34. E.Fiorini, Riv. Nuovo Cim. **2**, 1 (1972).
35. D.Bryman and C.Picciotto, Rev. Mod. Phys. **50**, 11 (1978).
36. Yu.G.Zdesenko, Sov. J. Part. Nucl. **11**, 542 (1980).
37. H.Primakoff and S.P.Rosen, Ann. Rev. Nucl. Part. Sci. **31**, 145 (1981).
38. W.C.Haxton and G.J.Stephenson Jr., Prog. Part. Nucl. Phys. **12**, 409 (1984).
39. M.G.Schepkin, Sov. Phys. Usp. **27**, 555 (1984).
40. M.Doi, T.Kotani, and E.Takasugi, Prog. Theor. Phys. Suppl. **83**, 1 (1985).
41. J.D.Vergados, Phys. Rep. **133**, 1 (1986).
42. F.T.Avignone III and R.L.Brodzinski, Prog. Part. Nucl. Phys. **21**, 99 (1988).
43. A.Faessler, Prog. Part. Nucl. Phys. **21**, 183 (1988).
44. K.Muto and H.V.Klapdor, in Neutrinos, ed. by H.V.Klapdor (Springer-Verlag, Berlin/Heidelberg/New York, 1988), p.183.
45. D.O.Caldwell, Int. J. Mod. Phys. A **4**, 1851 (1989).
46. M.K.Moe, Nucl. Phys. B (Proc. Suppl.) **19**, 158 (1991).
47. F.Boehm and P.Vogel, Physics of Massive Neutrinos 2-nd ed., (Cambridge University Press, Cambridge, 1992).
48. M.K.Moe, Int. J. Mod. Phys. E **2**, 507 (1993).
49. V.I. Tretyak and Yu.G. Zdesenko, At. Data and Nucl. Data Tables 61, 43 (1995).

50. H.V. Klapdor-Kleingrothaus, Proc.Int. Workshop on Double Beta Decay and Related Topics, Trento, Italy, April 24 - May 5, 1995 (World Sci. Publ. Co., 1996) p.3.
51. R.N. Mohapatra, Proc.Int. Workshop on Double Beta Decay and Related Topics, Trento, Italy, April 24 - May 5, 1995 (World Sci. Publ. Co., 1996) p.44.
52. J.W.F. Walle, Proc.Int. Workshop on Double Beta Decay and Related Topics, Trento, Italy, April 24 - May 5, 1995 (World Sci. Publ. Co., 1996) p.69.
53. W.C.Haxton, G.J.Stephenson Jr., and D.Strottman, Phys. Rev. Lett. **47**, 153 (1981).
54. W.C.Haxton, S.P.Rosen, and G.J.Stephenson Jr., Phys. Rev. D **26**, 1805 (1982).
55. T.Tsuboi, K.Muto, and H.Horie, Phys. Lett. B **143**, 293 (1984).
56. W.C.Haxton, Nucl. Phys. B (Proc. Suppl.) **31**, 88 (1993).
57. P.Vogel and M.R.Zirnbauer, Phys. Rev. Lett. **57**, 3148 (1986).
58. K.Grotz and H.V.Klapdor, Nucl. Phys. A **460**, 395 (1986).
59. O.Civitarese, A.Faessler, and T.Tomoda, Phys. Lett. B **194**, 11 (1987).
60. T.Tomoda and A.Faessler, Phys. Lett. B **199**, 475 (1987).
61. J.Engel, P.Vogel, and M.R.Zirnbauer, Phys. Rev. C **37**, 731 (1988).
62. K.Muto, E.Bender, and H.V.Klapdor, Z. Phys. A **334**, 177; 187 (1989).
63. G.Pantis et al., J. Phys. G: Nucl. Part. Phys. **18**, 605 (1992).
64. A.Staudt, K.Muto, and H.V.Klapdor-Kleingrothaus, Europhys. Lett. **13**, 31 (1990).
65. J.Suhonen and O.Civitarese, Phys. Lett. B **308**, 212 (1993).
66. G.Pantis and J.D.Vergados, Phys. Rep. **242**, 285 (1994).
67. D.B.Stout and T.T.S.Kuo, Phys. Rev. Lett. **69**, 1900 (1992).
68. S.Stoica and W.A.Kaminski, Phys. Rev. C **47**, 867 (1993).
69. C.R.Ching and T.H.Ho, Comm. Theor. Phys. **10**, 45 (1988).
70. C.R.Ching, T.H.Ho, and X.Wu, Phys. Rev. C **40**, 304 (1989).
71. M.Gmitro and F.Simkovic, Izv. AN SSSR, ser. fiz. **54**, 1780 (1990).
72. X.R.Wu et al., Phys. Lett. B **272**, 169 (1991).
73. X.R.Wu et al., Phys. Lett. B **276**, 274 (1992).
74. M.Hirsch et al., Phys. Rep. **242**, 403 (1994).
75. J.Engel, W.C.Haxton, and P.Vogel, Phys. Rev. C **46**, 2153 (1992).
76. O. Castanos et al., Nucl. Phys A **571**, 276 (1994).
77. J.H. Hirsch et al., Nucl. Phys A **582**, 124 (1995).
78. J.D. Vergados, Nucl. Phys B **250**, 618 (1985).
79. M. Lusignoli and M.G. Schepkin, Nucl. Phys B **261**, 143 (1985).
80. J.Suhonen, S.B. Khadkikar and A. Faessler, Phys. Lett. B **237**, 8 (1990).
81. J.Suhonen, S.B. Khadkikar and A. Faessler, Nucl. Phys. A **529**, 727 (1991); Nucl. Phys. A **529**, 727 (1991).
82. K.Grotz and H.V.Klapdor, The Weak Interaction in Nuclear, Particle and Astrophysics (Adam Hilger, Bristol/Philadelphia/New York, 1990).
83. K.Muto, E.Bender, and H.V.Klapdor-Kleingrothaus, Z. Phys. A **339**, 435 (1991).

84. W.C.Haxton, Nucl. Phys. B (Proc. Suppl.) **31**, 88 (1993).
85. H.V. Klapdor-Kleingrothaus, Nucl. Phys. B (Proc. Suppl.) **48**, 216 (1996).
86. A. Balysh et al., Phys. Lett. B **322**, 176 (1994).
87. C.E. Aalseth et al., Nucl. Phys. B (Proc. Suppl.) **48**, 223 (1996).
88. Yu.G.Zdesenko et al., Proc.Int.Conf. Neutrino-82,Balatonfured, Hungary, 1982, ed. by A.Frenkel and L.Jenik (Budapest, 1982), p.209.
89. S.I.Vasilyev et al., JETP Lett. **51**, 622 (1990).
90. Yu.Zdesenko, J. Phys. G: Nucl. Part. Phys. **17**, s243 (1991).
91. F.A. Danevich et al., JETP Lett. **49**, 476 (1989).
92. F.A. Danevich et al., Nucl. Phys. B (Proc. Suppl.) **48**, 232 (1996).
93. A.Alessandrello et al., Nucl. Phys. B (Proc. Suppl.) **35**, 366 (1994).
94. A.Alessandrello et al., Nucl. Phys. B (Proc. Suppl.) **48**, 238 (1996).
95. J. Busto et al., Nucl. Phys. B (Proc. Suppl.) **48**, 251 (1996).
96. T.A. Girard et al., Nucl. Instrum. Meth. A **316**, 44 (1992).
97. A.G.Prokopets, S.Sasaki, and M.Miyajima, KEK Proc. **94-7**, 60 (Tsukuba, 1994).
98. M.K.Moe, Phys. Rev. C **44**, 931 (1991).
99. M.Miyajima, S.Sasaki, and H.Tawara, Proc. 6-th Workshop on Rad. Detectors and Their Uses. Tsukuba, Japan, Jan. 29-30, 1991 (Tsukuba, 1991), p.19.
100. A.A.Klimenko et al., Nucl. Instrum. Meth. B **17**, 445 (1986).
101. M.K.Moe, M.A.Nelson, and M.A.Vient, Prog. Part. Nucl. Phys. **32**, 247 (1994).
102. V.A. Artemiev et al., Nucl. Phys. B (Proc. Suppl.) **48**, 241 (1996).
103. F.A. Danevich et al., Nucl. Phys. B (Proc. Suppl.) **48**, 235 (1996).
104. A. Morales, Proc. of TAUP'89, Aquilla, September 25-28, 1989, ed. A. Bottino and P. Monacelli (Editions Frontieres, Gif-sur-Yvette, 1989) p.97.

NEUTRINO MASS MODELS WITH AN ABELIAN FAMILY SYMMETRY

Stéphane Lavignac

Laboratoire de Physique Théorique et Hautes Energies
Université Paris-Sud, Bât. 210
F-91405 Orsay Cédex, France

Introduction

Fermion masses are one of the most fundamental problems of particle physics. While the origin of the observed hierarchy between quark and charged lepton masses remains unexplained in the Standard Model and most of its extensions, the question of whether the neutrinos are massive or not is still open. On the theoretical side, since neutrino masses are not protected by any fundamental symmetry[1], there is no reason to expect them to be zero. Now, if the neutrinos are massive, the rather unnatural suppression of their masses relative to the quarks and charged leptons of the same family has to be explained. On the phenomenological side, massive neutrinos could solve in a natural way several astrophysical and cosmological problems.

Family symmetries, which have been first introduced in order to explain the quark mass and mixing hierarchies, provide a predictive framework for neutrino mass models. We present two seesaw models based on an abelian family symmetry, the one leading to a hierarchical mass spectrum, the other yielding two quasi-degenerate neutrinos with a large mixing angle.

Neutrino oscillations

There is no direct laboratory evidence for non-zero neutrino masses, but some experimental data suggest neutrino oscillations. Before discussing them, let us briefly review neutrino oscillations in the simple case of two flavours. Neutrinos oscillate when the weak eigenstates $\nu_{\alpha,\beta}$ are mixtures of the mass eigenstates $\nu_{1,2}$:

$$\begin{pmatrix} \nu_\alpha \\ \nu_\beta \end{pmatrix} = \begin{pmatrix} \cos\theta & \sin\theta \\ -\sin\theta & \cos\theta \end{pmatrix} \begin{pmatrix} \nu_1 \\ \nu_2 \end{pmatrix} \tag{1}$$

[1]Neutrino masses are only protected by lepton number symmetry, which is an accidental global symmetry of the Standard Model, and turns out to be violated in most of its extensions.

Masses of Fundamental Particles: Cargèse 1996
Edited by M. Lévy *et al.*, Plenum Press, New York, 1997

Suppose a weak eigenstate ν_α is produced at $t = 0$: $\nu(0) = \nu_\alpha = \cos\theta\, \nu_1 + \sin\theta\, \nu_2$. At t it will have evolved into $\nu(t) = \cos\theta e^{-iE_1 t}\nu_1 + \sin\theta e^{-iE_2 t}\nu_2$. Assuming relativistic neutrinos, the probability of detecting the other weak eigenstate is:

$$P(\nu_\alpha \to \nu_\beta) = | < \nu_\beta|\nu(t) > |^2 = \sin^2 2\theta \, \sin^2\left(\frac{\Delta m^2 L}{4E}\right) \tag{2}$$

where E is the neutrino energy, L the distance travelled by the neutrino between the source and the detector, and $\Delta m^2 = m_2^2 - m_1^2$ is the squared mass difference between the two mass eigenstates. Note that the oscillation probability, which is a function of the ratio L/E, is characterized by the parameters $\sin^2 2\theta$ and Δm^2.

The strongest indication in favour of neutrino oscillations comes from the solar neutrino deficit. All solar neutrino experiments have observed a suppression of the ν_e flux relative to the predictions of the standard solar models. The most convincing explanation of this deficit is provided by the Mikheyev-Smirnov-Wolfenstein (MSW) [1] conversion of the electron neutrino into another specie inside the sun. There are two allowed regions in the ($\sin^2 2\theta$, Δm^2) plane, one with a small mixing angle ($\Delta m^2 \sim 6.10^{-6} eV^2$ and $\sin^2 2\theta \sim 10^{-3} - 10^{-2}$), the other with a large mixing angle ($\Delta m^2 \sim 10^{-5} - 10^{-4} eV^2$ and $0.2 \leq \sin^2 2\theta \leq 0.9$). Another hint for neutrino oscillations is the atmospheric neutrino anomaly. Some experiments have measured a significantly lower ratio of the ν_μ to the ν_e flux than predicted, which could be a signature of ν_μ oscillations into another flavour with a large mixing angle ($\Delta m^2 \sim 10^{-2} eV^2$ and $\sin^2 2\theta \geq 0.5$). However, this anomaly has not been observed by all experiments, and some uncertainties remain. Finally, there may be indications in favour of $\bar{\nu}_\mu \rightleftharpoons \bar{\nu}_e$ oscillations with a small mixing angle ($\sin^2 2\theta \sim$ a few 10^{-3} for Δm^2 in the few eV^2 region) from the LSND experiment. This interpretation needs to be confirmed.

Massive neutrinos are also interesting for cosmology. Structure formation requires, in addition to cold dark matter, a small amount of hot dark matter, which could be composed of a neutrino with mass between 1 and $10\, eV$, or several degenerate neutrinos in the few eV range.

Models of neutrino masses

Generalities

There are numerous models of neutrino mass. All of them need an extension of the particle content of the Standard Model. A Dirac mass term ($\mathcal{L}_m = -\, m_D\, \bar{\nu}_L N_R + h.c.$) requires the introduction of a right-handed (RH) neutrino N_R in addition to the standard left-handed (LH) neutrino ν_L. Since N_R is a $SU(2)_L$ singlet, such a mass term violates the weak isospin by $\Delta I_W = 1/2$, and must therefore be generated by a Yukawa coupling to a Higgs doublet: $\mathcal{L}_{Yuk} = -\, h_\nu\, (\bar{\nu}_L \bar{e}_L)\, H_2\, N_R + h.c.$, with $m_D = h_\nu < H_2^0 >$. A Dirac neutrino is then like other fermions, but its Yukawa coupling h_ν has to be unnaturally small. A Majorana mass term ($\mathcal{L}_m = -1/2\, m_M\, \bar{\nu}_L \nu_R^c + h.c.$), which violates lepton number by two units ($\Delta L = 2$), involves a transition from the standard ν_L ($I_W = +1/2$) to its CP conjugate ν_R^c ($I_W = -1/2$). Such a mass term has $\Delta I_W = 1$ and must therefore originate from a Yukawa coupling to a weak Higgs triplet (Gelmini-Roncadelli model) or from an effective interaction.

In the presence of a RH neutrino, both Dirac and Majorana mass terms can be present, as well as a $\Delta I_W = 0$ Majorana mass term for the RH neutrino, $\mathcal{L}_m = -1/2\, M_R\, \bar{N}_L^c N_R + h.c.$. The full mass term takes then the following form (in the absence of a Higgs triplet):

$$-\frac{1}{2}\left(\bar{\nu}_L \bar{N}_L^c\right)\begin{pmatrix} 0 & m_D \\ m_D & M_R \end{pmatrix}\begin{pmatrix} \nu_R^c \\ N_R \end{pmatrix} + h.c. \tag{3}$$

The physical neutrino states, which are two Majorana neutrinos, are obtained from the diagonalization of this 2x2 matrix. Particularly interesting is the seesaw limit [2], $m_D \ll M_R$, in which one eigenstate has a mass far below the weak scale[2]:

$$\begin{cases} m_1 \simeq m_D^2/M_R \\ m_2 \simeq M_R \end{cases} \quad \text{and} \quad \tan\theta \simeq m_D/M_R \tag{4}$$

Since the mixing angle is small, the light eigenstate is mainly the Standard Model neutrino.

Seesaw models

The seesaw mechanism is very popular, because it naturally generates neutrino masses much lighter than the weak scale. Moreover, it can be easily implemented in numerous extensions of the Standard Model, like $SO(10)$ GUT's or string models, where such Standard Model singlets as N_R with masses in the phenomenologically interesting range (typically, $M_R \sim 10^{12} - 10^{16}\ GeV$, which corresponds to a light neutrino mass $m_\nu \sim 10^{-3} - 10\ eV$) can be present. Assuming one RH neutrino per family, the Dirac (Majorana) mass in (3) is replaced by a 3x3 matrix in family space $\mathcal{M}_D$ ($\mathcal{M}_M$), and the light neutrino spectrum is obtained by diagonalizing the symmetric matrix:

$$\begin{aligned} \mathcal{M}_\nu &= -\mathcal{M}_D\, \mathcal{M}_M^{-1}\, \mathcal{M}_D^T \\ &= R_\nu\, Diag(m_{\nu_1}, m_{\nu_2}, m_{\nu_3})\, R_\nu^T \end{aligned} \tag{5}$$

Note that the mixing angles relevant for neutrino oscillations are given by the analog of the CKM matrix, which also involves the charged lepton sector[3]:

$$V_L = R_e^L R_\nu \tag{6}$$

In general, the entries of both the Dirac and the Majorana matrices are free parameters, and one has to choose a specific ansatz in order to make any definite prediction in the neutrino sector. It is often assumed that the Dirac mass matrix has the same structure than the up quark mass matrix[4]: $\mathcal{M}_D \sim \mathcal{M}_U$. For the Majorana matrix, however, no such simplifying assumption can be done, and it is necessary to assume a specific form. It follows that the neutrino spectrum of a given model depends on the ansatz that has been chosen[5], which is not very satisfactory.

Alternatively, one can try to derive the structures of the Dirac and the Majorana matrices from a symmetry. This symmetry has to act in a different way on the three neutrino families, otherwise the matrices would be unconstrained. Such a symmetry is called a *family symmetry.* This approach has proved to be successful in the quark sector, where, following the original idea by Froggatt and Nielsen [3], several groups [4, 5, 6, 7, 8, 13] have shown that an abelian family symmetry can reproduce the observed mass and mixing hierarchies.

---

[2]The Dirac mass m_D, which is protected by the electroweak symmetry, is expected to be of the order of the breaking scale $M_{weak} = 246\ GeV$, whereas the Majorana mass M_R, being not constrained by any symmetry, can be much larger than M_{weak}. Thus $m_D^2/M_R \ll M_{weak}$.

[3]The charged lepton mass matrix is in general not hermitian, so it is diagonalized by two unitary matrices: $Diag(m_1, m_2, m_3) = R_e^L \mathcal{M}_e R_e^{R\,\dagger}$.

[4]This arises naturally in Standard Model extensions with a quark/lepton symmetry, like $SO(10)$.

[5]For example, if there is no significant hierarchy between the heavy Majorana masses, the light neutrino masses are expected to scale as the up quark squared masses: $m_{\nu_i} \sim m_{u_i}^2/M_R$.

Neutrino mass models with a $U(1)$ family symmetry

The class of models we consider are extensions of the Minimal Supersymmetric Standard Model (MSSM) with: (i) a gauge group $SU(3)_C \times SU(2)_L \times U(1)_Y \times U(1)_X$, where $U(1)_X$ is an abelian family symmetry; (ii) a SM singlet field θ with X-charge $X_\theta = -1$, which is used to break $U(1)_X$ and to generate fermion masses; and (iii) three RH neutrinos $\bar{N}_i$ (i is a family index), in addition to the MSSM spectrum, which are needed to generate neutrino masses by the seesaw mechanism. We require that the family symmetry reproduce the experimental data on quarks and charged leptons, which forces it to be anomalous [6, 13], and that its anomalies be compensated for by an appropriate mechanism (the Green-Schwarz mechanism[6]).

In the following, we concentrate on the neutrino sector. We denote the lepton doublets by L_i (with their $I_W = +1/2$ components ν_i), and the right-handed neutrinos by $\bar{N}_i$, and their charges under $U(1)_X$ respectively by l_i and n_i. We note h_1 and h_2 the X-charges of the two MSSM Higgs doublets H_1 and H_2.

Dirac and Majorana matrices

Let us show how the Dirac ($\mathcal{M}_D$) and Majorana ($\mathcal{M}_M$) matrices are constrained by the family symmetry. Each Dirac mass term $L_i \bar{N}_j H_2$ carries an X-charge $p_{ij} = l_i + n_j + h_2$. If $p_{ij} \neq 0$, the coupling is forbidden by $U(1)_X$, and the corresponding entry of $\mathcal{M}_D$ is zero. However, if the excess charge p_{ij} is positive, one can write non-renormalisable interactions involving the chiral singlet θ:

$$L_i \bar{N}_j H_2 \left(\frac{\theta}{M} \right)^{p_{ij}} \tag{7}$$

where M is a large scale characteristic of the underlying theory (typically $M \sim M_{Planck}$ or M_{GUT}). When θ acquires a vev, $U(1)_X$ is spontaneously broken and effective Dirac masses are generated:

$$(\mathcal{M}_D)_{ij} \sim v_2 \left(\frac{< \theta >}{M} \right)^{p_{ij}} \tag{8}$$

where $v_2 =< H_2 >$. Since $U(1)_X$ is broken below the scale M, $\epsilon \equiv < \theta > / M$ is a small parameter. Thus the Dirac matrix obtained has a hierarchical structure[7], with the order of magnitude of its entries fixed by their excess charges under $U(1)_X$.

The entries of the Majorana matrix $\mathcal{M}_M$ are generated in the same way, with non-renormalizable interactions of the form:

$$M_R\, \bar{N}_i \bar{N}_j \left(\frac{\theta}{M} \right)^{q_{ij}} \tag{9}$$

giving rise to effective Majorana masses

$$(\mathcal{M}_M)_{ij} \sim M_R \left(\frac{< \theta >}{M} \right)^{q_{ij}} \tag{10}$$

[6]Remarkably enough, the observed fermion mass hierarchy, through this mechanism, fixes the Weinberg angle [9] to its standard value at the unification scale, $\sin^2 \theta_W = 3/8$. This, together with the anomalous character of $U(1)_X$, suggests a superstring origin to the model.

[7]Remember that the first motivation for introducing a family symmetry was to understand the quark and charged lepton mass hierarchies.

provided that $q_{ij} = n_i + n_j$ is a positive integer (otherwise $(\mathcal{M}_M)_{ij} = 0$). Thus, the light neutrino mass matrix $\mathcal{M}_\nu = -\mathcal{M}_D \mathcal{M}_M^{-1} \mathcal{M}_D^T$, and consequently the neutrino masses and mixing angles, is determined by the charges of the leptons under $U(1)_X$. *No particular ansatz for $\mathcal{M}_D$ nor $\mathcal{M}_M$ is required.* Note, however, that each of the entries is determined only up to an arbitrary factor of order one by the family symmetry.

Of course, there is a large variety of models, depending on the charges one assigns to the lepton fields. Contrary to the quark charges, whose possible values are strongly restricted by the experimental data on quark masses and CKM angles, the lepton charges are poorly constrained. In the following, we present two classes of models. The first one leads to a hierarchical mass spectrum [10, 11, 12, 13], the second one has two quasi-degenerate neutrinos with a large mixing angle [14].

Model 1: hierarchical mass spectrum

By analogy with the quark and charged lepton mass matrices, we start from a matrix with only one coupling allowed by the family symmetry:

$$\mathcal{M}_\nu = m_{\nu_3} \begin{pmatrix} 0 & 0 & 0 \\ 0 & 0 & 0 \\ 0 & 0 & 1 \end{pmatrix} \tag{11}$$

The breaking of $U(1)_X$ fills in the zero entries with powers of the small parameter ϵ, leading to a hierarchical spectrum. Assuming that (a) the X-charges of all mass terms are positive ($p_{ij} \geq 0$, $q_{ij} \geq 0$) and (b) the dominant entry of each mass matrix is the (3,3) entry, one automatically obtains pattern (11). The light neutrino masses are then:

$$m_{\nu_1} \sim \frac{m_3^2}{M_3}\, \epsilon^{2(l_1-l_3)} \qquad m_{\nu_2} \sim \frac{m_3^2}{M_3}\, \epsilon^{2(l_2-l_3)} \qquad m_{\nu_3} \sim \frac{m_3^2}{M_3} \tag{12}$$

The ν_τ mass is given by the usual seesaw formula (M_3 is the mass of the heaviest RH neutrino, m_3 the largest Dirac mass), whereas the other neutrino masses are suppressed relative to m_{ν_τ} by powers of the small breaking parameter ϵ. Note that the hierarchy depends only on the X-charges of the lepton doublets L_i. The lepton mixing matrix is:

$$V_L \sim \begin{pmatrix} 1 & \epsilon^{|l_1-l_2|} & \epsilon^{|l_1-l_3|} \\ \epsilon^{|l_1-l_2|} & 1 & \epsilon^{|l_2-l_3|} \\ \epsilon^{|l_1-l_3|} & \epsilon^{|l_2-l_3|} & 1 \end{pmatrix} \tag{13}$$

This model has several remarkable features. First, the neutrino mass and mixing hierarchies do *not* depend on the particular form of the Majorana matrix. This is a great difference with most seesaw models. The reason for this is that the dependences of $\mathcal{M}_D$ and $\mathcal{M}_M$ on the heavy neutrino charges compensate for each other in $\mathcal{M}_\nu$. Secondly, the mass spectrum obtained is naturally hierarchical[8], without hierarchy inversion: $m_{\nu_e} \ll m_{\nu_\mu} \ll m_{\nu_\tau}$. Finally, the mixing angles and the mass ratios are related by:

$$\sin^2\theta_{ij} \sim \frac{m_{\nu_i}}{m_{\nu_j}} \tag{14}$$

These relations, which are common to numerous seesaw models, show that small mixing angles are associated with mass hierarchies. They also imply $V_{e\nu_\mu} V_{\mu\nu_\tau} \sim V_{e\nu_\tau}$ in lepton charged current, in analogy with $V_{us} V_{cb} \sim V_{ub}$ in quark charged current.

[8]The possibility of mass degeneracies will be discussed in the next section.

The experimental data on solar neutrinos and atmospheric neutrinos put constraints on the parameters of the model. For example, if one wants to explain simultaneously the solar neutrino deficit by MSW $\nu_e \to \nu_\mu$ transitions, and the atmospheric neutrino anomaly by $\nu_\mu \rightleftharpoons \nu_\tau$ oscillations, one must choose:

$$l_1 - l_3 = 3 \qquad l_2 - l_3 = 1 \tag{15}$$

which leads to the following spectrum:

$$\begin{cases} m_{\nu_e} \sim 10^{-5} & eV \\ m_{\nu_\mu} \sim 5.10^{-3} & eV \\ m_{\nu_\tau} \sim 0.1 & eV \end{cases} \qquad \begin{cases} sin^2 2\theta_{e\mu} \simeq 2.10^{-3} - 4.10^{-2} \\ sin^2 2\theta_{\mu\tau} \simeq 5.10^{-2} - 0.8 \end{cases} \tag{16}$$

The uncertainties in the mixing angles are due to the fact that the mass matrix entries are determined only up to a factor of order one by the family symmetry . It is quite difficult to obtain a large mixing angle with a hierarchical spectrum [see (14)], as required by the atmospheric neutrino data. Furthermore, the tau neutrino is too light to be a good candidate for hot dark matter. However, if one ignores the atmospheric neutrino problem, it is possible to obtain a cosmologically relevant tau neutrino and to account for the solar neutrino deficit at once.

Model 2: quasi-degenerate neutrinos

As suggested above, in the context of abelian family symmetries, large mixing angles are naturally related to mass degeneracies. It thus seems rather difficult for the previous model to account for the atmospheric neutrino anomaly, or to accommodate the large angle branch of the MSW effect. Yet mass degeneracies and large mixing angles are not excluded: according to (12) and (13), $l_2 = l_3$ yields $m_{\nu_\mu} \sim m_{\nu_\tau}$ and $V_{\mu\nu_\tau} \sim V_{\tau\nu_\mu} \sim 1$. But the presence of unconstrained factors of order one in each entry of $\mathcal{M}_\nu$ can upset these formulae. Also, an accurate mass degeneracy requires fine-tuning of these factors. This leads us to consider another class of family symmetries, allowing for two equal[9] couplings, e.g.:

$$\mathcal{M}_\nu \;=\; m_{\nu 3} \begin{pmatrix} 0 & 0 & 0 \\ 0 & 0 & 1 \\ 0 & 1 & 0 \end{pmatrix} \tag{17}$$

Such a matrix has two degenerate eigenvalues[10], $m_{\nu_2} = -\, m_{\nu_3}$, with a maximal mixing angle, $\sin^2 2\theta = 1$. This degeneracy is slightly lifted by the breaking of $U(1)_X$. In order to reproduce pattern (17), we choose the following assignment of lepton charges:

$$\begin{cases} l_1 = l' > l \\ l_2 = -l_3 = l \end{cases} \quad \text{and} \quad 0 \leq n_1 < (-n_3) \leq l \leq n_2 \tag{18}$$

and, for simplicity, we assume $h_1 = h_2 = 0$. The light neutrino mass matrix is then:

$$\mathcal{M}_\nu \sim m_{\nu_3} \begin{pmatrix} \epsilon^{2l'} & \epsilon^{l'+l} & \epsilon^{l'-l} \\ \epsilon^{l'+l} & \epsilon^{2l} & 1 \\ \epsilon^{l'-l} & 1 & \epsilon^{2l} \end{pmatrix} \tag{19}$$

[9]The equality of the couplings in (17) follows from the symmetry of Majorana mass terms.
[10]The relative sign simply means that the mass eigenstates have opposite CP parities.

As expected, the spectrum contains two heavy, strongly degenerate neutrinos

$$m_{\nu_1} \sim \epsilon^{2l'} \ll m_{\nu_2} \simeq - m_{\nu_3} \tag{20}$$

with a squared mass difference determined by the family symmetry:

$$\Delta m^2_{23} \sim m^2_{\nu_3} \epsilon^{2l} \tag{21}$$

Taking into account the charged lepton sector, one obtains the lepton mixing matrix:

$$V_L \sim \begin{pmatrix} 1 & \epsilon^{l'-l} & \epsilon^{l'+l} \\ \epsilon^{l'-l} & \frac{1}{\sqrt{2}} & \frac{1}{\sqrt{2}} \\ \epsilon^{l'-l} & -\frac{1}{\sqrt{2}} & \frac{1}{\sqrt{2}} \end{pmatrix} \tag{22}$$

The quasi-degenerate neutrinos are almost maximally mixed, while the third neutrino, which is lighter, has small mixings with the other ones.

Such a mass and mixing pattern can account for the hot dark matter of the Universe, and simultaneously explain the atmospheric neutrino deficit in terms of $\nu_\mu \rightleftharpoons \nu_\tau$ oscillations. This requirement, together with the experimental limits on $\bar{\nu}_\mu \rightleftharpoons \bar{\nu}_e$ oscillations and the charged lepton masses, fixes the parameters of the model to be:

$$l = 2 \qquad l' = 5 \qquad m_{\nu_3} = 2 - 3\, eV \tag{23}$$

Thus, the mu and tau neutrinos are in the relevant mass range for hot dark matter:

$$m_{\nu_e} \sim (5 - 8).10^{-7}\, eV \qquad m_{\nu_\mu} \simeq m_{\nu_\tau} = 2 - 3\, eV \tag{24}$$

and, being strongly degenerate, they can oscillate with the parameters needed in order to solve the atmospheric neutrino problem:

$$\Delta m^2_{\mu\tau} \sim (0.9 - 2).10^{-2}\, eV^2 \quad \text{and} \quad \sin^2 2\theta_{\mu\tau} \simeq 1 \tag{25}$$

Furthermore, $\bar{\nu}_\mu \rightleftharpoons \bar{\nu}_e$ oscillations are found to be in the domain of sensitivity of the LSND and KARMEN experiments:

$$\Delta m^2_{e\mu} = 4 - 9\, eV^2 \quad \text{and} \quad \sin^2 2\theta_{e\mu} \sim 10^{-3} \tag{26}$$

Finally, let us note that the solar neutrino problem cannot be solved by this model, unless a sterile neutrino ν_s is added.

Conclusion

We have presented two seesaw models based on an abelian family symmetry, the one leading to a hierarchical mass spectrum, the other yielding an accurate mass degeneracy and a large mixing angle between the two heaviest neutrinos. In such models, the neutrino masses and mixing angles are determined in terms of the lepton charges under the family symmetry. Also, the squared mass difference between quasi-degenerate neutrinos is predicted. No ansatz for the Dirac nor the Majorana matrix is needed. Furthermore, the fact that the same symmetry is able to explain the observed fermion mass hierarchy and simultaneously constrains the neutrino spectrum sets an interesting connection between two fundamental problems in particle physics. Unfortunately, the lepton charges, though constrained by experimental data, are not fully determined by the model, which leads us to consider different classes of models, corresponding to different mass patterns. However, some generic properties of abelian family symmetries suggest that the model may originate from a more fundamental theory, which would fix all of its parameters.

Acknowledgements

This talk is based on work done in collaboration with P. Binétruy and P. Ramond, and with P. Binétruy, S. Petcov and P. Ramond. This work has been supported in part by the Human Capital and Mobility Programme, contract CHRX-CT93-0132.

References

[1] S.P. Mikheyev and A.Yu. Smirnov, Yad. Fiz. 42, 1441 (1985) [Sov. J. Nucl. Phys. 42, 913 (1985)]; Il Nuovo Cimento C 9, 17 (1986); L. Wolfenstein, Phys. Rev. D 17, 2369 (1978); Phys. Rev. D 20, 2634 (1979).

[2] M. Gell-Mann, P. Ramond, and R. Slansky in Sanibel Talk, CALT-68-709, Feb 1979, and in *Supergravity* (North Holland, Amsterdam 1979). T. Yanagida, in *Proceedings of the Workshop on Unified Theory and Baryon Number of the Universe*, KEK, Japan, 1979.

[3] C. Froggatt and H. B. Nielsen Nucl. Phys. B147 (1979) 277.

[4] M. Leurer, Y. Nir, and N. Seiberg, Nucl. Phys. B398 (1993) 319, Nucl. Phys. B420 (1994) 468.

[5] L. Ibáñez and G. G. Ross, Phys. Lett. B332 (1994) 100.

[6] P. Binétruy and P. Ramond, Phys. Lett. B350 (1995) 49.

[7] V. Jain and R. Shrock, Phys. Lett. B352 (1995) 83.

[8] E. Dudas, S. Pokorski and C.A. Savoy, Phys. Lett. B356 (1995) 45.

[9] L. Ibáñez, Phys. Lett. B303 (1993) 55.

[10] H. Dreiner, G.K. Leontaris, S. Lola, G.G. Ross and C. Scheich, Nucl. Phys. B436 (1995) 461; G.K. Leontaris, S. Lola, C. Scheich and J.D. Vergados, Phys. Rev. D53 (1996) 6381.

[11] Y. Grossman and Y. Nir, Nucl. Phys. B448 (1995) 30.

[12] P. Ramond, 25th Anniversary Volume of the Centre de Recherches Mathématiques de l'Université de Montréal.

[13] P. Binétruy, S. Lavignac and P. Ramond, preprint LPTHE-ORSAY 95/54, UFIFT-HEP-96-1, hep-ph/9601243 (to be published in Nuclear Physics).

[14] P. Binétruy, S. Lavignac, S. Petcov and P. Ramond, in preparation.

DARK MATTER

Joseph Silk

Departments of Astronomy and Physics
and Center for Particle Astrophysics
University of California, Berkeley, CA 94720

The search for dark matter constitutes an ongoing effort that is at the core of modern astronomy. The dark matter problem was first posed more than fifty years ago, and it remains largely unresolved. We have learnt much about its distribution in the universe, yet remain abysmally ignorant as to its nature. These lectures will address two questions: where is the dark matter, and what is its nature? In the first lecture, I will, after reviewing the history of dark matter, describe the status of dark matter probes over scales that range from the solar neighborhood to galaxy halos, clusters, and superclusters up to the horizon scale. The second lecture will focus on the composition of the dark matter. I will describe the cases for both baryonic and non-baryonic matter, and review the various dark matter candidates.

DARK MATTER — AN OVERVIEW

Dark matter accounts for at least 90 percent, and possibly for 99 percent of the matter in our universe. Elucidating its nature is one of the most urgent problems in astrophysics that has attracted the attention of astronomers and particle theorists. Its existence is predicated on our theory of gravitation. Various attempts have been made to tinker with Einstein's theory, modifying the law of gravitation to be able to account for, say, flat rotation curves without introducing dark matter. However, such modified laws of gravity are generally *ad hoc*, ugly, and not self-consistent when applied to strong gravity fields or in the relativistic limit.

However, while Einstein's theory of gravity is elegant, appealing, and accounts for all known observations, it is sobering to recall how Newton's theory of gravity, developed for the terrestrial environment, coped with the challenge of the solar system. The prediction of the existence of Neptune to account for the orbital anomalies of Uranus was a great success, but the search for Pluto as an explanation for anomalies

Masses of Fundamental Particles: Cargèse 1996
Edited by M. Lévy *et al.*, Plenum Press, New York, 1997

in Neptune's orbit was misguided, although it may have culminated in Pluto's discovery. Pluto's mass, we now know, is inadequate to have any influence on Neptune's orbit. The saga of Vulcan should be sobering: a planet interior to Mercury's orbit was postulated to account for the precession of the perihelion of Mercury's orbit. Vulcan does not exist, and Einstein's theory of general relativity superceded Newtonian gravity to feature the explanation of the precession of Mercury's perihelion as one of its major successes.

One should constantly be open to the possibility of a new theory of gravity that may dispense with the need for dark matter. However in the absence of even a remotely compelling theory, I will proceed with the more conventional approach to dark matter that takes the observational evidence at face value.

The dark matter problem was posed by Fritz Zwicky in 1933. He noted that observation of galaxy velocities in the Coma cluster of galaxies implied a mass-to-light ratio of about 400 $M_\odot/L_\odot$, as compared to the more modest value of a few that seemed appropriate for galaxies composed of Sun-like stars. Remarkably, Zwicky's estimate has not changed significantly in sixty years, despite an increase in the distance scale by an order of magnitude.

The first estimates of dark mass in galaxy halos were made by Vera Rubin in the 1950's. Much modern work has confirmed the view that the mass-to-light ratio of galaxies is 50 – 100 $\mathrm{M}_\odot/\mathrm{L}_\odot$ out to a scale of 50 – 100 kpc. Theorists became enthusiastic about dark matter after the argument, now largely discounted, was developed that dark halos are required in order to stabilize disks against bar formation. The culmination of theoretical arguments for dark matter came in 1980 with the emergence of inflationary cosmology. This theory argued that the size and homogeneity of the universe was explained by a cosmological model which predicted that a period of exponential inflation at the grand unification scale resulted in a density of dark matter almost precisely equal to the critical value for closure, $3H_0^2/8\pi G$. The implied mass-to-luminosity ratio is $1500\,h$, where $h \equiv H_0/100\,\mathrm{kms}^{-1}\,\mathrm{Mpc}^{-1}$ explicitly shows the scaling with the Hubble constant. This argument also is now largely discounted: there are many inflationary spin-offs, and open models are possible.

THE SOLAR NEIGHBORHOOD: <100 PC

The velocity dispersion of a uniform population of luminous stars such as K giants for which a scale height can be measured gives the mass surface density in a cylinder near the Sun that is perpendicular to the galactic plane. One finds a value of about 70 $M_\odot\,\mathrm{pc}^{-2}$. Most of this can be accounted for in terms of known stars, stellar remnants and interstellar matter. Brown dwarfs, objects below the minimum mass of $0.08\,M_\odot$ for a star to ignite hydrogen and undergo thermonuclear fusion, account for at most a small fraction of the local mass density in the galactic disk.

LUMINOUS REGIONS OF GALAXIES: <10 KPC

Studies of the inner regions of galaxies, defined to be within the half-light radius, utilize either rotation curves for disks or velocity dispersions for spheroids to infer the mass-to-light ratio. Typical values range from $\sim 2\ \mathrm{M}_\odot/\mathrm{L}_\odot$ for Sc galaxies to $\sim 7\ \mathrm{M}_\odot/\mathrm{L}_\odot$ for Sa's, and $\sim 10\ \mathrm{M}_\odot/\mathrm{L}_\odot$ for S0's and ellipticals. The old star population is characterized by the larger values of M/L in galaxies. Globular star clusters, which also contain exclusively old stars, appear to have undergone substantial dynamical relaxation and mass loss in order to have attained their observed $M/L \sim 2\ M_\odot/L_\odot$.

A simple explanation of $M/L \sim 10\ M_\odot/L_\odot$ is that the typical star is from an old stellar population and has a mass of $\sim 0.5\,M_\odot$. Since on the hydrogen-burning main sequence, L increases rapidly with M, approximately as M^4, for individual stars, one can actually account for the range in M/L by varying the mean age of the stellar population but retaining the same initial (that is, birth) mass function. A younger galaxy is more luminous than an older galaxy, for the same total mass in stars and initial stellar mass function. Metallicity adds a further degree of complication in evaluating the age of an old stellar population, as the degeneracy between age and metallicity cannot be resolved without adequate spectral data[1].

HALO DARK MATTER: <100 KPC

Rotation curves of spiral galaxies utilize a mass tracer such as Hα emission from HII regions or 21 cm emission from interstellar HI that can be observed out to 2 or more half-light radii. In fact, in the best studied systems, one can trace the HI out to about 15 disk scale lengths. The typical rotation curve first rises, more or less as in solid body rotation, then flattens after $\sim$ 3 kpc and remains flat as far out as it can be measured. In the case of the Milky Way, both globular clusters and satellite galaxies have been used to probe the halo beyond the measured extent of the rotation curve. One concludes that M ($\propto$ R) increases approximately linearly with radius R out to the measurement limit that is typically $\sim$ 50 kpc.

Diffuse gas is not a substantial component of the halo in gas-rich dwarf galaxies. One finds that the HI surface density is about 10 percent of the surface density of dark matter. Interestingly, this ratio appears to be independent of radius within the dark halo.

Simulations of galaxy rotation curves have had mixed success. Tidal torques between neighboring asymmetric fluctuations in the expanding universe provide the initial angular momentum at the onset of collapse. The baryonic component dissipates and collapses, torquing against a non-collapsing dark halo to form a disk with a flat rotation curve and with amplitude in the observed range. However, the scale length of the disk is found to be too small[2]. Moreover the rotation curve stays flat close to the center of the galaxy, in contradiction with rotation curves measured for dwarf galaxies[3,4]. The resolution of these problems may lie in including non-gravitational physics, notably energy feedback from stars into the gas, that will help increase the disk scale length. Dwarf galaxies are sufficiently dark-matter dominated that enlarging the central core may require a baryonic component of dark matter: dissipation and early star formation would result in a core that was larger than that inferred from cold dark matter simulations (*e.g.*, reference 5).

Galaxy halos around elliptical galaxies are studied by mapping the diffuse x-ray emission from halo gas. The gas is at $\sim 10^7$ K in a typical luminous elliptical galaxy, and must be confined by a massive dark halo. The pressure gradient of the gas gives a direct measure of the halo mass. In rare cases, shells are found well outside the half-light radius of the elliptical. These are remnants of mergers that occurred several Gyr ago. Dynamical modelling of the merging provides another estimate of the halo mass. As with spirals, a mass-to-light ratio of 50–100 $M_\odot/L_\odot$ is found out to a radius of 50–100 kpc.

GALAXY CLUSTERS: <3 MPC

Clusters are the largest bound objects that one can study in detail. The dark matter content of clusters is mapped by three independent techniques: mapping of the X-ray emission from the cluster gas; measurements of the galaxy velocity dispersion; gravitational lensing of background galaxies. Both the X-ray mapping and global velocity dispersion techniques assume that hydrostatic equilibrium applies both to the gas and galaxies out to the virial radius, $\sim 3M_{15}V_{1000}^{-2}$ Mpc, where the cluster mass is $10^{15}M_{15}$ $M_\odot$ and the velocity dispersion is $10^3 V_{1000}$ km/s. Within the virial radius, both X-ray maps and the global velocity dispersion enable the dark matter content to be deduced, whereas the gravitational lensing method makes no assumption about virial equilibrium. The X-ray method requires measurement of the X-ray emissivity with sufficient spectral resolution to yield the gas temperature as well as reconstruction of the gas density, and the hydrostatic support assumption then yields the total mass. The X-ray flux is proportional to the square of the gas density and is biased by density fluctuations. One can attempt to circumvent this problem in part with Sunyaev-Zeldovich mapping at radio frequencies, since the Sunyaev-Zeldovich effect due to Compton scattering of cosmic microwave background photons is proportional to the gas pressure. The optical method applies the virial theorem, and one has to make some assumptions about the isotropy of the galaxy velocity distribution. A third technique, which makes no assumptions about virial equilibrium, is that of studying the gravitational lensing of background galaxies via the image distortions. Arcs and many arclets are found in typical rich clusters in the central cores, and weak shear can be measured in the outer regions.

One complication that results in a major source of systematic error is that many clusters are still in the process of formation. Clusters are anisotropic and inhomogeneous. The galaxy distribution is clumpy, and there are large temperature gradients in the gas that are presumably due to large-scale shocks. Cluster structure is consistent with hierarchical formation, in which progressively larger lumps of self-gravitating matter aggregate together. Correlations of field galaxies give a typical non-linear scale of $\sim 5h^{-1}$ Mpc or a mass scale of $\sim 10^{14}\Omega h^{-1}M_\odot$, which is similar to the typical scale of substructure in rich clusters. Clusters condense from a comoving scale of $\sim 10M_{15}^{1/3}\,\Omega^{-1/3}h^{-2/3}$ Mpc.

The mass-to-light ratio of a typical rich cluster is found to be about $400\,h$ $M_\odot/L_\odot$, to be compared with $\sim 10h$ $M_\odot/L_\odot$ for the mass-to-light ratio of the typical cluster E or $S0$ galaxy, as evaluated at the galaxy's half-light radius. A significant fraction of the cluster mass is gas: the gas fraction is measured to be[6] $\sim 2-10h^{-3/2}$ percent, out to a radius of ~ 1 Mpc. In rich clusters, the ratio of gas to total mass appears to be constant at larger radii, more or less out to the virial radius.

SUPERSTRUCTURES AND BEYOND: < 100 MPC

On scales larger than $\sim 5h^{-1}$ Mpc, one is in the linear regime for fluctuations in luminous matter. The nonlinear role of the dominant dark matter fluctuations cannot exceed the correlation length scale, otherwise there would be conflict with observations. On scales of $\sim 1h^{-1}$ Mpc, pairwise velocities of galaxies constrain the dark matter component that is clumped on these scales. On $\sim 10h^{-1}$ Mpc scales, collapse and formation of clusters probes the dark matter distribution. If mass traces light in a universe at critical density, one would produce an order of magnitude more rich clusters than are seen in either optically or x-ray selected samples. In this way, one can limit the parameter $\sigma_8\Omega^{0.6}$ to lie in the range 0.5 – 0.6, where σ_8 is the ratio of dark matter to luminous matter fluctuations on a scale of $8\,h^{-1}$ Mpc, and the $\Omega^{0.6}$ factor is an approximation to the density fluctuation growth rate[7]. This limit applies to COBE-normalized primordial fluctuation spectra.

Next consider a supercluster, such as the Local Supercluster, which is centered on the Virgo cluster of galaxies at $\sim 10\,h^{-1}$ Mpc distance. This supercluster is not self-gravitating. At the distance of the Milky Way from Virgo, the excess overdensity of the local supercluster retards the Hubble recession of Virgo relative to us. From the Hubble flow distortion around us, which amounts to about a 10 percent diminution toward Virgo, one can deduce a mass-to-light ratio of $(400 \pm 200)\,h$ $M_\odot/L_\odot$.

The peculiar velocities of galaxies around us can be measured because distances to the galaxies can be directly determined via the Tully-Fisher effect: an empirical relation is found for spiral galaxies according to which the optical luminosity is approximately proportional to the fourth power of the maximum rotational velocity. This relation is calibrated for several nearby galaxies with Cepheid variable distances. Since the measured rotational velocity is distance-independent, one can apply the Tully-Fisher relation to distant galaxies, provided one can obtain accurate rotation curves. Comparison of the Tully-Fisher distance with distance inferred from the galaxy redshifts via Hubble's law then yields the pecular velocity, to an accuracy of about 15 percent per galaxy.

Peculiar velocities are used in two distinct ways to probe Ω. With essentially the only assumption being that peculiar velocities are gaussian distributed, studies of peculiar velocities around voids and of the skewness of the velocity divergence field directly give a measure of Ω, which is claimed to exceed 0.3 at 95 percent confidence. A more indirect approach combines peculiar velocity measurements with IRAS and optically-selected galaxy redshift catalogs.

The bias factor now enters as an additional parameter: this is the ratio of luminous to matter density fluctuation amplitudes. The combined technique has been applied to samples of galaxies extending to a depth of $\sim 100h^{-1}$ Mpc. Elliptical galaxy samples can also be used, since the so-called fundamental plane of surface brightness, velocity dispersion and half-light radius also provides an apparently universal correlation that yields direct, redshift-independent, distances to ellipticals. Studies of the pecular velocity field over $\sim 10h^{-1}$ — $80h^{-1}$ Mpc indicate that there is an *rms* component to galaxy peculiar motions of about 400 km s^{-1}. There is also a bulk flow, amounting to about 350 km s^{-1} on a scale of 50 h^{-1} Mpc. To explain this high pecular velocity dispersion and bulk flow over such large scales requires a universe that is near the critical density. From the measured radial velocity component, one derives a velocity potential, and then computes the three-dimensional velocity field, an exact procedure in the linear regime for irrotational flows. From this, one evaluates the divergence of the velocity field, and infers the density field. The linear theory prediction is that the velocity divergence is proportional to the product of dark (and luminous) matter density field with a function of Ω that varies approximately as $\Omega^{0.6}$. Comparison with, say, the IRAS caredshift survey then yields the quantity $\beta_I \equiv \Omega^{0.6}/b_I$, where b_I is the IRAS bias factor. Recent analyses find that β_I varies from about ~ 1 at $\sim 10h^{-1}$ Mpc to ~ 0.5 at $\sim 3h^{-1}$ Mpc, showing that the tracing of mass to light is scale-dependent.

With a smoothing scale, for example of $\sim 12h^{-1}$ Mpc, one obtains[8] $\sigma_8\Omega^{0.6} = 0.8 \pm 0.2$. If systematics are not contaminating this result, one finally concludes that, on the largest scales where one can map out galaxy velocities, mass presumably traces light and there is considerable dark matter that, conservatively, requires $\Omega > 0.4$ or $M/L > 600h$. One is still short of the simple inflationary prediction that $\Omega = 1$.

To probe dark matter at even larger scales, one has to resort to measurements of the geometry of the universe. These involve a combination of Ω and the deceleration parameter. Such approaches include most notably the use of distant SNIa supernovae as standard candles. Cosmic microwave background fluctuations potentially provide our best hope of measuring Ω and q_0. Before addressing this issue, which depends sensitively on the choice of the dark matter, I turn to the question of what comprises the dark matter.

POSSIBILITIES FOR DARK MATTER

Dark matter candidate masses span a range of some 80 orders of magnitude. One can divide this into two domains, those of elementary particle physics and astrophysics. The particle candidates range from axions, weighing in near 10^{-4} eV, to neutralinos, stable supersymmetric relics of mass near the SUSY symmetry-breaking scale of $\sim$ 100 GeV. The generic name for such particles is WIMPs, at least in the more massive end of the range, for weakly interacting massive particles. The astrophysical candidates are generically referred to as MACHOS (massive astrophysical compact halo objects), in part because diffuse gas is not a significant contributor to the dark matter content of the universe, and in part because the logical environment for dark baryons is in galaxy halos, surrounding the luminous baryons.

MACHOS

At least two distinct arguments point to the existence of a substantial amount of baryonic dark matter. Primordial nucleosynthesis of such light elements as ^{2}H, ^{4}He, ^{3}He, amd ^{7}Li requires an amount of baryons that significantly exceeds the mass in luminous galaxies, in the form of stars and interstellar matter. Studies of quasar absorption lines give a hydrogen mass fraction at redshift $\sim 3-4$ that is comparable to the low redshift mass fraction in luminous matter. Quasar absorption line studies measure intergalactic gas clouds, both of high hydrogen column density (damped Lyman alpha clouds) and low column density (Lyman alpha forest). The former systems consist of atomic hydrogen, and directly give a hydrogen mass fraction that exceeds, at redshift $\sim$ 3, the low redshift mass fraction in luminous matter. Lyman alpha forest clouds are predominantly ionized hydrogen, and provide a measure of the ratio of the *rms* gas density to the square root of the ionizing photon flux, this being the quantity that enters into the ionization balance for determing the observed subdominant HI column. The ionizing photon flux can be estimated from the observed quasar luminosity function and quasar spectra.

These arguments provide evidence for a baryonic fraction in the range

$$\Omega_B = (0.01 - 0.03)\,h^{-2},$$

where the baryon fraction in luminous galaxies is

$$\Omega_{lum} \approx 0.005\,.$$

The former condition is on baryon density n_B (where $\Omega_B \propto n_B/\rho_{cr} \propto n_B h^{-2}$), whereas the latter constraint specifies the product of luminosity density ρ_{lum} ($\propto h$)

and mass-to-light ratio ($\propto h$) relative to ρ_{cr} ($\equiv 3H_0^2/8\pi G$). Hence one can conservatively estimate that there is at least a factor 3 more dark baryonic matter than luminous baryonic matter.

The obvious candidates for MACHOs are known varieties of dark baryonic objects that are relatively common in the inner galaxy. One can argue that the uncertainties in star formation permit a systematic change in the initial mass function that would allow MACHOs to dominate in the halo, where the protogalactic environment was very different from that, say, in the vicinity of the Sun. Plausible MACHO candidates are therefore brown dwarfs (which are presumed to be common), white dwarfs (which constitute up to a third of the mass of the oldest star clusters in our galaxy, the globular star clusters), neutron stars (these are thought to amount to only a tenth of a percent of the mass of a globular cluster), and black holes. While black holes are extremely rare as stellar relics, one may also have a substantial number of primordial black holes.

MACHOs have been detected in the galactic halo. The experimental technique is that of gravitational microlensing. From the galactic rotation curve, one knows the local dark matter density in the halo. MACHOs occasionally cross the line-of-sight to a distant star, the light from which is amplified as the gravity field of the MACHO acts like a gravitational lens. The best target is the Large Magellanic Cloud, which provides many source stars for which one can estimate the lensing probability. If the halo consists entirely of MACHOs, the optical depth to microlensing is $(V_{rot}/c)^2 \sim 10^{-6}$, where V_{rot} is the galactic rotation velocity. Hence one has to monitor of order 10^7 stars to have a reasonable expectation of detecting an amplification event. There is also image splitting, and for a directly aligned observer–lens–source, the source star images into a ring (the Einstein ring) of radius $R_E \sim (R_G R_{os})^{1/2}$, where R_G is the Schwarzschild radius of the MACHO, $2Gm/c^2$, and R_{os} is the observer-source distance as long as the observer–lens or source–lens distance is small compared with the distance from observer to source. However the splitting is typically a millisecond for subsolar mass MACHOs, and is not directly measurable.

The duration of the lensing event is $t_L \sim R_E/v$, where v is the component of the MACHO velocity transverse to the line of sight. For LMC star microlensing, one finds that

$$t_L \sim 100(M/M_\odot)^{\frac{1}{2}} \quad \text{days.}$$

Two experiments (EROS and MACHO) have monitored some 10^7 LMC stars: EROS mostly photographically (using Schmidt plates), and also CCD detectors, for one year, MACHO digitally (using CCD's) for 3 years. Both experiments have detected microlensing events. The experiments are performed in two colors. The expected signal is achromatic, and time-symmetric for a single MACHO–single source star lensing event. Most of the results have come to date from the analysis of 2 years of data by the MACHO collaboration, who report some 8 microlensing events[9]. At least six of these are due to halo (as opposed to LMC) MACHOs. From the event duration, the MACHO mass is inferred to be $0.1 - 1\ M_\odot$, and from the event rate, one concludes that between 20 and 80 percent of our halo, between us and the LMC (at 50 kpc distance), consists of MACHOs. The longest duration events, being due to more massive MACHOs, dominate the mass fraction, and the experiment is progressively more incomplete at durations above $\sim$ 100 days. Hence the currently derived mass fraction is likely to represent a lower bound on the actual halo mass fraction in MACHOs.

What are the MACHOs? Neutron stars seem an unlikely bet, as the supernovae that generated the neutron stars would be prolific producers of heavy elements. One estimates that at a rate of one SN of type II per 100 yr, there are $\sim 10^8$ neutron stars in our galaxy. Allowing a factor of 10 for enhanced early star formation, one would obtain oas much as a percent of the stellar content of the Milky Way in MACHOs. The associated early supernovae produced all of the oxygen, silicon, etc., seen in stars like the Sun. Hence if one were to generate enough supernovae to fill the halo of mass$\sim 10^{12}\, M_\odot$ with neutron stars, the heavy element yields would be enhanced by some 3 orders of magnitude. This seems difficult to hide, for example by ejection from the galaxy.

White dwarfs carry a weaker nucleosynthetic signature. The number of white dwarfs in the Milky Way is equal to the number of stars of mass between ~ 1 and $8\, M_\odot$ that have formed. With a mass function $dN/dm \propto M^{-2.5}$ approximately describing the mass distribution between $\sim 0.3\, M_\odot$, below which the number of stars converges, and $\sim 100\, M_\odot$, one infers that white dwarfs constitute about 10 – 20 percent of the number of stars in the galaxy. Since the mean stellar mass ($\sim 0.5\, M_\odot$) is also the typical white dwarf mass, about 10 – 20 percent of the mass in stars may be in white dwarfs. White dwarf precursors produce most of the carbon in the galaxy. Hence even a white dwarf halo of $\sim 10^{12}\, M_\odot$ would have overproduced carbon by about 2 orders of magnitude.

It is possible that stars of sufficiently low metallicity ($\lesssim 10^{-4} Z_\odot$) do not undergo helium flashes and so dredge-up of carbon does not occur during the final stages of stellar evolution. Helium production and ejection is inevitable however, and a halo mass of white dwarfs would imply helium production that would be comparable to the primordial nucleosynthesis amount. However, the helium abundance is difficult to measure, especially in the intergalactic medium, where quasar absorption line studies probe abundances. If the stellar debris during halo formation was ejected via a galactic wind, although the energetics for this is questionable, into the intergalactic environment, one might be able to allow a white dwarf halo. Such a model requires an initial stellar mass function that is finely tuned so that only stars in the mass range between 2 and 8 $M_\odot$ were present in the protohalo. This seems implausible, but cannot be excluded given our theoretical ignorance about the origin of the initial mass function[10].

Our lack of any fundamental understanding of the IMF is of especial concern in more extreme environments than nearby regions of star formation. Black holes in the mass range 0.1 – 1 $M_\odot$ would not have formed via conventional stellar evolution, and hence are not candidates for the MACHO detectors. Primordial black holes, however, are an intriguing possibility. Indeed the only way to form subsolar mass black holes would have been in the very early universe. One requires large primordial density fluctuations on small scales. Then black hole formation on the horizon scale, when $\delta\rho/\rho$ was of order unity, would have been inevitable.

Producing black holes is only part of the problem. One must also not overproduce black holes, which do not redshift away in the radiation-dominated era, so that only a fraction $(1 + z_{eq})/(1 + z_{bh})$ of the universe could have formed primordial black holes. Here z_{eq} is the redshift of matter-radiation equality and z_{bh} ($\gg z_{eq}$) is the epoch of black hole formation. One has $z_{eq} \approx 2 \times 10^4 \Omega h^2$, and it is likely that $z_{bh} \approx 10^{12}$, corresponding to the quark-hadron phase transition era at $T \sim 100$ MeV, when the horizon contained approximately a solar mass. Theoretical constraints on the primordial fluctuation spectrum on these scales are non-existent. All one can say is that on a scale of $\sim 10^6\, M_\odot$, corresponding to the horizon at nucleosynthesis ($T \sim 0.1$ MeV), the universe could not have been highly inhomogeneous, otherwise light element abundances would have been unacceptably perturbed.

WIMPS

Supersymmetry provides strong, if not completely compelling or unambiguous, motivation for the existence of WIMPs. While SUSY signatures remain elusive (for example neither anomalies in Z^0 decays nor missing energy events have been confirmed), the standard arguments, such as convergence of the running of the coupling strengths of electroweak and strong interactions at the GUT scale, remain unchallenged. SUSY doubles the number of particles, providing massive partners, fermions to known bosons, and vice versa. SUSY particles include the squark ($\tilde{q}$, spin 0), slepton ($\tilde{\mu}$, spin 0), photino ($\tilde{\gamma}$, spin 1/2), zino ($\tilde{Z}$, spin 1/2), wino ($\tilde{W}^{\pm}$, spin 1/2), and higgsino ($\tilde{H}^{\pm,0}$, spin 1/2).

The lightest supersymmetric particle (or LSP) is expected to be a stable relic, frozen out when SUSY is broken above ~ 100 GeV. Stability of the LSP is guaranteed for SUSY particles by conservation of R-parity, defined by $(-1)^{3B+2L+S}$, where B is baryon number, L is lepton number, and S is spin. The LSP is likely to be a mixture of photino and neutral higgsino, generically called a neutralino. If the SUSY scale is indeed of order the $W^{\pm}$ mass, then the LSP interacts with a cross-section of order the weak nuclear cross-section. Moreover it is expected to have a mass of order the SUSY breaking scale. The LSP provides the ideal WIMP candidate.

In fact, one can make predictions of the LSP mass by appealing to cosmology. The relic densities of LSPs is determined by the LSP self-annihilation cross-section. LSPs are majorana particles, and hence annihilate via $\tilde{\chi}\tilde{\chi} \to f\bar{f}$. Freeze-out occurs at $kT \sim m_{\tilde{\chi}}c^2/20$, where

$$\langle \sigma v \rangle_{ann}\, n_{\tilde{\chi},fr}\, t_{fr} \sim 1.$$

Here $\langle \sigma v \rangle_{ann}$ is the annihilation rate at freeze-out, $n_{\tilde{\chi},fr}$ is the density of LSP's at freeze-out, and t_{fr} is the freeze-out epoch. Annihilations occur via weak interactions (Z^0 exchange for $\tilde{H}\tilde{H} \to f\bar{f}$ or $\tilde{f}$ exchange for $\tilde{\gamma}\tilde{\gamma} \to f\bar{f}$), and the annihilation rate for $\tilde{\chi}\tilde{\chi}$ is proportional to $m_{\tilde{f}}^4/m_{\tilde{\chi}}^2$. Accelerator lower bounds on $m_{\tilde{f}}$ are used to get a lower bound on $\langle \sigma v \rangle_{ann}$, and hence an upper bound on Ω_{χ} at specified $m_{\tilde{\chi}}$. Conversely, knowledge of Ω_{χ} bounds $m_{\tilde{\chi}}$.

Consider first massive particles that freeze out when non-relativistic, namely, stable SUSY relics. For $m_{\tilde{\chi}} < \frac{1}{2}m_z$, $\Omega_{\chi} \propto m_{\tilde{\chi}}^{-2}$, since $\sigma \propto m_{\chi}^2/m_z^4$. However for $m_{\tilde{\chi}} > \frac{1}{2}m_z$, $\Omega_{\chi} \propto m_{\tilde{\chi}}^2$, since $\sigma \propto m_{\chi}^{-2}$. This leads to the following bounds on Ω_{χ} (if the dark matter consists of the LSP). The cosmology bound $\Omega_{\chi}^2 \lesssim 1$ translates to $m_{\chi} \gtrsim 3\,\mathrm{GeV}$. As $m_{\tilde{\chi}}$ increases, Ω_{χ} decreases as $m_{\tilde{\chi}}^{-2}$, reaching a minimum value at $m_z \sim 100\,\mathrm{GeV}$, above which annihilations are slow. Now $\Omega_{\tilde{\chi}}$ increases, approximately as $m_{\tilde{\chi}}^{-2}$, and reaches unity again at $m_{\tilde{\chi}} \sim 1\,\mathrm{TeV}$. Use of LEP1 and LEP1.5 limits on $m_{\tilde{f}}$ plus the refined cosmological bound $\Omega_{\chi} < 0.25$ requires the lower bound on the neutralino mass to be about 20 GeV in minimal SUSY[11]. A more robust bound that does not appeal to cosmology, utilizing chargino and neutralino searches at LEP1 and LEP1.5 plus $e^+e^- \to \gamma +$ *nothing* at TRISTAN and KEK, requires the neutralino mass to exceed 10 GeV in minimal SUSY. A more refined estimate of the upper limit to avoid excessively large Ω_{χ} is likely to be about 5 TeV.

Consider next the more conventional case of particles possibly carrying mass that freeze out when relativistic, for which the prototype is the neutrino. Suppose that the dark matter Ω_ν consists of massive neutrinos. In this case, $\Omega_\nu \propto m_\nu$, since the relic number density of neutrinos is specified by cosmology, and specifically satisfies

$$\Omega_\nu h^2 = (m_\nu/100\,\mathrm{ev}).$$

Since neutrinos freeze out relativistically, their abundance is not suppressed: for $m_\nu > 100\,\mathrm{eV}$, neutrinos would overclose the universe. Indeed for $m_\nu > 10\,\mathrm{eV}$, the mass fraction in hot dark matter would contradict large-scale structure observations. For example, the velocity dispersion of massive neutrinos, which become non-relativistic at $kT < m_\nu c^2$, is $\sigma_\nu = 10^{-5}c(1+z)(30\,\mathrm{eV}/m_\nu)$. At $T \sim 3000\,\mathrm{K}$, $z \sim 1000$, when

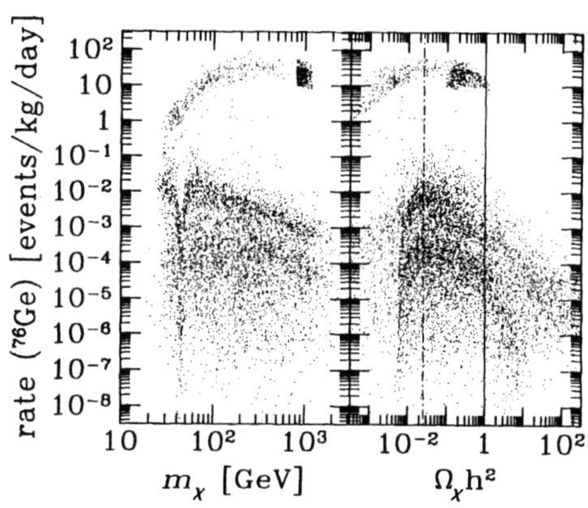

Figure 1. Dark matter constraints from direct detection experiments for the LSP from a grid of SUSY models[12,13]. Current bounds are about 1 event per kg per day. Lower Ω_χ generally means higher annihilation cross-section, higher scattering cross-section, and higher direct detection rate, but the dispersion is large.

structure formation can first commence, $\sigma_\nu \sim 3000\,\mathrm{km/s}$ and would inhibit galaxy-mass fluctuations which can only condense if $\sigma_\nu \lesssim 300\mathrm{km/s}$, the typical virial velocity of a galaxy, from growing. Structure formation consequently is top-down, massive clusters forming first. This is in conflict with the observed hierarchy of structure: galaxies are old whereas clusters are young.

COLD DARK MATTER

The stable LSP is a natural motivation for cold dark matter-dominated cosmology. Freeze-out occurred early at $kT > 1\,\mathrm{GeV}$. By matter-domination, at $kT \sim 10\,\mathrm{eV}$, the particle velocity dispersion is negligible relative to that of any virialization velocity on scales of interest for structure formation. Hence bottom-up formation occurs. The primordial fluctuation spectrum is expected to be nearly scale-invariant, as a consequence of inflation-amplified quantum fluctuations. Cold dark matter allows growth of fluctuations on sub-horizon scales at a logarithmic rate during radiation-domination, and at a rate proportional to $(1+z)^{-1}$ during matter-domination. Consequently, the fluctuation amplitude at horizon crossing, a measure of the critical value of the primordial metric fluctuations, is required to be approximately 10^{-5} in order to form structure by the present.

There is a fundamental problem with cold dark matter, if one assumes that light traces mass and that $\Omega = 1$. Galaxy count fluctuations have unit amplitude over spheres of radius $8h^{-1}$Mpc. One therefore expects bulk motions on scales that are just non-linear to be of order

$$v \approx \left(\frac{4\pi}{3} G \delta\rho(R) R^2\right)^{\frac{1}{2}}$$
$$\approx \Omega^{\frac{1}{2}} HR |\delta\rho/\rho|^{\frac{1}{2}}_{rms} \approx 10^3 \Omega^{\frac{1}{2}} \text{ km/s}.$$

However bulk flows, as measured by pairwise galaxy velocities and *rms* galaxy pecular motions, over scales that span the range 1–$50h^{-1}$ Mpc, are about 400 km s^{-1}. The excess in predicted power amounts to a factor of $\sim$ 4.

The discrepancy can be resolved by introducing the concept of biasing. If the mass distribution is more smoothly distributed than that of luminous matter, one can lower the predicted pecular velocity field amplitude. If one defines σ to be the *rms* value of $\delta\rho/\rho$, and σ_8 to be σ evaluated at $8h^{-1}$ Mpc, then the bias factor b can be defined as σ_8^{-1}, the ratio of luminous to matter density fluctuation amplitudes at $8h^{-1}$ Mpc.

In linear theory, $\delta v \approx HR\,|\delta\rho/\rho| \propto \sigma(t_0)$. As with the galaxy cluster abundance, large-scale velocities effectively measure the product $\sigma_8\Omega^{0.6}$ because of the dependence on linear growth rate ($\lesssim \Omega^{0.6}$) of the COBE-normalized primordial fluctuation spectrum. The abundance of rich galaxy clusters requires that $\sigma_8\,\Omega^{0.6} = 0.6 \pm 0.2$. Reconciliation of CDM with the large-scale structure data therefore implies that $b \approx 2$ if $\Omega = 1$ but $b \approx 1$ if $\Omega \approx 0.2$ (*e.g.*, Kolatt and Dekel 1997).

DARK MATTER EXPERIMENTS

One may classify dark matter detection strategies as being either direct or indirect. Direct detection experiments typically utilize nuclear recoil in a cryogenic detector, or ionization in a liquid scintillator, beginning with the hypothesis that the halo dark matter density near the Sun is about 0.03 GeV cm^{-3}. The interaction cross-section is simply calculated from the low temperature limit of the annihilation cross-section. The flux at Earth of $3 \times 10^5 (m_\chi/1\,\text{GeV})^{-1}\text{cm}^{-2}\,\text{s}^{-1}$ leads to a signal on the order of one event per day per kg of detector material, for cryogenic Ge or Si detectors that search for recoil signatures via phonon and ionization signatures.

A different type of experiment utilizes annihilations of WIMPs in the Earth or the Sun. Again, the current, low temperature annihilation cross-section is determined once Ω_χ, and therefore the annihilation rate at freeze-out, is specified. High energy neutrinos, of typical energy $\sim 0.1\,m_{\tilde{\chi}}$, are produced that can be observed in underground, under-ice, or underwater neutrino detection experiments, typically via Cerenkov radiation in water (or ice).

Neutrino masses provide another avenue for hunting dark matter. Direct limits are not useful, except for the electron neutrino ($m_{\nu_e} < 5\,\mathrm{eV}$, $m_{\nu_\mu} < 250\,\mathrm{keV}$; $m_{\nu_\tau} < 30\,\mathrm{MeV}$). However there is evidence for neutrino oscillations, both in the atmosphere and in the Sun. These require a mass difference. For example, $m_{\nu_e} - m_{\nu_\mu} \approx 0.1\,\mathrm{eV}$ and $m_{\nu_\mu} - m_{\nu_e} \approx 10^{-3}\,\mathrm{eV}$, respectively, would indicate the possibility that either m_{ν_μ} might be heavy, or that all three neutrinos $m_{\nu_e} \approx m_{\nu_\tau} \approx m_{\nu_e}$, are heavy. The cosmologically desired mass, preferably in three species, is about 6 eV, to give a model that is about 20 percent hot and 75 percent cold dark matter, the remainder being both luminous and (mostly) dark baryons.

Large-scale structure on scales of $\sim$ 100 Mpc provides a potentially important dark matter probe. It is here that degree-resolution cosmic microwave background maps probe the primordial fluctuations, and ongoing deep redshift surveys probe the galaxy distribution back to z $\sim$ 0.2. The Anglo-Australian 2 degree field survey will obtain 250,000 galaxy redshifts, and the Sloan Digital Sky Survey will generate 10^6 redshifts, within the next five years. This data will provide full sampling of the universe over 10 Mpc – 100 Mpc scales. The shape of the linear power spectrum will be derived. It will be possible to probe various dark matter variants, since the alternative models that reconcile CDM with large-scale structure generally differ in $P(k)$ over this range.

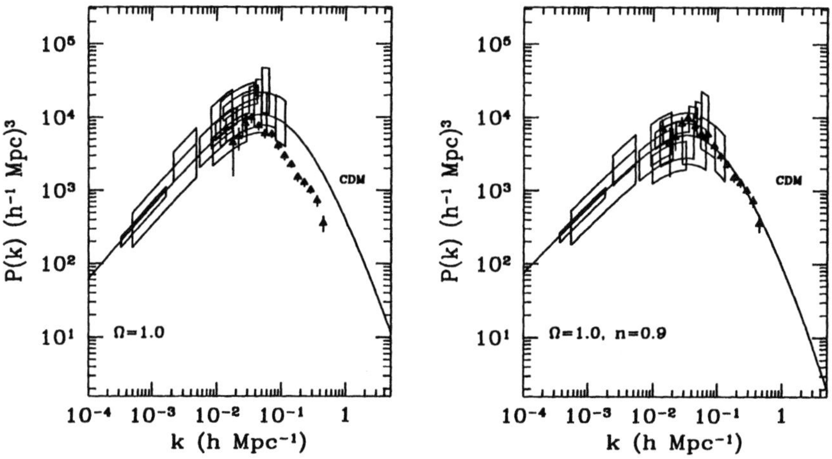

Figure 2. The matter power spectrum, $P(k)$, as reconstructed from large–scale structure and cosmic microwave background data (updated from reference 14). Boxes are $\pm 1\sigma$ values of $P(k)$ inferred from CMB measurements, assuming CDM ($\Omega_0 = 1$, $h = 0.5$, $\Omega_B = 0.03$) with $n = 1$ and $n = 0.9$. The overall amplitude of the LSS data (triangles), taken from a recent compilation[15], is uncertain to $\sim$20%.

Below 10 Mpc, one studies $P(k)$ via its implications for the epoch of galaxy formation. Here, redshift surveys with very large telescopes will obtain large samples of galaxies at $z = 1$ and possibly beyond. Indeed with HST, redshift 3 galaxies are becoming accessible. Mixed dark matter, for example has less small-scale power than, say, tilted cold dark matter, and will produce fewer high redshift galaxies.

The ultimate dark matter probe will come with the forthcoming generation of satellites designed to map the cosmic background radiation. One can represent primordial density fluctuations in the universe prior to matter-radiation decoupling as gigantic sound waves. At decoupling of matter and radiation the sound speed drops by many orders of magnitude, from near light speed to about 3 km s^{-1}. The sound waves consequently become gravitationally unstable fluctuations, and the growing mode develops into large-scale structure at low redshift. The cosmic microwave background provides a snapshot of the universe at the last scattering epoch, and one directly views the imprint of the sound waves as temperature fluctuations. An inflationary origin for the fluctuations guarantees that the sound waves are coherent, at any given wavelength, or equivalently, angular scale at which they are viewed. This is because the superhorizon amplitude, specified by inflation, of the growing mode fixes the phase. With allowance for radiation diffusion damping of the waves, one finds that several peaks (and rarefactions) of the wave are observable, corresponding to harmonics of the sound horizon at last scattering. The wavelength range probed extends from the maximum sound horizon scale to about one-fifth of the sound horizon scale.

These acoustic features show up as peaks in $|\delta T/T|_{rms}$, and are sensitive to various cosmological parameters[16]. The amplitude increases both as Ω_B and ρ_γ/ρ_m increase, because of enhanced driving of the oscillations in the coupled baryon-photon fluid prior to decoupling. Enhancment of Ω_B also suppresses the higher frequency peaks because of damping, and reduction of Ω_0 (in the open model) modifies the geometry and projects all of the features to smaller angular scales. There is sufficient breaking of the parameter degeneracy that one can hope to disentangle all of the key cosmological parameters with measurement of the first three peaks. In particular, it should be possible to reconstruct the primordial matter power spectrum. A first attempt at such a reconstruction is shown in Figure 2, where the reconstruction from CMB experiments is compared with the power spectrum inferred from deep redshift surveys for two CDM models. COBE-normalized standard CDM overproduces power on scales below $\sim$ 100 Mpc, whereas introduction of a slight tilt from primordial spectral index $n = 1$ to $n = 0.9$ goes a long way towards alleviating the problem.

Experiments now in preparation should be capable of accomplishing this goal of solving the outstanding issues in cosmology. Long duration ballooning will map up to 2 or 3 percent of the sky at 10' – 30' resolution, and interferometers will map smaller regions of the sky at a resolution of several arc-minutes. The ultimate CMB mapping experiment will be carried on the MAP satellite, to be launched by NASA in 2000 with 20' resolution and 20 μK sensitivity at frequencies up to 90 GHz, and the COBRAS-SAMBA satellite, to be launched by ESA in 2004 with 7' resolution, 5 μK sensitivity, and frequencies spanning 40 – 300 GHz.

With the all-sky coverage afforded by satellite experiments, one will be able to measure individual Fourier components of the CMB fluctuations to a precision of 1 or 2 percent. Up to 1,000 Fourier components ($\phi \lesssim 1000$) are measurable. Since each component is independent (for a Gaussian signal), one will be able to measure cosmological parameters to comparable precision after the foregrounds are subtracted.

Possible non-Gaussianity of the cosmological signal will complicate matters. However all the non-Gaussian models constructed to date, involving topological defects such as cosmic strings or textures, look remarkably Gaussian on degree scales. If the forthcoming experiments provide the claimed precision, one will know within a few years precisely how much dark matter there is in the universe. As to its nature, while there is a CMB signal that distinguishes a massive neutrino component, and even distinguishes the number of species carrying mass, from cold dark matter, one has to look to direct detection experiments for the ultimate discriminant.

ACKNOWLEDGEMENTS

I thank Eric Gawiser and Paulo Gondolo for providing the two figures, and for reading the manuscript.

REFERENCES

1. Charlot, S., Worthey, G. and Bressan A., *Astrophys. J.* **457** (1996) 625.
2. Navarro, J., Frenk, C. S. and White, S. D. M., *Mon. Not. R. astr. Soc.* **275** (1995) 56.
3. Flores, R. and Primack, J., *Astrophys. J.* **427** (1994) L1.
4. Moore, B. 1994 *Nature* **370** 629.
5. Navarro, J., Eke, V. R. and Frenk, C. S. *Mon. Not. R. astr. Soc.* (1996) in press.
6. Loewenstein, M., & Mushotzky, R.F., *Astrophys. J.* **471** (1996) L83.
7. Eke, V.R., Cole, S., & Frenk, C.S., astro-ph 9601088 (1996).
8. Kolatt, T., & Dekel, A. , *Astrophys. J.* (1997) in press.
9. Alcock, C. *et al.*, astro-ph 9606165, *Astrophys. J.* (1997) in press.
10. Fields, B. D., Mathews, G. A. and Schramm, D. S., *Astrophys. J.* (1997) in press.
11. Ellis, J., Falk, T., Olive, K. A. and Schmitt, M., CERN Preprint TH/96-102, hep-ph/9606292, *Phys. Lett. B.* (1996) in press.
12. Bergstrom, L. and Gondolo, P., hep-ph/9510252, *Astropart. Phys.* (1996) in press.
13. Gondolo, P., in *Dark Matter in Cosmology, Quantum Measurements, Experimental Gravitation*, Editions Frontières, Gif-sur-Yvette (1996) in press.
14. Scott, D., Silk, J. and White, M., *Science* **268** (1995) 829.
15. Peacock, J. A. & Dodds, D. D., *Mon. Not. R. astr. Soc.* **267** (1994) 1020.
16. Hu, W., Sugiyama, N. and Silk, J. *Nature* (1996) in press.

ASPECTS OF DARK MATTER DIRECT DETECTION

L. Baudis[a], V. Bednyakov[b], J. Hellmig[a], G. Heusser[a], M. Hirsch[a], H.V. Klapdor-Kleingrothaus[a], S. Kovalenko[b], A. Müller[a], H. Päs[a], F. Petry[a], Y. Ramachers[a1], H. Strecker[a]

[a]Max-Planck-Institut für Kernphysik
P.O.Box 10 39 80, D-69029 Heidelberg, Germany
[b]Joint Institute for Nuclear Research, Dubna, Russia

INTRODUCTION

The purpose of this article is to give a short introduction into the main aspects of a fast evolving field of research, namely the direct detection of possible dark matter particles [1, 2, 3], the so called WIMPs (Weakly Interacting Massive Particles).

Under the assumption that WIMPs constitute the dark halo of our galaxy they should in principle be detectable here on earth. Up to now no evidence for WIMPs has been detected in any experiment. Consequently, experiments give limits on the cross section for the underlying physical process, i.e. WIMP-nucleus elastic scattering, as results. The aim is so far the exclusion of possible candidate particles. The most favorable candidate is the LSP (Lightest Supersymmetric Particle), usually assumed to be the neutralino, within the MSSM (Minimal Supersymmetric Standard Model).

The plan is to show the underlying foundations of the direct detection of dark matter particles with which the current results and perspectives, summarized in the Figs. 3,4, are obtained. As an example for an experiment specialized to give stringent cross section limits for WIMPs, the HDMS (Heidelberg Dark Matter Search) Experiment will be introduced [10]. Finally, we give some remarks about calculations for dark matter within the MSSM in order to understand the typical scatter plots for expectations of WIMP events in direct detection experiments [2, 11, 12, 13, 14].

FOUNDATIONS OF DARK MATTER DIRECT DETECTION

Direct detection of dark matter particles is defined by the attempt to measure counts in a given detector induced by an elastic scattering event of a particle with a detector nucleus. Most existing experiments measure in addition the deposited energy

[1]Talk presented at the NATO ASI on Masses of Fundamental Particles (Cargèse, August 5-17, 1996) by Y. Ramachers

Masses of Fundamental Particles: Cargèse 1996
Edited by M. Lévy *et al.*, Plenum Press, New York, 1997

transferred by the recoiling nucleus. Other possible observables are pulse shapes of the events, the element specific splitting of the deposited energy into ionization and lattice vibrations (phonons) and also the recoil direction [3–10].

The main ingredients that enter the description of dark matter events can be shown in a schematic rate equation:

$$R \propto \rho_0 \langle v \rangle \frac{\sigma}{m_\chi m_N}, \tag{1}$$

where R is the total rate of events in counts per unit time and mass. The input from astrophysics is described by the local halo density of WIMPs, ρ_0, and $\langle v \rangle$, the average WIMP halo velocity. These parameters can in principle be determined in astronomy and are usually assumed to take values of $\rho_0 \approx 0.3$ GeV/cm^3, $\langle v \rangle \approx 300$ km/s. Note that ρ_0 is unsure by a factor of two and $\langle v \rangle$ depends on the geometry of the halo, the solar rotation velocity and a lot of other parameters with non-negligible errorbars.

Particle Physics enters the equation by the WIMP–nucleus elastic scattering cross section σ and the WIMP mass m_χ. Both are unknown and highly dependent on the assumed nature of the particle candidate. The nucleus mass m_N is mainly determined by practical reasons of availability and usefulness as a detection material. Typical materials are for instance Germanium, Sodium, Iodine, Fluorine, Xenon, Sapphire (Al_2O_3) and some other.

The procedure to get limits or exclusion curves (*c.f.* Figs. 3,4) is in principle quite simple. Take an energy sensitive detector and let it measure for a reasonable time (to get a sufficient statistic) a spectrum of counts in the low energy region, *i.e.* from the threshold to about 100 keV. Calculate with the differential rate equation

$$\frac{dR}{dQ} = N_T \frac{\rho_0 \sigma_0}{2 m_\chi m_r^2} F(Q) \int_{v_{min}(E_{thr})}^{v_{max}} \frac{f(v)}{v} dv \quad , \tag{2}$$

(dR/dQ differential rate in counts/(day kg keV); N_T number of target nuclei; F(Q) nuclear form factor (exponential decaying function); E_{thr} threshold energy of the detector; m_r reduced mass of the WIMP–nucleus system; f(v) velocity distribution of the WIMPs in the galactic halo) several theoretical WIMP spectra by varying the unknown WIMP mass and taking σ_0 as free parameter. Then fit the experimental spectrum to the theoretical spectra, depending on m_χ, where σ_0 is the fit parameter. This is how to get points in a σ–m_χ–diagram which give the exclusion curve. The cross sections above these curves are excluded to a given confidence level (included into the fit procedure) usually to 90%. There are some corrections to add to the described procedure [1] but they do not spoil the outline given above.

From an experimental point of view there are just three main ingredients to be optimized: the background of the detector (remember that you look for extremely rare events), the threshold energy (a decrease raises the sensitivity for WIMP events) and the target mass (increases the sensitivity; high masses are necessary to look for a WIMP signature). For the design of the experiment introduced in the next section the aim is to reduce the background drastically with a sufficiently low threshold.

A NEW EXPERIMENT

In this section we introduce a new idea for a dark matter detector using Ge–semiconductor technology [10]. The idea is as follows: A small Ge–crystal (natural Ge for the prototype, isotopically enriched Ge for the second stage) is surrounded by a well–type Ge–crystal and the configuration is run in anticoincidence mode (see Fig. 1). Both crystals are mounted into a common cryostat system. We will use our experience with Ge detectors from the Heidelberg–Moscow Experiment [15, 16] to build the cryostat system and the surrounding. A thin insulator (about 1mm) is placed between the two crystals in order to shield leakage currents on the surfaces.

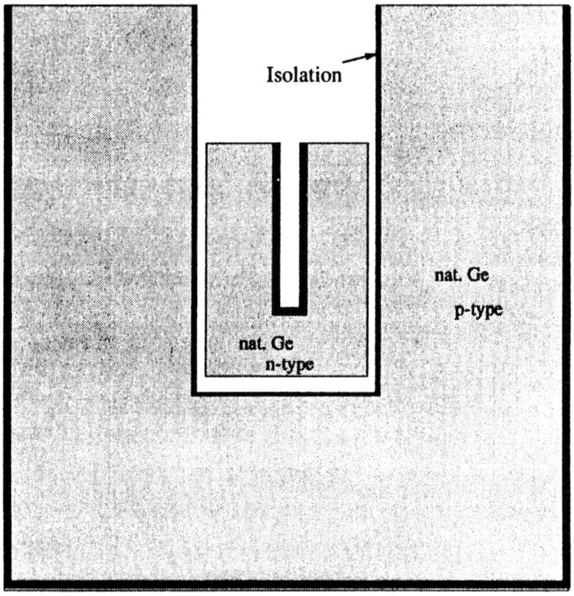

Figure 1. Schematic geometry of our new anticoincidence detector design. The well-type Ge crystal is planned as a p–type semiconductor, the small measurement Ge crystal as n–type. The approximate diameter of the well–type will be 90mm, height 90mm; the small measurement crystal will have about 30mm diameter and height 40mm.

The outer well–type Ge detector will be a very effective anticoincidence counter. This efficiency for different photon energies has been simulated in a Monte–Carlo simulation using the GEANT [17] code for the schematic geometry from Fig. 1. The results are expressed as suppression factors as functions of the initial photon energies. The suppression factor is defined by the ratio of the number of unsuppressed counts in the interesting energy interval 0–120 keV over the number of suppressed counts in the same energy interval measured by the small inner detector crystal.

In Fig. 2 we show the suppression factors for the simulated γ–ray radioactive impurities located in four different volumes: i) a cylinder volume surrounding the well–type Ge detector, ii) a volume above the top and iii) under the bottom side of the well–type and iv) inside the insulator material between the small and the well–type detector. In addition to the simulated discrete photon energies we simulated a typical cosmogenically induced background contribution from ^{60}Co and the natural decay chain of ^{228}Ac. The results are comparable or even better (^{60}Co decays with two coincident photons which increases the anticoincidence efficiency) than the suppression factors of the 2614 keV photons.

The efficiency of the anticoincidence is lowest for photons starting in the insulator material between the crystals. Therefore it is essential to use selected radiopure material for this structural part and we are currently testing selected materials in the Gran Sasso underground laboratory. However, even a specific activity comparable to that of the selected copper surrounding the crystals would result in a negligible contribution to the background due to the low mass needed for the insulation material. Suppression factors for photons started from the top of the well detector (open side, see Fig. 1) are lower than for photons from the bottom or the cylinder shell. They are more likely to escape the anticoincidence by Compton backscattering thereby just leaving a signal inside the small detector crystal.

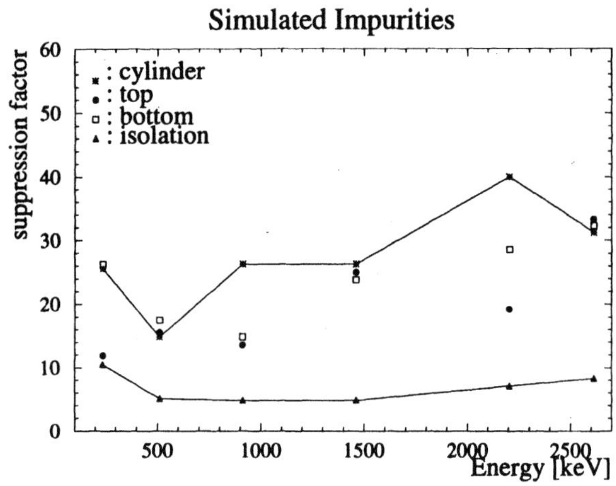

Figure 2. The results of the Monte–Carlo simulation for the efficiency of the anticoincidence (suppression factor in the energy interval 0–120 keV) between the two Ge crystals from Fig. 1. Shown is the dependence on different photon energies for different photon source locations. The simulated geometry is shown in Fig. 1, photons have been started from the cylinder shell surrounding the well–type, from the open top, the bottom and from inside the isolation material between the crystals.

From this simulation we roughly estimate a background reduction of a factor of 20 due to the anticoincidence operation mode. Note that this is still a conservative estimate due to our experience with low–level Ge–detectors. It gives us the opportunity to start from a background level at least as low as already achieved in the Heidelberg–Moscow experiment [16]. The main background contribution always originates from materials nearest to the measurement crystals. In our case this would be one of the radiopurest materials we know, a second Ge–crystal. Therefore we think to give a conservative estimation when we prospect the WIMP cross section limit for our new detector design in Figs. 3,4. From this prospect we conclude to reach WIMP cross section limits comparable to planned cryogenic experiments [6, 7] with our new Ge–semiconductor dark matter experiment which will start in 1997.

CALCULATIONS WITHIN THE MSSM FOR DARK MATTER

In principle calculations within the MSSM for dark matter can be divided in two blocks: The first block of calculations deals with assumptions to reduce the number of free parameters of the theory. The aim is to get as a result a completely specified supersymmetric (SUSY) particle model. This model is then used in the second block to calculate interesting quantities like amplitudes for neutralino–nucleus elastic scattering.

The assumptions can be stated in the form of the unifications for the soft SUSY breaking parameters and gauge coupling constants at a high energy scale M_X (SUSY–GUT–scale) typically higher than 10^{16} GeV. Current efforts [13] stress relaxation of these assumptions in the sector of soft scalar masses, so called non–universal scalar mass unification. Note however, that already in the earlier paper ref. [12] a similar scenario was used. It implies universal soft sfermion masses not unified with the soft Higgs masses. As a result of these assumptions the SUSY model predictions depend on only six parameters (five for complete GUT unification conditions) compared to more than sixty for a generic low-energy supersymmetric model. These six (or five) parameters can not be varied arbitrarily. First, the demand for internal consistency of the theory has to be fulfilled and second there exist a lot of constraints on the parameter space from experiments (accelerator experiments, measured branching ratios where supersymmetry could contribute etc.) as well as a cosmological constraint (no overproduction of relic matter). All of these have to be taken into account [14].

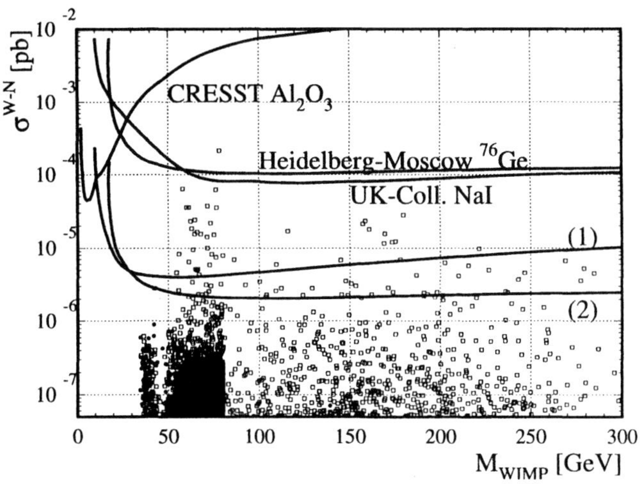

Figure 3. Speculative view for obtainable cross section limits on WIMP–neutron (spin–independent interaction) elastic scattering for three future experiments, CDMS (Label 2) and CRESST [6,7] and our proposal [10] (Label 1). For comparison the present lowest–limit results, the Heidelberg–Moscow result [4] and the UK–Coll. result for NaI [9], are also shown together with expected cross sections for neutralinos, first for a SUSY–GUT scenario (filled dots) and second for a SUSY–GUT scenario without universal scalar mass unification [14] (open boxes).

Neutralino–nucleus elastic scattering

Having specified the SUSY model ,*i.e.* the sparticle mass spectrum, couplings etc., one can calculate the cross section of neutralino–nucleus elastic scattering. This cross section can be divided in factors depending on properties of the detector nucleus and factors depending on the nucleon structure and the specific SUSY model [18]:

$$\frac{d\sigma}{dq^2}(v^2, q^2) = \frac{8\,G_F}{v^2}\,(a_0^2\,\mathcal{F}_{00}^2 + a_0\,a_1\,\mathcal{F}_{10}^2 + a_1^2\,\mathcal{F}_{11}^2 + c_0^2\,\mathcal{F}_S^2), \tag{3}$$

with $\mathcal{F}_{ij}^2$ being the nuclear form factors and the coefficients a_i, c_0 depending on the nucleon structure as well as on the given SUSY model. Terms with the coefficients a_i describe the spin–dependent interaction of neutralinos with nuclei, the term with c_0 the spin–independent interaction. From eqn. (3) it should be clear that expectations for WIMP events depend on the type of interaction either dominantly spin–dependent or spin–independent.

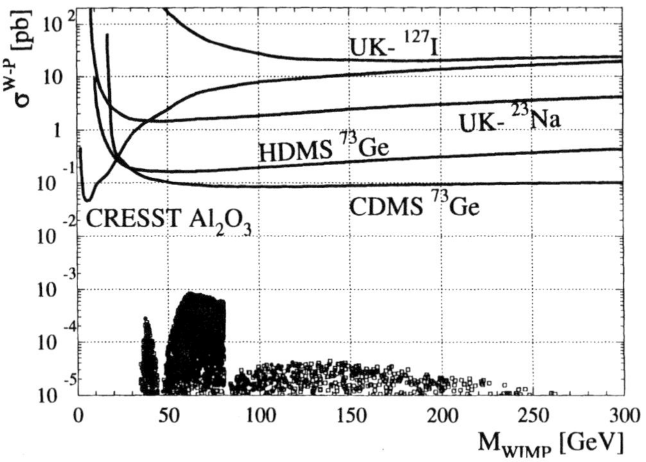

Figure 4. The same as Fig.3 but for WIMP-proton (spin-dependent interaction) elastic scattering. The difference between the SUSY GUT and the relaxed SUSY GUT scenario is nearly invisible.

Now one can calculate expected event rates usually represented in scatter–plots. Every dot a complete SUSY model. As can be seen in Figs. 3,4 the dominantly spin–dependent interaction is suppressed compared to the spin–independent as was shown by [12].

CONCLUSION

The direct detection of dark matter particles is a large field of research with connections to many different areas in physics and astrophysics. New or upgraded experiments just start operation or will start in the near future. To conclude this short survey it should be stressed that up to now no evidence for WIMPs has been found but the next years will show an interesting development for this specific technique to detect dark matter in the universe.

References

[1] P.F. Smith and J.D. Lewin, Phys. Rep. **187** (1990) 203

[2] G. Jungman, M. Kamionkowski and K. Griest, Phys. Rep. **267** (1996) 195

[3] R. Bernabei, Riv. Nuov. Cim. **18** (1995) 1

[4] M. Beck *et.al.*, Phys. Lett. **B 336** (1994) 141

[5] W. Seidel *et.al.*, J. Low Temp. Phys. **93** (1993) 797; Phys. Lett. **B 323** (1994) 95; P.D. Barnes *et.al.*, J. Low Temp. Phys. **93** (1993) 791; D. Reusser *et.al.*, Phys. Lett. **B 255** (1991) 143

[6] P.D. Barnes *et.al.*, NIM **A370** (1996) 233

[7] M. Bühler *et.al.*, NIM **A370** (1996) 237

[8] C.J. Martoff *et.al.*, Phys. Rev. Lett. **76** (1996) 4882

[9] P.F. Smith *et.al.*, Phys. Lett. **B 379** (1996) 299

[10] L. Baudis *et.al.*, submitted to Nuclear Instruments and Methods, Section A (1996)

[11] M. Drees and M.M. Nojiri, Phys. Rev. **D 48** (1993) 3483

[12] V. Bednyakov, H.V. Klapdor–Kleingrothaus and S.G. Kovalenko, Phys. Rev. **D50**(1994) 7128

[13] V. Berezinsky, A. Bottino, J. Ellis, N. Fornengo, G. Mignola and S. Scopel, Preprint HEP–PH–9508249

[14] V. Bednyakov, H.V. Klapdor–Kleingrothaus, S.G. Kovalenko and Y. Ramachers, Preprint HEP–PH–9606261

[15] HEIDELBERG–MOSCOW–COLLABORATION, Phys. Lett. **B356** (1995) 450

[16] H.V. Klapdor–Kleingrothaus, in Double Beta Decay and Related Topics, eds. H.V. Klapdor–Kleingrothaus and S. Stoica, World Scientific, Singapore 1996; HEIDELBERG–MOSCOW–COLLABORATION, submitted to Phys. Rev. **D**

[17] R. Brun, GEANT3 User's Guide, CERN Report No. DD/EE/84-1 (1987)

[18] V. Bednyakov, H.V. Klapdor–Kleingrothaus and S. Kovalenko, Preprint HEP–PH–9510292; Proceedings of Double Beta Decay and Related Topics, ECT Workshop, Trento 24. April–5. May 1995, (World Scientific, Singapore 1996), eds. H.V. Klapdor–Kleingrothaus and S. Stoica

MASSES BEYOND THE STANDARD MODEL

G.G. Ross

Department of Physics, Theoretical Physics
University of Oxford, 1 Keble Road, Oxford, OX1 5DX

1 Introduction

If we are to understand the pattern of fermion masses and mixing angles it is necessary to go beyond the Standard Model. One obvious possibility is that there is further structure, Grand Unification, strings etc., at a high scale which determines the Yukawa couplings responsible for the masses. However attempts to implement such ideas have to explain why the electroweak breaking scale is much less than the unification scale. Low energy supersymmetry provides a way of protecting such an hierarchy of mass scales and if supersymmetry is broken by a gaugino condensate it can even explain the origin of the hierarchy.

Within the context of supersymmetry, unification has had considerable success in determining the parameters of the Standard Model. The measured values of the gauge couplings are consistent with simple unification values with a unification scale of $O(10^{16}GeV)$ provided the Standard Model spectrum is extended to that of the minimal supersymmetric model (the MSSM). In addition the pattern (and magnitude) of spontaneous breakdown of the Standard Model follows naturally from the structure of radiative corrections in the MSSM provided there is some unification of the supersymmetry breaking masses at the unification scale. This simplicity in the parameters of the (supersymmetric) Standard Model at high scales appears to extend to the Yukawa couplings involved in determining the heavy fermion masses. The measured values of the bottom quark and the τ lepton masses are consistent with their equality at the unification scale [1]. Indeed this simplicity might extend to the top quark sector for it is a viable possibility that the top, bottom and tau Yukawa couplings are all equal and governed by the infra-red structure of the theory[2, 3] However the pattern of light masses and mixing angles appears at first sight to be too complicated to be explained by such simple unification ideas. Here we wish to show that this is not the case and that very simple extensions of the Standard Model are capable of generating both the qualitative and quantitative structure of the full set of fermion masses and mixing angles.

We start with a review of the measurements of the quark masses and mixing angles and their implications for the underlying structure of the quark mass matrices. We then turn to the question whether the observed hierarchical structure for the masses and mixing angles can be explained by a family symmetry. It is shown how even the simplest

Masses of Fundamental Particles: Cargèse 1996
Edited by M. Lévy *et al.*, Plenum Press, New York, 1997

Table 1. Experimental data on quark masses and mixing angles.

$$|V_{i,CKM}| \approx \begin{bmatrix} 1 & 0.219-0.222 & 0.0025-0.0035 \\ V_{cd} \approx V_{us} & 1 & 0.0365-0.0421 \\ V_{td} = (1.4-4.6)V_{ub} & V_{ts} \approx V_{cb} & 1 \end{bmatrix}$$

$$M_d = (4-6)MeV \qquad M_u = (3-5)MeV$$
$$M_s = (0.11-0.23)GeV \qquad M_c \sim 1.35GeV$$
$$M_b \sim 5.2GeV \qquad M_t = 175 \pm 6GeV$$

such symmetry can reproduce the hierarchy of masses and generate "texture zeros" in the mass matrices in a manner consistent with observation. The constraints following from anomaly cancellation are reviewed showing how they may lead to a prediction for the ratio of gauge coupling constants without Grand Unification. Although the general hierarchical structure of masses follows simply from a family symmetry it is necessary to have further structure if the absolute magnitude of the masses and mixing angles are to be determined. We consider two possible routes to making such predictions. We first consider the possibility that the infra-red structure of the theory fixes the masses, particularly those of the third generation, through infra-red-fixed points. We then consider the possibility that the masses are fixed through further symmetry. This leads us to a discussion of Grand Unification and Supersymmetric unification. The unification predictions for gauge couplings are reviewed and extended to the other parameters of the supersymmetric Standard Model.

2 Quark Masses and Mixings

As the subject of these lectures is a discussion of our attempts to determine in a field theory description of the fundamental interactions the masses and mixing angles of quarks (and leptons) it is perhaps appropriate first to summarise what we know about them.

In Table 1 I present the data for the quark masses and mixing angles as currently known [4]. Note that what we know are the eigenvalues of the mass matrix and the CKM matrix which is a combination of the rotations matrices needed to diagonalize the left-handed components of the up and down quarks.

$$\mathrm{U}_{CKM} = V_L^{u\dagger} V_L^d \tag{1}$$

However, the full mass matrix requires a knowledge of both the left and right handed rotation matrices

$$M^{u,d} = V_R^{u,d^{T\dagger}} M_{Diag}^{u,d} V_L^{u,d} \tag{2}$$

Thus measurement of the CKM matrix and the diagonal mass entries is insufficient to construct the full mass matrix. This is an embarrasment when trying to develop a theory of masses because masses in a field theory arise through the matrix of couplings of quarks to the Higgs boson. Upon spontaneous breakdown this leads to a mass matrix as the fundamental starting point. What then do we know about this mass matrix?

Bjorken [5] has introduced a systematic way to determine the mass matrix from the observed masses and mixing angles for the case of small off diagonal matrix elements. Although not inevitable, this is the preferred case for it leads to small mixing angles as are observed without requiring a cancellation between up and down rotations. In this case it is easy to show that the CKM matrix elements determine those elements of the mass matrix above the diagonal. Those below the diagonal are only weakly constrained. To see this we write the mass matrix in the form

$$M = M_D + \Delta M + m \tag{3}$$

in which M_D is a diagonal matrix with the mass eigenvalues along the diagonal, m refers to the small off diagonal elements and ΔM refers to the correction to the diagonal elements on going to the non-diagonal form. The mass matrix M

$$V_L^+ M M^+ V_L = M_D^2 \qquad (V = V^{u,d} \quad M = M^{u,d}) \tag{4}$$

Substituting the form of eq(3) and premultiplying by V_L gives

$$(M_D + \Delta M + m)(M + \Delta M + m^+)V_L = V_L M_D^2 \tag{5}$$

Since the rotation matrices are ordered in powers of the small parameters m_{ij} by

$$V_{ij} = O(m) \quad V_{ii} = 1 + O(m^2) \tag{6}$$

we may solve eq(5) perturbatively to obtain two equations in leading non-vanishing order

$$2M_i \, \Delta M_i + \sum_j |m_{ij}|^2 + \sum_j (m_{ij} M_j + M_i m_{ji}^x) V_{ji} = 0 \tag{7}$$

and

$$M_i^2 V_{ij} + \sum_k (m_{ik} M_k + M_i m_{ki}^*) V_{kj} + \sum_k m_{ik} m_{jk}^\tau = V_{ij} M_j^2 \tag{8}$$

These may be solved for the off diagonal elements of the CKM matrix giving

$$V_{ij} = -\frac{(m_{ij}M_j + m_{ji}^* M_i)}{(M_i^2 - M_j^2)} + \frac{(m_{ik}M_k + m_{ki}^* M_i)(m_{kj}M_j + m_{jk}^* M_k)}{(M_i^2 - M_j^2)(M_k^2 - M_j^2)} - \frac{m_{ik} m_{jk}^*}{(M_i^2 - M_j^2)} + O(m^3) \tag{9}$$

From eq(9) we may determine the contributions to the CKM coming from diagonalisation of the up and the down quarks. Combining these we have

$$\begin{aligned} V_{12}^{CKM} &= \frac{m_{12}^d}{M_s} - \frac{m_{12}^u}{M_c} + \frac{m_{23}^{d*} m_{13}^u}{M_b M_t} + \frac{m_{21}^{d*}}{M_s} \frac{M_d}{M_s} + \ldots \\ V_{12}^{CKM} &= -\frac{m_{12}^{d*}}{M_s} + \frac{m_{12}^{u*}}{M_c} + \frac{m_{13}^{d*} m_{23}^u}{M_b M_t} + \ldots . \end{aligned} \tag{10}$$

This form immediately leads to a constraint between the matrix elements. To leading order we have

$$|V_{12}| \simeq |V_{21}| \tag{11}$$

This is in excellent agreement with the measurements $|V_{12}| = 0.218 - 0.224$, $|V_{21}| = 0.218 - 0.224$ and supports the initial assumption of small off diagonal mass matrix elements. Note that to leading order the magnitude of the matrix elements is determined only by the mass matrix entries above the diagonal.

The same procedure applied to V_{23}, V_{32} gives

$$\begin{aligned} V_{23}^{CKM} &= \frac{m_{23}^d}{M_b} - \frac{m_{23}^u}{M_t} + \ldots \\ V_{32}^{CKM} &= -\frac{m_{23}^{d*}}{M_b} + \frac{m_{23}^{u*}}{M_t} + \ldots \end{aligned} \tag{12}$$

so

$$|V_{23}| \simeq |V_{32}| \tag{13}$$

again in good agreement with experiment $|V_{23}| = 0.032 - 0.048$, $|V_{32}| \simeq 0.030 - 0.048$.

Finally one finds

$$\begin{aligned} V_{13} &= \frac{m_{13}^d}{M_b} - \frac{m_{13}^u}{M_t} + \frac{m_{12}^u}{M_c}\left(\frac{m_{23}^u}{M_t} - \frac{m_{23}^d}{M_b}\right) + \ldots \\ V_{31} &= -\frac{m_{13}^{d*}}{M_b} + \frac{m_{13}^{u*}}{M_t} - \frac{m_{12}^{d*}}{M_s}\left(\frac{m_{23}^{u*}}{M_\tau} - \frac{m_{23}^{d*}}{M_b}\right) + \ldots \end{aligned} \tag{14}$$

Again if the leading term dominates we have approximate equality of V_{13} and V_{31} in magnitude, but this is in disagreement with the observation (*cf* Table 1). If, on the other hand, we assume that the leading terms are zero *ie* $m_{13}^d = m_{13}^u = 0$, so called "texture zeros", then the next term in eq(14) dominates and leads to

$$\begin{aligned} |V_{13}| &= \sqrt{\frac{M_u}{M_c}}\,|V_{23}| \\ |V_{31}| &= \sqrt{\frac{M_d}{M_s}}\sqrt{\frac{M_c}{M_u}}\,|V_{13}| \end{aligned} \tag{15}$$

Substituting the measured values given in Table 1 one finds these relations are in remarkably good agreement with experiment.

2.1 Texture Zeros

We have seen that the assumption of the vanishing (in leading order) of two elements of the mass matrix ("texture zeros") leads immediately to a prediction for a CKM matrix elements in excellent agreement with experiment. Are there any other possible texture zeros? Suppose the diagonal element $(M + \Delta M)_{11} = 0$ and further assume that the $(1,2)$, $(2,1)$ elements are symmetric. The sub-matrix for either the up or the down quarks has the form

$$\begin{pmatrix} 0 & m_{12} \\ m_{12} & m_{22} \end{pmatrix}$$

This has now only two parameters (the phase can be absorbed in a redefinition of the fermion fields). Hence there is a relation between its two mass eigenvalues and the rotation diagonalising it. One may readily see that this implies

$$\begin{aligned} |V_{us}| &= \left|\sqrt{\frac{M_D}{M_s}} - \sqrt{\frac{M_u}{M_c}}\,e^{i\sigma}\right| \\ c.f.(0.218 - 0.224) &= |(0.16 - 0.33) - (0.047 - 0.07)e^{i\sigma}| \end{aligned} \tag{16}$$

Here σ is a phase that enters when combining the up and the down rotations. The experimental comparison is also shown and one may see this prediction is in good agreement with experiment (due to the smallness of the up quark contribution the σ dependence is quite small).

Thus we see that experiment is in good agreement with the hypothesis of four texture zeros (counting pairs of off-diagonal zeros as one zero since we have assumed a symmetric form). We shall return to a more systematic discussion of texture zeros shortly. Note that as was the case for V_{12}, V_{21} the other CKM matrix elements principally depend on the elements of m *above* the diagonal so we are largely ignorant of the elements below the diagonal. In fact the only hint for the magnitude of these off-diagonal lower elements comes from the success of eq(16) which applies only if $m_{21} = m_{12}$. In what follows we will often follow this hint and assume that the full mass matrices are symmetric.

To summarise, we have found the data is consistent with four texture zeros plus the assumption that the off-diagonal matrix elements are small. This simple structure determines 5 of the 6 of diagonal CKM matrix elements (*ie* three of the four independent elements of a unitry matrix)

The mass matrix has the form

$$M^{d(u)} = \begin{pmatrix} 0 & m_{12} & 0 \\ m_{12} & m_{22} & m_{23} \\ 0 & m_{33} & m_{33} \end{pmatrix} \tag{17}$$

The one remaining matrix element left to be determined is V_{cb} which is related to the mass matrix elements as in eq(9)

$$V_{cb} = \frac{m^d_{23}}{M_b} - \frac{m^u_{23}}{M_t} \tag{18}$$

We will return to a discussion of how such a structure might emerge from a field theory description of the fundamental interactions in a later section. First however we turn to a more systematic discussion of the possible texture zeros in the mass matrix.

2.2 General texture zero structure

The appearance of texture zeros in the mass matrix strongly suggests that there is some new symmetry which, when exact, enforces the zero of the mass matrix element. We wish to use this "bottom-up" approach, starting with low-energy observables, to explore the possible appearance of such new symmetries as an indication of a further stage of unification of the fundamental interactions. As we shall discuss there is some indication of unification of the gauge couplings at a very high scale of $O(10^{16}GeV)$. In this case the analysis we have presented should be applied to the masses and mixing angles continued up to the unification scale before looking for texture zeros. The reason is that the mass matrix is scale dependent and zeros at one scale will be filled in by radiative corrections at another scale. To do this one uses the renormalisation group to continue the measured masses and mixings to the high scale and looks for simplicity appearing at the unification scale. In the case of gauge couplings, simplicity is equality between the couplings. In the case of the mass matrices simplicity is simple ratios of quark and lepton masses and the appearance of "texture" zeros. However the analysis of masses so far presented falls short of the ideal "bottom-up" approach for it assumes a particular set of texture zeros rather than just starting with measured values and seeing the zeros emerge.

The major difficulty in implementing the ideal program is that, as we have just discussed, laboratory measurements only determine the masses, i.e. the diagonal mass matrix and the CKM mass matrix, *not* the full mass matrices, If we are to determine $\mathbf{V}^{u,d}_{L,R}$, and hence the mass matrices separately, a simplifying assumption is needed. A plausible simplifying assumption that has been explored in detail is that the mass matrices are symmetric (as we have seen the success of the (1,1) texture zero lends some support to this hypothesis). We can analyse the most general symmetric mass matrix case leading to five or six texture zeroes, because there are just 6 possible forms of symmetric mass matrix with an hierarchy of the three (non-zero) eigenvalues and three texture zeroes (at least one of the up or down quark mass matrices must have three of the texture zeroes). Allowing for the redefinition of the quark fields to absorb phases, these matrices involve just three real parameters. This allows us to determine, up to the six fold discrete ambiguity, the diagonalising matrix V^u (or V^d) in terms of the masses. Hence, using eq(2), we may compute $\mathbf{M}^d$ (or $\mathbf{M}^u$) in terms of the CKM matrix and hence find M^d (or $\mathbf{M}^u$). Further texture zeroes in $\mathbf{M}^u$, $\mathbf{M}^d$ will result in predictions for the mixing angles of the CKM matrix.

The advantage of this technique is that it allows a determination of the down quark mass matrix using experimentally measured quantities without prejudicing the result by the choice of a specific texture. Thus the general problem of searching for (five or six texture zero) structure in mass matrices may be solved with the assumption of symmetric mass matrices[1] and the predictive ansatz that *either* the up or the down quark mass matrices have three texture zeros and thus contain just three real parameters giving a mass matrix with non-zero determinant..

Using this method one may find all structures with 6 or 5 texture zeroes. Encouragingly there are solutions, consistent with the hoped for simplicity in the mass matrices, but the number of possible solutions is limited; no solutions with 6 texture zeroes were found but there are several 5 texture zero solutions, given in Table 2. Altogether five solutions are found for the Yukawa matrices at $Q = M_X$ which are consistent with the low energy fermion masses and CKM matrix elements. In table 2, rather than list the numeric values, we have presented approximations to the matrix elements in powers of $\lambda(\simeq 0.22)$, the small parameter in eqn (11).

In these solutions, after using the freedom to choose the 9 independent quark phases, there is one phase left which we take to be the phase of the (complex) (1,2) matrix element of the down quark matrix (This is not displayed in 2). In addition, solution 2 has a phase associated with the (2,3) element.

To me structure of the type displayed in Table 2 strongly suggests a new symmetry of the type extensively explored by Froggart and Nielsen and many others[6, 7]. In this the origin of the hierarchy follows because the symmetry allows only one non-zero entry (by definition the (3,3) entry). Subsequent entries occur after the symmetry is broken their magnitude being determined by how many of the (small) symmetry breaking parameter insertions are needed to generate the term. In this context a texture zero need not be absolutely zero but merely sufficiently small that it does not affect the mass eigenvalues and mixing angles to leading order in the symmetry breaking parameter expansion. For example in the matrix of Table 2 the texture zero in the (1,1) position need only be zero to $O(\lambda^8)$.

3 Family Symmetries

How difficult is it to find a family symmetry capable of generating an acceptable fermion mass matrix? The surprising answer is "Not at all difficult" as I will illustrate by a very simple example utilising an Abelian family symmetry group[8, 9]. As we shall discuss this automatically gives texture zeros simultaneously in the (1,1) and (1,3) positions

[1] In fact the analysis applies to hermitian matrices too for a general 3×3 symmetric mass matrix may be transformed to an hermitian matrix through the freedom to redefine the nine phases of the three left-handed doublets and six right-handed singlets of quark fields. These 9 phases may be used to make both the up and down quark mass matrices hermitian (up to an overall irrelevant phase) since it is always possible to choose a basis in which either the top or the bottom mass matrix is diagonal.

Table 2. Approximate forms for the five-zero symmetric textures.

Solution	$\mathbf{Y}_u$	$\mathbf{Y}_d$
1	$\begin{pmatrix} 0 & \sqrt{2}\lambda^6 & 0 \\ \sqrt{2}\lambda^6 & \lambda^4 & 0 \\ 0 & 0 & 1 \end{pmatrix}$	$\begin{pmatrix} 0 & 2\lambda^4 & 0 \\ 2\lambda^4 & 2\lambda^3 & 4\lambda^3 \\ 0 & 4\lambda^3 & 1 \end{pmatrix}$
2	$\begin{pmatrix} 0 & \lambda^6 & 0 \\ \lambda^6 & 0 & \lambda^2 \\ 0 & \lambda^2 & 1 \end{pmatrix}$	$\begin{pmatrix} 0 & 2\lambda^4 & 0 \\ 2\lambda^4 & 2\lambda^3 & 2\lambda^3 \\ 0 & 2\lambda^3 & 1 \end{pmatrix}$
3	$\begin{pmatrix} 0 & 0 & \sqrt{2}\lambda^4 \\ 0 & \lambda^4 & 0 \\ \sqrt{2}\lambda^4 & 0 & 1 \end{pmatrix}$	$\begin{pmatrix} 0 & 2\lambda^4 & 0 \\ 2\lambda^4 & 2\lambda^3 & 4\lambda^3 \\ 0 & 4\lambda^3 & 1 \end{pmatrix}$
4	$\begin{pmatrix} 0 & \sqrt{2}\lambda^6 & 0 \\ \sqrt{2}\lambda^6 & \sqrt{3}\lambda^4 & \lambda^2 \\ 0 & \lambda^2 & 1 \end{pmatrix}$	$\begin{pmatrix} 0 & 2\lambda^4 & 0 \\ 2\lambda^4 & 2\lambda^3 & 0 \\ 0 & 0 & 1 \end{pmatrix}$
5	$\begin{pmatrix} 0 & 0 & \lambda^4 \\ 0 & \sqrt{2}\lambda^4 & \frac{\lambda^2}{\sqrt{2}} \\ \lambda^4 & \frac{\lambda^2}{\sqrt{2}} & 1 \end{pmatrix}$	$\begin{pmatrix} 0 & 2\lambda^4 & 0 \\ 2\lambda^4 & 2\lambda^3 & 0 \\ 0 & 0 & 1 \end{pmatrix}$

suggesting that we may be on the right track in concentrating on the texture zero structure.

The symmetry giving rise to Froggart Nielsen structure may be an additional gauge symmetry, a global symmetry or a discrete symmetry. In the case of a global symmetry there may be problems associated with the appearance of Goldstone bosons when the symmetry is broken and with possibly large gravitational corrections due to wormhole effects. The latter are known to be absent if the symmetry is a local gauge symmetry, either continuous or discrete. Here I will only consider the continuous possibility further restricted to the simplest possibility namely that there is an additional Abelian gauge symmetry beyond the Standard Model gauge group and that it is broken at a very high scale so that the phenomenology of the Standard Model is essentially unchanged. The example applies to the Supersymmetric version of the Standard Model (if the reader is unfamiliar with this, read on. It is covered in a later section) although the techniques apply also to non-supersymmetric schemes.

It turns out to be remarkably easy to construct a model with one of these mass matrices through the introduction of an abelian gauge symmetry, $U(1)$ (such additional symmetries abound in string theories). The most general charge assignment of the Standard Model states is given in Table 3. This follows since the need to preserve

Table 3. $U(1)_{FD}$ symmetries.

	Q_i	u_i^c	d_i^c	L_i	e_i^c	ν_i^c	H_2	H_1
$U(1)_{FD}$	α_i	α_i	α_i	a_i	a_i	a_i	$-2\alpha_1$	$-2\alpha_1$

$SU(2)_L$ invariance requires (left-handed) up and down quarks (leptons) to have the same charge. This plus the requirement of symmetric matrices then requires that all quarks (leptons) of the same i-th generation transform with the same charge $\alpha_i(a_i)$. If the light Higgs, H_2, H_1, responsible for the up and down quark masses respectively have $U(1)$ charge so that only the (3,3) renormalisable Yukawa coupling to H_2, H_1 is allowed, only the (3,3) element of the associated mass matrix will be non-zero as desired. The remaining entries are generated when the $U(1)$ symmetry is broken. We assume this breaking is spontaneous via Standard Model singlet fields, θ, $\bar{\theta}$, with $U(1)_{FD}$ charge -1, +1 respectively, which acquire vacuum expectation values (vevs), $<\theta>$, $<\bar{\theta}>$,along a "D-flat" direction (This is discussed in more detail shortly). After this breaking all entries in the mass matrix become non-zero. For example, the (3,2) entry in the up quark mass matrix appears at $O(\epsilon^{|\alpha_2-\alpha_1|})$ because U(1) charge conservation allows only a coupling $c^c t H_2(\theta/M_2)^{\alpha_2-\alpha_1}$, $\alpha_2 > \alpha_1$ or $c^c t H_2(\bar{\theta}/M_2)^{\alpha_1-\alpha_2}$, $\alpha_1 > \alpha_2$ and we have defined $\epsilon = (<\theta>/M_2)$ where M_2 is the unification mass scale which governs the higher dimension operators. As discussed in reference[8] one may expect a different scale, M_1 for the down quark mass matrices (it corresponds to mixing in the H_2, H_1 sector with M_2, M_1 the masses of heavy H_2, H_1 fields). Thus we arrive at mass matrices of the form

$$\frac{M_u}{m_t} \approx \begin{pmatrix} h_{11}\rho_{11}\epsilon_a^{|2+6a|} & h_{12}\rho_{12}\epsilon_b^{|3a|} & h_{13}\rho_{13}\epsilon_a^{|1+3a|} \\ h_{21}\rho_{21}\epsilon_b^{|3a|} & h_{22}\rho_{22}\epsilon^2 & h_{23}\rho_{23}\epsilon^1 \\ h_{31}\rho_{31}\epsilon_a^{|1+3a|} & h_{32}\rho_{32}\epsilon^1 & h_{33} \end{pmatrix} \tag{19}$$

$$\frac{M_d}{m_b} \approx \begin{pmatrix} k_{11}\sigma_{11}\bar{\epsilon_a}^{|2+6a|} & k_{12}\sigma_{12}\bar{\epsilon_b}^{|3a|} & k_{13}\sigma_{13}\bar{\epsilon_a}^{|1+3a|} \\ k_{21}\sigma_{21}\bar{\epsilon_b}^{|3a|} & k_{22}\sigma_{22}\bar{\epsilon}^{2} & k_{23}\sigma_{23}\bar{\epsilon}^{1} \\ k_{31}\sigma_{31}\bar{\epsilon_a}^{|1+3a|} & k_{32}\sigma_{32}\bar{\epsilon}^{1} & k_{33} \end{pmatrix} \tag{20}$$

where $\bar{\epsilon} = (\frac{<\theta>}{M_1})^{|\alpha_2-\alpha_1|}$, $\epsilon = (\frac{<\theta>}{M_2})^{|\alpha_2-\alpha_1|}$, and $a = (2\alpha_1 - \alpha_2 - \alpha_3)/3(\alpha_2 - \alpha_1)$. For $-3a > 1$ $\epsilon_a = \epsilon_b = \epsilon$ and $\bar{\epsilon}_a = \bar{\epsilon}_b = \bar{\epsilon}$. In this case it is easy to check that there are *no* texture zeros because all matrix elements contribute at leading order to the masses and mixing angles. For $1 > -3a > 0$, ϵ_a, $\bar{\epsilon}_a$ change and are given by $\bar{\epsilon_a} = (\frac{<\bar{\theta}>}{M_1})^{|\alpha_2-\alpha_1|}$, $\epsilon_a = (\frac{<\bar{\theta}>}{M_2})^{|\alpha_2-\alpha_1|}$. In this case texture zeros in the (1,1) and (1,3) positions *automatically* appear for small $< \bar{\theta} >$. However the (1,2) matrix element is too large (cf Table 2). For $a > 0$ however $\bar{\epsilon}_{a,b} = (\frac{<\bar{\theta}>}{M_1})^{|\alpha_2-\alpha_1|}$, $\epsilon_{a,b} = (\frac{<\bar{\theta}>}{M_2})^{|\alpha_2-\alpha_1|}$, the texture zeros in the (1,1) and (1,3) positions persist, and the (1,2) matrix element can be of the correct magnitude.

Thus we see that the simplest possibility of an additional $U(1)$ gauge family symmetry *requires* texture zeros in the phenomenologically desirable positions for a large range of the single relevant free parameter, a. In addition it generates Froggart Nielsen structure for the other matrix elements which can duplicate the required hierarchical structure of masses and mixing angles. After mixing the light Higgs states are given by $H_{33}^2 + \sum \rho_{ij} H_{ij}^2 \epsilon_{a,\,,b}^{n_{ij}}$ and $H_{33}^1 + \sum \sigma_{ij} H_{ij}^1 \bar{\epsilon}_{a,\,,b}^{n_{ij}}$ where the powers n_{ij} are those appearing in eq(20) and ρ, σ are related to Yukawa couplings in the Higgs sector in a manner discussed below. These and the Yukawa couplings h_{ij}, k_{ij} are all assumed to be of $O(1)$.

As discussed above, for $a > 0$, there are two approximate texture zeros in the (1,1) and (1,3), (3,1) positions. These give rise to excellent predictions for two combinations of the CKM matrix. The magnitude of the remaining matrix elements is sensitive to the magnitude of a and the values of the expansion parameters. Then choosing $a = 1$ the remaining non-zero entries have magnitude in excellent agreement with the measured values. From Table 2 we see that to a good approximation we have the relation [8]

$$\epsilon = \bar{\epsilon}^2 \tag{21}$$

which also implies that $M_2 > M_1$.

The charged lepton mass matrix may similarly be determined. In [8] we restricted the lepton charges by requiring the good relation $m_b = m_\tau$ at unification scale $\alpha_1 = a_1$ giving

$$\frac{M^l}{m_\tau} \approx \begin{pmatrix} l_{11}\sigma_{11}\bar{\epsilon}^{|2+6a+2b|} & l_{12}\sigma_{12}\bar{\epsilon}^{|3a|} & l_{13}\sigma_{13}\bar{\epsilon}^{|1+3a+b|} \\ l_{21}\sigma_{21}\bar{\epsilon}^{|3a|} & l_{22}\sigma_{22}\bar{\epsilon}^{|2(1+b)|} & l_{23}\sigma_{23}\bar{\epsilon}^{|1+b|} \\ l_{31}\sigma_{31}\bar{\epsilon}^{|1+3a+b|} & l_{32}\sigma_{32}\bar{\epsilon}^{|1+b|} & l_{33} \end{pmatrix} \tag{22}$$

where $b = (a_2 - \alpha_2)/(\alpha_2 - \alpha_1)$ and again the Yukawa couplings, l_{ij}, are assumed of $O(1)$. At this stage b is not determined but for any value of $b > 0$ there are texture zeros in the $(1,1)$ and $(1,3)$ positions giving rise to the phenomenologically successful prediction $Det(M^l) \approx Det(M^d)$. To proceed further a value of b must be chosen and two viable choices were explored. For $b = 0$ the lepton charges are the same as the down quark sector, and so the structure of the down quark and lepton mass matrices are identical. In order to explain the detailed difference between down quark and lepton masses it is necessary in this case to assume that the constants of proportionality determined by Yukawa couplings which we have so far taken to be equal (and of O(1)) differ slightly for the lepton case. A factor 3 in the (2,2) entry is sufficient to give excellent charged lepton masses. An alternative which does not rely on different Yukawa couplings is to choose b half integral. For $a = 1$, $b = 1/2$ we found excellent agreement for the charged

Table 4. Anomaly-free $U(1)_{FI}$ symmetries.

	Q	u	d	L	e	H_2	H_1
$U(1)_H$	0	0	0	0	0	1	-1
$U(1)_{XX}$	0	0	1	1	0	0	0
$U(1)_X$	1	1	0	0	1	0	0

Table 5. Anomaly-free $U(1)$ symmetries.

$U(1)$	$\alpha_i + x$	$\alpha_i + x$	$\alpha_i + y$	$a_i + y$	$a_i + x$	z-$2\alpha_1$	-z+$w\alpha_1$

lepton masses (in this case there is a Z_2 symmetry forcing the $(1,3)$, $(3,1)$, $(2,3)$, $(3,2)$ matrix elements to vanish). The extension of these ideas to include neutrino masses is straightforward[10].

3.1 Anomaly Cancellation and the prediction of the weak mixing angle

To complete the discussion we turn to a consideration of the anomaly structure of the model. Although the choice of charges for the quarks and letons given above has no $SU(3)^2U(1)_{FD}$, $SU(3)^2U(1)_{FD}$ or $U(1)^2U(1)_{FD}$ anomalies it is clear that the Higgs charges are not anomaly free. However in theories derived from a string theory there is a significant new possibility for a non-vanishing anomaly associated with a new $U(1)$ gauge factor can be cancelled by the Green-Schwarz (GS) anomaly cancellation mechanism[11, 12]. In the 4-D version of the GS mechanism one cancels the anomalies of a single $U(1)$ by an appropriate shift of the axion present in the dilaton multiplet of four-dimensional strings. This happens because such an axion has a direct coupling to $F\tilde{F}$.

For the GS mechanism to be possible, the coefficients A_i, $i = 3, 2, 1$ of the mixed anomalies of the $U(1)$ with $SU(3)$, $SU(2)$ and $U(1)_Y$ have to be in the ratio $A_3 : A_2 : A_1 = k_3 : k_2 : k_1$. Here k_i are the Kac-Moody levels of the corresponding gauge factors and they determine the boundary condition of the gauge couplings at the string scale by the well-known equation $g_3^2 k_3 = g_2^2 k_2 = g_1^2 k_1^2$. The usual (e.g. GUT) canonical values for these normalization factors (corresponding to the successful result $sin^2(\theta_W) = 3/8$) yield $k_3 : k_2 : k_1 = 1 : 1 : 5/3$ and hence the GS mechanism can only work in this case if the mixed anomalies of the $U(1)$ with the SM gauge factors are in the ratio $A_3 : A_2 : A_1 = 1 : 1 : 5/3$[11].

One can easily convince oneself that there are only two $U(1)$ symmetries with anomalies consistent with this ratio of gauge coupling constants, namely $U(1)_X$ and $U(1)_{XX}$ given in Table 4 .

Thus, in a supersymmetric SM coming from a string the most general family-independent anomaly-free $U(1)$ consistent with canonical gauge coupling unification is given by:

$$U(1)_{FI} \;=\; z\, U(1)_H \;+\; x\, U(1)_X \;+\; y\, U(1)_{XX} \,. \tag{23}$$

As the $U(1)_{FI}$ is family blind one may use any combination of the $U(1)_{FD}$ and the $U(1)_{FI}$ currents while still maintaining the structure for the fermion masses discussed above. he full charges of this $U(1)$ factor may now be determined using Tables 3 and 4 and give the charges shown in Table 5.

One can easily check that for $z = -2x$ one gets the results of 19 for the u-quark mass matrix. If one further has $3x + y = -4\alpha_1$ (and $w = -2$) one gets the results of eq.20 for the d-quark masses. Note, however, the choice of the flavour-independent component allows for further possibilities for the d-quark matrices. Thus we conclude that the generic problem raised by anomaly cancellation may naturally be solved in the context of string based models.

It is important to recall that the $U(1)$s whose anomalies are cancelled through a GS mechanism are *necessarily spontaneously broken not far below the string scale.* The reason for this is that the piece in the Lagrangian cancelling the anomalies has a supersymmetric counterpart which is a sort of field-dependent Fayet-Iliopoulos term for the $U(1)$. This term forces $U(1)$ symmetry breaking in a natural way at a scale of order $1/\sqrt{192\pi} M_{string}$[11, 12]. Thus the present scheme also explains why the extra $U(1)$ symmetry required to generate the fermion mass patterns does not survive down to low energies.

So far we have assumed that $sin^2(\theta_W) = 3/8$ at the string unification scale. In fact an acceptable pattern of fermion masses actually *requires* this value! To see this let us compute the mixed anomalies A_i for the $U(1)_{FD}$ symmetry of table 3 with $SU(3), SU(2)$ and $U(1)_Y$. One finds respectively:

$$\begin{aligned} A_3 &= 2\sum_{i=1}^{3}\alpha_i \\ A_2 &= \frac{3}{2}\sum_{i=1}^{3}\alpha_i + \frac{1}{2}\sum_{i=1}^{3}a_i + \frac{\alpha_1}{2}(w-2) \\ A_1 &= \frac{11}{6}\sum_{i=1}^{3}\alpha_i + \frac{3}{2}\sum_{i=1}^{3}a_i + \alpha_1(w-2) \end{aligned} \tag{24}$$

where, as we are working with the full $U(1)$ charges, we no longer have $\alpha_3 = -(\alpha_1+\alpha_2)$ and $a_3 = -(a_1+a_2)$ as the $U(1)_{FI}$ piece adds a family independent term to α_i and also to a_i. However, to maintain the result $m_b = m_\tau$, this term is the same for α_i and for a_i and so

$$\sum_{i=1}^{3}\alpha_i = \sum_{i=1}^{3}a_i \tag{25}$$

Now, in order that $A_3 = A_2$ so that the $SU(3)$ and $SU(2)$ couplings are unified eq(24) gives

$$w = 2 \quad or\ \alpha_1 = 0 \tag{26}$$

$$and\ A_3 : A_2 : A_1 = 1 : 1 : \frac{5}{3} \tag{27}$$

The second equation requiring $sin^2(\theta_W) = 3/8$ may be seen as a *consequence* of an acceptable mass matrix structure. The first condition in eq(27) requires $\alpha_1 = 0$ since $w = 2$ is no good as it leads to very bad results for the d-quark masses.

Thus we see that a simple $U(1)_{FD}$ extension of the standard model generates much of the quark and lepton mass matrix structure through the constraints of anomaly cancellation, but that the full $U(1)$ extension including the Higgs doublets needed for the MSSM is only anomaly free in the context of string theory via a Green Schwarz

Table 6. Anomaly-free $U(1)$ gauge symmetry giving rise to the textures in eqs(A,B,C)

$U(1)_F$	Q	u	d	L	e	H_1	H_2	θ	$\bar{\theta}$
3rd generation	0	0	0	0	0	0	0	1	-1
2nd generation	1	1	1	-1/2	-1/2				
1st generation	-4	-4	-4	-5/2	-5/2				

term. The charges of each individual particle with respect to this anomaly-free $U(1)$ are shown in Table 4.

The simplicity of the assignments is remarkable. It is also worth to emphasise that this $U(1)$ symmetry may be made anomaly free through the GS mechanism *if and only if the normalization of the coupling constants is the canonical one* $g_3^2 = g_2^2 = 5/3g_1^2$ yielding the successful prediction $sin^2\theta_W = 3/8$. Thus the present scheme not only predicts a successful pattern of fermion masses and mixings but also predicts $sin^2\theta_W = 3/8$ even without any grand unification group.

To summarise we have found it very straightforward to construct an Abelian family symmetry capable of generating both the texture zeros and the required hierarchical structure for the remaining mass matrix elements. For the simple choice of the simple Abelian family symmetry discussed above the mass matrix is given by

$$\frac{M^{u,d}}{m_{t,b}} = \begin{pmatrix} 0 & c\varepsilon_{u,d}^3 & o \\ c'\varepsilon_{u,d}^3 & b\varepsilon_{u,d}^2 & a\varepsilon_{u,d} \\ 0 & a'\varepsilon_{u,d} & 1 \end{pmatrix} \tag{28}$$

where the non-zero entries have been expanded in a power series in the expansion parameters $\varepsilon_u \approx \varepsilon_d^2 \approx 0.04$. The quantities a, b and c are constants of order 1.

Essentially all attempts to explain the pattern of light, quark and lepton masses and mixing angles rely on some spontaneously broken family symmetry of this type to organise the magnitude of the mass matrix elements in the manner given in eq 28, most involving an Abelian family symmetry [6, 8, 9]. In string theories such symmetries abound and have been used to generate fermion mass structure [13]. Such a symmetry can generate an hierarchy of masses and mixing angles in powers of $< \theta >$ but does not determine the coefficients a, b and c of order one which are needed if one is to obtain quantitative predictions for masses. Two ways of going beyond this have been investigated. One possibility follows if there is "strong unification" for then the Yukawa couplings and hence the coefficients a, b and c may be determined by the infra red fixed point structure of the low-energy theory [2, 3] A second possibility is to invoke a stage of Grand Unification relating the quark and lepton coefficients of a given generation [14]. This may be extended to a non-Abelian family symmetry to try to relate the coefficients involving different generations. I will consider both of these possibilities in turn.

4 Infra-red-fixed-points

Just as the gauge coupling "run" on going from the unified scale to low energies so too do the Yukawa couplings responsible for fermion masses. It is reasonable therefore to ask if there is some systematic behaviour in this running. To illustrate this let us consider the Standard Model in the simple limit where we keep only the top Yukawa coupling and the gauge couplings

$$8\pi^2\frac{dln(h_t)}{dt} = 8g_3^2 + \frac{3}{4}(3g_2^2 + g_1^2) + \frac{2}{3}g_1^2 - \frac{9}{2}h_t^2 \tag{29}$$

where $t = ln(\frac{\mu_0^2}{\mu^2})$ and μ_0 and μ are the initial and final scales at which the couplings are determined. In the limit where h_t is much smaller than the gauge couplings we see that only the positive terms on the RHS remain and these cause the mass to grow as one goes down in energy from the unification scale. This is the case relevent to light quarks and was used in the phenomenologically successful prediction of m_b/m_τ in $SU(5)$ where there gauge corrections split the predicted degeneracy of the b and the τ at the unification scale. However in the case of the top quark the Yukawa coupling cannot be neglected relative to the gauge couplings and something very interesting happens in the flow to the infra-red. To simplify the discussion let us keep only the largest (QCD) gauge coupling. In this case the equation has a fixed point structure which relates the top Yukawa coupling to the QCD coupling g_i. To see this we keep only the largest (QCD) gauge coupling and write a renormalisation group equation for the *ratio* of the Yukawa to the gauge coupling

$$8\pi^2\frac{dln(\frac{h_t}{g_3})}{dt} = g_3^2 - \frac{9}{2}h_t^2 \tag{30}$$

This clearly has an infra-red stable fixed point (IRFP) at [15]

$$(h_t^2)^* = \frac{2}{9}g_3^2 \tag{31}$$

for which the RHS of eq(30) vanishes and thus the ratio remains unchanged. This then suggests that the Yukawa coupling is predicted in termsof the gauge coupling irrespective of the intal value at the unification scale because, as may be seen from eq(30), the ratio is always driven towards the IRFP as one goes to lower energy. However, as stressed by Chris Hill [16], this fixed point value is not reached for large initial values of the top quark coupling because the range in t as μ varies between between the Planck scale and the electroweak scale is too small to cause the trajectories closely to approach the fixed point. Rather Hill showed a "Quasi fixed point" governs the value of h_t for large initial values of h_t. To exhibit this it is useful to provide a complete analytic solution to eq(30).

4.1 Analytic solution of the renormalisation group equations

We first solve the general form of the renormalisation group equations for the case of a single dominant Yukawa coupling and several gauge couplings

$$\begin{aligned}\frac{dg_i^2}{dt} &= -\frac{b_i g_i^4}{(4\pi)^2}\\ \frac{dY_t}{dt} &= Y_t(\sum_i r_i\tilde{\alpha}_i - sY_t)\end{aligned} \tag{32}$$

where

$$\tilde{\alpha_i} = \frac{g_i^2}{(4\pi)^2}$$

$$Y_t = \frac{h_t^2}{(4\pi)^2}$$
$$\beta_i = \tilde{\alpha}_i(0) b_i \tag{33}$$

The solution to these equations is [17]

$$\tilde{\alpha}_i(t) = \frac{\tilde{\alpha}_i(0)}{(1+\beta_i t)}$$
$$Y_t(t) = \frac{Y_t(0)E_1(t)}{1+sY_t(0)F_1(t)} \tag{34}$$

where

$$E_1(t) = \Pi_i(1+\beta_i t)^{(B_i-1)}$$
$$F_1(t) = \int_0^t E_1(t')dt' \tag{35}$$

and

$$B_i = \frac{r_i}{b_i} + 1 \tag{36}$$

4.2 Fixed point structure

In the Standard Model if we keep only the dependence on the largest gauge coupling, g_3, the equations have an infra-red-fixed point. In this instance we have

$$E_1(t) = (1+\beta_3 t)^{(B_3-1)}$$
$$F_1(t) = \frac{(1+\beta_3 t)^{B_3}}{B_3\beta_3} - \frac{1}{B_3\beta_3} \tag{37}$$

The solutions presented in the previous section do not make explicit the fixed point structure. To do this it is necessary to eliminate the functions E_1 and F_1 in eq(34) giving

$$\frac{Y_t(t)}{\tilde{\alpha}_3(t)} = \frac{Y_t(0)}{\tilde{\alpha}_3(0)} \frac{(\frac{\alpha_i(0)}{\alpha_i(t)})^{B_3}}{1+\frac{sY_t(0)}{B_3\beta_3}((\frac{\alpha_i(0)}{\alpha_i(t)})^{B_3}-1)} \tag{38}$$

After some algebra this reduces to

$$\frac{Y_t(t)}{\tilde{\alpha}_3(t)} = \left(\frac{Y_t}{\tilde{\alpha}_3}\right)^* \frac{1}{1+(\frac{\alpha_i(t)}{\alpha_i(0)})^{B_3}(\frac{Y_t}{\tilde{\alpha}_3})^*(\frac{\tilde{\alpha}_3(0)}{Y_t(0)}-(\frac{\tilde{\alpha}_3}{Y_t})^*)} \tag{39}$$

where the star superscript denotes the fixed point value

$$\left(\frac{Y_t}{\tilde{\alpha}_3}\right)^* = \frac{B_3 b_3}{s} \tag{40}$$

In this form it is clear that as $t \to \infty$ the ratio of the Yukawa coupling to the gauge coupling tends to its fixed point value because $(\frac{\alpha_i(t)}{\alpha_i(0)})^{B_3} \to 0$. In the case of the Standard Model

$$b_3 = -7$$
$$r = 8$$
$$s = \frac{9}{2} \tag{41}$$

giving $(\frac{Y_t}{\tilde{\alpha}_3})^*_{SM} = \frac{2}{9}$ and $B_3 = -\frac{1}{7}$. The smallness of the power B_3 in eq(39) together with the slow evolution of the QCD coupling in the range M_X to m_t means that in practice the fixed point is not reached. If we take $M_X = O(10^{16}GeV)$ then $(\frac{\alpha_i(t)}{\alpha_i(0)})^{B_3} \approx 0.8$ so we are quite far away from the true fixed point. This means that, in general, one will be sensitive to the initial value of the top Yukawa coupling. For example if $\frac{Y_t(0)}{\tilde{\alpha}_3(0)} = (\frac{Y_t}{\tilde{\alpha}_3})^*$ then it will remain at its fixed point. On the other hand if $\frac{Y_t(0)}{\tilde{\alpha}_3(0)} >> (\frac{Y_t}{\tilde{\alpha}_3})^*$ then we may see from eq(39) that the value at low scales are relatively insensitive to the initial value giving a "Quasi-fixed-point" value which is really just the fixed point value radiatively corrected by known gauge boson contributions

$$\left(\frac{Y_t}{\tilde{\alpha}_3}\right)^{QFP} = \frac{(\frac{Y_t}{\tilde{\alpha}_3})^*}{(1 - (\frac{\alpha_i(t)}{\alpha_i(0)})^{B_3})} \tag{42}$$

Again using $M_X = O(10^{16}GeV)$ we find the Quasi-fixed-point gives a value for the top quark Yukawa coupling and hence the top quark mass approximately twice the true fixed point value, 220GeV rather than 110Gev[2]. In addition to the QCD gauge corrections just discussed which correct for the fact that the RG flow is over a relatively small distance there are also significant corrections due to the $SU(2) \otimes U(1)$ gauge interactions. In this case the renormalisation group equations eq(32) do not have an exact infra-red-fixed point and we must use the full solution to the renormalisation group equations eq(34). Including these effects the quasi fixed point for the top pole mass is $240GeV$.

In the case of the minimal supersymmetric standard model (MSSM) we have [17, 19]

$$\begin{aligned} b_3 &= -3 \\ r &= \frac{16}{3} \\ s &= 6 \end{aligned} \tag{43}$$

resulting in a larger value for $(\frac{Y_t}{\tilde{\alpha}_3})^*_{MSSM} = \frac{7}{18}$ and B_3, $B_3 = -\frac{7}{9}$. In this case $(\frac{\alpha_i(t)}{\alpha_i(0)})^{B_3} \approx 0.46$ giving a somewhat closer approach to the true fixed point; the Quasi-fixed-point is approximately a third greater than the true fixed point at $155 sin\beta$ where $tan\beta$ is the ratio of the two Higgs vacuum expectation values of the MSSM. Of course the interesting question is whether the top mass is determined by this (quasi) fixed point behaviour. Until the value of $\sin\beta$ is determined this is not known but clearly the large value predicted by the quasi-fixed-point is needed to accommodate a top mass of $O(174GeV)$.

5 Grand Unification

The original suggestion that there might be an underlying Grand Unified gauge field theory beyond Standard Model was triggered by the observation of Georgi Quinn and Weinberg[20] that, although the strong electromagnetic and weak interactions are quite different at low energies the couplings, if continued to high energies, approach each other. This could be explained if there was an underlying simple gauge group with a single coupling constant from which the Standard Model emerged after symmetry breaking -Grand Unification[21].

[2] All masses quoted here are pole masses, related to running masses by a simple correction [18].

5.1 $SU(5)$

The prototype Grand Unified Theory is based on the group $SU(5)$ a rank 5 group with just enough neutral generators to accommodate those of the rank 5 Standard Model[22]. In $SU(5)$ the states of a single family are accommodated in just 2 representations $\bar{5}$ and a 10. Thus the representaion content of the Standard Model is simplified. The assignment to $\bar{5}$ and 10 are shown in eqn(44)

$$SU(5) \supset SU(3) \otimes SU(2) \otimes U(1)$$

$$\psi_{\bar{5}} = \begin{bmatrix} d_1^c \\ d_2^c \\ d_3^c \\ e^- \\ \nu_e \end{bmatrix}_L$$

$$\chi^{10} = \frac{1}{\sqrt{2}} \begin{bmatrix} 0 & u_3^c & -u_2^c & -u^1 & -d^1 \\ -u_3^c & 0 & u_1^c & -u^2 & -d^2 \\ u_2^c & -u_1^c & 0 & -u^3 & -d^3 \\ u^1 & u^2 & u^3 & 0 & e^+ \\ d^1 & d^2 & d^3 & e^+ & 0 \end{bmatrix}_L \tag{44}$$

where c denotes the charge conjugate and the numerical indices are $SU(3)$ indices. It may be seen that the quarks and leptons belong to the same multiplet. Of course $SU(5)$ must be broken and it is important to consider the pattern of the symmetry breaking leading to the Standard Model. In the case of $SU(5)$, this is shown in eqn(45).

$$SU(5) \underset{<\Sigma_{24}^A>}{\longrightarrow} SU(3) \otimes SU(2) \otimes U(1) \underset{<H_5>}{\longrightarrow} SU(3) \otimes U(1) \tag{45}$$

where one can see that the single 24-dimensional adjoint representation, Σ, leads to the breaking of $SU(5)$ immediately to the Standard Model $SU(3) \otimes SU(2) \otimes U(1)$ and subsequently the electroweak breaking is triggered by a doublet contained in a 5-dimensional Higgs representation, H_5. The most interesting prediction following from $SU(5)$ follows because there is just a single gauge coupling. Thus the weak mixing angle which is given in terms of the ratio of the two coupling constants associated with $SU(2) \otimes U(1)$ is determined. The prediction is[3]

$$\sin\theta_W{}^2 = \frac{T_r(T_{3L}^2)}{T_r(Q^2)} = 3/8 \tag{46}$$

but this applies at the unification scale, M_X. As we shall discuss including radiative corrections to the couplings to run them down to low scales one finds that the 3/8 initial value leads to quite acceptable values at laboratory energies for the weak mixing angle but only in a supersymmetric GUT.

The Yukawa couplings of the theory are also restricted by $SU(5)$ in a way that gives predictions for fermion masses. This may be seen from the equation

$$L_Y = h\bar{\psi}_{5R}^p \chi_{p,qL}^{10} H_5^q \Rightarrow m_e = m_d = h < h_5^5 > \tag{47}$$

The mass of electron is equal to the mass of down quark at the unification scale. The same is true for the heavier generations. As we will discuss in the last lecture

[3] I leave it to the reader to derive this form!

these predictions are not realistic apart from that of the third generation, so some modification is needed. It is possible to generate such a modifications for example through the coupling

$$L'_Y = h'\bar{\psi}^p_R H^{sq}_{45p}\chi_{sqL} \Rightarrow m_q = 3m_e \tag{48}$$

where H_{45} is a 45 dimensional representation of Higgs fields. The Yukawa coupling generates a relation between quark masses which says that the electron would be 1/3 of the down quark mass a much more acceptable relation.

6 Supersymmetric Unification

Grand Unification offers a very plausible way to relate quarks and leptons and to make predictions for couplings in the Standard Model which may shed light on the pattern of quark and lepton masses. However it suffers from a very serious problem. Because it introduces a very high scale of unification it introduces an hierarchy problem in that it is difficult to separate the electroweak scale from the GUT scale. Even if this is arranged at tree level the problem is reintroduced through radiative corrections. The only way to solve this problem is to introduce a new symmetry, supersymmetry, which protects the hierarchy.

I find it instructive to introduce the need for supersymmetry simply starting from the assumption that the Standard Model can be viewed as an effective field theory following from some unified theory at a high scale, M_X, the Grand Unified scale or perhaps the Planck scale. Following from this very simple assumption, one can immediately derive various properties about the effective theory. Firstly the theory should be built from a local gauge invariance principle for only such a symmetry can guarantee the existence of light vector gauge bosons. This is just the structure that is found in the Standard Model. Secondly the theory should have chiral invariance in the sense that the left-handed state of any fermions in the theory should have different symmetry properties from the right-handed fermions. In this case the lightness of the fermion is guaranteed because fermion mass terms can arise only at the scale at which the symmetry is broken. Again such a structure is observed in the Standard Model; the left-handed states transform as doublets under the SU(2) gauge group while the right-handed state are singlets. The fermion masses arise only at the stage of SU(2) breaking. Finally one should have a principle which guarantees that the Higgs scalars needed for electroweak symmetry breaking remain light. However, in the Standard Model, mass terms for Higgs scalars are allowed and one would naively expect them to be at the highest possible scale, M_X. In order to maintain light Higgs in a natural way one has to extend the symmetries of the Standard Model to include supersymmetry. Of course we know that the supersymmetry can not be exact because otherwise, for example, the supersymmetric partners of quarks, squarks, would have been seen. However, we also know that these new superpartners cannot be too heavy. In particular radiative corrections will generate a mass squared for the Higgs-boson proportional to the difference between the super partner mass squared and that of the normal state, the constant of proportionality involving the appropriate coupling squared. Therefore the condition that the Higgs scalar should be of the order of the electroweak scale places a condition on the mass difference. This is the only evidence we have for low energy supersymmetry. Quantifying this we arrive at the values of Table 7 showing the naturalness bounds (i.e. assuming no large "unnatural" cancellations occur in the calculation) on the mass of SUSY states [23, 24] together with the most "natural" value [24].

Once one has extended the Standard Model to include low energy supersymmetry,

Table 7. Naturalness bounds on the mass of SUSY states, the lightest and next-to-lightest (LSP and NLSP), the chargino, $C^{\pm}$, the gluino, $\tilde{g}$, and the lightest squark the stop $\tilde{t}$. The upper bound corresponds to a fine tuning by a factor 10 [1]. Other measures of naturalness are ($1\frac{1}{2}$ - 2) times weaker [2].

SUSY state	Upper bound	Most "natural"
LSP	50GeV	40 GeV
NLSP	90GeV	
$C^{\pm}$	90GeV	55GeV
$\tilde{g}$	370GeV	270 GeV
$\tilde{t}$	300GeV	220 GeV

Table 8. Flowchart showing the steps involved in the construction of a supersymmetric extension of the Standard Model.

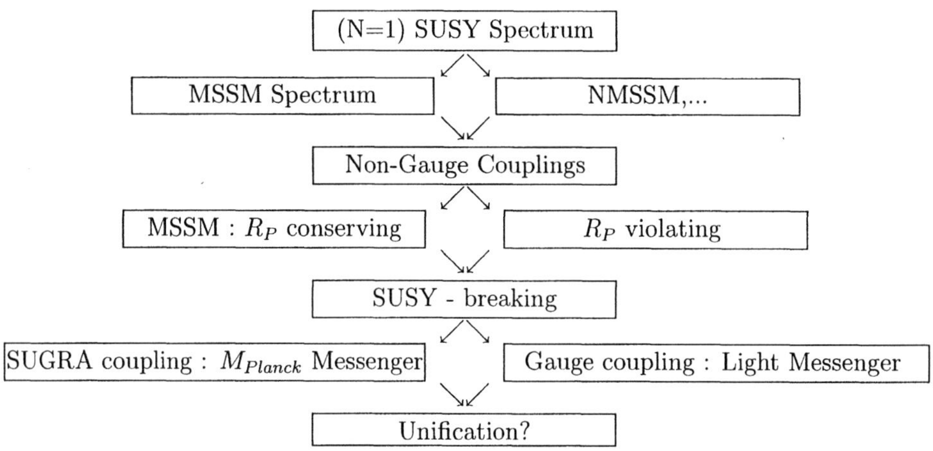

$$\begin{pmatrix} g_\mu^{a=1..8} \\ \widetilde{g}^{a=1..8} \end{pmatrix} + \begin{pmatrix} W_\mu^{i=1..3} \\ \widetilde{W}^{i=1..3} \end{pmatrix} + \begin{pmatrix} B_\mu \\ \widetilde{B} \end{pmatrix}$$

$$\begin{pmatrix} q_L \\ \widetilde{q}_L \end{pmatrix} + \begin{pmatrix} u_L^c \\ \widetilde{u}_L^c \end{pmatrix} + \begin{pmatrix} d_L^c \\ \widetilde{d}_L^c \end{pmatrix} + \begin{pmatrix} l_L \\ \widetilde{l}_L \end{pmatrix} + \begin{pmatrix} E_L^c \\ \widetilde{E}_L^c \end{pmatrix}$$

$$\begin{pmatrix} \widetilde{H}_1 \\ H_1 \end{pmatrix} + \begin{pmatrix} \widetilde{H}_2 \\ H_2 \end{pmatrix}$$

Table 9: The MSSM multiplet content. Standard Model states have no tilde. The first line shows the gauge bosons and their fermionic partners the gauginos. The second line shows the quarks and leptons together with their scalar partners the squarks and sleptons. Finally the Higgs sector is shown with two Higgs doublets together with their fermion partners the Higgsinos

one is able to build a fully unified theory at the very high scale without introducing any fine tuning problems. One hopes that such a scheme will be capable of explaining the origin of the symmetries, the multiplet structure and the couplings of the Standard Model leading to the final "Theory of Everything".

6.1 Supersymmetric Extensions of the Standard Model

As an introduction I show in Table 8 the steps involved in building a supersymmetric extension of the Standard Model [25]. There are various choices to be made. First of all the states of Standard Model must be assigned to supermultiplets. The choice of multiplet structure in the minimal supersymmetric Standard Model (the MSSM), is given in Table 9. In this table you will see that the gauge bosons have fermion partners, the gauginos. The quarks and leptons have scalar (spin 0) partners, the squarks and sleptons. The Higgs boson has a fermionic partner, the Higgsino. Moreover, due to the necessity to cancel anomalies it is impossible to work with a single Higgs doublet. A second Higgs doublet with opposite hypercharge has to be introduced with its Higgsino counterpart. Of course this minimum multiplet structure may not be the one realised in nature. For example, one of the fundamental couplings in the theory, the so-called μ term, shown in eq (49), may arise as a result of a vacuum expectation value for a new singlet field $\widetilde{N}$, the scalar component of the chiral superfield N.

$$\begin{aligned} [\mu \quad H_1 \quad H_2]_F \rightarrow [N \quad H_1 \quad H_2]_F \\ \mu = < \widetilde{N} > \end{aligned} \tag{49}$$

In this case MSSM is extended to include this additional Higgs fields and its Higgsino counterpart.

Once a multiplet structure is chosen, the gauge boson and gaugino couplings are specified by the gauge principle. In addition there are associated quartic scalar interactions. Of these the most interesting is that involving the Higgs giving a quartic coupling of the form (for the neutral components)

$$\frac{g_1^2 + g_2^2}{8}(|H_1^0|^2 - |H_2^0|^2)^2 \tag{50}$$

In the Standard Model this coupling is not predicted. In the supersymmetric extension, as we shall discuss, the fact that it is determined in terms of known gauge couplings leads to an important prediction for the Z mass when studying the electroweak symmetry breaking pattern.

Having specified the gauge sector, the next step in Table 8 is the specification of the non-gauge couplings. Again there is more than one choice. In the MSSM the couplings are those shown in eq(51)

$$\begin{aligned} l^{Yuk}_{MSSM} = & h^u_{ij}\left[q_i H_2 u^c_j\right]_F + h^d_{ij}\left[q_i H_1 d^c_j\right]_F \\ & + h^l_{ij}\left[l_i H_1 E^c_j\right]_F + \mu\left[H_1 H_2\right]_F \end{aligned} \tag{51}$$

Each of these terms corresponds to a series of couplings involving the primary states. For example the first term leads to the three separate Yukawa couplings shown in eq(52)

$$\begin{aligned} & h_{ij}(\overline{u} \quad \overline{d})_i \begin{pmatrix} H_2^0 \\ H_2^- \end{pmatrix} u_j + h^u_{ij}(\overline{u} \quad \overline{d})_i \\ & \begin{pmatrix} \widetilde{H}_2^0 \\ \widetilde{H}_2 \end{pmatrix} \widetilde{u}_j + h^u_{ij}(\overline{\widetilde{u}} \quad \overline{\widetilde{d}})_i \begin{pmatrix} \widetilde{H}_2^0 \\ \widetilde{H^-}_2 \end{pmatrix} u_j \end{aligned} \tag{52}$$

plus "F" terms involving quartic couplings [25]. The minimal set of couplings in the MSSM respect what is known as R- parity (a Z_2) symmetry given by eq(53).

$$R_p \equiv (-1)^{3B+L+2S} \tag{53}$$

Under this all the new SUSY states are all odd so (cf eq(52)) they appear quadratically in the Lagrangian with the important phenomenological implication that the super partners of Standard Model states can only be produced in pairs and that the lightest supersymmetric particle will be absolutely stable and indeed a candidate for dark matter. However, the terms of the eq(53) are not the only ones which are allowed. In particular one may have the additional gauge invariant terms [26] of the form given in eq(54).

$$\begin{aligned} L^{Yuk}_{Rpviolating} = & \lambda_{ijk}\underbrace{\left[l_i \quad l_j \quad E^c_k\right]_F}_{\Delta L=1} + \lambda'_{ijk}\underbrace{\left[l_i \quad q_j \quad d^c_k\right]_F}_{\Delta L=1} \\ & + \lambda''_{ijk}\underbrace{\left[u^c_i \quad d^c_j \quad d^c_k\right]_F}_{\Delta B=-1} \end{aligned} \tag{54}$$

These have the property that they violate R-parity, allowing single production of supersymmetric states and allowing the decay of the lightest supersymmetric particle. However, they also have property that they violate Baryon number and/or Lepton number and all the terms of eq(54) can not be simultaneously present if proton decay is not to proceed at an unacceptably fast rate.

Having specified the non-gauge couplings all that is left to complete the model is to specify the mechanism leading to supersymmetry breaking. Unfortunately this is rather poorly understood at present and the best one can do is to parameterise this breaking by a series of soft supersymmetry breaking terms which do not disturb the solution to the hierarchy problem. They are shown in eq(55) .

$$\begin{aligned} L_{SB} = & -\frac{1}{2}\sum_{\widetilde{A}} M_{\widetilde{A}}\lambda_{\widetilde{A}}\lambda_{\widetilde{A}}\widetilde{q}|^2 - m^2_{\widetilde{u}}|\widetilde{u}|^2 \\ & -m^2_{\widetilde{d}}|\widetilde{d}|^2 - m^2_{\widetilde{l}}|\widetilde{l}|^2 - m^2_{\widetilde{E}}|\widetilde{E}| - m_0(h^u A^u \widetilde{q}\widetilde{u}^c H_2 \\ & + h^d A^d \widetilde{q}\widetilde{d}^c H_1 + h^l A^l \widetilde{l}\widetilde{E}^c H_1) - m_0(B\mu H_1 H_2) + h.c. \end{aligned} \tag{55}$$

It is clear that one is introducing a large number of additional parameters. However there are some strong constraints following from the need to suppress flavour changing neutral currents. For example graphs such as those shown in Fig 1 lead to the following constraints [27]

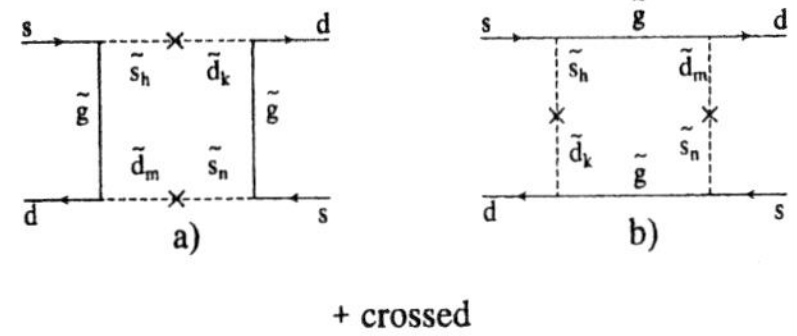

Figure 1. The gluino box diagrams for $\Delta S = 2$ transitions.

$$\begin{aligned} \frac{m_{\tilde{d}}^2 - m_{\tilde{s}}^2}{m_{\tilde{d}}^2} &\lesssim 6.10^{-3} \left(\frac{m_{\tilde{d}}}{Tev}\right)\left(\frac{\sin\tilde{\theta}_c}{\sin\theta_c}\right)^2 \\ \frac{m_{\tilde{e}}^2 - m_{\tilde{\mu}}^2}{m_{\tilde{e}}^2} &< 10^{-1} \left(\frac{m_{\tilde{e}}}{Tev}\right)^2 \left(\frac{\sin\tilde{\theta}_e}{\sin\theta_c}\right)^2 \end{aligned} \tag{56}$$

where the tilde refers to the relative mixing angle between the Standard Model state and its SUSY partner. One of the questions anyone studying supersymmetry breaking must answer is why there should be such a close degeneracy between squark states or between slepton states? In order to achieve an acceptable pattern of supersymmetry breaking in which the lightest squark or slepton states are heavier than their Standard Model partners it is necessary that supersymmetry be broken in an "hidden" sector and supersymmetry be communicated to the MSSM states only via radiative corrections. The different supersymmetry breaking mechanisms can be characterised by the mass of the messenger which couples the supersymmetry breaking sector to the visible sector.

Most explored has been the case where the supersymmetry breaking sector is coupled by a Planck scale messenger via super gravity (SUGRA)[25]. In this case the usual assumption is that the gravity is flavour blind and at the Planck scale the gaugino masses are all equal to one common scale, $M_{1/2}$, and squark, slepton and Higgs masses are equal to another common scale, M_0 (These are both expected to be of order the gravitino mass $m_{3/2} = \Lambda^2/M_{Planck}$ where Λ is the supersymmetry breaking scale). Radiative corrections split these degeneracies at lower energy scales; for example for gauginos one has [28]

$$m_{\lambda_i}(\mu) = \frac{\alpha_i(\mu)}{\alpha_i(M_{Planck})} m_{1/2} \tag{57}$$

In this class of super-symmetry breaking the lightest supersymmetric particle is a combination of neutralino and higgsino. It may be questioned whether the universality assumption which is necessary for agreement with eq(56) is reasonable. In string theories in which the communication from the hidden sector is ony via Planck scale messengers it is not automatically the case that universality occurs although for a specific symmetry breaking mechanism it does happen [29].

A second possibility is that the messenger is a gauge non-singlet field coupling the supersymmetry breaking sector to the visible sector. This was explored in the first realisations of the supersymmetric extensions of the Standard Model [25], but subsequently was neglected in favour of the supergravity messenger. However, this year there has been resurgence of interest in this mechanism[30]. One big of advantage of this scheme arises because the communication is via gauge interactions that automatically lead to a degenerate gaugino and a degenerate squark and slepton sector and hence obey the constraint of eq(56). The expectation for the gaugino and scalar masses in the simplest model are [31]

$$
\begin{aligned}
m_{\tilde{\lambda}_i}(\mu) &= c_i N \alpha_i \frac{\Lambda^2}{M} >> m_{3/2} \\
m_{\phi_i}^2 &= 2\Lambda^2 N \sum_i C_i \alpha_i^2 >> m_{3/2}
\end{aligned} \tag{58}
$$

where $M << M_{Planck}$ is the messenger mass. In this scheme the lightest supersymmetric particle is the gravitino whose mass is given above. The next to lightest supersymmetric particle is expected to be the neutralino, heavier than the gravitino because the messenger mass is less than the Planck mass. Because of this the signals in the gauge-non-singlet messenger scheme are characteristically different as neutralino production ultimately leads to the production of a photon in its decay to the gravitino. Thus one may expect that signals will involve missing energy plus a displaced γ vertex or a charged lepton or the products of a neutral Higgs decay. The CDF event [?], $p\bar{p} \to e^+e^- +$ 2 photons + missing energy, is thus a prime candidate as a signal for the gauge non-singlet messenger scheme [31, 32, 33].

6.2 The String Effective Lagrangian

String theory is the only candidate we have for a unification of all the fundamental interactions *including* gravity. Of course at low energies one is looking at the four dimensional effective field theory which results from string theory defined at the Plack scale. Often this involves a stage of compactification of some of the space-time dimensions and in the low energy theory these act as internal degrees of freedom. Let me summarise the common features of compactified string theory [34]:

- The symmetries are specified by the underlying string. The resulting gauge group can easily accommodate the Standard Model but also can be larger. For example it can incorporate the Grand Unified groups SU(5), SO(10) or E_6. The only symmetries that are known to descend from the string are gauge symmetries, either continuous or discrete. Unfortunately there are many possible vacua in the compactified 4 dimensional theory and we do not know how to select between them.

- The multiplet structure is specified in a given string vacua. Usually this includes some number n_g of chiral families + some number n_v of vector like states which come in complex conjugate pairs. The latter are expected to acquire mass at a high scale through a stage of symmetry breaking below the compactification scale.

- String theory has only one parameter, the string scale which can be related to the Planck scale. All other parameters are determined in terms of the vacuum expectation values (vevs) of moduli fields, Φ_M. For example the string coupling constant is determined by

$$g_i^{1/2}(M_{string}) = \frac{k_i}{g_{string}^2} \tag{59}$$

where k_i is a parameter associated with the particular string construction and

$$g_{string}^2 = \frac{1}{< S >}. \tag{60}$$

Here S is a moduli field known as the dilaton. In the absence of supersymmetry breaking the moduli fields have no potential and any value for their vacuum expectaion value gives a viable string theory. Once supersymmetry is broken, the moduli's vev will be fixed and the coupling will be determined as in eq(60). Similarly the Yukawa couplings will be determined as functions of additional moduli known as complex structure moduli. Unfortunately we are still far from an understanding of the string vacuum structure and hence from a prediction of the couplings of the theory.

7 SUSY Unification Predictions

To date we have no unambiguous evidence for supersymmetry, although the departure of R_b from the Standard Model prediction could be accommodated in a supersymmetric theory [35]. However, it was pointed out that it is very special property of supersymmetry that as the superpartners become heavy one recovers the Standard Model predictions (this is not necessarily true of theories beyond the Standard Model, for example technicolour theories tend to give very large corrections). Thus supersymmetric theories are at least consistent with the precision tests without the need for fine tuning of parameters. In this decoupling limit one would not have expected to see the new SUSY partners (cf Table 7).

Despite the present lack of low energy indications of supersymmetry we do have some hints that supersymmetry is indeed correct. These are the so-called unification predictions. The most famous of these is gauge coupling unification. If one uses the known radiative corrections of the Standard Model to continue the gauge couplings up in energy they come together [36]. However, comparing with the new precision measurements of these couplings one finds the unification fails by some 10 standard deviations. However, if one allows for a low energy stage of supersymmetry, the new states that one adds change the evolution of couplings in a known way [37] and, remarkably, the couplings come together provided the supersymmetry breaking scale is low, of the order of a Tev or less[38].

The second unification prediction is what I call "mass unification". As discussed above this is the assumption that, at the unification scale, the scalar masses and the gaugino masses are separately degenerate (a similar universality assumption may be made for the other soft parameters). At low energy radiative corrections splits these degeneracies. For example the squarks and gluinos will become heavier than the winos and leptons simply because the gauge interactions causes the masses to rise and QCD is the strongest gauge interaction. Lacking information about the mass spectrum of the supersymmetric particles this of course does not lead us to any testable predictions. However, when one applies the same ideas to the Higgs boson something very interesting happens. Due to the large Yukawa coupling of the Higgs to the top quark the Higgs boson mass squared is actually driven down at low energies and at some stage will become *negative,* triggering symmetry breaking [39]. Thus this very simple mass-unification picture leads naturally to the idea of electroweak symmetry breaking and indeed to the prediction that the group $SU(3) \otimes SU(2) \otimes U(1)$ is broken to $SU(3) \otimes U(1)$. Further, since supersymmetry predicts the quartic Higgs couplings, once we have the soft mass of the Higgs determined in terms of the parameters of high scale we can determine the vacuum expectation value of the Higgs and hence the Z-mass. Thus one ends up with a prediction for the Higgs and Z-masses in terms of soft supersymmetry breaking parameters.

The third unification prediction is Yukawa unification. For example, if the bottom and τ Yukawa couplings are equal at the unification scale as is predicted in some Grand Unified theories such as $SU(5)$, then at that scale the bottom and τ masses are equal. Continuing down in energy, one finds radiative corrections split this degeneracy forcing the bottom quark to be heavier and indeed going in the right direction to fit the observed spectrum at low scales [1]. Once again the prediction for m_b and m_τ depends on the soft supersymmetry breaking parameters.

Finally there is also a prediction for the dark matter abundance made up by the LSP, the lightest supersymmetric particle which is stable in the MSSM. Since the spectrum of the supersymmetric state is determined by the soft supersymmetric breaking parameters and the radiative corrections, one can determine the character of the LSP and hence, knowing its interactions, determine its annihilation rate and thus its present day abundance.

The remarkable thing is that all four of these unification predictions work well for a common choice of the soft supersymmetry breaking parameters. Much work had been done on this topic over the years. Here I present recent results by De Boer et al [40] on the constrained MSSM. This is the MSSM in which the soft supersymmetry breaking parameters are constrained by a simultaneous fit to gauge coupling constant unification, to the value of Z mass coming from radiative electroweak breaking, to the Yukawa coupling constant unification, $m_b = m_\tau$, to the dark matter abundance. In addition one requires consistency with phenomenological constraints, namely the SUSY spectrum should be consistent with the lower limits on SUSY states established experimentally and also with the branching ratio $b \rightarrow s\gamma$ (this turns out to be quite sensitive to radiative corrections involving supersymmetric states). Two regions of parameter space have been identifed where all of these constraints can be simultaneously satisfied. They correspond to low and high values of tan β. The allowed regions of the m_0, $m_{1/2}$ plane are shown in Fig 2 together with the reach of LEP, showing that almost all the allowed parameter space is accessible to LEP.

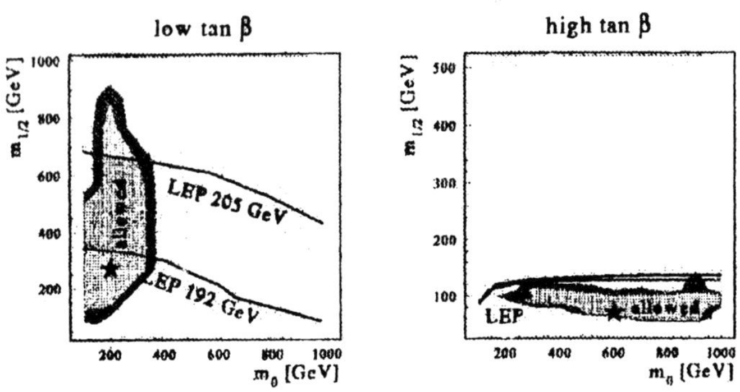

Figure 2. The allowed regions in the m_0 $m_{\frac{1}{2}}$ plane for the CMSSM fit discussed in the text. Also shown is the future LEP sensitivity.

In the low tan β solution the top Yukawa coupling is large, close to its infra-red fixed point value, $M_t = 206 \sin\beta$ Gev. The Higgs mass is relatively light $M_{H<}110$ Gev (cf the 132 GeV absolute upper bound in the MSSM) with 81 Gev the best fit value. The LSP abundance is $\Omega h^2 = 0.42$, consistent with it being the dominant component of dark matter with $M_{LSP} = 80$ Gev and composed mainly (99%) of bino. The large tan β solution extends the equality of the bottom and τ Yukawa couplings to equality with the top Yukawa coupling, giving $m_b = m_\tau = m_\tau$, as is suggested by many SO(10) grand unified models. Again the lightest Higgs state is relatively light, $M_H = 110GeV$, and again the dark matter abundance is large, $\Omega h^2 = 0.19$, consistent with the LSP being the dominant dark matter component. An interesting corrolary is the implication of the constrained MSSM for dark matter searches. The conclusion is that imposing all of the constraints including $b \to s\gamma$ does *not*, as had previously been thought, reduce the dark matter rate significantly so current experiments should be sensitive to SUSY dark matter.

7.1 Gauge Unification : Threshold effects

Given the remarkable success of the unification of gauge couplings, it is worth exploring in some detail the sensitivity of the gauge unification predictions for the measured couplings to the details of the light spectrum and to the GUT thresholds. The sensitivity of the strong coupling to the SUSY spectrum is summarised by the formula of eq(61) [41]

$$\alpha_s(M_z)^{-1} = \alpha_s^{SUSY}(M_z)^{-1} + \frac{19}{28\pi}\ln(\frac{T_{SUSY}}{M_z}) \tag{61}$$

where T_{SUSY} contains the effect of the SUSY thresholds and is given by

$$T_{SUSY} = m_{\tilde{H}}(\frac{m_{\tilde{H}}}{m_{\tilde{H}}})^{\frac{28}{19}}[(\frac{m_H}{m_{\tilde{H}}})^{\frac{3}{19}}(\frac{m_{\tilde{H}}}{m_{\tilde{H}}})^{\frac{4}{19}}(\frac{m_{\tilde{l}}}{m_{\tilde{l}}})^{\frac{3}{19}}] \tag{62}$$

Using this one arrives at the following value for the strong coupling at the Z mass

$$\begin{aligned}\alpha_s(M_Z) =& 0.1268 + 1.21.10^{-4}(M_t - 165) + 3.36.10^{-7}\\ &(M_t - 165)^2 - 2.16.10^{-5}(m_h - 100)\\ &+9.6.10^{-8}(m_h - 100)^2 \pm 0.009|^{T_s=15TeV}_{T_s=15GeV}\end{aligned} \tag{63}$$

where the latter error summarises the SUSY uncertainty for extreme choices of the SUSY spectrum. For the weak mixing angle one finds

$$\begin{aligned}\sin^2\theta_w(M_Z) \simeq& 0.2326 - 0.25(\alpha_5(M_Z) - 0.123)\\ &\pm 0.0020|^{T_s=1TeV}_{T_s=15GeV}\end{aligned}$$

The effect of GUT thresholds on the gauge couplings is given by [42]

$$\alpha_i^{-1}(M_G) = \alpha_G^{-1} + \frac{1}{2\pi}\sum_{Heavies,X} b_i^X \ln|\frac{M_X}{M_G}| \tag{64}$$

In order to apply this formula one needs to know the "Heavy" spectrum with masses at the GUT scale. For example in the "realistic" theory [48] discussed in Section 4.1 the dominant correction comes from a "light" triplet Higgs with mass $10^{13}GeV$ giving a GUT threshold correction $\delta\ \alpha_s = -0.003$.

7.2 Implications of Coupling Unification

One of the parameters of gauge unification is the unification scale, M_X, which turns out to be between 1 and 3 times $10^{16}GeV$. In Grand Unified theories this scale is not determined. In string theories however there is only one fundamental scale, the string scale of order the Planck mass. The gauge unification scale, M_X, is determined in terms of the string scale[43]. In the heterotic string this scale turns out to be some 5 x $10^{17}GeV$, not so far away from the gauge unification scale. It is difficult to overemphasise the potential importance of this result. Our belief that there is a stage of Grand Unification of the strong, weak and electromagnetic interactions rests largely on the quantitative success of the unification of the associated couplings. The prediction of the unification scale would be the first indication of unification with gravity. While the prediction within a factor of 20 is encouraging, the residual discrepency leaves the question open. What could be the source of this discrepancy? This has been examined in some detail. There are various effects that can cause the prediction of the Grand Unified scale to vary. One should remember that the Grand Unified scale is the argument of the log and thus, in order to get it correctly, one has to work to very high precision. Possible problems in determining this scale are :

- SUSY threshold effects. It seems that although these could make the factor of 20 difference this only can be achieved with a very peculiar supersymmetry spectrum at low energies in which the gluino is lighter than the Wino [44].
- A second possibility are threshold effects the Grand Unified scale [42, 45]. One can easily arrange that small corrections add up to give a change in the prediction for the Grand Unified scale by a factor of 20. However, in my opinion this is rather unsatisfactory for the moment one adds extra states at high energy the predictive powers is reduced and the remarkable agreement of gauge unification looks less convincing.
- Another possibility that has been explored are string threshold effects which in a given string theory are calculable[46] and amount to including the effect of the heavy Kaluza Klein modes which are themselves split when the Grand Unified theory is broken. Although it is possible to construct theories in which they can also give you a change in the unification scale by a factor of 20, there is only a small class of theories in which this is happens and they look rather unnatural.

Thus we are left with somewhat unsatisfactory situation in which the prediction for the unification scale following from string theory is close in the sense that it is within 2 orders of magnitude of the prediction coming from evolution of coupling, but not precise enough to give some confidence that this is indeed what is happening. Recently there has been a new contribution to this debate which I think is potentially very important [47]. The calculation which lead to a prediction of the string scale assumed that the four dimensional gauge coupling and the 10 dimensional gauge coupling in the decompactified theory are both in the perturbative domain. However, this is not necessarily true. For example it is possible that the 4 dimensional coupling is small, as suggested by the value found continuing the couplings from low energies, while the 10 dimensional coupling is large. In this case the calculation breaks down and, apparently, becomes intractable due to the non-perturbative effects. However recent advances in calculating at strong coupling have raised the possibility that one may calculate the string scale even when the 10 dimensional coupling becomes strong. The reason is that the strongly coupled heterotic string is dual to another theory in the weakly couple domain. This theory is known as M theory and for this theory that prediction the string unification scale changes. In fact it depends upon new parameters and, although these are not known one is able to bound the maximum valueof the string unification scale.

This is given by

$$M_{string} \leq 5.10^{17} (\frac{4\pi}{\alpha_{GUT}^{\frac{1}{3}}})^{-1} GeV \tag{65}$$

We see that if this bound is saturated the string unification scale will be reduced by a factor sufficient to bring it into agreement with the gauge unification scale.

I find the study of M_X in string theories a very exciting development because it raises the possibility that the bottom up test of gauge unification is giving us the first indication of unification of all of the fundamental interactions including gravity.

7.2.1 SO(10)⊗U(1)^5

In this example the spontaneously broken family symmetry comes from five Abelian U_1 factors [48]. Other possibilities have been considered [14], for example an (SO(10)⊗Δ(48)⊗ $U(1)$) model has recently been considered in which a discrete symmetry [49], Δ(48), provides the spontaneously broken family symmetry. Returning to the SO(10)⊗U(1)^5 model, the multiplet structure is given by

$$3.16 + (16 + \overline{16}) + 4.45 + 2.54 + 4.1 \tag{66}$$

Following from this the structure for the up down Lepton mass matrices is given by

$$m^{u,d,l} = \begin{pmatrix} 0 & z_a^1 C & 0 \\ z_a C & y_a E_e^{i\Phi} & x_a' B \\ 0 & x_a B & A \end{pmatrix} \tag{67}$$

The relative coefficients x_a, y_a and z_a for a=1,2,3 are determined by the SO(10) Clebsch Gordon coefficients and relate quark and lepton masses of a given generation. However, because the spontaneously broken family symmetry is Abelian, there is no prediction for the *relative* magnitudes of x_a, y_a and z_a. The magnitude of the coefficients A, B, C and E are determined by working out at what order in the symmetry breaking vevs one generates a given entry. The mechanism involves the mixing with heavy states as pioneered by Froggatt and Nielsen [6]. An example is given in Fig 3 giving for the Yukawa coupling responsible for the (2,3) mass the structure

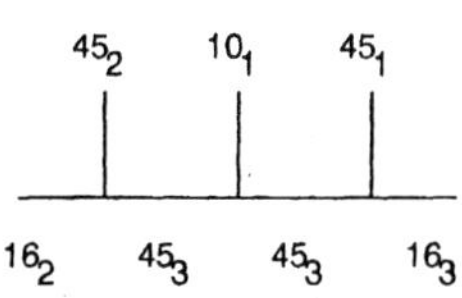

Figure 3. Tree graph generating intergenerational mixing.

$$16_2 \frac{<45_2>}{<45_3>} 10_1 \frac{<45_1>}{<45_3>} 16_3 \tag{68}$$

In this the $< 45_{1,2} >$ vevs are responsible for breaking the symmetry and inserting these in tree graphs generates off diagonal elements in the mass matrices. The model has been explored in detail in [50]. They find a good fit to all low energy data. The SO(10) symmetry generates a Yukawa unification with $h_t = h_b = h_\tau$ the splitting in the top and bottom masses coming from a large asymmetry in the Higgs vevs (a large tan β solution). At the unification scale $m_b = m_\tau$, while $m_\mu = 3m_s$ the factor of 3 coming from the SO(10) factor [51] associated with the coupling generating that entry (the 3 is just the number of colours). Being a Grand Unified theory based on SO(10), the proton is unstable. Its dominant decay mode is generated by a heavy Higgs triplet, H_T, exchange giving

$$P \to K^+\nu^-, \ \tau_P \gtrsim 10^{32} yrs,$$
$$(M_{H_T} = 4.10^{19} Gev, M_{H_D} = 10^{13} Gev) \tag{69}$$

(H_D is the associated Higgs doublet and must be strongly split to give acceptable gauge unification.) This is a very interesting example of a Grand Unified Theory of fermion masses and represents an almost complete theory. The sector generating the spontaneous breakdown of the theory has not been completely specified. One reservation about such unification schemes is that although the parameters determining the fermion masses are much reduced there are always a large number of additional parameters associated with the GUT symmetry breaking sector. In addition there may be problems with flavour changing nutral currents due to radiative corrections at the GUT scale; these have not yet been computed. However, it is clear that the combination of Grand Unification and family symmetry is able quite closely to reproduce a pattern of fermion masses. Improvements in the measurement of the fermion masses and mixing angles will be able to test such a theory.

7.2.2 Strong Unification

In order to relate masses of different generations the parameters a, b, c, ... of $O(1)$ in eq(28) must be determined. Grand Unification without a family group only relates masses within a given family. Attempts to extend thiem via a non-Abelian family symmetry have not been very successful. However there is another possibility if there is "strong unification" which I would like to review here. The point is that if the gauge coupling at unification is large then there will be rapid flow to the infra red fixed points as one comes down in energy from the unification scale towards the electroweak scale [?, 2, 52, 3]. The renormalisation group equations have the form 30

$$\frac{d\ln(\frac{y}{\alpha})}{dt} = a\alpha - by \tag{70}$$

where $y = h^2/4\pi$ and h is a Yukawa coupling. The ratio $\frac{y}{\alpha}$ has an infra-red fixed point given by $(\frac{y}{\alpha})^* = \frac{a}{b}$ relating the Yukawa coupling to the gauge coupling . The soft supersymmetry breaking masses may also be determined by the infra-red structure. Interestingly this gives a new explanation for the degeneracy of scalar masses needed to avoid flavour changing neutral currents. The point is that the soft scalar mass terms satisfy a renormalisation group equation of the form given by

$$\frac{d(m_{Q_i}^2 - m_{Q_j}^2)}{dt} \propto y(m_{Q_i}^2 - m_{Q_j}^2) \tag{71}$$

At the fixed point the mass differences tend to zero. Therefore if there is rapid flow towards the fixed point there is a dynamical reason for universality.

To explore this possibility let us suppose there is some extension of the Standard Model at a scale M. It is plausible to identify this with the scale, M_X, at which the gauge couplings unify. At first sight it seems very improbable that the IRFP is relevent because there is only a small gap between the gauge unification scale $O(10^{16}GeV)$ and the compactification or Planck scale there seems little room for logarithmic corrections to drive couplings to their fixed points. However the analytic solution presented above shows that the couplings may run to the fixed point much faster than in the Standard model. This happens if, beyond the gauge unification scale the theory is not asymptotically free and the couplings start to rise. Once they get near the strong coupling limit the evolution becomes rapid. This means that flowing in the reverse direction one rapidly reaches the IRFP even though the flow is over only one or two decades in energy.

Is there a case in which strong unification occurs without spoiling the phenomenological success of gauge unfication? The answer is yes for if the new states responsible for driving the couplings large at high scales come in complete SU(5) representations they will not in affect the unification predictions at the one loop order. However, in the models studied two loop effects increase M_X and simultaneously reduce the value of α_s needed for gauge unification. This is welcome for the MSSM value of eq(**??**) is somewhat large compared to the latest values for α_s $\alpha_s = 0.118 \pm 0.004$. A further advantage of this scheme is that it may answer one of the problems involving the string moduli for a dilaton vacuum expectation value corresponding to weak coupling is unstable [53]. However, if unification occurred at strong coupling the dilaton potential will be driven by non-perturbative effects and it has been argued this is necessary to provide a stable minimum.

In fact string theories offer a strong motivation for considering just this type of multiplet structure. There are several advantages in starting with a compactified string theory in which the gauge group is just that of the Standard Model (possibly with additional Abelian family factors). In particular this can solve the doublet triplet splitting problem for, without Grand Unification, there is no symmetry reason requiring the troublesome coulour triplet partners of the Higgs doublets. However, without Grand Unification there is no obvious reason why the observed quarks and leptons should fill out complete $SU(5)$ multiplets. The class of compactified string theories based on the heterotic string *do* however provide just such a motivation. In these theores there is an underlying (E_6) symmetry which is broken on compactification by Wilson loops. Such breaking does not change the Euler characteristic (equal to twice the number of generations) and so the multiplets in the compactifed theory must make up complete E_6 multiplets even though the gauge group is just the Standard Model gauge group. Some of the E_6 states can pair up to form large $SU(3) \otimes SU(2) \otimes U(1)$ invariant masses but there necessarily is left light three generations of representations filling out $\bar{5} + 10$ representations of $SU(5)$. The remaining ambiguity in the string prediction for the low-energy spectrum is that there may be, before Wilson line breaking, N copies of $27 + \bar{27}$ E_6 representations. Wilson line breaking may leave these massless, or may give mass to a specific subset of the states. The latter possibility is utilised to solve the doublet triplet splitting problem giving just Higgs doublets. The former possibility gives additional vectorlike states in complete $SU(5)$ representations, just as we argued were desirable for strong unification without spoiling the phenomenological success of gauge unification.

There are several attractive features of strong unification. In the first place the predictions rely only on a knowledge of the dynamics of the theory, which follow from the symmetries, and do not require knowledge of the initial conditions. As a result the uncertainties associated with the final "Theory of Everything"[4] are avoided or reduced. Secondly the flavour changing problems associated with a family symmetry are under control in a class of theories with a family independent fixed point structure[54]. Indeed we may take this as a strong hint that the IRSFP structure plays an important role for it seems likely that there *must* be new family dependent interactions if the pattern of fermion masses is to be understood and then it is essential to explain why flavour changing processes are small. Finally the known structure of the quark and lepton masses and mixing angles may be used to infer the symmetry structure needed for the

[4]For example the uncertainties associated with the moduli vevs in string theories.

theory beyond the Standard Model. Given the symmetries the dynamics of the theory and thus the renormalisaton group (RG) equations are determined and so one may try to develop a systematic and *quantitative* investigation of the idea that IRSFP structure determines the masses and mixing angles.

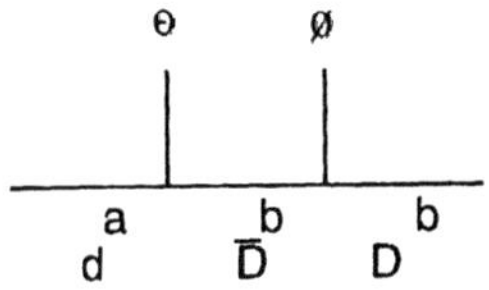

Figure 4. Tree graph mixing d quarks of different $\tilde{U}(1)$ charges.

The theory with strong unification is largely determined by its symmetries and multiplet content alone for many of the couplings of the theory will be determine by the fixed point structure. As an example consider a model [2] with the symmetry $SU(3) \otimes SU(2) \otimes U(1) \otimes \tilde{U}(1)$ and multiplet structure 3 families, f, $+3(f+\overline{f})$+3(1). Assigning the $\tilde{U}(1)$ charges (-4,1,0) and (4,-1,0) to the three generations of left-handed quark components and right-handed quark components respectively and 0 charge to the Higgs field immediately leads to the form of eq(28) through an expansion analagous to eq(??) with $\epsilon_{u,d} =< \theta > /M_{u,d} =< \bar{\theta} > /M_{u,d}$ where θ and $\bar{\theta}$ are standard model singlet fields with $\tilde{U}(1)$ charge ± 1 whose vevs spontaneously break the symmetry along a D-flat direction. The origin of the couplings involving the θ fields is through diagrams similar to Fig 3. For example if the $3(f+\overline{f})$ additional fields contain SU(2) singlet down quarks with charge 0 and -1 then the light d quark will be a mixture of different $\tilde{U}(1)$ charge singlet fields

$$d_{light} = d_1 + \lambda_d \bar{\epsilon} D_0 + \lambda'_d \bar{\epsilon}^2 D_{-1} \tag{72}$$

For example the second term arises through the mixing of the type shown in Fig 4. Here $\bar{\epsilon} = \frac{<\theta>}{<\Phi>}$ and λ_d is a coupling of O(1) determined by the Yukawa couplings involved in Fig 4. With this light state it is easy to see it will couple to the Higgs field generating the (2,2) and (2,3) entries of eq(28). Further mixing will fill in the remaining entries. The only requirement is that there should be an appropriate quark or Higgs field amongst the $3(f+\overline{f})$ additional fields to allow for the mixing. Once the multiplet structure is determined the allowed mixing terms follow. In addition in the case of "strong coupling" the Yukawa couplings such as λ and λ' are all determined by the fixed point structure of the theory leading to a prediction of the constants a, a', b, c, c'. For example, if the entries in the (2,2), (3,2) and (2,3) positions are generated by mixing amongst the quarks while the elements in the (1,3) and (3,2) are generated by the Higgs mixing, a very promising form results [55] as the couplings evolve into the infra red. One finds

$$\begin{aligned}
m_t &= 181 GeV \\
tan\beta &= 46 \\
m_b &= 0.95\ m_\tau \\
m_s &= 1.25(\frac{\alpha(M_x)}{\alpha(M_{Planck})})^{0.55} m_\mu \\
m_d m_s &= 1.2\ m_e m_\mu \\
V_{cb} &\simeq [1-(\frac{\alpha(M)}{\alpha(M_{Planck})})] \sqrt{\frac{m_d}{m_s}}
\end{aligned}$$

$$\delta \approx \frac{\pi}{2}$$
$$sin\theta_W = 3/8 \tag{73}$$

The remarkable thing about this is that the relationship such as $m_b = 0.95\ m_\tau$ apply in a theory without Grand Unification. Similarly, as we discussed above, the prediction for the weak mixing angle, which is the same as is found in $SU(5)$, follows from the need to cancel the anomalies in the new $\tilde{U}(1)$ and does not require a Grand Unified group [8]. Thus we see there is a an alternative in strong coupling theories to Grand Unification which preserve all the successful GUT relations. In addition there is a dynamical origin for the difference between the muon and strange quarks following from the infrared structure. This should be compared to the group theoretic difference introduced in the previous section.

It is encouraging that the rather complicated pattern of fermion masses and mixings may have a relatively simple interpretation in terms of a spontaneosly broken symmetry. Improvements in the accuracy with which the CKM matrix elements and the quark masses are known will allow us to distinguish between the various models proposed and one may hope that we may be able ultimately be able to determine the mechanism responsible for fermion masses and mixing angles.

8 Summary

Supersymmetry is a necessary ingredient if one is to embed the Standard Model in a more unified structure at some very high scale. The support for such an hypothesis comes from the quantitative success of gauge unification. The unification relations have been extended to include a prediction for the pattern and magnitude of breaking of the Standard Model group and also for fermion masses and the dark matter abundance. Remarkably all can be made consistent for a choice of soft supersymmetry breaking parameters. It now seems possible that the unification with gravity as is implemented in string theories can be also quantitatively tested through the prediction of the gauge coupling unification scale and the initial results are encouraging. The structure of light quark masses and their mixing angles has also been extensively studied in SUSY models and various schemes based on spontaneously broken symmetries proposed which can explain the hierarchical pattern observed and even provide absolute predictions for the masses and mixing angles. It may be hoped that as the accuracy of measurement of the quark masses and mixings improve we will be able to distinguish between the various models and maybe move closer to identifying the "Theory of Everything".

References

[1] A. Buras, J.Ellis, M.K.Gaillard and D.V.Nanopoulos, Nucl. Phys. B135 (1978) 66; D.V.Nanopoulos and D.A.Ross, Nucl. Phys. B157 (1970) 273; H.Arason, D.J.Castano, B.E.Keszthelyi, S.Mikaelian, E.J.Piard, P.Ramond and B.D.Wright, Phys. Rev. Lett.67 (1991) 2933; L.Ibanez and C.Lopez, Nucl. Phys. B233 (1984) 511; S.Dimopoulos, L.Hall and S.Raby, Phys. Rev. Lett.68 (1992) 1984; Phys. Rev. D45 (1992) 4192.

[2] G.G. Ross, Phys. Lett. B364 (1995) 216; M.Lanzagorta and G.G.Ross, Phys Lett. B364 (1995) 163; ibid B349 (1995)319.

[3] A related approach is discussed in "Gauge-Yukawa Unification in SO(10) SUSY GUTS", J. Kubo, M. Mondragon, S.Shoda, George Zoupanos, contributed paper PA05-108; Warsaw International conference on High Energy Physics, 1996; ibid

"Unification of Gauge and Yukawa Couplings without Symmery", J. Kubo, M. Mondragon, M. Olech, George. Zoupanos, contributed paper PA05-109.
[4] For a recent review and further references see A. Ali, D.London, DESY Preprint DESY-96-140, hep-ph/9607392; H.Leutwyler, Bern Preprint HEPPH-9609467, hep-ph/9609467 and Phys. Lett.B378(313)1996.
[5] B.J. Bjorken, lectures presented at Oxford 1996, available on Oxford WWW home page.
[6] M.E. Machacek and M.T. Vaughn, Phys. Lett. **B103** (1981) 427;
C. Wetterich, Nucl. Phys. **B261** (1985) 461; Nucl. Phys. **B279** (1987) 711;
J. Bijnens and C. Wetterich, Phys. Lett. **B176** (1986) 431; Nucl. Phys. **B283** (1987) 237; Phys. Lett. **B199** (1987) 525;
P. Kaus and S. Meshkov, Mod. Phys. Lett. **A3** (1988) 1251;
C.D.Froggat and H.B. Nielsen, Origin of symmetries, World Scientific (1991);
S. Dimopoulos, L. J. Hall and S. Raby, Phys. Rev. Lett. 68(1992)1984; Phys. Rev. D45(1992)4195;
H. Arason, D. J. Castaño, P. Ramond and E. J. Piard, Phys.Rev.D47(1993)232;
G. F. Giudice, Mod. Phys. Lett. **A7** (1992)2429. K.S.Babu and R.N.Mohapatra, Univ. of Maryland preprint, UMD-PP-95-57; M.Bando, K.-I. Izawa and T. Takahashi, Kyoto Univ. preprint, KUNS 1252.
[7] P.Ramond, R.G.Roberts and G.G.Ross, Nucl.Phys. B406(1993)19.
[8] L.Ibanez, G.G.Ross, Phys.Lett.B332:100-110,1994.
[9] P.Binetruy and P.Ramond, Phys. Lett. B350(1995)49;
V. Jain and R.Schrock, Phys. Lett. B352(1995);
E.Dudas, S.Pokorski and C.A.Savoy, Phys. Lett. B356(1995)45;
Y.Nir, Phys. Lett. B345(1995)107.
[10] E.Papageorgiu, Orsay preprint, LPTHE Orsay 40/94; H. Dreiner, G. K. Leontaris, S. Lola, G. G. Ross and C. Scheich, CERN-TH.7311/94 preprint, (to appear in Nucl. Phys. B).
[11] L. Ibanez, Phys. Lett. B 303 (1993) 55.
[12] For a review of string theories, see M. Green, J. Schwarz and E. Witten, Superstring Theory, Cambridge University Press, 1987.
[13] J.L.Lopez and D.V.Nanopoulos, Phys. Lett. B268(1991)359; A.E.Faraggi, Phys. Lett. B278(1992)131; Princeton preprint, IASSN-HEP-94/31; A.E.Faraggi and E.Halyo, Nucl. Phys. B416(1994)63.
[14] For a review and extensive references see S.Raby, Ohio State Preprint OHSTPY-HEP-T-95-001.
[15] B. Pendleton, G. G. Ross, Phys. Lett. 98B, 291(1981).
[16] C. T. Hill, Phys. Rev. D24, 691, (1981); C. T. Hill, C. N. Leung, S. Rao, Nucl. Phys. B262, 517(1985); J. Kubo, K. Sibold, W. Zimmermann, Phys. Lett. B222, 191(1989);Nucl. Phys. B259, 331(1985).
[17] L. Ibanez, C. Lopez, Nucl. Phys. B233, 545 (1984).
[18] J.Gasser and H.Leutwyler, Phys. Rep. 87C(1982) 77;
S.Narison, Phys. Lett B216 (1989) 191;
N.Gray, D.J.Broadhurst, W.Grafe and K.Schilcher, Z.Phys. C48 (1990) 673.
[19] L.Alvarez-Gaume, J. Polchinski and M.B.Wise, Nucl. Phys. B221 (1983) 495;
J.Bagger, S.Dimopoulos and E.Masso, Phys. Lett. 156B (1985) 357.
[20] H.Georgi,H.R.Quinn,S.Weinberg,Phys.Rev.Lett.33(1974)451.
[21] For general reviews of GUTs and the original references see:
P. Langacker, Physics Reports 72 (1981) 185;
G. G. Ross, Grand Unified Theories, Benjamin/Cummings (1984);

R.N.Mohapatra, Unification and Supersymmetry, Springer-Verlag,(1992).
[22] H. Georgi and S.L.Glashow, Phys.Rev.Lett.32(1974)438. Nucl.Phys.B193(1981)150.
[23] J. Ellis, K. Enqvist, D.V. Nanopoulos and F. Zwirner, Mod. Phys. Lett. A1 (1987) 57; R.Barbieri and G.F. Giudice, Nucl. Phys. B306 (1988) 63; S. Dimopoulos and G.F. Giudice, Phys. Lett. B357 (1995)573.
[24] G. Anderson and D.J. Castano, Phys. Lett. B347 (1995) 300 and Phys. Rev. D52 (1995) 1693
[25] For reviews, see: P. Fayet, Unification of the Fundamental Particle Interactions, eds. S. Ferrara, J. Ellis and P. Van Nieuwenhuizen (Plenum Press, New York, 1980), p. 587; H.P. Nilles, Phys. Rev. 110C (1984) 1; G.G. Ross, Grand Unified Theories, Benjamin (New York 1984); H.E. Haber and G.L. Kane, Physics Reports 117C (1985) 75.
[26] L.J. Hall and M. Suzuki, Nucl. Phys. B231 (1984) 419; F. Zwirner, Phys. Lett. B132 (1983) 103.
[27] F. Gabbiani and A. Masiero, Nucl. Phys. B322 (1989) 235; J.Hagelin, S. Kelley and T. Tanaka, Nucl. Phys. B 415 (1994) 293; Mod. Phys.Lett. A 8 (1993) 2737; D. Choudhury, F. Eberlein, A. Konig, J. Louis and S. Pokorski, Phys. Lett. B342 (1995), 80; F. Gabbiani, E. Gabrielli, A. Masierio and L. Schwestrini, Rome preprint ROM2F/96/21; hep-ph/9604387.
[28] K.Inoue, A. Kakuto, H.Komatsu and S. Takeshta, Progr.Theor. Phys. 68(1982)927, and 71 (1984) 348.
[29] L.E. Ibanez and D.Lust, Nucl. Phys. B382 (1992) 305; V.S. Kaplunovsky and J.Louis, Phys. Lett. B306 (1993) 269; A.Bringnole, L.E. Ibanez and C. Munoz, Nucl. Phys. B422 (1994) 125 [Erratum: B436 (1995) 747; S. Ferrara, C. Kounnas and F. Zwirner, Nucl. Phys. B429 (1994) 589 [Erratum: B433 (1995) 255]; T. Kobayashi, D. Suematsu, K. Yamada and Y. Yamagishi, Phys. Lett. B348 (1995) 402; P. Brax and M. Chemtob, Phys. Rev. D51 (1995) 6550; E. Dudas, S. Pokorski and C.A. Savoy, Phys. Lett. B369 (1996) 255.
[30] M. Dine, A.E. Nelson and Y. Shirman, Phys. Rev. D 51 (1995) 1362; M. Dine, A.E. Nelson, Y. Nir and Y. Shirman, SCIPP-95-32, hep-ph/9507378.
[31] S. Dimopoulos et al., Phys. Rev. Lett. 76, 3494 (1996).
[32] "Supersymmetric Analysis and Predictions Based on the CDF ee $\gamma\gamma^+$ missing E_T Event", S. Ambrosanio, G.L. Kane, contributed paper PA10-008; S. Ambrosinis et al Phys Rev Lett 76(1996)3498.
[33] "A Search for Supersymmetry with a Light Gravitino at the Fermilab Tevatron and CERN LEP Collider", S. Ambrosanio , G.L. Kane, contributed paper PA11-008.
[34] For a review of string theories, see M. Green, J. Schwarz and E. Witten, Superstring Theory, Cambridge University Press, 1987.
[35] For a review and extensive references see S. Pokorski, plenary talk, IHEP conference, Warsaw, 1996. . See also "Implication of the Reported Deviations from the Standard Model for Gamma ($Z \rightarrow b\bar{b}$) and $\alpha_s(m_{Z^2})$", J.D.Wells, G.L.Kane, contributed paper PA11-006, IHEP conference, Warsaw, 1996.
[36] H. Georgi, H.R. Quinn and S. Weinberg, Phys. Rev. Lett. 33 (1974) 451
[37] S.Dimopoulos,S.Raby,F.Wilczek,Phys.Rev.D24(1981)1681; L.E.Ibanez,G.G.Ross,Phys.Lett.105B(1982)439; M.Einhorn,D.R.T.Jones,Nucl.Phys.B196(1982)475.
[38] J.Ellis, S.Kelley, D.V.Nanopoulos, Phys. Lett. 260B(1991)131; U.Amaldi, W.deBoer, H.Furstenau, Phys. Lett.260B(1991)447; P.Langacker,M.Luo,Phys.Rev.D44(1991)817; G.G. Ross and R.G. Roberts, Nucl. Phys. B377(1992)571; F. Anselmo, L. Ciafarelli,

A. Peterman and A. Zichichi, Nuovo Cim. 104A (1991) 1817.
[39] L.E.Ibanez and G.G.Ross, Phys Lett, ; K. Inoue, A. Kakuto, H. Komatsu and S. Takeshita, Progr. theor. Phys. 67(1982) 1889; L.Alvarez-Gaume, J Polchinski and M.B. Wise, Nucl. Phys. B221 (1993) 495.
[40] For the latest fits and extensive references to earlier work see W. de Boer, G. Burkart, R. Ehret, J. Lautenbacher, W. Oberschulte-Beckmann, U. Schwickerath, V.Bednyakov, D.I.Kazakov and S.G.Kovalenko, Univ. of Karlsruhe preprint IEKP-KA/95-07.
[41] For a recent review and extensive references see M. Carena, Proceedings International conference on High Energy Physics, World Scientific (1995) 481.
[42] S.Weinberg, Phys. Lett. 91B 51 91980); L.J. Hall, Nuc. Phys. B178 75 (1981); C.H. Llewellyn-Smith, G.G.Ross, J. Wheater, Nucl. Phys. B177 (1981) 263; R. Barberi and L. Hall, Phys. Rev. Lett. 68 (1992) 752.
[43] V. Kaplunovsky, Nucl. Phys. B 307 (1988) 145; for a recent review see K.Dienes,
[44] L.Roszkowski and M.Shifman, Phys.Rev. D53(1996)404
[45] "Non-Universalities and Planck Scale Effects on Grand Unification", Pran Nath, R. Arnowitt, contributed paper PA11-014.
[46] V. Kaplunovsky, Nucl. Phys. B 307 (1998) 145; 382 (1992) E436; L.J. Dixon, v.s. Kaplunovsky and J.Louis, Nucl. Phys. B 329 (1990) 27; 355 (1991) 649; J.P. Derendinger, S. Ferrara, C. Kounnas and F. Zwirner, Nucl. Phys. B372 (1992) 145; Phys. Lett. B271 (1991) 307; G. Lopez Cardoso and B.A. Ovrut, Nucl. Phys. B 369 (1992) 351; I. Antoniadis, K.S. Narain and T.R. Taylor, Phys. Lett. B 267 (1991) 37; I. Antoniadis, E. Gava and K.S. Narain, Phys. Lett. B 283 (1992) 209; 393 (1992) 93; I. Antoniadis, E.Gava, K.S. Narain and T.R. Taylor, preprint NUB-3057 (1992); L.E. Ibanez, D. Lust and G.G. Ross, Phys. Lett. B 272 (1991) 251; L. Ibanez and D. Lust, Nucl. Phys. B 382 (1992) 305; I. Antoniadis, J. Ellis, R. Lacaze and D.V. Nanopoulos, Phys. Lett. B 268 (1991) 188; G. Cardoso, D. Lust and T. Mohaupt, HUB-IEP-95/50, hep-th/9412209.
[47] E. Witten, Nucl Phys. B433, 85 (1995); P. Horava and E. Witten, Nucl. Phys. B460, 506 (1996); E.Caceras, V.Kaplunovsky, I.Michael Mandelberg, UTTG-07-96, hep-th/9606036.
[48] V.Lucas and S.Raby, Ohio State Preprint OHSTPY-HEP-T-95-028
[49] "Realistic SUSY GUT for CP violation, neutrino oscillation, fermion masses and mixings", Yue-Liang Wu, K.C.Chou, contributed paper PA08-002.
[50] "SUSY GUT's and global fits to low energy data", Tomas Blazek, Marcela Carena, Stuart Raby, contributed paper PA11-029.
[51] H.Georgi and C.Jarlskog, Phys.Lett.B86(1979)297; H.Georgi and D.V.Nanopoulos, Nucl.Phys. B159(1979)16; J.Harvey, P.Ramond and D.B.Reiss, Phys.Lett. B92(1980)309; Nucl.Phys. B199(1982)223.
[52] P.M. Ferreira, I. Jack and D.R. T. Jones Phys. Lett. B357 91995) 359
[53] M. Dine and N.Seiberg, Phys. Lett. 162 B, 299 (1985); T.Banks and M.Dine, Phys. Rev. D50(1994)7454; M.Dine and Y.Shirman, hep-th/9601175; M.Dine,hep-th/9508085
[54] M. Lanzagorta and G. G. Ross, CERN-TH.95-162
[55] G.G.Ross, in preparation.

ELECTROWEAK BARYOGENESIS AND THE MINIMAL SUPERSYMMETRIC STANDARD MODEL *

D. Delépine†

Institut de Physique Théorique
Université catholique de Louvain
B-1348 Louvain-la-Neuve, Belgium

ABSTRACT

In principle, the baryon asymmetry of the Universe can be generated at the electroweak phase transition but the experimental lower limit on the Higgs mass seems to rule out a Standard Model scenario. However, it has been shown recently that in the Minimal Supersymmetric Standard Model, the electroweak phase transition can be a strong enough first order one for baryogenesis if the mass of one top squark is close to or smaller than the top mass.

INTRODUCTION

Why the Universe is dominated by matter and not by antimatter is one of the most intriguing characteristic of the Nature[1]. Astrophysical observations imply in fact a very small value for the Baryon Asymmetry of the Universe (BAU) :

$$\frac{n_B - n_{\bar{B}}}{n_\gamma} \simeq 4 - 7\ 10^{-10} \tag{1}$$

with n_B,$n_{\bar{B}}$ and n_γ are respectively the densities of baryons, antibaryons and photons.

In 1967, Sakharov [2] was the first to point out that the BAU could be explained in terms of high energy physics. He showed that,in order to generate it, a particle theory has to satisfy 3 conditions:

*Work supported in part by the EEC Science Project SC1-CT91-0729.
†Research assistant of the National Fund for the Scientific Research

- Baryon number cannot be conserved.
- C and CP(C for Conjuguaison Charge and P for parity) have to be violated. Otherwise, the rate of reactions with particles and the rate of reactions with antiparticles should be the same.
- Departure from thermal equilibrium is needed. At some stage of its history, the Universe had to be out of thermal equilibrium. During these phases, the state of Universe was non-stationary and some macroscopic variables as the baryonic charge was time-dependent.

This letter is divided in 2 parts. In the first, properties of the Standard Model are discussed.In the second, it will be shown that in a simplified scenario described by the Minimal Supersymmetric Standard Model (MSSM) the baryogenesis constraint can be satisfied. The allowed range of the parameters of the model is consistent with the present experimental bounds [8].

THE STANDARD MODEL

In the electroweak Standard Model (SM), the Sakharov's conditions are fullfilled.The SM has 2 sources of CP-violation. The first is the CP-violation coming from the phase (δ_{CKM}) in the Cabibbo-Kobayashi-Maskawa quark mixing matrix [9] with 3 generations of quarks. A basis-invariant measure of this is given by the Jarlskog's determinant [10]:

$$\Delta_{CP} = det[M_u M_u^\dagger, M_d M_d^\dagger] \tag{2}$$

where M_u and M_d are respectively the 3×3 up and down quark mass matrices.

Some recent attempts to calculate the BAU using the CKM CP-violating phase led to the conclusion that the δ_{CKM} is not efficient enough to produce the right order of magnitude [3].

The second source of CP-violation in the SM is the θ_{strong}-angle. The most general QCD lagrangian contains the following 4-dimensional term:

$$\mathcal{L}_\theta = \theta_{QCD} \frac{g_s^2 N_f}{32\pi^2} G_a^{\mu\nu} \tilde{G}_{\mu\nu}^a \tag{3}$$

with N_f is the flavor number,g_s the QCD gauge coupling, $G_a^{\mu\nu}$ is the gluon field strength tensor and $\tilde{G}_{\mu\nu}^a$, its dual. This term is T-violating (T for time reversal) and its related to axial $U(1)$ anomaly.

Indeed, at the classical level, the QCD lagrangian for N_f massless quarks is invariant under the chiral $U(N_f)_L \times U(N_f)_R$ symmetry. But at the quantum level, the flavor-singlet axial current $J^{\mu 5}(\bar{q}\gamma^\mu\gamma^5 q)$ is not conserved.

$$\partial_\mu J^{\mu 5} = \frac{g_s^2 N_f}{16\pi^2} G_a^{\mu\nu} \tilde{G}_{\mu\nu}^a \tag{4}$$

In the procedure to diagonalize the arbitrary mass matrices of the standard electroweak model, a chiral redefinition of the right-handed fields is necessary.

$$q_R \rightarrow e^{-i\frac{\theta}{N_f}} q_R \tag{5}$$

with $\theta = argdet M_u M_d$.

The anomalous effect of this chiral transformation is to induce the following modification to the lagrangian:

$$\delta\mathcal{L}_\theta = \theta\frac{g_s^2 N_f}{32\pi^2}G_a^{\mu\nu}\tilde{G}_{\mu\nu}^a \tag{6}$$

Therefore,the physical θ is the sum of both contributions: one from QCD (Eq.[3]) and the other from the mass matrices (Eq.[6]).

The strongest constraint on the physical θ is coming from the neutron electric dipole momentum [11]:

$$\theta_{physical} < 10^{-10} \tag{7}$$

It is interesting to note that this limit is of the same order of magnitude than the BAU.

The out of equilibrium condition will be satisfied if the Electroweak Phase transition (EWPT) is a first order one.It means that the vacuum of the symmetric phase is metastable.The Phase Transition (PT) proceeds by nucleation. In such a case, 3 temperatures can be defined: one when both vacua are degenerated (T_1), a second when the PT occurs (T_c) and the last one is when the potential is flat at the origin(T_0). These temperatures follow this hierarchy:

$$T_0 < T_c < T_1 \tag{8}$$

In the SM, the first order phase transition is induced by the weak gauge bosons. It is a consequence of the cubic term in the finite temperature bosonic effective potential[5].

$$V_{bosons}(m,T) = -\frac{T^4\pi^2}{90} + \frac{m^2T^2}{24} - \frac{m^3T}{12\pi} + ... \tag{9}$$

$$V_{fermions}(m,T) = -\frac{7}{180}\pi^2T^4 + \frac{m^2T^2}{12} + \frac{m^4}{16\pi^2}ln\frac{m^2}{T^2} + ... \tag{10}$$

with m is the field-dependent mass of the bosons or the fermions.

As we can see, the first order character of the EWPT in the SM is proportional to the weak gauge coupling.The phase transition is not expected to be strong enough for baryogenesis as we shall see later.

The last Sakharov's condition is the non-conservation of the baryonic charge. In the SM, only the B-L current is conserved (B and L are respectively the Baryonic and the Leptonic currents) but the divergence of the B+L current is given by the electroweak anomaly induced by the chiral structure of the weak gauge symmetry.

$$\partial_\mu J^\mu_{B+L} = \frac{g_w^2 N_g}{16\pi^2}W_a^{\mu\nu}\tilde{W}_{\mu\nu}^a \tag{11}$$

with N_g is the generation number,g_w the weak gauge coupling, $W_a^{\mu\nu}$ is the weak field strength tensor and $\tilde{W}_{\mu\nu}^a$, its dual.

But at zero temperature, the rate of the anomalous B-violating reactions (Γ_B) is strongly suppressed by an exponential factor [12]:

$$\Gamma_B \propto e^{\frac{-1}{\alpha_w}} \approx 10^{-100} \tag{12}$$

This suppression can be avoided at high temperature. In that case, Γ_B is proportionnal to a Boltzman factor and the B-violating transition is induced by an unstable solution of the equation of motion called "Sphaleron" [13].

$$\Gamma_B \propto e^{\frac{-M_{sph}}{T}} \tag{13}$$

where $M_{sph} = 4\pi v(T)/g_w B(\lambda/g^2)$ is the sphaleron mass, B is a constant which in the standard model ranges between $1.5 \leq B \leq 2.7$ for $0 \leq \lambda/g^2 < \infty$ and $v(T)$ is the Higgs expectation value at the temperature T.

To avoid a wash-out of the BAU by the sphalerons after the phase transition, the B-violating processes have to be out of equilibrium. A way to impose this property is to ask that Γ_B has to be smaller than the expansion rate of the Universe (Γ_H):

$$\Gamma_B < \Gamma_H \tag{14}$$

Using Eq.[13] and the expression of the M_{sph}, the last condition can be written as follows:

$$\frac{v(T_c)}{T_c} \geq 1 \tag{15}$$

In the SM, this baryogenesis constraint implies an upper bound on the Higgs mass[4]:

$$m_H \leq 60 GeV \tag{16}$$

This value is experimentally ruled out[14].

In conclusion of this first part, we have to mention that the SM effectively fills the 3 Sakharov's conditions but it fails on 2 main points. First, even if the CP-violating processes producing the BAU are not well known and understood , the SM CP-violation seems to be too small. Secondly, at the Electroweak Phase Transition, the jump in the Higgs expectation value is too weak. So, in order to explain the BAU, we need to go beyond the SM.

THE MINIMAL SUPERSYMMETRIC STANDARD MODEL (MSSM)

A simple extension of the SM is the MSSM which not only predicts a light Higgs boson but also contains new CP-violating phases. In the MSSM, there are 2 complex scalar doublets. But we shall assume that at low-energy, only one neutral scalar remains light while all the other Higgs bosons and supersymmetric partners of the SM particles have a mass of the order of the global supersymmetry breaking scale. The tree level scalar potential for the real component h of the lightest Higgs boson reads

$$V_{\text{tree}} = \frac{1}{2}\mu^2 h^2 + \frac{1}{32}\tilde{g}^2 \cos^2 2\beta \; h^4 \tag{17}$$

where $h = h_1 \cos\beta + h_2 \sin\beta$, $\tilde{g}^2 = (g_y^2 + g_w^2)$ and $g_{y,w}$ are the $U(1)$ and $SU(2)$ gauge couplings respectively.

In order to simplify our discussion, we shall assume that the stop masses are given by the following relations:

$$m^2_{\tilde{t}_{L,R}} = m^2_{L,R} + m^2 \tag{18}$$

with $m \equiv \frac{g_t h}{\sqrt{2}}$. We have neglected the D-term contribution to the stop masses as well as the left-right mixing effects. The top and the stop loops are the dominant contributions to the effective 1-loop potential. For the Higgs mass,one obtains

$$m_h^2 = m_Z^2 \cos^2 2\beta + \frac{3}{4\pi^2}\frac{m_t^4}{v_0^2} \ln\left[\left(1+\frac{m_L^2}{m_t^2}\right)\left(1+\frac{m_R^2}{m_t^2}\right)\right] \tag{19}$$

where $m_Z = \tilde{g}v_0/2$ is the Z-boson mass and $m_t = g_t v_0/\sqrt{2}$ is the top quark mass, $v_0 = 246$ GeV. As we can see from Eq.[9], the strength of the phase transition can be enhanced by the right- or left-handed stop field contribution if m_L or m_R is close to zero. A scenario with $m_R \ll m_L$ is naturally implemented even if at GUT scale $m_R = m_L$ (universality of the soft masses)[6].This effect is due to the 3:2:1 hierarchy in the renormalisation group equations for the Higgs scalar h_2, right-handed and left-handed squared masses respectively[7]. Finally, assuming $m_R \ll m_L$ and $m_t \ll m_L$ and keeping only the relevant terms in the effective potential, we obtain

$$V(h,T) = M^2(T)h^2 - \delta(T)h^3 - a(T)(h^2+b^2)^{3/2} + \lambda(T)h^4 \ , \tag{20}$$

where

$$\begin{aligned} M^2(T) &= -\frac{1}{4}m_Z^2\cos^2 2\beta - \frac{3}{16\pi^2}\frac{m_t^2}{v_0^2}\left\{m_t^2 \ln\left[\left(1+\frac{m_L^2}{m_t^2}\right)\left(1+\frac{m_R^2}{m_t^2}\right)\right]\right. \\ &+ \left. m_L^2 \ln\left(1+\frac{m_t^2}{m_L^2}\right) + m_R^2 \ln(m_t^2+m_R^2) + \frac{1}{2}m_R^2\right\} \\ &+ \frac{m_t^2}{2v_0^2}T^2 + \frac{3}{16\pi^2}\frac{m_t^2}{v_0^2}m_R^2(2\ln T + C_B) \ , \end{aligned} \tag{21}$$

$$\delta(T) = \frac{2m_W^3 + m_Z^3}{6\pi v_0^3}T \ , \tag{22}$$

$$a(T) = \frac{m_t^3 T}{2\pi v_0^3} \ , \quad b = \frac{m_R v_0}{m_t} \ , \tag{23}$$

$$\lambda(T) = \frac{1}{8}\frac{m_Z^2}{v_0^2}\cos^2 2\beta - \frac{3}{16\pi^2}\frac{m_t^4}{v_0^4}\left(\ln\frac{T}{m_L} - C_F - \frac{1}{2}C_B - \frac{3}{4}\right) \ , \tag{24}$$

C_B and C_F are constants coming from the high temperature expansion [5] ($C_B = 5.41$ and $C_F = -2.64$). The form of the potential is simple and can be analytically studied as done in [8].As said before, the phase transition occurs at a temperature T_c between T_0 and T_1 and these relations remain valid for the $\frac{v(T)}{T}$:

$$\frac{v(T_1)}{T_1} \lesssim \frac{v(T_c)}{T_c} \lesssim \frac{v(T_0)}{T_0} \ . \tag{25}$$

To conclude, the $\frac{v(T)}{T}$ ratios are plotted as a function of tanβ for $m_{\tilde{t}_R} = m_t$ and $m_L = 500$ GeV in Fig.1. We can see that from the point of view of the baryogenesis the favourite value for tanβ is between 0.5 and 1.5. The maximum value for the ratio is reached for tanβ=1.This corresponds to the lower value for the Higgs mass (≈ 60 GeV) which is consistent with the present experimental data on MSSM [14].

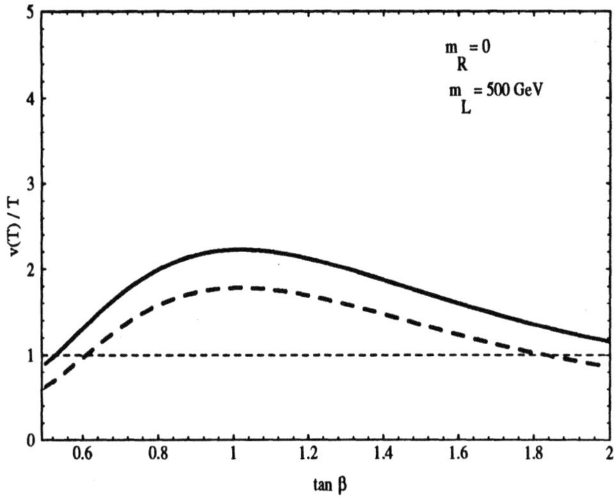

Figure 1. The curves $v(T_0)/T_0$ (solid line) and $v(T_1)/T_1$ (bold-dashed line) as functions of $\tan\beta$ for $m_R = 0$, $m_L = 500$ GeV and $m_t = 174$ GeV.

Under the assumption that the right-handed stop mass is close to or smaller than the top mass, we have shown that the baryogenesis constraint (Eq. [15]) can be satisfied for low value of tanβ in the MSSM. In [15], similar results were obtained using a numerical analysis of the effective potential. A 2-loop numerical analysis of this potential [16]and lattice calculations [17] confirm the enhancement of the first order phase transition in the range of the MSSM parameters studied in this letter.

ACKNOWLEDGEMENT

These results were obtained in a fruitful collaboration with J.-M. Gérard, R. Gonzalez Felipe and J. Weyers.

References

[1] See *e.g.* the reviews: A.G. Cohen, D.B. Kaplan and A.E. Nelson, Annu. Rev. Nucl. Part. Sci. **43** (1993) 27; V.A. Rubakov and M.E. Shaposhnikov, preprint CERN-TH/96-13 (1996) and references therein.

[2] A.D. Sakharov , JETP Lett. **5** (1967) 24.

[3] M.B. Gavela, P. Hernández, J. Orloff, O. Pène and C. Quimbay, Mod. Phys. Lett. **9** (1994) 795; Nucl. Phys. **B 430** (1994) 382; P. Huet and E. Sather, Phys. Rev. **D 51** (1995) 379.

[4] M.E. Shaposhnikov, JETP Lett. **44** (1986) 465; Nucl. Phys. **B287** (1987) 757;Nucl. Phys. **B299** (1988) 797.

[5] L. Dolan and R. Jackiw, Phys. Rev. **D 9** (1974) 3320; S.Weinberg, Phys. Rev. **D 9** (1974) 3357.

[6] M. Carena and C.E.M. Wagner, Nucl. Phys. **B 452** (1995) 45.

[7] H.P.Nilles,Phys.Rep **110** (1984) and references therein.

[8] D. Delépine,J.-M. Gérard, R. Gonzalez Felipe, J. Weyers, UCL-IPT-96-05, hep-ph 9604440 (to appear in Phys.Lett.B.).

[9] N.Cabibbo, Phys. Rev. Lett. **10** (1963) 531;M.Kobayashi, T. Maskawa Progr. Theor. Phys. **49** (1973) 652.

[10] C. Jarlskog Phys. Rev. Lett. **55** (1985) 1039; Z.Phys.C **29** (1985) 491.

[11] R.J. Crewther, P.D. Vecchia, G.Veneziano Phys. Lett. **88 B** (1979) 123; E.Witten, Phys. Lett. **91 B** (1980) 487.

[12] G. 't Hooft, Phys. Rev. Lett. **37** (1976) 8; Phys. Rev. **D 14** (1976) 3432.

[13] N.S. Manton, Phys. Rev. **D 28** (1983) 2019; F.R. Klinkhamer and N.S. Manton, Phys. Rev. **D 30** (1984) 2212.

[14] *Review of Particle Properties*, Phys. Rev. **D 49** (1994) 1173.

[15] M. Carena, M. Quiros and C.E.M. Wagner, Phys. Lett. **B 380** (1996) 81.

[16] J.R. Espinosa, preprint DESY-96-064, hep-ph 9604320 (1996)

[17] J.M. Cline and K. Kainlulainen, preprint McGill-96-20, hep-ph 9605235 (to appear in Nucl. Phys.**B** (1996);M. Laine, preprint HD-THEP-96-13, hep-ph9605283 (1996).

BPS STATES IN SUPERSTRINGS WITH SPONTANEOUSLY BROKEN SUSY

Costas Kounnas

Theory Division, CERN
CH-1211 Geneva 23, Switzerland

Abstract and Disclaimer

We show the existence of a supersymmetry-breaking mechanism in string theory, where $N = 4$ supersymmetry is broken *spontaneously* to $N = 2$ and $N = 1$ with moduli-dependent gravitino masses. The BPS spectrum of the theory with lower supersymmetry is in one-to-one correspondence with the spectrum of the heterotic $N = 4$ string. The mass splitting of the $N = 4$ spectrum depends on the moduli as well as the three R-symmetry charges. In the case of $N = 4 \to N = 2$, the perturbative $N = 2$ prepotential is determined by the perturbative $N = 4$ BPS states. This observation led us to suggest a method that determines the exact non-perturbative prepotential of the effective $N = 2$ supergravity using the shifted spectrum of the non-perturbative BPS states of the underlying $N = 4$ theory.

Introduction

When a local symmetry is spontaneously broken, the physical states can be classified in terms of the unbroken phase spectrum and in terms of a well-defined mass splitting given in terms of vacuum expectation values of some fields, weighted by the charges of the broken symmetry. In the case of gauge symmetry breaking, the fields with non-zero vev's are physical scalar fields, while in the case of supersymmetry breaking they are auxiliary fields. In extended supersymmetric theories (local or global), the supersymmetric vacua are degenerate, with zero vacuum energy for any vev of the moduli fields (S, T^i). For instance, in the case of $N = 4$ supergravity based on a gauge group $U(1)^6 \times G$, the space of the moduli fields is given in terms of $2 + 6r$ physical scalars, which are coordinates of the coset space [1,2]

$$\left[\frac{SU(1,1)}{U(1)}\right]_S \times \left[\frac{SO(6,r)}{SO(6) \times SO(r)}\right]_T . \tag{1}$$

r is the rank of the gauge group G.

In an arbitrary point of the moduli space the gauge symmetry G is broken down to $U(1)^r$ while at some special points of the moduli space the gauge symmetry is extended

Masses of Fundamental Particles: Cargèse 1996
Edited by M. Lévy *et al.*, Plenum Press, New York, 1997

to some non–Abelian gauge group of the same rank due to the presence of some extra gauge multiplets that become massless at the special points of the moduli space.

In the heterotic $N = 4$ superstring solution obtained by T^6 compactification of the $10-d$ superstring, the rank of the group r has a fixed value, $r = 22$ [3-6]. In an arbitrary point of the moduli space the gauge group is $U(1)^r$ and in special points the symmetry is extended as in field theory. There is however a fundamental difference between the field theory Higgs phenomenon and the string theory one. Indeed, if in an $N = 4$ field theory the gauge group is $G = U(1)^6 \times SO(32)$ at any given point of the moduli space, then at any other point the remaining gauge symmetry G_{T_i} is always a subgroup of G with smaller dimensionality $\dim(G_{T_i}) \leq \dim(G)$. On the contrary, in the string Higgs phenomenon, owing to the existence of winding states, we can connect gauge groups which are not subgroups of a larger gauge group. For instance, it is possible to connect continuously $G = U(1)^6 \times SO(32)$ with $G = U(1)^6 \times E_8 \times E_8$, as well as with the most symmetric of same rank, namely $G = SO(44)$. Indeed, starting from a $10 - d$ $N = 1$ supergravity theory with $G = SO(32)$ or $G = E_8 \times E_8$ after compactification in four dimensions the only possible $N = 4$ supergravity effective theories are based either to $G = U(1)^6 \times SO(32)$ or $G = U(1)^6 \times E_8 \times E_8$ (and their subgroups obtained with Higgs phenomenon). In string theory the gauge group can be further extended due to the existence of extra gauge bosons with non-zero winding numbers, which can become massless in special points of the moduli space.

When some auxiliary fields of the supergravity theories have non–vanishing vev, some (or all) of the supersymmetries are spontaneously broken [10–12]. There is a consistent class of $N = 1, 2$ and $N = 4$ models defined in flat space-time in which all supersymmetries are broken or partially broken [4–12]. The most interesting case for our purposes is that in which there is one of the supersymmetries left unbroken. In that case we know that it is possible, in general, to have chiral representations of matter scalar multiplets which can describe the quarks and leptons of the supersymmetric standard model. All previous examples about the partial breaking of $N = 2$ to $N = 1$ supersymmetry was done at the field theory level [13]. In this work we will first show the extension of the partial spontaneous breaking at the perturbative string level and then we will generalize our result to the non-perturbative level using as a tool the heterotic–type II string duality of the $N = 4$ 4d-superstrings [14,15].

Perturbative $N = 4$ Mass Spectrum

Our starting point is a four-dimensional heterotic $N = 4$ superstring solution. From the world-sheet viewpoint, these theories are constructed by the following left- and right-moving degrees of freedom:

- Four left-moving non-compact supercoordinates, X^μ, Ψ^μ
- Six left-moving compactified supercoordinates, Φ^I, Ψ^I
- The left-moving super-ghosts, b, c and β, γ
- Four right-moving coordinates, $\bar{X}^\mu$
- Six right-moving compactified coordinates, $\bar{\Phi}^I$
- 32 right-moving fermions, $\bar{\Psi}^A$
- The right-moving ghosts, $\bar{b}, \bar{c}$

In order to obtain a consistent (without ghosts) $N = 4$ solution the left-moving fermions Ψ^μ, Ψ^I and the β, γ ghosts must have the same boundary conditions. In that case the global existence of the left-moving spin-3/2 world-sheet supercurrent

$$T_F = \Psi^\mu \partial X^\mu + \Psi^I \partial \Phi^I \tag{2}$$

implies periodic boundary conditions for the compact and non-compact left-moving coordinates, Φ^I, X^μ. Modular invariance implies the right-moving coordinates $\bar{\Phi}^I$, $\bar{X}^\mu$ to be periodic as well. The solution with $G = U(1)^6 \times SO(32)$, is when the right-moving fermions Ψ^A have the same boundary conditions (periodic or antiperiodic), while the solution with $G = U(1)^6 \times E_8 \times E_8$ is when the $\bar{\Psi}^A = (\bar{\Psi}^{A_1}, \bar{\Psi}^{A_2})$ are in two groups of sixteen with the same boundary conditions. Starting either from the $G = U(1)^6 \times E_8 \times E_8$ solution or from the $G = U(1)^6 \times SO(32)$ we can obtain all others by deforming the momentum lattice of compactified bosons together with the charge lattice of the 32 fermions $\bar{\Psi}^A$.

The partition function of the heterotic $N = 4$ solutions in a generic point of the moduli space is well known and has the following expression [3]:

$$Z(\tau,\bar{\tau}) = \frac{\mathrm{Im}\tau^{-1}}{\eta^2(\tau)\ \bar{\eta}^2(\bar{\tau})} \times \frac{1}{2}\sum_{\alpha,\beta}(-)^{\alpha+\beta+\alpha\beta}\ \frac{\vartheta^4[^\alpha_\beta](\tau)}{\eta^4(\tau)}\ \frac{\Gamma_{(6,22)}(\tau,\bar{\tau})}{\eta^6(\tau)\bar{\eta}^{22}(\bar{\tau})}\ , \tag{3}$$

where $\Gamma_{(6,22)}(\tau,\bar{\tau})$ denotes the partition function due to the compactified coordinates $\Phi^I, \bar{\Phi}^A$ and due to the sixteen right-moving $U(1)$ currents constructed with the fermions $\bar{\Psi}^I$

$$\bar{J}^k = \bar{\Psi}^{2k-1}\bar{\Psi}^{2k}, \qquad k = 1, 2, ..., 16. \tag{4}$$

$\Gamma_{(6,22)}(\tau,\bar{\tau})$ has in total 6×22 moduli parameters which correspond to (1,1) marginal deformations of the world-sheet action.

$$\delta S^{2d} \ = \ \delta T_{IJ}\ \partial\Phi^I\ \bar{\partial}\bar{\Phi}^J\ +\ Y_I^k\ \partial\Phi^I\ \bar{J}^k. \tag{5}$$

In terms of the six-dimensional backgrounds of the compactified space, the T_{IJ} moduli are related to the internal background metric G_{IJ} and the internal antisymmetric tensor B_{IJ}; $T_{IJ} = G_{IJ} + B_{IJ}$. The Y_I^k moduli are the six-dimensional internal gauge field backgrounds which belong in the Cartan subalgebra of the ten-dimensional gauge group (either $E_8 \times E_8$ or $SO(32)$). From the four-dimensional viewpoint the moduli T_{IJ} and Y_I^k correspond to the vev's of massless scalar fields, members of the $N = 4$ vector supermultiplets. The explicit form of the $N = 4$ heterotic partition function [3,12] $\Gamma^{SO(32)}_{(6,22)}[T_{IJ}, Y_I^k]$ is:

$$\Gamma^{SO(32)}_{(6,22)}(T,Y)(\tau,\bar{\tau}) = \mathrm{Im}\tau^{-3}(\mathrm{det}G_{IJ})^3 \times \sum_{m^I,n^I}\ \exp\left[-\pi\ T_{IJ}\frac{(m^I+\tau n^I)(m^J+\bar{\tau}n^J)}{\mathrm{Im}\tau}\right] \times$$
$$\frac{1}{2}\sum_{\gamma,\delta}\prod_{k=1}^{16}\exp\left[\ i\pi\frac{1}{4}(n^I\ Y_I^kY_J^k\ m^J + 2\ \delta\ Y_I^k\ n^I)\right] \times\ \bar{\vartheta}\begin{bmatrix}\gamma+n^I\ Y_I^k\\ \delta+m^I\ Y_I^k\end{bmatrix}(\bar{\tau}) \tag{6}$$

When all Y-moduli are zero $Y_I^k = 0$ then the gauge group is extended from $G = U(1)^{22}$ to $G = U(1)^6 \times SO(32)$.

An alternative representation of $\Gamma_{(6,22)}$ is the one in which for $Y_I^k = 0$ the extended gauge symmetry is $G = U(1)^6 \times E_8 \times E_8$ instead of $U(1)^6 \times SO(32)$:

$$\Gamma^{E_8\times E_8}_{(6,22)}(T,Y)(\tau,\bar{\tau}) = \mathrm{Im}\tau^{-3}(\mathrm{det}G_{IJ})^3\ \times \sum_{m^I,n^I}\ \exp\left[-\pi\ T_{IJ}\frac{(m^I+\tau n^I)(m^J+\bar{\tau}n^J)}{\mathrm{Im}\tau}\right]$$
$$\times\frac{1}{2}\sum_{\gamma_1,\delta_1}\prod_{k=1}^{8}\quad \exp\left[\ i\pi\frac{1}{4}(n^I\ Y_I^k\ Y_J^k\ m^J + 2\delta_1\ Y_I^k\ n^I)\right] \times \bar{\vartheta}\begin{bmatrix}\gamma_1+n^I\ Y_I^k\\ \delta_1+m^I\ Y_I^k\end{bmatrix}(\bar{\tau})$$
$$\times\frac{1}{2}\sum_{\gamma_2,\delta_2}\prod_{k=9}^{16}\quad \exp\left[\ i\pi\frac{1}{4}(n^I\ Y_I^k\ Y_J^k\ m^J + 2\delta_2\ Y_I^k\ n^I)\right] \times \bar{\vartheta}\begin{bmatrix}\gamma_2+n^I\ Y_I^k\\ \delta_2+m^I\ Y_I^k\end{bmatrix}(\bar{\tau}) \tag{7}$$

Both the $SO(32)$ and $E_8 \times E_8$ representations are connected continuously by marginal deformations with the $G = SO(44)$ maximal gauge symmetry point [4, 5]:

$$\Gamma^{SO(44)}_{(6,22)}(\tau, \bar{\tau}) = \frac{1}{2} \sum_{\gamma,\delta} \vartheta^6 \left[{}^{\gamma}_{\delta}\right] (\tau) \bar{\vartheta}^{22} \left[{}^{\gamma}_{\delta}\right] (\bar{\tau}). \tag{8}$$

Another useful representation of the $\Gamma_{(6,22)}$ is that of the lorenzian left- and right-momentum even self dual lattice. This representation is obtained by performing Poisson resummation on m^I, using either $\Gamma^{SO(32)}_{(6,22)}(T, Y)$ or $\Gamma^{E_8 \times E_8)}_{(6,22)}(T, Y)$:

$$\Gamma_{(6,22)}\left(P_I, \bar{P}_I, Q^k\right) = \sum_{m_I, n^I, Q^k} \exp i\pi \left[\frac{\tau}{2} P_I\, g^{IJ} P_J - \frac{\bar{\tau}}{2} \bar{P}_I g^{IJ} \bar{P}_J - \bar{\tau} \hat{Q}^k \hat{Q}^k\right] \tag{9}$$

with

$$\frac{1}{2}\, P_I\, g^{IJ}\, P_J \; - \; \frac{1}{2}\, \bar{P}_I\, g^{IJ}\, \bar{P}_J \; - \; \hat{Q}^k \hat{Q}^k \; = 2\, m_I\, n^I \; - \; Q^k\, Q^k \; = \; \text{even integer}. \tag{10}$$

In the above equations g^{IJ} is the inverse of G_{IJ}; the lattice momenta P_I, $\bar{P}_I$, and the left–charges $\hat{Q}^k$ are qiven in terms the moduli parameters G_{IJ}, B_{IJ} and Y^k_I and in terms of the charges (m_I, n^I, Q^k):

$$P_I = m_I + Y^k_I Q^k + \frac{1}{2} Y^k_I Y^k_J\, n^J + B_{IJ} n^J + G_{IJ} n^J$$

$$\bar{P}_I = m_I + Y^k_I\, Q^k + \frac{1}{2} Y^k_I Y^k_J\, n^J + B_{IJ} n^J - G_{IJ}\, n^J$$

$$\hat{Q}^k \; = \; Q^k \; + \; Y^k_I\, n^J. \tag{11}$$

All $N = 4$ heterotic solutions are defined in terms of the vev's of the moduli fields (T_{IJ}, Y^k_I) and thus different solutions are connected to each other by a string-Higgs phenomenon. The heterotic $N = 4$ spectrum is invariant under the target-space duality group $SO(6, 22; Z)$, e.g the generalization of the $R \to 1/R$ spectrum symmetry in a compactification on S^1 [16]. At some special points of the moduli space we have extensions of the gauge group as in the effective $N = 4$ supergravity theories. In string theories further extensions can take place, since due to the non-zero winding charges (n^I) can become massless in special points of the moduli space. Thus, in string theory, a large class of disconnected $N = 4$ supergravities are continuously related among themselves due to the existence of the winding states. This precise fact is the origin of the perturbative string unification of all interactions in string theories.

$N = 4 \to N = 2$ spontaneous SUSY breaking

One of the defining characteristics of the $N = 4$ theories is that the states are classified by their transformation properties under the R-symmetry group which, for $N = 4$ supersymmetry is $G_R = SU(4) \sim SO(6)$. In the gravitational multiplet the gravitinos are in **4** representation of G_R, the graviphotons are in **6**, while the graviton, the dilaton and the antisymmetric tensor field are singlets. The degrees of freedom of a massless $N = 4$ vector multiplet are also in definite representations of G_R: the scalars are in **6**, the gauginos are in **4**, while the gauge bosons are singlets. In the heterotic string G_R is constructed in terms of the six–left moving compactified supercoordinates, (Φ^I, Ψ^I). The world-sheet fermion bilinears $\Psi^I\ \Psi^J$ form an $SO(6)_{k=1}$ Kac–Moody algebra. In the light-cone picture, the full spectrum of the theory is classified in representations of

$SO(6)_{k=1}$ and in terms of the $U(1)_0$ helicity charge $q^0 = \oint j^0$, $j^0 = \Psi^\mu \Psi^\nu$, $\mu, \nu = 3, 4$. In the $N = 4$ spectrum the three internal helicity charges $q^i = \oint j^i$, $j^i = \Psi^{2k-1}\Psi^{2k}$, $k = 1, 2, 3$ and q^0 are all simultaneously integers for space-time bosonic states and simultaneously half-integers for the fermionic states:

$$q^i = \text{half integers for space-time fermions}$$
$$q^i = \text{integers for space-time bosons.} \tag{12}$$

Furthermore all physical states have odd total q^i charge (GSO-projection)

$$q^0 + q^1 + q^2 + q^3 = \text{odd integer.} \tag{13}$$

The last condition remains valid for supersymmetric solutions with less than four supersymmetries. In order to have a lower number of supersymmetries, the q^i's must not be simultaneously integers or half-integers. It is then necessary to modify the world-sheet action S^{2d}, adding background fields that can change the individual values of the q^i's, keeping however their total q^i charge:

$$\Delta S^{2d} = \int dz d\bar{z}\, F^a_{IJ}\, (\Psi^I\, \Psi^J - \Phi^I\, \overleftrightarrow{\partial}\, \Phi^J)\, \bar{J}^a, \tag{14}$$

where $\bar{J}^a$ denotes any dimension (0,1) operator. The part of the left–moving operator $(\Phi^I\, \overleftrightarrow{\partial}\, \Phi^J)$ is necessary to ensure the $N = (1,0)$ super-reparametrization of the 2-d action. From a higher–dimensional point of view, the F^a_{IJ} denote non-trivial gauge or gravitational ($\mathcal{R}^{(KL)}_{IJ}$) field backgrounds. In four dimensions they give rise to non-vanishing auxiliary fields. The permitted values of F^a_{IJ} ($\mathcal{R}^{(KL)}_{IJ}$) are not arbitrary. Only those for which

$$U_L(F) = \exp\, [\int dz\, F^a_{IJ}\, (\Psi^I\, \Psi^J - \Phi^I\, \overleftrightarrow{\partial}\, \Phi^J)] \tag{15}$$

commutes with the 2-d super-current ($T_F = \Psi^\mu\, \partial\Phi^\mu + \Psi^I\, \partial\Phi^I$) are allowed. This restriction generates a quantization of the permitted F^a_{IJ} ($\mathcal{R}^{(KL)}_{IJ}$) backgrounds.

A partial $N = 4 \rightarrow N = 2$ breaking is possible [10] when $F^a_{3,4} = -F^a_{5,6} = H$ is not zero (self-duality condition). Indeed, in that case the q^2 and q^3 charges are shifted, preserving the total q^i charge. In order to define the full deformation of the spectrum it is necessary to find a representation of the partition function in which the bosonic charges

$$Q^B_2 = \oint dz \Phi^3\, \overleftrightarrow{\partial}\, \Phi^4 \text{ and } Q^B_3 = \oint dz \Phi^5\, \overleftrightarrow{\partial}\, \Phi^6 \tag{16}$$

are well defined. As a starting point we fermionize the four internal bosonic coordinates

$$\partial\Phi^I = y^I w^I, \text{ and } \bar{\partial}\bar{\Phi}^I = \bar{y}^I\, \bar{w}^I,\ I = 3, 4, 5, 6. \tag{17}$$

In this representation the 2d supercurrent is [4],

$$T_F = \Psi^\mu\, \partial\Phi^\mu + \sum_{I=1}^{2} \Psi^I\, \partial\Phi^I + \sum_{I=3}^{6} \Psi^I\, y^I\, w^I. \tag{18}$$

We will now perform the following Z_4 transformation:

$$\Psi^3 \rightarrow \Psi^4,\quad y^3 \rightarrow y^4,\quad \Psi^5 \rightarrow -\Psi^6,\quad y^5 \rightarrow -y^6,$$

$$\Psi^4 \rightarrow -\Psi^3,\quad y^4 \rightarrow -y^3,\quad \Psi^6 \rightarrow \Psi^5,\quad y^6 \rightarrow y^5,$$

$$w^3 \to w^4, \qquad w^4 \to w^3, \qquad w^5 \to w^6, \qquad w^6 \to w^5, \tag{19}$$

which leaves (18) invariant. The above transformation corresponds to a $\pi/2$ rotation on the complex fermion basis:

$$\begin{aligned}
\chi_1 &\to e^{2i\pi\phi}\,\chi_1 \quad \text{with} \quad \chi_1 = \frac{\Psi^3 + i\Psi^4}{\sqrt{2}} \\
\chi_2 &\to e^{-2i\pi\phi}\,\chi_2 \quad \text{with} \quad \chi_2 = \frac{\Psi^5 + i\Psi^6}{\sqrt{2}} \\
Y_1 &\to e^{2i\pi\phi}\,Y_1 \quad \text{with} \quad Y_1 = \frac{y^3 + iy^4}{\sqrt{2}} \\
Y_2 &\to e^{-2i\pi\phi}\,Y_2 \quad \text{with} \quad Y_2 = \frac{y^5 + iy^6}{\sqrt{2}} \\
W_- &\to e^{4i\pi\phi}W_- \text{ with } W_- = \frac{(w^3 - w^4) + i(w^5 - w^6)}{2} \\
W_+ &\to W_+ \quad \text{with} \quad W_+ = \frac{(w^3 + w^4) + i(w^5 + w^6)}{2}
\end{aligned} \tag{20}$$

Similarly for the right–moving degrees of freedom ($\bar{\Psi}^I$, $\bar{y}^I$, $\bar{w}^I$, $I = 3,4,5,6$). The above transformation is a symmetry only if the rotation angle is a multiple of $\pi/2$ or $\phi = k/4$, with k integer.

Observe that with the help of the word sheet fermions we can classify the $N = 4$ string spectrum in terms of a left and right $U(1)$ charges $Q_L = \oint j_L$ and $Q_R = \oint j_R$, where

$$\begin{aligned}
j_L &= \chi_1\chi_1^\dagger - \chi_2\chi_2^\dagger + Y_1Y_1^\dagger - Y_2Y_2^\dagger + 2W_-W_-^\dagger\,, \\
Q_L &= q_{\chi_1} - q_{\chi_2} + q_{Y_1} - q_{Y_2} + 2q_{W_-}
\end{aligned} \tag{21}$$

and

$$\begin{aligned}
j_R &= \bar{\chi}_1\bar{\chi}_1^\dagger - \bar{\chi}_2\bar{\chi}_2^\dagger + \bar{Y}_1\bar{Y}_1^\dagger - \bar{Y}_2\bar{Y}_2^\dagger + 2\bar{W}_-\bar{W}_-^\dagger, \\
Q_R &= q_{\bar{\chi}_1} - q_{\bar{\chi}_2} + \bar{q}_{Y_1} - \bar{q}_{Y_2} + 2\bar{q}_{W_-}
\end{aligned} \tag{22}$$

We are now in a position to switch on non-vanishing F^a_{IJ} by performing a boost among the fermionic charge lattice and the $\Gamma_{(2,n)}$ lattice:

$$\begin{aligned}
q_{\chi_1} &\to q_{\chi_1} + h_i n^i, & q_{\chi_2} &\to q_{\chi_2} - h_i n^i \\
q_{\bar{\chi}_1} &\to q_{\bar{\chi}_1} + h_i n^i, & q_{\bar{\chi}_2} &\to q_{\bar{\chi}_2} - h_i n^i \\
q_{Y_1} &\to q_{Y_1} + h_i n^i, & q_{Y_2} &\to q_{Y_2} - h_i n^i \\
q_{\bar{Y}_1} &\to q_{\bar{Y}_1} + h_i n^i, & q_{\bar{Y}_2} &\to q_{\bar{Y}_2} - h_i n^i \\
q_{W_-} &\to q_{W_-} + 2h_i n^i, & q_{W_+} &\to q_{W_+} \\
q_{\bar{W}_-} &\to q_{\bar{W}_-} + 2h_i n^i, & q_{\bar{W}_+} &\to q_{\bar{W}_+} \\
P_i^L(h_i) &= P_i^L - h_i\,(Q_L - Q_R,), \\
P_i^R(h_i) &= P_i^R - h_i\,(Q_L - Q_R)
\end{aligned} \tag{23}$$

with

$$P_i^L = m_i \ + \ Y_i^a \ Q^a \ + \ \frac{1}{2} Y_i^a \ Y_j^a \ n^j \ + B_{ij} \ n^j \ + \ G_{ij} \ n^j$$

$$P_i^R = m_i \ + \ Y_i^a \ Q^a \ + \ \frac{1}{2} Y_i^a \ Y_j^a \ n^j \ + B_{ij} \ n^j \ - \ G_{ij} \ n^j \tag{24}$$

Y_i^a $i = i, 2$, $a = 1, 2, ..., 18$ are the Wilson-line moduli of the $\Gamma(2, 18)$ lattice. Owing to the non-zero h_i shift, two of the $N = 4$ gravitinos become massive, with mass proportional to $|q_{\chi_1} - q_{\chi_2}|$. The $N = 4$ gravitinos have vanishing $m_i, n^i, q_{Y_i}, q_{\bar{Y}_i}, q_{W_i}, q_{\bar{W}_i}$ charges. The two of them remain massless since $|q_{\chi_1} - q_{\chi_2}| = 0$, while for the other two become massive since $|q_{\chi_1} - q_{\chi_2}| = 1$:

$$(m_{3/2}^2)_{1,2} = 0, \quad (m_{3/2}^2)_{3,4} = \frac{|F|^2}{4\text{Im}T\text{Im}U}, \tag{25}$$

with $F \ = \ h_1 \ + U \ h_2$, T and U are usual complex moduli of the $\Gamma_{(2,2)}$ lattice:

$$T = i\sqrt{\det G_{ij}} \ + \ B_{12},$$

$$U = \frac{(i\sqrt{\det G_{ij}} \ + \ G_{12})}{G_{22}}. \tag{26}$$

The global existence of the supercurrent implies in this case the quantization condition: $4h_i$ = integer. The $N = 2$ partition function $\mathbb{Z}^{4\to 2}(F)$ is obtained from that of $N = 4$ by shifting the lattice momenta P_i and the R–charges q_i as above. Performing a Poisson resummation on m_i, we obtain the following expression:

$$\gamma = 2h_i n^i. \quad \delta = 2h_i m^i, \quad F = h_1 + U \ h_2$$

$$\mathbb{Z}^{4\to 2}(F) = \frac{(\text{Im}\tau)^{-1}}{\eta^2\bar{\eta}^2} \sum_{m^i,n^i} \frac{1}{2} \sum_{\alpha,\beta} (-)^{\alpha+\beta+\alpha\beta} \times \frac{\vartheta^2[^\alpha_\beta]}{\eta^2} \frac{\vartheta[^{\alpha\ +\ \gamma}_{\beta\ +\ \delta}]}{\eta} \frac{\vartheta[^{\alpha\ -\ \gamma}_{\beta\ -\ \delta}]}{\eta} \frac{\Gamma_{(2,2)}[^{n^i}_{m^i}]}{|\eta|^4} Z_{(4,4)}[^\gamma_\delta]$$

$$\times \sum_{\bar{\alpha},\bar{\beta}} \frac{1}{2} \sum_{\bar{\alpha},\bar{\beta}} \frac{\bar{\vartheta}^6[^{\bar{\alpha}}_{\bar{\beta}}]}{\bar{\eta}^6} \frac{\bar{\vartheta}[^{\bar{\alpha}\ +\ \bar{\gamma}}_{\bar{\beta}\ +\ \bar{\delta}}]}{\bar{\eta}} \frac{\bar{\vartheta}[^{\bar{\alpha}\ -\ \bar{\gamma}}_{\bar{\beta}\ -\ \bar{\delta}}]}{\bar{\eta}} \sum_{\epsilon,\zeta} \frac{1}{2} \frac{\bar{\vartheta}^8[^\epsilon_\zeta]}{\bar{\eta}^8}, \tag{27}$$

where

$$\Gamma_{(2,2)}[^{n^i}_{m^i}] \ = \ \sqrt{\det\, G_{ij}}\,(\text{Im}\tau)^{-1} \times \exp\,[-\pi G_{ij} \frac{(m^i + n^i\tau)(m^j + n^j\ \bar{\tau})}{\text{Im}\tau} + 2i\pi B_{ij} m^i n^j] \tag{28}$$

and

$$Z_{(4,4)}[^\gamma_\delta] \ = \frac{1}{2} \sum_{a,b} \frac{|\vartheta[^a_b]|^2}{|\eta|^2} \frac{|\vartheta[^{a\ +\ 2\gamma}_{b\ +\ 2\delta}]|^2}{|\eta|^2} \frac{|\vartheta[^{a\ +\ \gamma}_{b\ +\ \delta}]|^2}{|\eta|^2} \frac{|\vartheta[^{a\ -\ \gamma}_{b\ -\ \delta}]|^2}{|\eta|^2}. \tag{29}$$

When $h_i \ = \ 0$ (γ, $\delta \ = \ 0$), $\mathbb{Z}^{4\to 2}(F \ = \ 0)$ corresponds to the $N \ = \ 4$ heterotic string solution based on a gauge group $U(1) \times U(1) \times SO(8) \times E_8 \times E_8$; the $SO(8)$ gauge group factor corresponds to the extended symmetry of the $\Gamma_{(4,4)}$ lattice at the fermionic point

$$Z_{(4,4)}[^0_0] \ = \frac{1}{2} \sum_{a,b} \frac{|\vartheta[^a_b]|^8}{|\eta|^8}. \tag{30}$$

The sum over m^i and n^i gives rise to the $\Gamma(2,2)$ lattice at an arbitrary point of the moduli space:

$$\sum_{m^i,n^i} \Gamma_{(2,2)}[^{n^i}_{m^i}] \;=\; \Gamma_{(2,2)}[T,U]. \tag{31}$$

When $h_i \neq 0$ $(\gamma,\ \delta) = (2h_i n^i,\ 2h_i m^i)$, then the $N = 4$ supersymmety is spontaneously broken to $N = 2$ and the gauge group is reduced to $U(1)^2 \times E_7 \times E_8$, as in orbifold models.

The important difference between the $N = 2$ model described above and the orbifold models [8] of order $\mathbf{N}$ is in the parameters γ and δ, which appear as arguments in ϑ-functions. In the model in which somme of the $N = 4$ the supersymmetries are broken spontaneously, $\gamma = 2h_i n^i$ and $\delta = 2h_i m^i$ are not independent but are given in terms of the h_i and in terms of the charges n^i, m^i of the $\Gamma(2,2)[^{n^i}_{m^i}]$ lattice. In the standard symmetric orbifolds of order $\mathbf{N}$, the arguments γ and δ, ($\gamma = 2l/\mathbf{N}$ and $\delta = 2k/\mathbf{N}$ with $l, k =$ 0,1,...,$\mathbf{N}$ -1), are independent arguments; their summation gives rise to the orbifold projections and to some additional states in the twisted sector:

$$\begin{aligned}
\mathbb{Z}^{N=2}_{\rm orb} = {} & \frac{(\mathrm{Im}\tau)^{-1}}{|\eta(\tau|^4}\ \frac{1}{\mathbf{N}} \sum_{\gamma,\ \delta=0}^{1}\ \sum_{\alpha,\ \beta} (-)^{\alpha+\beta+\alpha\beta} \\
& \times \frac{\vartheta^2[^\alpha_\beta](\tau)}{\eta^2(\tau)}\frac{\vartheta[^{\alpha\,+\,\gamma}_{\beta\,+\,\delta}](\tau)}{\eta(\tau)}\frac{\vartheta[^{\alpha\,-\,\gamma}_{\beta\,-\,\delta}](\tau)}{\eta(\tau)}\frac{\Gamma_{(2,2)}}{|\eta(\tau)|^4}\ Z_{(4,4)}[^\gamma_\delta] \\
& \times \frac{1}{2}\sum_{\bar\alpha,\bar\beta}\frac{\bar\vartheta^6[^{\bar\alpha}_{\bar\beta}](\bar\tau)}{\bar\eta^6(\bar\tau)}\frac{\bar\vartheta[^{\bar\alpha\,+\,\gamma}_{\bar\beta\,+\,\delta}](\bar\tau)}{\bar\eta(\bar\tau)}\frac{\bar\vartheta[^{\bar\alpha\,-\,\gamma}_{\bar\beta\,-\,\delta}](\bar\tau)}{\bar\eta(\bar\tau)} \times \frac{1}{2}\sum_{\epsilon,\ \zeta}\frac{\bar\vartheta^8[^\epsilon_\zeta](\bar\tau)}{\bar\eta^8(\bar\tau)}.
\end{aligned} \tag{32}$$

In the language of orbifolds, the spontaneously broken theory, $\mathbb{Z}^{4\to 2}$, corresponds to a *freely acting orbifold.* Indeed, using the quantization condition,

$$\mathbf{N}\ h_i \;=\; \text{integer} \tag{33}$$

and the mod 2 periodicity properties of the ϑ-functions in the arguments

$$\vartheta[^{a+2k}_{b+2l}] \;=\; \vartheta[^a_b]\ e^{i\pi la}, \tag{34}$$

it is possible to write the $\mathbb{Z}^{4\to 2}$ theory in orbifold language. In order to make this correspondence explicit, we must first redefine the lattice charges $n^i = \mathbf{N}\hat{n}^i + \gamma^i$ and $m^i = \mathbf{N}\hat{m}^i + \delta^i$. Thanks to the property (34), the above lattice charge redefinition makes the arguments of the ϑ-functions independent of $\hat{n}^i$ and $\hat{m}^i$; they depend only on $\hat\gamma = 2h_i\gamma^i$ and $\hat\delta = 2h_i\delta^i$. Performing now a Poisson resummation on $\hat{m}^i$, we obtain the orbifold representation for $\mathbb{Z}^{4\to 2}$ theory:

$$\begin{aligned}
\mathbb{Z}^{4\to 2}(F) = {} & \frac{(\mathrm{Im}\tau)^{-1}}{\eta^2(\tau)\bar\eta^2(\bar\tau)}\frac{1}{\mathrm{N}}\sum_{\gamma^i\ \delta^i}\frac{1}{2}\sum_{\alpha,\beta}(-)^{\alpha+\beta+\alpha\beta} \times \frac{\vartheta^2[^\alpha_\beta](\tau)}{\eta^2(\tau)}\frac{\vartheta[^{\alpha\,+\,\hat\gamma}_{\beta\,+\,\hat\delta}](\tau)}{\eta(\tau)}\frac{\vartheta[^{\alpha\,-\,\hat\gamma}_{\beta\,-\,\hat\delta}](\tau)}{\eta(\tau)} \\
& \times \sum_{\bar\alpha,\bar\beta}\frac{1}{2}\sum_{\bar\alpha,\bar\beta}\frac{\bar\vartheta^6[^{\bar\alpha}_{\bar\beta}](\bar\tau)}{\bar\eta^6(\bar\tau)}\frac{\bar\vartheta[^{\bar\alpha\,+\,\hat\gamma}_{\bar\beta\,+\,\hat\delta}](\bar\tau)}{\bar\eta(\bar\tau)}\frac{\bar\vartheta[^{\bar\alpha\,-\,\hat\gamma}_{\bar\beta\,-\,\hat\delta}](\bar\tau)}{\bar\eta(\bar\tau)} \\
& \times Z^{SO(8)}_{(4,4)}[^{\hat\gamma}_{\hat\delta}]\ \frac{\Gamma_{(2,2)}[^{\gamma^i}_{\delta^i}]}{|\eta(\tau)|^4}\sum_{\epsilon,\zeta}\frac{1}{2}\frac{\bar\vartheta^8[^\epsilon_\zeta](\bar\tau)}{\bar\eta^8(\bar\tau)},
\end{aligned} \tag{35}$$

where

$$\Gamma_{(2,2)}\ [^{\gamma^i}_{\delta^i}] = \sum \exp i\pi\left[\frac{2\delta^i\hat{m}_i}{\mathrm{N}} + \tau\frac{1}{2}P^L_i g^{ij} P^L_j - \bar\tau\frac{1}{2}P^R_i g^{ij} P^R_j\right], \tag{36}$$

with

$$P_i^L = \hat{m}_i + \left(\hat{n}^j + \frac{\gamma^j}{\mathrm{N}}\right) G_{ij} \text{ and } P_i^R = \hat{m}_i - \left(\hat{n}^j + \frac{\gamma^j}{\mathrm{N}}\right) G_{ij}. \tag{37}$$

The connection of $\mathbb{Z}^{4\to 2}$ with the freely acting orbifolds gives us the way to switch all the moduli of $Z_{(4,4)}$ and so to move out of the $SO(8)$ extended symmetry point. This extension can be done by replacing the $SO(8)$ characters of $Z_{(4,4)}^{SO(8)}[^{\gamma}_{\delta}]$, which are defined at the fermionic point

$$Z_{(4,4)}^{SO(8)}[^{\gamma}_{\delta}] = \frac{1}{2}\sum_{a,b} \frac{|\vartheta[^a_b]|^2}{|\eta(\tau)|^2} \frac{|\vartheta[^{a+2\gamma}_{b+2\delta}]|^2}{|\eta(\tau)|^2} \frac{|\vartheta[^{a+\gamma}_{b+\delta}]|^2}{|\eta(\tau)|^2} \frac{|\vartheta[^{a-\gamma}_{b-\delta}]|^2}{|\eta(\tau)|^2} \tag{38}$$

by the characters of the Z_{N}-orbifold, $Z_{(4,4)}^{(twist)}[^{\gamma}_{\delta}]$.

• When $(\gamma,\ \delta) = (0,0)$, $Z_{(4,4)}^{SO(8)}[^0_0]$ must be replaced by the "untwisted" orbifold partition function, which depends on the T_{IJ} moduli

$$Z_{(4,4)}^{SO(8)}[^0_0] \to Z_{(4,4)}[^0_0][T_{IJ}] = \frac{\Gamma_{(4,4)}[T_{IJ}]}{|\eta(\tau)|^8}, \tag{39}$$

• When $(\gamma,\ \delta) \neq (0,0)$, $Z_{(4,4)}^{SO(8)}[^{\gamma}_{\delta}]$ *no modification is necessary* since the "twisted" orbifold partition function remains the same at any point of the moduli space:

$$Z_{(4,4)}^{SO(8)}[^{\gamma}_{\delta}] \to Z_{(4,4)}^{twist}[^{\gamma}_{\delta}] = Z_{(4,4)}^{(SO(8)}[^{\gamma}_{\delta}] \quad (\gamma,\ \delta) \neq (0,0). \tag{40}$$

The models described above are special cases of a general class of models having the interpretation of freely acting orbifods of the $N=4$ heterotic string theory. They are obtained in the following way. Consider $\Gamma(6,22)$ and set the appropriate moduli to special values, so that it factorizes as

$$\Gamma_{(6,22)} \to \Gamma_{(2,18)}\,\Gamma_{(4,4)} \tag{41}$$

Now consider the orbifold that acts as a Z_{N} rotation on $\Gamma_{(4,4)}$ and as a translation by an N-th lattice vector $\vec{\varepsilon}/\mathrm{N}$ with $\vec{\varepsilon} = (\vec{\varepsilon}_L; \vec{\varepsilon}_R, \vec{\zeta})$, on $\Gamma_{(2,18)}$. $\vec{\varepsilon}_{L,R}$ are two–dimensional vectors while $\vec{\zeta}$ is a sixteen dimensional vector. Owing to the accompanying translation on $\Gamma_{(2,18)}$, this is a freely acting orbifold.

The two types of constructions of $N=4 \to N=2$ theories we have presented above, have complementary features. In the first approach of using a specific generalized boost at the fermionic point, it is evident that there is a *one-to-one correspondence* of states between the original $N=4$ supersymmetric theory and the final *spontaneously broken* $N=4 \to N=2$ theory. This is what should be expected during spontaneous symmetry breaking. In the second, *freely acting orbifold* approach, we have a clear geometrical intuition about the spontaneously broken theory, which will be very useful for the identification of the heterotic and type II dual theories.

Inspection of the standard $N=4$ gravitino vertex operators shows that two of them are invariant while the other two transform, one with a phase $e^{2\pi i/\mathrm{N}}$ and the other with $e^{-2\pi i/\mathrm{N}}$. In order for them to survive in the spectrum they have to pair up with a state of the (2,18) lattice carrying momentum $p = (\vec{m}; \vec{n}, \vec{Q})$ but no oscillators (these will shift the mass to the Planck scale). Since such a lattice state picks up a phase $e^{2\pi i\varepsilon\cdot p/\mathrm{N}}$, one of the two massive gravitinos will have momentum p_1 with the property that $p_1 \cdot \varepsilon = 1 \mod \mathrm{N}$ while the other p_2 with $p_2 \cdot \varepsilon = -1 \mod \mathbf{N}$. The mass formulae given in (25) are special cases of the above.

There is an essential difference between the models with spontaneous breaking of the $N = 4 \longrightarrow N = 2$ and the standard $N = 2$ orbifold models.

• First, in the spontaneously broken case, one expects an effective restoration of the $N = 4$ supersymmetry in a corner of the moduli space T, U, where the two massive gravitinos become light, $m_{3/2} \longrightarrow 0$.

• Second, in the standard obifolds there is no restoration of the $N = 4$ supersymmetry at any point of the moduli space.

If there is an effective restoration of the $N = 4$ supersymmetry in the spontaneously broken case, then one must find zero higher-genus corrections to the coupling constants of the theory in the $N = 4$ restoration limit $m_{3/2} \longrightarrow 0$. This restoration phenomenon has been checked in ref.[17, 18] where the one-loop corrections of the coupling constants were performed for a class of Z_2 models based on $E_8 \times E_7 \times SU(2) \times U(1)^2$ gauge group. A more detailed discussion of the general heterotic models and their type II duals will appear in ref. [19]. Here I will restrict myself to the case of Z_2 freely acting orbifolds with $F = h_1 + Uh_2 = 1/2$. The $m_{3/2} \longrightarrow 0$ limit in this class of models corresponds to the corner of the moduli space $\text{Im}T\ \text{Im}U \longrightarrow \infty$, which implies an effective decompactification of one of the two coordinates of $\Gamma_{2,2}(T,U)$, ($R_1 \longrightarrow \infty$ and R_2 arbitrary; $\text{Im}T \sim R_1 R_2$, $\text{Im}U \sim R_1/R_2$). In this limit, T, $U \longrightarrow \infty$, one expects vanishing corrections to the coupling constants due to the effective $N = 4$ restoration. Using the explicit results of ref. [17],

$$\Delta^{freely}_{(8,7)} = \frac{16\pi^2}{g^2_{E_8}} - \frac{16\pi^2}{g^2_{E_7}} = \delta b\ \log\left[|\mu|^2\ \text{Im}\ T\text{Im}\ U|\vartheta_4(T)\ \vartheta_4(U)|^4\right], \qquad (42)$$

where $\delta b = b_8 - b_7$ and b_i are the β-function coefficients due to massless particles. When T and U are large, $\text{Im}T\ \text{Im}U \gg 1$, due to the asymptotic behaviour of $\vartheta_4(T) = 1 + \mathcal{O}(e^{-i\pi\ T})$:

$$\Delta^{freely}_{(8,7)} \longrightarrow \delta b \log |\mu|^2\ \text{Im}T\ \text{Im}U\ . \qquad (43)$$

The logarithmic contribution is an artefact due to the infrared divergences. In fact by turning on Wilson lines appropriately (e.g. small Higgs vev's of the vector multiplets), we can arrange that there are no charged states with masses $\mu_W^2 \sim |W|^2/\text{Im}T\ \text{Im}U$ below $m^2_{3/2}$. In this case the logarithmic term becomes:

$$\delta b \log |\mu|^2\ \text{Im}T\ \text{Im}U \longrightarrow \delta b \log \frac{\mu_W^2}{m^2_{3/2} + \mu_W^2} \sim \mathcal{O}\left(\frac{m^2_{3/2}}{\mu_W^2}\right); \qquad (44)$$

the logarithmic divergence thus disappears and the thresholds vanish, which shows the restoration of the $N = 4$ supersymmetry in the light massive gravitino limit as expected. In the calculation of individual couplings, there is an extra contribution $Y(T,U)$, which is "universal" for g_{E_8} and g_{E_7}; the explicit calculation in ref. [17–19] shows that $Y(T,U)$ behaves like

$$Y(T,U) \longrightarrow \frac{m^2_{3/2}}{M_s^2} \quad \text{as} \quad m_{3/2} \longrightarrow 0. \qquad (45)$$

Thus individual couplings also vanishe in the limit $m_{3/2} \longrightarrow 0$.

In the standard orbifold with $N = 2$ space-time supersymmetry, the corrections to the coupling constants have a different behaviour for T, $U \gg 1$ [20, 18]

$$\Delta^{orb}_{(8,7)} = \delta b\ \log\left[\mu^2\ \text{Im}T\text{Im}U|\eta(T)\eta(U)|^4\right]. \qquad (46)$$

When T, U is large, $ImTImU \gg 1$,

$$\Delta^{orb}_{(8,7)} \longrightarrow \delta b \left[\frac{\pi}{3}(\text{Im}T + \text{Im}U)\ +\ \log \frac{|W|^2}{M_s^2}\right] + \text{finite terms.} \qquad (47)$$

So, in the standard orbifolds, the correction to the coupling constants grows linearly with the five–dimensional volume. This shows that the $N = 2$ supersymmetry is "not extended" in the decompactification limit $R_1 \rightarrow \infty$. On the other hand there is an extension of the supersymmetry in the freely acting orbifold case.

In the opposite limit $\mathrm{Im}T\mathrm{Im}U \rightarrow 0$, the situation is different:

i) In the freely acting orbifolds the two massive gravitinos becomes superheavy: $m_{3/2} \rightarrow \infty$ in the limit $\mathrm{Im}T\mathrm{Im}U \rightarrow 0$.

ii) In the standard orbifolds, thanks to the duality symmetry $R_i \rightarrow 1/R_i$ the behaviour $T, U \rightarrow 0$, is identical to the dual model with $T' = -1/T,\ U' = -1/U \ \rightarrow \infty$ and thus

$$\Delta^{orb}_{(8,7)}(T, U, W) = \Delta^{orb}_{(8,7)}(T', U', W') \rightarrow \delta b \left[\frac{\pi}{3}(\mathrm{Im}T' + \mathrm{Im}U') + \log \frac{|W'|^2}{M_s^2}\right] + \text{finite terms} \tag{48}$$

In the freely acting orbifolds, the $SO(2,2;Z)$ duality symmetry is reduced to a smaller subgroup due to the Z_2 action on the lattice. Thus one expects non-restoration of the $N = 4$ supersymmetry in this limit ($T,\ U \rightarrow 0\ m_{3/2} \rightarrow \infty$):

$$\Delta^{freely}_{(8,7)} \ = \ \delta b' \ \left[\ \log|\vartheta_2(T')\ \vartheta_2(U')|^4 + \log\frac{|W'|^2}{M_s^2}\right]. \tag{49}$$

In the above equation we have used the ϑ-identity

$$\mathrm{Im}T|\vartheta_4(T)|^4 = \mathrm{Im}T'|\vartheta_2(T')|^4, \quad T' = -\frac{1}{T}. \tag{50}$$

Using the asymptotic behaviour for $T', U' \gg\ 1$ of $\log\ |\vartheta_2(T')\ \vartheta_2(U')|^4$ one obtains:

$$\Delta^{freely}_{(8,7)} \rightarrow \delta b' \left[\frac{\pi}{3}(\mathrm{Im}\ T' + \mathrm{Im}\ U') + \log \frac{|W'|^2}{M_s^2}\right] + \text{finite terms.} \tag{51}$$

It is interesting to observe that the $m_{3/2} \rightarrow \infty$ limit [17] of the freely acting orbifolds corresponds to a corner in the moduli space of T, U where the two classes of theories (the freely and non-freely acting orbifolds) "touch" each other. Both theories are effectively five-dimensional. Thus the five-dimensional standard $N = 2$ orbifolds can be viewed as an $m_{3/2} \rightarrow \infty$ limit of some spontaneously broken $N = 4$ models.

$N = 4 \rightarrow N = 1$ spontaneous SUSY breaking

Using the connection between the freely acting orbifolds and the spontaneous breaking $N = 4 \rightarrow N = 2$, we can proceed to further break the supersymmetry to $N = 1$. We will restrict ourselves to the case where the possible quantized parameters are of order N=2 $2|h^i| = 1$. In that case the spontaneously broken $N = 4 \rightarrow N = 1$ theory is strongly connected to $Z^2 \times Z^2$ freely acting orbifolds; the $Z^2 \times Z^2$ acts simultaneously as a rotation on the coordinates $\Phi^I,\ \Phi^J$ and $\Psi^I,\ \Psi^J$ of the two complex planes and as a translation on the third complex plane Φ^L. Denoting by $\Phi_A,\ A = 1, 2, 3$, the complex internal coordinates and by $\chi_A,\ A = 1, 2, 3$, the three complex fermionic world-sheet superpartners, the non-trivial actions of the orbifold are:

1) $\Phi_1 \rightarrow \Phi_1 + 2\pi h_1, \quad (\Phi_2,\ \chi_2) \rightarrow e^{\ i2\pi h_1}(\Phi_2,\ \chi_2), \quad (\Phi_3,\ \chi_3) \rightarrow e^{-i2\imath h_1}(\Phi_3,\ \chi_3).$

2) $\Phi_2 \rightarrow \Phi_2 + 2\pi h_2, \quad (\Phi_1,\ \chi_1) \rightarrow e^{i2\pi h_2}(\Phi_1,\ \chi_1), \quad (\Phi_3,\ \chi_3) \rightarrow e^{-i2\pi h_2}(\Phi_3,\ \chi_3).$

$3)\Phi_3 \longrightarrow \Phi_3 + 2\pi h_3, \quad (\Phi_1,\ \chi_1) \ \longrightarrow e^{i2\pi h_3}(\Phi_1,\ \chi_1), \quad (\Phi_2,\ \chi_2) \ \longrightarrow e^{-i2\pi h_3}(\Phi_2,\ \chi_2).$

In order to obtain the partition function and define the theory, we need to introduce the "shifted" and "twisted" characters of the three complex coordinates. We denote by $(\gamma_A,\ \delta_A)$ the translation shifts and by $(H_A,\ G_A)$ the rotation twists. When the "twist" is zero $(H_A,\ G_A) = (0,\ 0)$:

$$Z_A \begin{bmatrix} \gamma_A \ ; \ 0 \\ \delta_A \ ; \ 0 \end{bmatrix} \ = \ \frac{\Gamma_{(2,2)} \left[\begin{smallmatrix} \gamma_A \\ \delta_A \end{smallmatrix} \right]}{|\eta|^4} \ \delta(H_A)\delta(G_A). \tag{52}$$

When the twist is non-zero $(H_A,\ G_A) \neq (0,\ 0)$:

$$Z_A \begin{bmatrix} \gamma_A \ ; \ H_A \\ \delta_A \ ; \ G_A \end{bmatrix} \ = \ \frac{1}{2} Z^{twist}_{(2,2)} \begin{bmatrix} H_A \\ G_A \end{bmatrix} \times \left[\ \delta(\gamma_A) \ \delta(\delta_A) \ + \ \delta(\gamma_A + H_A) \ \delta(\delta_A + G_A) \right]. \tag{53}$$

In the above equation, the $Z^{twist}_{(2,2)} \begin{bmatrix} H_A \\ G_A \end{bmatrix}$ can be written either in terms of a twisted boson or in terms of 2d-fermionic characters with shifted boundary conditions:

$$Z^{twist}_{(2,2)} \begin{bmatrix} H_A \\ G_A \end{bmatrix} = \frac{4|\eta|^2}{\vartheta\binom{1\ +\ H_A}{1\ +\ G_A}\vartheta\binom{1\ -\ H_A}{1\ -\ G_A}|} = \frac{1}{2} \sum_{a,b} \frac{|\vartheta^2\binom{a}{b}\vartheta\binom{a\ +\ H_A}{b\ +\ G_A}\vartheta\binom{a\ -\ H_A}{b\ -\ G_A}|}{|\eta|^4}$$
$$\text{if} \quad (H_A,\ G_A) \neq (0,0) \tag{54}$$

The world-sheet modular properties of $Z_A \begin{bmatrix} \gamma_A \ ; \ H_A \\ \delta_A \ ; \ G_A \end{bmatrix}$ are the same for any point of the moduli space and thus at the $SO(4)_A$ fermionic point, which takes the following form:

$$Z_A \begin{bmatrix} \gamma_A \ ; \ H_A \\ \delta_A \ ; \ G_A \end{bmatrix} |_{T_0^A, U_0^A} \ = \frac{1}{2} \sum_{a,b} \frac{|\vartheta^2\binom{a}{b}\vartheta\binom{a+H_A}{b+G_A}\vartheta\binom{a-H_A}{b-G_A}|}{|\eta|^4} e^{i\pi(a\delta^A + b\gamma^A + \gamma^A\delta^A)}. \tag{55}$$

The above expression makes the world-sheet modular properties of $Z_A \begin{bmatrix} \gamma_A \ ; \ H_A \\ \delta_A \ ; \ G_A \end{bmatrix}$ more transparent under $SL(2, Z)_\tau$; it also makes clear the connection to the fermionic models. The role of the phase factor $e^{i\pi \ (\ a \ \delta^A \ + \ b \ \gamma^A \ + \ \gamma^A \ \delta^A)}$ is of main importance, since it clarifies the way we had to choose the coefficient of the fermionic characters.

We are now in a position to construct consistent $N = 4 \rightarrow N = 1$ models using the fermionic construction algorithm [4]. Although these constructions are at special points of the moduli space $(T_0^A,\ U_0^A)$, the generalization of them for arbitrary moduli is automatic by a simple replacement of the fermionic "twisted" characters with the characters of the "shifted" and "twisted" bosonic coordinates:

$$Z_A \begin{bmatrix} \gamma_A \ ; \ H_A \\ \delta_A \ ; \ G_A \end{bmatrix} |_{T_0^A, U_0^A} \ \rightarrow Z_A \begin{bmatrix} \gamma_A \ ; \ H_A \\ \delta_A \ ; \ G_A \end{bmatrix} |_{T,U}. \tag{56}$$

Many models can be constructed in this way. We will display below the partition function of a model with one unbroken and three spontaneously broken supersymmetries, $N = 4 \rightarrow N = 1$ (the unbroken gauge group of this example is $E_8 \times E_6 \times U(1)^2$:

$$\mathbb{Z}^{4\rightarrow 1}(F_i) = \frac{(\mathrm{Im}\tau)^{-1}}{\eta^2\bar{\eta}^2} \times \frac{1}{4} \sum_{h_i, g_i} Z_1 \begin{bmatrix} h_1 \ ; \ h_2 \\ g_1 \ ; \ g_2 \end{bmatrix} \ Z_2 \begin{bmatrix} h_2 \ ; \ h_3 \\ g_2 \ ; \ g_3 \end{bmatrix} \ Z_3 \begin{bmatrix} h_3 \ ; \ h_1 \\ g_3 \ ; \ g_1 \end{bmatrix}$$
$$\times \frac{1}{2} \sum_{\alpha,\beta} (-)^{\alpha+\beta+\alpha\beta} \ \frac{\vartheta[^\alpha_\beta]}{\eta} \frac{\vartheta[^{\alpha+h_2}_{\beta+g_2}]}{\eta} \frac{\vartheta[^{\alpha+h_3}_{\beta+g_3}]}{\eta} \frac{\vartheta[^{\alpha+h_1}_{\beta+g_1}]}{\eta}$$
$$\times \frac{1}{2} \sum_{\bar{\alpha},\bar{\beta}} \frac{\bar{\vartheta}^5[^{\bar{\alpha}}_{\bar{\beta}}]}{\bar{\eta}^5} \frac{\bar{\vartheta}[^{\bar{\alpha}+h_2}_{\bar{\beta}+g_2}]}{\bar{\eta}} \frac{\bar{\vartheta}[^{\bar{\alpha}+h_3}_{\bar{\beta}\ +\ g_3}]}{\bar{\eta}} \frac{\bar{\vartheta}[^{\bar{\alpha}+h_1}_{\bar{\beta}+g_1}]}{\bar{\eta}} \times$$

$$\delta(h_1 + h_2 + h_3)\ \delta(g_1 + g_2 + g_3)\frac{1}{2}\sum_{\epsilon,\zeta}\frac{\bar{\vartheta}^8[^{\epsilon}_{\zeta}]}{\bar{\eta}^8}. \tag{57}$$

The existence of one unbroken supersymmetry is ensured because of the relations $h_1 + h_2 + h_3 = 0$ and $g_1 + g_2 + g_3 = 0$; these relations guarantee the existence of an $N = 2$ superconformal symmetry on the world-sheet and thus the existence of $N = 1$ space-time supersymmetry[21].

It is easy to see that the partition function $\mathbb{Z}^{4\to 1}$ can be decomposed in four sectors:

• **The $N = 4$ sector**, with no rotations or translations in all three complex planes $((h_A, g_A) = (0, 0))$

• **Three $N = 2$ sectors**, with a non-zero translation in one of the complex planes and opposite non-zero rotations in the remaining two complex planes.

The contribution to the partition function of the $N = 4$ sector is one quarter of the $N = 4$ partition function with lattice momenta in the reduced $\Gamma(2,2)^3$ lattice. The contribution of the other three $N = 2$ sectors are equal sector by sector to the corresponding $N = 4 \to N = 2$ partition function divided by a factror of 2. The untwisted complex plane lattice momenta correspond to the shifted $\Gamma_{(2,2)}\ [^{\gamma_A}_{\delta_A}]$ lattice. The moduli-dependent corrections to the gauge couplings can be easily determined by combining the results of the individual $N = 2$ sectors.

$$\frac{16\pi^2}{g^2_{E_8}} - \frac{16\pi^2}{g^2_{E_6}} = \Delta_{(8,6)} = \frac{1}{2}\sum_{A=1}^{3}\Delta^A_{(8,7)}, \tag{58}$$

where the expressions of the $\Delta^A_{(8,7)}$ are given in (42).

As we mentioned in the $N = 4 \to N = 2$ spontaneous breaking, one expects a restoration of the $N = 4$ supersymmetry in the limit in which the massive gravitinos become massless; in order to prove the $N = 4$ restoration in the $N = 4 \to N = 1$ defined above as a $Z^2 \times Z^2$ freely acting orbifold, we need to identify the three massive gravitinos and express their masses in terms of the moduli fields and the three R-symmetry charges q_i $(i = 1, 2, 3)$:

$$m^2_{3/2}(q_i) = \frac{|q_2 - q_3|^2}{4\ \text{Im}\ T_1\ \text{Im}\ U_1} + \frac{|q_3 - q_1|^2}{4\ \text{Im}\ T_2\ \text{Im}\ U_2} + + \frac{|q_1 - q_2|^2}{4\ \text{Im}\ T_3\ \text{Im}\ U_3} \tag{59}$$

with $|q_0 + q_1 + q_2 + q_3| = 1$ and $|q_i| = |q_0| = \frac{1}{2}$ where q_0 is the left-helicity charge. Using the above expression, one finds the desired result:

$$(m^2_{3/2})_1 = \frac{1}{4\ \text{Im}\ T_2\ \text{Im}\ U_2} + \frac{1}{4\ \text{Im}\ T_3\ \text{Im}\ U_3},$$

$$(m^2_{3/2})_2 = \frac{1}{4\ \text{Im}\ T_3\ \text{Im}\ U_3} + \frac{1}{4\ \text{Im}\ T_1\ \text{Im}\ U_1},$$

$$(m^2_{3/2})_3 = \frac{1}{4\ \text{Im}\ T_1\ \text{Im}\ U_1} + \frac{1}{4\ \text{Im}\ T_2\ \text{Im}\ U_2},$$

$$(m^2_{3/2})_0 = 0. \tag{60}$$

The three massive gravitinos become massless in the decompactification limit $\text{Im}\ T_I\ \text{Im}\ U_I \to \infty$, $I = 1, 2, 3$, with ratios $\text{Im}\ T_I/\text{Im}\ U_I$ fixed. Thus the full restoration of the $N = 4$ effectively takes place in seven dimensions. Partial restoration of an $N = 2$ supersymmetry can happen in six dimensions when $\text{Im}T_I\ \text{Im}\ U_I \to \infty$, $I = 1, 2$; In this limit $(m^2_{3/2})_0 = 0$ and $(m^2_{3/2})_3 \to 0$.

$N = 2 \to N = 1$ **spontaneous SUSY breaking**

Using similar techniques as before, it is possible to construct $N = 2$ models with one of the supersymmetries to be spontaneously broken, $N = 2 \to N = 1$. In this class of models the restoration of $N = 2$ takes place in six dimensions. No further restoration of supersymmetry is possible. Examples can be obtained as in $(T^2 \otimes K_3)/Z_f^2$ orbifold compactification in which the Z_f^2 is freely acting. A representative example of this class of models is the one in which the K_3 compactification is chosen to be at the orbifold point $T^4/Z_o^2 \sim K_3$ (we denote by Z_o^2 the orbifold group and by Z_f^2 that which corresponds to the freely acting orbifold). We will give below the exact partition function that corresponds to this construction. From the explicit expression we can directly verify the effective restoration of $N = 2$ supersymmetry in the large-volume limit of K_3. Using the $Z_o^2 \otimes Z_f^2$ orbifold notation, the partition function of the $(T^2 \otimes T^4/Z_o^2)/Z_f^2$ model is:

$$\begin{aligned} \mathbb{Z}^{2\to 1}(F_i) = {} & \frac{(\mathrm{Im}\tau)^{-1}}{\eta^2\bar\eta^2} \times \frac{1}{2}\sum_{h_f,g_f} \frac{1}{2}\sum_{h_o,g_o} Z_1\left[{}^{0;h_f}_{0;g_f}\right] Z_2\left[{}^{h_f;h_o}_{g_f;g_o}\right] Z_3\left[{}^{h_f;-h_f-h_o}_{g_f;-g_f-g_o}\right] \\ & \times \frac{1}{2}\sum_{\alpha,\beta}(-)^{\alpha+\beta+\alpha\beta} \frac{\vartheta[^\alpha_\beta]}{\eta} \frac{\vartheta[^{\alpha+h_f}_{\beta+g_f}]}{\eta} \frac{\vartheta[^{\alpha+h_o}_{\beta+g_o}]}{\eta} \frac{\vartheta[^{\alpha-h_f-h_o}_{\beta-h_f-h_o}]}{\eta} \\ & \times \frac{1}{2}\sum_{\bar\alpha,\bar\beta} \frac{\bar\vartheta^5[^{\bar\alpha}_{\bar\beta}]}{\bar\eta^5} \frac{\bar\vartheta[^{\bar\alpha+h_f}_{\bar\beta+g_f}]}{\bar\eta} \frac{\bar\vartheta[^{\bar\alpha+h_o}_{\bar\beta+g_o}]}{\bar\eta} \frac{\bar\vartheta[^{\bar\alpha-h_f-h_o}_{\bar\beta-g_f-g_o}](\bar\tau)}{\bar\eta} \times \frac{1}{2}\sum_{\epsilon,\zeta} \frac{\bar\vartheta^8[^\epsilon_\zeta]}{\bar\eta^8}. \end{aligned} \tag{61}$$

In the above expression, the parameters $(h_f,\ g_f)$ and $(h_o,\ g_o)$ correspond to Z_f^2 and Z_o^2 respectively. The unbroken gauge group of this model is the $E_8 \otimes E_6 \otimes U(1)^2$. Switching on continuous or discrete Wilson lines, we can construct a large class of models with different gauge group but with a universal behaviour with respect to the $N = 2$ restoration at the large moduli limit; the massive gravitino of the broken $N = 2$ becomes massless when (Im T_2 ImU_2 and Im T_3 Im U_3 large).

$$(m^2_{3/2})_1 \;=\; \frac{1}{4\ \mathrm{Im}\ T_2\ \mathrm{Im}\ U_2} + \frac{1}{4\ \mathrm{Im}\ T_3\ \mathrm{Im}\ U_3},\ (m^2_{3/2})_0 \;=\; 0. \tag{62}$$

An easy way to view this ground-state is as an orbifold of the original $N = 4$ theory by the following non-trivial $Z_2 \times Z_2$ elements: $(1, r, r)$, (r, rt, t), (r, t, rt), (r stands for "π-rotation" and t for 1/2–lattice translation); $(1, r, r)$ has four fixed planes while the others have none. Because of the $N = 2$ restoration phenomenon, we expect that the only non-vanishing corrections to the gauge coupling constants are those that correspond to the $N = 2$ sector with $(h_f,\ g_f) = (0,\ 0)$ and $(h_o,\ g_o) \neq (0,\ 0)$. Indeed in this sector the Z_o^2 acts trivially on the $\Gamma_{(2,2)}(T_1,\ U_1)$ lattice as in the usual orbifolds. On the other hand, in the remaining two $N = 2$ sectors,

i) $(h_o, g_0) = (0, 0)$, $(h_f, g_f) \neq (0, 0)$

ii) $(h_o, g_0) + (h_f, g_f) = (0, 0)$, $(h_f, g_f) \neq (0, 0)$.

In both sectors the corresponding Z^2 acts without fixed points because of the simultaneous non-trivial shift $(h_f,\ g_f)$ on the corresponding $\Gamma_{(2,2)}(T_A,\ U_A)$, $A = 2, 3$, lattice.

The moduli-dependent corrections to the gauge couplings can be easily determined by combining the results of the individual $N = 2$ sectors.

$$\Delta_{(8,6)} \;=\; \frac{16\pi^2}{g^2_{E_8}} - \frac{16\pi^2}{g^2_{E_6}} \;=\; \frac{1}{2}\left(\ \Delta^1_{(8,7)} + \Delta^2_{(8,7)} + \Delta^3_{(8,7)}\right), \tag{63}$$

where the $\Delta^A_{(8,7)}$ are the threshold corrections of the three $N = 2$ sectors:

$$\Delta^1_{(8,7)} = (b^1_8 - b^1_7) \log \left[|\mu|^2 \mathrm{Im}T_1 \mathrm{Im}U_1 |\eta(T_1)\eta(U_1)|^4\right]$$
$$\to (b^1_8 - b^1_7) \left[\frac{\pi}{3}(\mathrm{Im}T_1 + \mathrm{Im}U_1) + \log|\mu|^2 \mathrm{Im}T_1 \mathrm{Im}U_1\right] \tag{64}$$

which corresponds to the threshold corrections of the standard orbifolds.

On the other hand $\Delta^A_{(8,7)}$ for $A = 2, 3$ will correspond to the threshold corrections of freely acting orbifolds which have different behaviour in the large-moduli limit:

$$\Delta^A_{(8,7)} = (b^A_8 - b^A_7) \log \left[|\mu|^2 \mathrm{Im}T_A \mathrm{Im}U_A \right] + (b^A_8 - b^A_7)\log \left[|\vartheta_4(T_A)\, \vartheta_4(U_A)|^4 \right]$$
$$\to (b_8 - b_7) \log |\mu|^2 \mathrm{Im}T_A \mathrm{Im}U_A \,. \tag{65}$$

Modulo the artificial sub-leading logarithmic contribution (due to the infrared divergences), the moduli contribution of the second and third plane T_A, U_A, $A = 2, 3$, is exponentially suppressed due to the asymptotic behaviour of $\vartheta_4(T_A)$, $\vartheta_4(U_A)$ for large T_A and U_A, $\vartheta_4(T_A) = 1 + \mathcal{O}(e^{-i\pi\, T_A})$.

There is a large class of such models obtained from $N = 2$ Z_2 orbifold compactifications by using D_4 type symmetries that act on the twist fields as well as the lattice.

Non-Perturbative BPS $N = 4$ Mass Formula

The low-energy effective $N = 4$ supergravity is manifestly invariant under the $SO(6, 22; R)$ group. The full string theory, however, is only invariant under the discrete subgroup $O(6, 22, Z)$. Furthermore the equations of motion and Bianchi identities are invariant under the $SL(2, R)$ transformation

$$S \to \frac{aS + b}{cS + d} \quad \text{with} \quad ad - bc = 1, \tag{66}$$

provided we perform an $SL(2, R)$ transformation on the "electric" and "magnetic" fields and charges. In particular, the transformation $S \to -1/S$ interchanges electric and magnetic charges. It has been conjectured [22–24] that a discrete subgroup $SL(2, Z)$ of this continuous symmetry of the equations of motion of the effective theory is a (non-perturbative) symmetry of the full theory.In the heterotic theory, only electrically charged states exist; these charges are the six quantized momenta and windings as well as the 16 $U(1)$ charges of $E_8 \times E_8$ or $SO(32)$, $(m_I,\ n^I,\ Q^k)$. Obviously the spectrum of the heterotic theory is not invariant under $SL(2, Z)$. For this to be true it is necessary to include in the theory non-perturbative states that carry both electric $(m_I,\ n^I,\ Q^k)$ and magnetic $(\tilde{m}_I,\ \tilde{n}^I,\ \tilde{Q}^k)$ charges [23,25–28]. Thanks to the $N = 4$ supersymmetric algebra and its central extension, one can write down an exact mass formula for all stable perturbative and non perturbative states, which preserves at least one of the four supersymmetries (BPS-states).

$$M^2_{BPS} = \frac{(P_I + S\, \Pi_I)\, g^{IJ}\, (P_J + S\, \Pi_J)}{4\mathrm{Im}S} + \frac{1}{2}\sqrt{(P_I\, g^{IJ}\, P_J)\,(\Pi_I\, g^{IJ}\, \Pi_J) - (P_I\, g^{IJ}\, \Pi_J)^2} \tag{67}$$

where the "electric" and the "magnetic" momenta P_I and Π_I are given in terms of the "electric" $(m_I,\ n^I,\ Q^k)$ and "magnetic" $(\tilde{m}_I,\ \tilde{n}^I,\ \tilde{Q}^k)$ charges:

$$P_I = m_I + Y^k_I Q^k + \frac{1}{2} Y^k_I Y^k_J n^J + B_{IJ} n^J + G_{IJ} n^J$$

$$\Pi_I = \tilde{m}_I + Y_I^k \tilde{Q}^k + \frac{1}{2} Y_I^k Y_J^k \tilde{n}^J + B_{IJ}\tilde{n}^J + G_{IJ}\tilde{n}^J \tag{68}$$

The square-root factor in the BPS mass formula is proportional to the difference of the two central charges squared: depending on whether this vanishes or not, the representation preserves 1/2 or 1/4 of the supersymmetries, (either short or intermediate supermultiplets). For the perturbative BPS states of the heterotic string ($\tilde{m}_I$, $\tilde{n}^I$, $\tilde{Q}^k$) = 0, and thus belong to short supermultiplets. Their mass reads

$$M^2_{BPS,\text{pert}} \;=\; \frac{1}{4\text{Im}S}\, P_I\, g^{IJ}\, P_J. \tag{69}$$

The factor of ImS is there because masses are measured in units of M_{Planck}. The BPS mass formula is manifestly invariant under $SL(2,Z)_S$.

$$S \to S+1; \qquad P_I \to P_I \,+\, \Pi_I, \qquad \Pi_I \to \Pi_I$$

$$S \to -\frac{1}{S}; \qquad \Pi_I \to P_I, \qquad P_I \to -\Pi_I. \tag{70}$$

Although the mass formula for non-perturbative BPS states is understood, we do not know a priori the multiplicities of all these states. From the $N=4$ heterotic string we know the multiplicities when $\Pi_I = 0$. Using $SL(2,Z)$ we also know the multiplicities of all states with $P_I\, g^{IJ}\, \Pi_J = \; 0$. To go further and learn more about the states with $P_I\, g^{IJ}\, \Pi_J \;\neq\; 0$ (namely intermediate multiplets) it is necessary to go beyond the string picture and learn more about the non-perturbative structure of the theory. The heterotic string on T^6 is supposed to be equivalent, in the strong coupling limit to the type II theory compactified on $K_3 \times T^2$. Moreover, there is a hypothetical 11-d theory (M-theory) that includes the non-perturbative dynamics of type IIA theory [15]. Thus compactification of M-theory on $K_3 \times T^3$ contains all the relevant non perturbative information about the heterotic $N = 4$ theory. This idea led to a conjecture on the multiplicities of dyonic BPS states in the 4-d $N=4$ theory [31]. This will be an important input, for our non-perturbative analysis of the spontaneously broken $N=4$ theory.

BPS states in models with partial SUSY breaking $N = 4 \to N = 2$

Let us consider an interesting question concerning the BPS spectrum of the theories where $N = 4$ is spontaneously broken to $N = 2$. In the original heterotic $N = 4$ theory, there are only short BPS multiplets in the perturbative spectrum. Their multiplicities can be easily counted by using helicity supertrace formulae [30]. In particular, the supertrace of helicity to the power 4 counts the multiplicities of $N = 4$ short (massless or massive) multiplets. From the partition function of the heterotic $N = 4$, we can construct the helicity-generating partition function:

$$\begin{aligned} Z^{het}_{N=4}(v,\bar v) &= \text{Str}[q^{L_0}\, \bar q^{\bar L_0} e^{2\pi i v\lambda_R - 2\pi i \bar v \lambda_L}] = \frac{1}{2}\sum_{\alpha\beta}(-1)^{\alpha+\beta+\alpha\beta}\, \frac{\vartheta[^\alpha_\beta](v)\vartheta^3[^\alpha_\beta]}{\eta^{12}\bar\eta^{24}}\; \xi(v)\; \bar\xi(\bar v)\; \frac{\Gamma_{(6,22)}}{\text{Im}\tau} \\ &= \frac{\vartheta_1^4(v/2)}{\eta^{12}\bar\eta^{24}}\xi(v)\bar\xi(\bar v)\frac{\Gamma_{(6,22)}}{\text{Im}\tau} \end{aligned} \tag{71}$$

The physical helicity in closed string theory λ is the sum of the left helicity λ_L and the right helicity λ_R:

$$\xi(v) \;=\; \prod_{n=1}^{\infty} \frac{(1-q^n)^2}{(1-q^n e^{2\pi i v})(1-q^n e^{-2\pi i v})} \;=\; \frac{\sin \pi v}{\pi}\,\frac{\vartheta_1'}{\vartheta_1(v)}, \qquad (\,\xi(v) = \xi(-v)\,), \tag{72}$$

which counts the contributions to the helicity due to the world-sheet bosons. If we define

$$Q = \frac{1}{2\pi i}\frac{\partial}{\partial v} \quad , \quad \bar{Q} = -\frac{1}{2\pi i}\frac{\partial}{\partial \bar{v}}, \tag{73}$$

then

$$B_4 = \langle \lambda^4 \rangle = (Q+\bar{Q})^4 Z^{het}_{N=4}(v,\bar{v})|_{v=\bar{v}=0} = \frac{3}{2}\frac{\Gamma_{6,22}}{\bar{\eta}^{24}}. \tag{74}$$

The numerator provides the mass formula while the denominator $1/\bar{\eta}^{24}$ provides the multiplicities. More precisely defined:

$$\frac{1}{\eta^{24}} = \sum_{N=-1}^{\infty} d(N) q^N = \frac{1}{q} + 24 + 324q + \mathcal{O}(q^2). \tag{75}$$

Then at the mass levels $M^2 = \frac{1}{4}P_L^2$, with

$$q_e^2 \equiv 2\vec{m}\cdot\vec{n} - \vec{Q}\cdot\vec{Q}, \tag{76}$$

the multiplicity is $d(q_e^2/2)$. The generalization of the string "electric" multiplicity to the non-perturbative dyonic states needs to assume a genus-2 generating function[31] $\Phi(\tau_{ij})$ $(\tau^{ij} = \tau^{ji})$:

$$\Phi(\tau_{ij}) \;=\; \sum_{N_{ij}} d(N_{ij}) \exp\left[\, 2i\pi\, N_{ij}\tau^{ij}\,\right], \tag{77}$$

where the levels N_{ij} of a dyonic state are characterized by the electric and magnetic charges

$$\vec{q}_1 \equiv \vec{q}_e = (\vec{m},\, \vec{n},\, \vec{Q}), \qquad \vec{q}_2 \equiv \vec{q}_m = (\vec{\tilde{m}}, \vec{\tilde{n}}, \vec{\tilde{Q}})$$

$$2N_{ij} = \vec{q}_i\cdot\vec{q}_i\ ; \tag{78}$$

the above equation generalizes the "electric" matching condition for the dyons and magnetic monopoles. The non-perturbative multiplicities are determined in terms of the electric and magnetic charges $d(\vec{q}_i\cdot\vec{q}_i/2))$. Thus the knowledge of the generating function $\Phi(\tau_{ij})$ determines the full spectrum of the perturbative and non-perturbative BPS states in terms of the moduli fields $(S,\ T_{IJ},\ Y_I^k)$ and the charges. In ref.[31], it was conjectured that the generating function $\Phi(\tau_{ij})$ is the genus-2 determinant of 24 bosons:

$$\Phi(\tau_{ij}) \;=\; \eta[\tau_{ij}]^{-24} \;=\; \left[\prod_{even} \vartheta[\tau_{ij}]\right]^{-2}. \tag{79}$$

Using the genus-2 interpretation of the non-perturbative multiplicities, we will present in section-**9** an algorithm that determines the non-perturbative BPS spectrum of the $N=2$ theories in terms of the shifted $N=4$ spectrum. Here we will restrict ourselves to the perturbative heterotic and type II cases. Some general facts, valid in $N=2$ theories, are in order:

• The $N=2$ massless multiplets M_0^λ have the following helicity content:

$$\pm\left(\lambda \pm \frac{1}{2}\right) + 2(\pm\lambda); \tag{80}$$

M_0^0 is the hypermultiplet, $M_0^{1/2}$ is the vector multiplet, while $M_0^{3/2}$ is the supergravity multiplet.

• The massive BPS multiplets have the following $SO(3)$ spin content

$$M^j \;:\; [j]\otimes([1/2]+2[0]) \tag{81}$$

and contain $2(2j+1)$ bosonic states and an equal number of fermionic ones.

• Finally the generic long-massive multiplet has the following SO(3) content

$$L^j \;:\; [j]\otimes([1]+4[1/2]+5[0]) \tag{82}$$

• The $N=2$ BPS states correspond to the short multiplets and are picked up by the supertrace of helicity squared, $B_2 = \langle\lambda^2\rangle$. We have

$$B_2(M_0^\lambda) = (-1)^{2\lambda}\;,\; B_2(L^j) = 0\;,\; B_2(M^j) = (-1)^{2j+1}(2j+1)/2. \tag{83}$$

For the perturbative heterotic theory, a direct computation determines B_2 in terms of the characters of the shifted $\Gamma_{2,18}[^h_g]$ and of the four twisted (right-moving) bosons $Z_{4,4}\left[^h_g\right]$. In what follows we will assume for simplicity the Z_2 freely acting orbifold defined in section-**3**:

$$\begin{aligned}
\tau_2\, B_2 = \tau_2\,\langle\lambda^2\rangle &= \Gamma_{2,18}[^0_1]\frac{\bar\vartheta_3^2\bar\vartheta_4^2}{\bar\eta^{24}} - \Gamma_{2,18}[^1_0]\frac{\bar\vartheta_2^2\bar\vartheta_3^2}{\bar\eta^{24}} - \Gamma_{2,18}[^1_1]\frac{\bar\vartheta_2^2\bar\vartheta_4^2}{\bar\eta^{24}}\\
&= \frac{\Gamma_{2,18}[^0_0]+\Gamma_{2,18}[^0_1]}{2}\bar F_1 - \frac{\Gamma_{2,18}[^0_0]-\Gamma_{2,18}[^0_1]}{2}\bar F_1\\
&- \frac{\Gamma_{2,18}[^1_0]+\Gamma_{2,18}[^1_1]}{2}\bar F_+ - \frac{\Gamma_{2,18}[^1_0]-\Gamma_{2,18}[^1_1]}{2}\bar F_-\,.
\end{aligned} \tag{84}$$

with

$$\bar F_1 = \frac{\bar\vartheta_3^2\bar\vartheta_4^2}{\bar\eta^{24}}\quad,\quad \bar F_\pm = \frac{\bar\vartheta_2^2(\bar\vartheta_3^2\pm\bar\vartheta_4^2)}{\bar\eta^{24}}. \tag{85}$$

For all $N=2$ heterotic theories B_2 has universal modular properties under $SL(2,Z)_\tau$:

$$\tau\to\tau+1\quad B_2\to B_2\quad \tau\to-\frac{1}{\tau}\quad B_2\to\tau^2\,B_2. \tag{86}$$

All functions $\bar F_i$ have positive coefficients and have the generic expansions

$$F_1 = \frac{1}{q}+\sum_{n=0}^{\infty} d_1(n)q^n = \frac{1}{q}+16+156q+\mathcal{O}(q^2) \tag{87}$$

$$F_+ = \frac{8}{q^{3/4}}+q^{1/4}\sum_{n=0}^{\infty} d_+(n)q^n = \frac{8}{q^{3/4}}+8q^{1/4}(30+481q+\mathcal{O}(q^2)) \tag{88}$$

$$F_- = \frac{32}{q^{1/4}}+q^{3/4}\sum_{n=0}^{\infty} d_-(n)q^n = \frac{32}{q^{1/4}}+32q^{3/4}(26+375q+\mathcal{O}(q^2)). \tag{89}$$

Also the lattice sums $(\Gamma_{2,18}[^h_0]\pm\Gamma_{2,18}[^h_1])/2$ have positive multiplicities.

The contribution of the generic massless multiplets is given by the constant coefficient of F_1 and agrees with the expectation, $16 = 20-4$, since we have the supergravity multiplet and 19 vector multiplets contributing 20, and 4 hypermultiplets contributing -4. Turning off all the Wilson lines and restoring the $E_7\times E_8$ group, the above result becomes

$$\langle\,\lambda^2\,\rangle = \Gamma_{2,2}[^0_1]\frac{\bar\vartheta_3^4\bar\vartheta_4^4(\bar\vartheta_3^4+\bar\vartheta_4^4)\bar E_4}{2\bar\eta^{24}} + -\Gamma_{2,2}[^1_0]\frac{\bar\vartheta_2^4\bar\vartheta_3^4(\bar\vartheta_2^4+\bar\vartheta_3^4)\bar E_4}{2\bar\eta^{24}} - \Gamma_{2,2}[^1_1]\frac{\bar\vartheta_2^4\bar\vartheta_4^4(\bar\vartheta_2^4-\bar\vartheta_4^4)\bar E_4}{2\bar\eta^{24}} \tag{90}$$

. Let us analyse the $N=4\to N=2$ BPS mass formulae; we denote by $\vec\varepsilon = (\vec\varepsilon_L;\vec\varepsilon_R,\vec\zeta)$ the shift vector of the $\Gamma_{(2,18)}$; $\vec\varepsilon$ must satisfy the modular-invariant constraint $\vec\varepsilon\cdot\vec\varepsilon\equiv 2\vec\varepsilon_L\cdot\vec\varepsilon_R-\vec\zeta\cdot\vec\zeta = 2(-1+\text{mod }4)$. The mass formula for the BPS states is:

• In the "untwisted" sector $h = 0$, $\vec{\varepsilon} = 0$

$$M^2(h=0) = \frac{|-m_1U + m_2 + Tn_1 + (TU - \frac{1}{2}\vec{W}^2)n_2 + \vec{W}\cdot\vec{Q}|^2}{4\,S_2\left(T_2U_2 - \frac{1}{2}\mathrm{Im}\vec{W}^2\right)}, \tag{91}$$

where $\vec{W}$ is the 16-dimensional complex vector of Wilson lines. When the integer

$$\rho = \vec{m}\cdot\vec{\varepsilon}_R + \vec{n}\cdot\varepsilon_L - \vec{Q}\cdot\vec{\zeta} \tag{92}$$

is even, these states are vector-like multiplets with multiplicity function $d_1(s)$ of (87) where s is :

$$s = \vec{m}\cdot\vec{n} - \frac{1}{2}\vec{Q}\cdot\vec{Q}. \tag{93}$$

When ρ is odd, these states are hypermultiplets-like with multiplicities $d_1(s)$.

• In the "twisted" sector $h = 1$, $\vec{\varepsilon} \neq 0$

$$M^2(h=1) = \frac{|-m_1'U + m_2' + Tn_1' + (TU - \frac{1}{2}\vec{W}^2)n_2' + \vec{W}\cdot\vec{Q}'|^2}{4\,S_2\left(T_2U_2 - \frac{1}{2}\mathrm{Im}\vec{W}^2\right)} \tag{94}$$

with

$$\vec{m}' \equiv \vec{m} + \frac{\vec{\varepsilon}_L}{2}, \quad \vec{n}' \equiv \vec{n} + \frac{\vec{\varepsilon}_R}{2}, \quad \vec{Q}' \equiv \vec{Q} + \frac{\vec{\zeta}}{2}. \tag{95}$$

When ρ is even (ρ' odd) the states are hypermultiplet-like with multiplicities $d_+(s')$, with

$$s' = \vec{m}'\cdot\vec{n}' - \frac{1}{2}\vec{Q}'\cdot\vec{Q}' \tag{96}$$

and

$$\rho' = \vec{m}'\cdot\vec{\varepsilon}_R + \vec{n}'\cdot\varepsilon_L - \vec{Q}'\cdot\vec{\zeta}, \tag{97}$$

When ρ is odd (ρ' even) the states are hypermultiplet- like with multiplicities $d_-(s')$.

Let us discuss here the gauge-symmetry enhancements in the presence of shift vectors. For simplicity we will ignore the charged sector coupled to the Wilson lines and focus on the $\gamma_{(2,2)}$ part. Let us fist consider the untwisted sector ($h = 0$). According to the above analysis, the masses are given by the unshifted mass formula (91) and they are vector multiplets when ρ is even and hypermultiplets when ρ is odd. Now the points where the standard $\Gamma_{(2,2)}$ mass vanishes are well known. At $T = U$, there are two configurations with zero mass, given by $m_1 = n_1 = \pm 1$, all the rest being zero. For both states, $|\rho| = |\varepsilon_L^1 + \varepsilon_R^1|$. Depending on it being even or odd, these states are either vector multiplets that enhance the gauge group $U(1)^2 \to SU(2)\times U(1)$ or hypermultiplets charged under one of the $U(1)$'s.

Let us now look for states becoming massless in the twisted ($h = 1$) sector at $T = U$. Again we obtain $m_1 + \varepsilon_L^1/2 = n_1 + \varepsilon_R^1/2$. Since $\varepsilon_L^1, \varepsilon_R^1$ are either 0 or 1, the previous condition can be satisfied only if both $\varepsilon_L^1 = \varepsilon_R^1 = \psi$, with $\psi = 0, 1$. Then $\rho = 2m_1\psi$ and is always even. The matching condition here for such a state becomes $s = 3/4$ when ρ is even (see (88)). Thus the condition on m_1 becomes

$$m_1^2 + \psi m_1 = \frac{3}{4} - \frac{\varepsilon^2}{8}. \tag{98}$$

From modular invariance we have $\varepsilon^2/2 = -1 \bmod 4 = 4k - 1$ $k \in Z$; then eq. (98) becomes

$$m_1^2 + \psi m_1 + k - 1 = 0 \tag{99}$$

and has either two solutions or none in the field of integers, depending on ψ and k. All such potentially massless states are hypermultiplets, come with multiplicity 8 and have equal and opposite charge under one of the 2-torus $U(1)$'s.

Heterotic-Type II dual pairs with partially broken SUSY $N=4 \to N=2$

The heterotic string compactified on T^4, with $N=2$ in $(6-d)$ space-time supersymmetry, has been conjectured to be dual to type II theory compactified on K_3 [29, 14]. This duality changes the sign of the dilaton, dualizes the field strength of the antisymmetric tensor and leaves the (4,20) gauge fields A^I_μ, the $SO(4,20)$ moduli and the Einstein metric invariant. Obviously this duality descends in four dimensions by compactifying both theories on an extra T^2. In four dimensions there are four extra gauge fields, two coming from the metric A^i_μ whose charges are the momenta of the T^2 and two coming from the antisymmetric tensor $B_{i,\mu}$, whose charges are the winding numbers of the T^2. Also, we have three extra scalars from the components of the metric on T^2, G_{ij} and one from the antisymmetric tensor B_{ij}. There are also $2\times$ 24 extra scalars, Y^i_I coming from the 6-d gauge bosons plus one more A, which is the four-dimensional dual of the antisymmetric tensor. If we denote heterotic variables by unprimed names and type II ones by primed names, then the heterotic-type II duality in four dimensions implies that

$$e^{-\phi} = \sqrt{\text{det}G'_{ij}} \quad , \quad e^{-\phi'} = \sqrt{\text{det}G_{ij}} \tag{100}$$

$$\frac{G_{ij}}{\sqrt{\text{det}G_{ij}}} = \frac{G'_{ij}}{\sqrt{\text{det}G'_{ij}}} \quad , \quad A'^i_\mu = A^i_\mu \tag{101}$$

$$e^{-\phi} g_{\mu\nu} = e^{-\phi'} g'_{\mu\nu} \quad \to \quad g^E_{\mu\nu} = g'^E_{\mu\nu} \tag{102}$$

$$M'_{4,20} = M_{4,20} \quad , \quad A^I_\mu = A'^I_\mu \quad , \quad Y^i_I = Y'^i_I \tag{103}$$

$$A = \frac{1}{2}\varepsilon^{ij} B'_{ij} \quad , \quad A' = \frac{1}{2}\varepsilon^{ij} B_{ij}. \tag{104}$$

Moreover, it effects an electric–magnetic duality transformation on the B^i_μ gauge fields

$$\frac{1}{2}\frac{\varepsilon_{\mu\nu}{}^{\rho\sigma}}{\sqrt{-\text{det}g}}\varepsilon^{ij} F^{B'}_{j,\rho\sigma} = \frac{\delta S^{het}}{\delta F^{B,\mu\nu}_i}. \tag{105}$$

On the electric and magnetic charges it acts on the T^2 charges and leaves the rest invariant.

For the configurations of moduli we are interested in, namely the factorization $(6,22) \to (2,18) \times (4,4)$, we proceed as follows. In the case of the heterotic string the complex moduli $T, U, \vec{W}$ are defined in terms of G_{ij}, B_{ij} and Y^k_i, $i,j=(1,2)$. However, for the type II string the situation is different. A careful analysis of the tree-level action shows that there is an analogue of the Green–Schwarz term $B\wedge F\wedge F$ at tree level; this appears at one loop at the heterotic side for 4-d descendants of both $B\wedge F^4$ and $B\wedge R^4$; the $B\wedge R\wedge R$ term appears at one loop in the type II side [32]. This term changes at tree level the definition of the type II S' field. There is an analogous phenomenon, which changes also at tree level the definition of the T' field. The correct formulae read:

$$S' = A' - \frac{1}{2}Y^I_1 Y^I_2 + \frac{U_1}{2} Y^I_2 Y^I_2 + i\left(\left(e^{-\phi'} + \frac{U_2}{2} Y^I_2 Y^I_2\right)\right. \tag{106}$$

$$T' = \sqrt{\det G'_{ij}} + iB' \tag{107}$$

where as usual

$$\frac{1}{\sqrt{\det\, G'_{ij}}} G'_{ij} = \frac{1}{U_2} \begin{pmatrix} 1 & U_1 \\ U_1 & |U|^2 \end{pmatrix}. \tag{108}$$

Thus (100)–(104) translate to

$$U = U' \quad , \quad \vec{W} = \vec{W}' \quad , \quad S = T' \quad , \quad T = S'. \tag{109}$$

Let us indicate how the $N = 4$ heterotic-type II duality works at the level the restricted $SO(2,18)$ BPS formula:

$$M^2_{BPS} = \frac{|P \;+\; S\,\Pi|^2}{\mathrm{Im}S(\mathrm{Im}T\mathrm{Im}U - \frac{1}{2}\mathrm{Im}\vec{W}\cdot\mathrm{Im}\vec{W})} \tag{110}$$

where P and Π are given in terms of the "electric" and "magnetic" charges and in terms of the complex moduli $T, U, \vec{W}$:

$$P = -m_1 + n_1 T + m_2 U + n_2(TU - \frac{1}{2}\vec{W}\cdot\vec{W}) + Q\cdot\vec{W}$$

and

$$\Pi = -\tilde{m}_1 + \tilde{n}_1 T + \tilde{m}_2 U + \tilde{n}_2(TU - \frac{1}{2}\vec{W}\cdot\vec{W}) + \tilde{Q}\cdot\vec{W} \tag{111}$$

We start first from the heterotic string not necessarily weakly coupled. We would like, however, to end up and compare with the weakly coupled type II string. Thus we must take the limit T_2 large in the mass formula and keep light states:

$$M^2_{het} = \frac{|-m_1 + m_2 U + \vec{W}\cdot Q + S(\tilde{m}_2 - \tilde{m}_1 U + \vec{W}\cdot\tilde{\vec{Q}})|^2}{4\; S_2(T_2U_2 - (\vec{W}_2)^2/2)} \tag{112}$$

However the terms containing the charged states are really absent; $\vec{m}\cdot\vec{n} = 0$ and $\tilde{\vec{m}}\cdot\vec{n} = 0$ and thus, there are no physical states with non-trivial $\vec{Q}, \tilde{\vec{Q}}$. Taking this into account, then using type II variables from (109), we can write (112) as

$$M^2_{pert-II} = \frac{|-m_1 + m_2 U' + T'(\tilde{m}_2) - \tilde{m}_1 U'|^2}{4\left(S'_2 - \frac{\vec{W}'^2_2}{2U'_2}\right) T'_2 U'_2} \tag{113}$$

which gives the correct tree-level type II mass formula in the large T'_2 limit, taking into account (106) and the duality map.

Owing to the adiabatic argument of ref.[33], we can obtain new dual heterotic–type II pairs by orbifolding both the $N = 4$ heterotic and $N = 4$ type II strings, by the same freely acting symmetry. Thus we would like to identify the duals of the heterotic models constructed in the previous sections with spontaneously broken supersymmetry.

For concreteness we will go to the Z_2 sub-manifold of K_3, where the conformal field theory is explicit, and we will map directly the heterotic to the type II string. The type II partition function on $K_3 \times T^2$ at the orbifold point is

$$Z^{II}_{N=4} = \frac{1}{\mathrm{Im}\tau|\eta|^4}\,\frac{1}{2}\sum_{h,g=0}^{1}\frac{\Gamma_{(2,2)}}{|\eta|^4}\, Z^{\mathrm{twist}}_{(4,4)}\left[{h \atop g}\right] \times \frac{1}{2}\sum_{\alpha,\beta=0}^{1}(-1)^{\alpha+\beta+\alpha\beta}\,\frac{\vartheta^2\left[{\alpha \atop \beta}\right]\vartheta\left[{\alpha+h \atop \beta+g}\right]\vartheta\left[{\alpha-h \atop \beta-g}\right]}{\eta^4}$$

$$\times \frac{1}{2} \sum_{\bar{\alpha},\bar{\beta}=0}^{1} (-1)^{\bar{\alpha}+\bar{\beta}+\bar{\alpha}\bar{\beta}} \frac{\bar{\vartheta}^2[^{\bar{\alpha}}_{\bar{\beta}}]\bar{\vartheta}[^{\bar{\alpha}+h}_{\bar{\beta}+g}]\bar{\vartheta}[^{\bar{\alpha}-h}_{\bar{\beta}-g}]}{\bar{\eta}^4} \tag{114}$$

Let us examine the massless bosonic spectrum of the $N = 4$ type II, and try to match it to that of the $N = 4$ heterotic string [34, 14].

• In the NS–NS ($\alpha = \bar{\alpha} = 0$) untwisted sector ($h = 0$), there are 32 degrees of freedom, corresponding to the graviton, 2 scalars (axion-dilaton), 4 vectors, and another 20 scalars (the $\Gamma_{2,2}$ and $Z^{\text{twist}}_{(4,4)}$ moduli). Two of the gauge bosons are graviphotons while the other two belong to $U(1)$ vector multiplets. Thus these four gauge bosons have lattice signature (2,2). Similarly the (2,2) moduli belong to these two vector multiplets while the (4,4) moduli are in multiplets with vectors coming from the R–R untwisted sector.

• In the NS–NS ($\alpha = \bar{\alpha} = 0$) twisted sector ($h = 1$), there are 16 Z_2 invariant states in the T^4/Z_2 part: H^I. There are in total 4×16 massless states; all of them are scalars in multiplets with vectors coming from the R-R twisted sector.

• In the R–R ($\alpha = \bar{\alpha} = 1$) untwisted sector there are 32 physical degrees of freedom. These correspond to 8 vectors and 16 scalars. The vectors have lattice signature (4,4) and four of them are graviphotons while the other four are in vector multiplets. The sixteen scalars complete the six vector multiplets.

• In the R-R, twisted sector, there are 4×16 massless states corresponding to 16 vectors and 32 scalars.

Here the gauge group is composed of $U(1)$'s, which implies that we are sitting at a generic point in the space of Wilson lines. The perturbative spectrum is charged under two of the graviphotons and two of the other gauge bosons with charges given by P_L, P_R of the T^2.

Consider now the freely acting orbifold on the heterotic side acting as a π rotation on the (4,4) part of the lattice and as a translation $\vec{\varepsilon}$ on the $\Gamma_{(2,18)}$ lattice. Again for simplicity we focus on the Z_2 case. On the type II side the Z_2 rotation on the (4,4) part changes the sign of the massless states coming from the untwisted R–R sector as well as the scalars coming from the twisted NS–NS sector. The effect of the (2,18) translation $\vec{\varepsilon} = (\vec{\varepsilon}_L; \vec{\varepsilon}_R, \vec{\zeta})$ is to give phases to massive charged states, but has no effect on the massless spectrum. Thus at the massless level the NS–NS twisted and R–R untwisted sectors have to be projected out. The projection in the type II case, which has the same effect as the (4,4) rotation in the heterotic side, is a combination of the right fermion number operator $(-1)^{F_R}$, which changes the sign of the right-moving Ramond sector, and the symmetry transformation $e = (-1)^h$, which acts on the twisted states of the orbifold with a minus sign and is inert on anything else.

The $\vec{\zeta}$ translation vector does not act in the perturbative type II string since the perturbative spectrum does not contains states charged under the 16 gauge bosons coming from the R-R twisted sector. However it will act on non-perturbative D-brane states carrying R–R charges. Finally the phase coming from the translation of the (2,2) piece is

$$(-1)^{\vec{m}\cdot\varepsilon_R+\vec{n}\cdot\varepsilon_L} \tag{115}$$

in the heterotic side. Under the type II–heterotic map, this becomes, in the type II side:

$$(-1)^{\vec{m}\cdot\vec{\varepsilon}_R+\vec{\tilde{m}}\hat{\times}\vec{\varepsilon}_L} \tag{116}$$

where $\vec{a}\hat{\times}\vec{b} = a_1 b_2 - a_2 b_1$. Thus the ε_L translation acts on the type II side on the magnetically charged states of the momentum-gauge fields of the two-torus; it is thus, not visible in type II perturbation theory.

The type II duals have 20 vector-multiplets and 4 hypermultiplets; thus they are "mirrors" of the type II models discussed in ref.[33] with 4 vector multiplets and 20 hypermultiplets. Therefore, the perturbative partition function of the type II models dual to the heterotic ones is

$$
\begin{aligned}
Z_{II}^{4\to 2} = & \frac{1}{\mathrm{Im}\tau|\eta|^4}\,\frac{1}{2}\sum_{h,g,\bar{h}\bar{g}=0}^{1} \frac{\Gamma_{2,2}^{\vec{\varepsilon}_R}\left[{\bar{h} \atop \bar{g}}\right]}{|\eta|^4} Z_{(4,4)}^{\mathrm{twist}}\left[{h \atop g}\right] \\
& \times \frac{1}{2}\sum_{\alpha,\beta=0}^{1}(-1)^{\alpha+\beta+\alpha\beta}\,\frac{\vartheta^2\left[{\alpha \atop \beta}\right]\vartheta\left[{\alpha+h \atop \beta+g}\right]\vartheta\left[{\alpha-h \atop \beta-g}\right]}{\eta^4} \times \frac{1}{2}\sum_{\bar{\alpha},\bar{\beta}=0}^{1}(-1)^{\bar{\alpha}+\bar{\beta}+\bar{\alpha}\bar{\beta}}\,\frac{\bar{\vartheta}^2\left[{\bar{\alpha} \atop \bar{\beta}}\right]\bar{\vartheta}\left[{\bar{\alpha}+h \atop \bar{\beta}+g}\right]\bar{\vartheta}\left[{\bar{\alpha}-h \atop \bar{\beta}-g}\right]}{\bar{\eta}^4} \\
& \times (-1)^{(\bar{\alpha}+h)\bar{g}+(\bar{b}+g)\bar{h}+\bar{g}\bar{h}}. \qquad (117)
\end{aligned}
$$

Here the reader might have noticed a potential puzzle. Consider a heterotic model defined by a translation vector with $\vec{\varepsilon}_R = \vec{0}$. In this model, in the limit Im $T \to 0$ $N = 2$ supersymmetry is restored to $N = 4$. Alternatively speaking, $m_{3/2} \sim$ Im T. Thus in weakly-coupled heterotic string we take $S \to \infty$ and also $T \to 0$. According to our duality map described above, there is no perturbative shift of the T^2 in the type II side. Thus, at the perturbative level, the type II $N = 2$ theory does not look like a spontaneously broken $N = 4$. However a look at (116) is sufficient to convince us that there are two gravitinos, with $m_{3/2} \sim$ Im S', which are light in the strong coupling region of the type II theory and certainly not visible in the weak coupling type II perturbation theory.

A similar phenomenon can happen in reverse. Consider a freely acting orbifold of the type II ($N = 4$) side, as in (117), where the (2,2) lattice translation acts on the windings of the two-torus with the phase $(-1)^{\vec{\varepsilon}_L\cdot\vec{n}}$. This is modular-invariant on the type II side. On the heterotic side the shift of the two-torus becomes non-perturbative via the heterotic-type II map, $(-1)^{\vec{m}\hat{\times}\vec{\varepsilon}_L}$. Thus, in heterotic perturbation theory, we only see the Z_2 rotation of the (4,4) torus. As it stands the heterotic $N = 2$ model is not modular-invariant. An extra shift in the gauge lattice is needed (not visible on the type II side). Thus the perturbative heterotic ground state has a $K_3 \times T^2$ structure (at the Z_2 orbifold point) and the supersymmetry $N = 4 \to N = 2$ looks explicitly broken in perturbation theory. Turning on all Wilson lines we find that the generic massless spectrum has 19 vector multiplets (including the dilaton) and 4 hypermultiplets. Moreover the $SL(2,Z)_S$ is broken to $\Gamma^-(2)_S$ as can easily be seen by following the fate of T-duality of the type II dual.

Another comment concerns the fate of the $SL(2,Z)_S$ electric–magnetic duality symmetry of the original N=4 theory, in the spontaneously broken phase. It is known that in the $N = 4$ case $SL(2,Z)_S$ is a corollary of heterotic–type II duality, since the T-duality of type II translates into the S-duality of the heterotic theory. Let us investigate what remains of the perturbative T duality in the broken type-II theory. We have argued above that the two-torus on the type II side gets a (perturbative) shift $(\vec{0};\vec{\varepsilon}_R)$ that amounts to the phase $(-1)^{\vec{m}\cdot\vec{\varepsilon}_R}$. The $SL(2,Z)_T$ acts on the two-torus charges as the set of matrices

$$
SL(2,Z)_T \;:\; \begin{pmatrix}\vec{m}\\ \vec{n}\end{pmatrix} \to \begin{pmatrix} a\,\mathbf{1} & b\, i\sigma^2 \\ -c\, i\sigma^2 & d\,\mathbf{1}\end{pmatrix}\begin{pmatrix}\vec{m}\\ \vec{n}\end{pmatrix}; \qquad (118)
$$

$$
ad - bc = 1 \;\;,\;\; a,b,c,d \in Z
$$

There are two subgroups of $SL(2,Z)$ that are relevant here; one is $\Gamma^+(2)$ defined by b even in (118); the other one is $\Gamma^-(2)$ defined by c even in (118). Thus when $\vec{\varepsilon}_R \neq \vec{0}$, $SL(2,Z)_T$ is broken to $\Gamma^+(2)_T$. Thus, the S-duality group is reduced to $\Gamma^+(2)_S$.

In the above discussion , it is obvious that there are non-perturbative ambiguities in the translation-related projections. The most general projection conceivable is determined by the "electric" translation vector $\vec{\varepsilon}$, but simultaneously by a "magnetic" translation vector $\tilde{\varepsilon}$ whose effects are not visible in the perturbative spectrum. Parts of these translations are never perturbatively visible either in the heterotic nor in the type II side. We will comment more on this issue in the next section.

One more remark is in order about the type II duals described above. Inspection shows that all of the $N = 2$ spacetime supersymmetry comes from the left side. Consequently, in these models the S field is in a vector multiplet [33]. Thus, as in the heterotic side, the vector-moduli space gets corrections while the hypermultiplet moduli space does not. At generic Wilson lines this class of models has a massless spectrum, which consists, apart from the supergravity and the dilaton vector multiplet, of 18 vector multiplets and 4 neutral hypermultiplets (the moduli of the four-torus). The non-perturbatively exact hypermultiplet quaternionic manifold is $SO(4,4)/SO(4) \times SO(4)$. The exactness of the hypermultiplet moduli space restricts the orbifolding possibilities on the type II side to the ones described in (117).

Non-perturbative BPS spectrum in partially broken SUSY $N = 4 \rightarrow N = 2$

Our conjecture for the non-perturbative multiplicities consists in generalizing the perturbative multiplicity functions (87)–(89) F_i in genus-2. First we rewrite F_i in a more convenient form:

$$F_1 = \frac{1}{\bar{\eta}^{24}}\, \bar{\chi}\left[{}^{0}_{1}\right], \quad F_{\pm} = \frac{1}{\bar{\eta}^{24}}\, \left(\bar{\chi}\left[{}^{1}_{0}\right] \pm \bar{\chi}\left[{}^{1}_{1}\right]\right) \tag{119}$$

where $\bar{\chi}\left[{}^{h}_{g}\right]$ are given in terms of the characters of four twisted 2d right-moving bosons:

$$\bar{\chi}\left[{}^{h}_{g}\right] \;=\; \frac{4(-)^h\, \bar{\eta}^6}{\bar{\vartheta}\left[{}^{1+h}_{1+g}\right]\, \bar{\vartheta}\left[{}^{1-h}_{1-g}\right]}, \tag{120}$$

where in the above equation $(h,g) \neq 0$. We can extend the validity of $\bar{\chi}\left[{}^{h}_{g}\right]$ for all (h,g) sectors using identities between right-moving, bosonic and fermionic, "twisted" characters:

$$\bar{\chi}\left[{}^{h}_{g}\right] \;=\; \frac{1}{8\,\bar{\eta}^6}\, \sum_{a,b} (-)^h\, \bar{\vartheta}^4\left[{}^{a+h}_{b+g}\right]\, \bar{\vartheta}^4\left[{}^{a-h}_{b-g}\right]\, \bar{\vartheta}\left[{}^{1+h}_{1+g}\right]\, \bar{\vartheta}\left[{}^{1-h}_{1-g}\right]. \tag{121}$$

In this expression, the absence of the $(h,g) = (0,0)$ sector is due to the vanishing of the odd-spin structures ($\bar{\vartheta}\left[{}^{1}_{1}\right]$ terms). In genus-2 h and g become $\vec{h} = (h,\ \tilde{h})$ and $\vec{g} = (g,\ \tilde{g})$ in correspondence with the "electric" and "magnetic" charge shifts. The generalization in genus-2 of the twisted characters consists in promoting the various ϑ-functions with characteristics in genus-2

$$\bar{\vartheta}\left[{}^{a\,+\,h}_{b\,+\,g}\right](\bar{\tau}) \quad \rightarrow \quad \bar{\vartheta}\left[{}^{\vec{a}\,+\,\vec{h}}_{\vec{b}\,+\,\vec{g}}\right](\bar{\tau}^{ij}). \tag{122}$$

Then, the proposed non-perturbative multiplicities will be generated by the genus-2 functions:

$$F\left[{}^{\vec{h}}_{\vec{g}}\right] \;=\; \Phi(\bar{\tau}^{ij})\, \bar{\chi}\left[{}^{\vec{h}}_{\vec{g}}\right](\bar{\tau}^{ij}), \tag{123}$$

where $\Phi(\bar{\tau}^{ij})$ is the $N = 4$ multiplicity function and $\bar{\chi}\left[{}^{\vec{h}}_{\vec{g}}\right](\bar{\tau}^{ij})$ are the genus–2 analogues of the genus-1 "twisted" characters $\bar{\chi}\left[{}^{h}_{g}\right](\bar{\tau})$ defined above.

Using the genus-2 multiplicity functions, we can construct weighted free-energy super-traces, which extend at the non-perturbative level the same perturbative quantities, e.g. the moduli dependence of the gauge and gravitational couplings. We define by $\mathcal{L}^{\mathcal{D}}$ the following quantity:

$$\mathcal{L}^{\mathcal{D}} = \int_{C} [dt \prod dX^{ij}] \sum_{h_i, g_i} \sum_{q_i} \mathcal{D}(\tau^{ij}) \, F \begin{bmatrix} \vec{h} \\ \vec{g} \end{bmatrix} (\bar{\tau}^{ij}) \times \exp \left[-2i\pi \, \mathrm{Re}\tau^{ij} \, (\vec{q}_i + \vec{\varepsilon}_i) \cdot (\vec{q}_j + \vec{\varepsilon}_j) \right]$$
$$\times \exp \left[-\pi \, t \, M^2_{BPS}(S; \, \vec{q}_i, \vec{\varepsilon}_i) \right], \tag{124}$$

where $M^2_{BPS}(S; \, \vec{q}_i, \vec{\varepsilon}_i)$ stands for the non-perturbative mass formula (110,111 with shifted charges; M^2_{BPS} depends on the shifted "electric" and "magnetic" charges, the moduli T, U, and $\vec{W}$ as well as the dilaton–axion moduli field S. The period matrix τ^{ij} of genus-2 in eq.(125), is constructed in terms of the parameters t, X^{ij} and S in the following way:

$$t = \sqrt{\det(\tau^{ij})}, \; X^{ij} = \mathrm{Re} \, \tau^{ij}, \text{ and } \frac{\tau^{ij}}{\sqrt{det \; \tau^{ij}}} = \frac{1}{\mathrm{Im}S} \begin{pmatrix} 1 & \mathrm{Re}S \\ \mathrm{Re}S & |S|^2 \end{pmatrix}. \tag{125}$$

The integration on X^{ij} in the domain $[-1/2, \; +1/2]$ would give rise to the non-perturbative matching conditions (78). The relevant multiplicities are generated by the functions $F \begin{bmatrix} \vec{h} \\ \vec{g} \end{bmatrix}$. This is a suggestive argument, and stands in a similar footing with the analogous τ_1 integration in the perturbative string. However we suggest that, like in the string case, the correct integration domain is the genus-two fundamental region. Thus we expect that the integration over t (in the fundamental domain of genus-2 with S fixed) gives rise to the non-perturbative quantity $\mathcal{L}^{\mathcal{D}}[S; \; T, \; U, \; \vec{W}]$ in terms of all moduli, S included.

The kernel $\mathcal{D}$ is the genus-2 analogue of a product of charge operators. In the perturbative string, this is given by a product of right-moving lattice vectors and contains also a "back-reaction" term [35]. There is an analogue of "right-moving" charges in the non-perturbative case when we also include the magnetic charges. The charge sum for the overall trace can be written in the perturbative case as a $\bar{\tau}$ derivative, which generalizes in the non-perturbative treatment to the $\partial_{\tau^{11}} + \partial_{\tau^{22}}$. The "back-reaction" term can be fixed since it has to restore the modular properties of the $\bar{\tau}^i j$ derivatives.

The physical interpretation of the summation over the "magnetic" charges reproduces the Euclidean space-time instanton corrections to the couplings.

The determination of the non-perturbative effective couplings constants (the gravitational one included) defines without any ambiguity the non-perturbative prepotential of the $N = 2$ effective theory. Therefore, the knowledge of $\mathcal{L}^{\mathcal{D}}_l$ determines at the non-perturbative level the $N = 2$ low-energy effective supergravity, which includes terms up to two derivatives.

Outlook

We have demonstrated the existence of partial spontaneous supersymmetry breaking in string theory, and gave several concrete examples in both the heterotic and type II theories. We have studied the issue of restoration of supersymmetry, at the classical and perturbative level. We have further analysed the consequences of heterotic–type II duality valid for the $N = 2$ models we presented. We have pointed out that in the dual theories the $N = 4 \rightarrow N = 2$ supersymmetries may look explicitly broken in their perturbation theory. This was also corroborated by our conjecture on the full non-perturbative structure of their effective theories. In some cases we can predict some novel non-perturbative (non-geometric) transitions between vacua of the type II string with (2,0) and (1,1) space-time supersymmetry.

An analysis of the perturbative BPS states of strings, with supersymmetry spontaneously broken $N = 4 \rightarrow N = 2$, and the underlying duality structure permit us to conjecture the full non-perturbative form of the effective field theory. This conjecture needs to be elaborated and tested in the context of explicit models. This will be the subject of future analysis.

Acknowledgements

Part of this work was done in collaboration with E. Kiritsis. We acknowledge useful discussions with B. Pioline and E. Verlinde. This work is supported in part by the TMR contract ERB–4061–PL–95–0789 in which the author is associated to ENS.

REFERENCES

1. M. de Roo, Nucl. Phys **B255** (1985) 515; Phys. Lett. **156** (1985) 331;
E. Bergshoeff, I.G. Koh and E. Sezgin, Phys. Lett. **B155** (1985) 71.
2. S. Ferrara, L. Girardello, C. Kounnas and M. Porrati, Phys. Lett. **B192** (1987) 368.
3. K.S. Narain, Phys. Lett. **B169** (1986) 41;
K.S. Narain, M.H. Sarmadi and E. Witten, Nucl. Phys. **B279** (1987) 369.
4. I. Antoniadis, C. Bachas C. Kounnas and P. Windey, Phys. Lett. **B171** (1986) 51;
I. Antoniadis, C. Bachas and C. Kounnas, Nucl. Phys. **B289** (1987) 87.
5. H. Kawai, D.C. Lewellen and S.-H.H. Tye, Nucl. Phys. **B288** (1987) 1.
6. W. Lerche, D. Lüst and A.N. Schellekens, Nucl. Phys. **B287** (1987) 477.
7. P. Candelas, G. Horowitz, A. Strominger and E. Witten, Nucl. Phys. **B258** (1985) 46;
8. L. Dixon, J. Harvey, C. Vafa and E. Witten, Nucl. Phys. **B261** (1985) 678;
K.S. Narain, M.H. Sarmadi and C. Vafa, Nucl. Phys. **B288** (1987) 551.
9. D. Gepner, Phys. Lett. **B199** (1987) 370; Nucl. Phys. **B296** (1988) 757;
L.E. Ibanez, H. P. Nilles and F. Quevedo, Phys. Lett. **B187** (1987) 25.
10. C. Kounnas and M. Porrati, Nucl. Phys. **B310** (1988) 355,
S. Ferrara, C. Kounnas, M. Porrati and F. Zwirner, Nucl. Phys. **B318** (1989) 75.
11. I. Antoniadis, Phys. Lett. **B246** (1990) 377;
I. Antoniadis and C. Kounnas, Phys. Lett. **B261** (1991) 369.
12. C. Kounnas and B. Rostand, Nucl. Phys. **B341** (1990) 641.
13. I. Antoniadis, H. Partouche and T.R. Taylor, Phys. Lett. **B372** (1996) 155;
S. Ferrara, L. Girardello and M. Porrati, Phys. Lett. **B376** (1996) 275;
J. Bagger and A. Galperin, hep-th/9608177.
14. C. Hull and P. Townsend, Nucl. Phys. **B438** (1995) 109, hep-th/9410167.
15. E. Witten, Nucl. Phys. **B443** (1995) 85, hep-th/9503124.
16. K. Kikkawa and M. Yamasaki, Phys. Lett. **B149** (1984) 357;
N. Sakai and I. Senda, Progr. Theor. Phys. **75** (1986) 692;
V.P. Nair, A. Shapere, A. Strominger and F. Wilczek, Nucl. Phys. **B287** (1987) 402;
B. Sathiapalan, Phys. Rev. Lett. **58** (1987) 1597;
R. Dijkgraaf, E. Verlinde and H. Verlinde, Commun. Math. Phys. **115** (1988) 649;
A. Giveon, E. Rabinovici and G. Veneziano, Nucl. Phys. **B322** (1989) 167;
A. Shapere and F. Wilczek, Nucl. Phys. **B320** (1989) 301;
M. Dine, P. Huet and N. Seiberg, Nucl. Phys. **B322** (1989) 301.
17. E. Kiritsis, C. Kounnas, M. Petropoulos and J. Rizos, Phys. Lett. **B385** (1996) 87, hep-th/9606087.
18. E. Kiritsis, C. Kounnas, M. Petropoulos and J. Rizos, hep-th/9605011; hep-th/9608034.
19. E. Kiritsis, C. Kounnas, M. Petropoulos and J. Rizos, CERN/TH.96/...to appear
20. L. Dixon, V. Kaplunovsky and J. Louis, Nucl. Phys. **B355** (1991) 649.
21. T. Banks and L. Dixon, Nucl. Phys. **B307** (1988) 93.
22. A. Font, L. Ibanez, D. Lüst and F. Quevedo, Phys. Lett. **B249** (1990) 35;
S.J. Rey, Phys. Rev. **D43** (1991) 256.
23. A. Sen, Phys. Lett. **B303** (1993) 22, **B329** (1994) 217 and Nucl. Phys. **B329** (1994) 217.

24. J. Schwarz and A. Sen, Phys. Lett. **B312** (1993) 105, and Nucl. Phys. **B411** (1994) 35.
25. M. Cvetic and D. Youm, Phys. Rev. **D53** (1996) 584; Phys. Lett. **B359** (1995) 87; hep-th/9508058; hep-th/9510098; hep-th/9512127;
Kwan-Leung Chan and M. Cvetic, Phys. Lett. **B375** (1996) 98.
26. M. Duff, J.T. Liu and J. Rahmfeld, Nucl. Phys. **B459** (1996) 125, hep-th/9508094.
27. M. Cvetic and A. Tseytlin, Phys. Lett. **B366** (1996) 95, hep-th/9510097; Phys. Rev. **D53** (1996) 5619, hep-th/9512031.
28. G. Lopes-Cardoso, G. Curio, D. Lüst, T. Mohaupt and S. J. Rey, Nucl. Phys. **B464** (1996) 18, hep-th/9512129.
29. M. Duff and R. Khuri, Nucl. Phys. **B411** (1994) 473, hep-th/9305142;
M. Duff and R. Minasian, Nucl. Phys. **B436** (1995) 507, hep-th/9406198.
30. C. Bachas and E. Kiritsis, hep-th/9611205.
31. R. Dijkgraaf, E. Verlinde and H. Verlinde, hep-th/9607026.
32. C. Vafa and E. Witten, Nucl. Phys. **B447** (1995) 261.
33. C. Vafa and E. Witten, hep-th/9507050.
34. S. Ferrara and C. Kounnas, Nucl. Phys. **B328** (1989) 406.
35. E. Kiritsis and C. Kounnas, Nucl. Phys. **B442** (1995) 472.

GAUGINO CONDENSATION, LOOP CORRECTIONS, AND S-DUALITY CONSTRAINT

Kamran Saririan

Theoretical Physics Group
Lawrence Berkeley National Laboratory
and
Department of Physics
Berkeley, CA 94720

Introduction

Amongst the candidates for fundamental unified theories, heterotic superstring theory with gauge group $E_8 \times E_8$ seems to be the most promising one. This is because the spectrum of the theory easily accomodates the Standard Model spectrum and gauge structure. In addition, the underlying gauge group contains an extra factor of E_8 which provides an 'hidden sector', which couples to the observable sector only through gravity, and, as will be discussed below, plays a crucial role in the mechanism of supersymmetry breaking. Furthermore, the effective theories that describe the heterotic string in 4 dimensions below the Planck scale are (nonrenormalizable) locally supersymmetric effective field theories. Indeed, requiring supersymmetry at energies well above M_W in order to stabilize the gauge hierarchy, in some sense forces one to consider locally supersymmetric theories: A unified field theory must include gravity. Within the framework of General Relativity, a supersymmetric theory has to be *locally* supersymmetric. This follows from the fact that the supersymmetry transformation on the metric, or on the vielbein must include general coordinate transformations. Supergravity theories are nonrenormalizable, but can be consistently viewed as low-energy effective field theories (LEEFT) for the massless modes of superstring.

A basic feature of superstring constructions in four dimensions is the presence of massless moduli in the effective field theory. These fields whose vevs parameterize the continuously degenerate string vacua, are gauge-singlet chiral fields; furthermore, they are exact flat directions of the low energy effective field theory (LEEFT) scalar potential. Generically, the moduli appear in the couplings of the LEEFT. For example, the tree level gauge couplings at the string scale depend on the dilaton, S, and the Yukawa couplings as well as the kinetic terms depend on the T-moduli (and S through

Masses of Fundamental Particles: Cargèse 1996
Edited by M. Lévy *et al.*, Plenum Press, New York, 1997

the Kähler potential) . There is mixing of the moduli beyond tree level, due to both string threshold corrections [1] and field-theoretical loop effects, as we shall dicuss.

Since the supersymmetric vacua of heterotic strings consist of continuously degenerate families (to all orders of perturbation theory), parameterized by the moduli vevs, the latter remain perturbatively undetermined. This degeneracy can only be lifted by a nonperturbative mechanism which would induce a nontrivial superpotential for moduli, and at the same time break supersymmetry. We shall assume that this nonperturbative mechanism takes place in the LEEFT and is not intrinsically stringy. This certainly appears to be the most "tractible" possibility. A popular candidate for such a mechanism has been gaugino condensation which is the focus of this talk.

As a specific model, we later consider gaugino condensation in a superstring-inspired effective field theory, with approximate S-duality invariance [2, 3] and exact T-modular invariance (generalization of the work in ref. [3]) and incorporate an intermediate scale M_I ($M_{\text{cond}} \ll M_I \ll M_{\text{string}}$), [4] in order to see how the intermediate-scale threshold corrections will affect gaugino condensation and supersymmetry breaking. This part of the talk is based on the work in ref. [4]. Incorporating the intermediate-scale threshold corrections into gaugino condensation is non-trivial in the sense that the field-theoretical threshold corrections at M_I are dilaton-dependent. Hence, these modifications can have non-trivial implications for supersymmetry breaking by gaugino condensation. Furthermore, *a priori*, nothing prohibits intermediate scales in the hidden sector.

The outline of this talk is as follows. In the next section, we review gaugino condensation, and of duality symmetries (modular and S-duality). We shall discuss our model in section 3.1, and give the renormalized Kähler potential including 1-loop threshold corrections at an intermediate mass, and constrained by duality symmetries. The issues related to the scalar potential, dilaton run-away, and supersymmetry breaking , as well as the role of the intermediate mass are discussed in section 3.2. Concluding remarks are given in section 4. Due to the significance of renormalization of the field-dependent gauge coupling in such models and its connection with modular anomaly cancellation in the effective theory, we give a review of these ideas in the Appendix.

Generalities

Gaugino Condensation

A possible mechanism for breaking supersymmetry within the framework of ($N = 1$, $D = 4$) LEEFT of superstring is gaugino condensation in the hidden sector. In this scenario, the nonperturbative effects arise from the strong coupling of the asymptotically free gauge interactions at energies well below M_{Pl}. Corresponding to this strong coupling is the condensation of gaugino bilinear $\langle\bar{\lambda}\lambda\rangle_{h.s.}$. Let us briefly remind the reader the overview of the development of gaugino condensation. It was recognized many years ago that gaugino condensation in globally supersymmetric Yang-Mills theories without matter does not break supersymmetry [6]. In fact, that dynamical supersymmetry breaking cannot be achieved in pure SYM theories was shown by topological arguments of Witten [7]. In the locally supersymmetric case the picture is rather different, namely, gaugino condensation can break supersymmetry [8], and the gauge coupling is itself generally field-dependent. When the gauge coupling becomes strong, it gives rise to gaugino condensation at the scale[1]

$$M_{cond} \sim M_{string}\langle \text{Re}T\rangle^{-1/2} e^{-\text{Re}S/2b_0} = M_{string}\langle \text{Re}T\rangle^{-1/2} e^{-1/b_0 g_{st}^2},$$

which breaks local supersymmetry spontaneously ($M_{cond}^3 \sim \langle\bar{\lambda}\lambda\rangle_{h.s.}$), and S is the dilaton/axion chiral field. Supersymmetry breaking in the obesrvable sector is induced by gravitational interactions which act as 'messenger' between the two otherwise decoupled sectors.

However, there are generally two problems associated with the above scenario. First, the destabilization of S — the only stable minimum of the potential in the S-direction being at $S \to \infty$; *i.e.,* in the direction where exact supersymmetry is recovered and the coupling vanishes! This is contrary to the expectation that the vacuum is in the strongly coupled, confining regime. This problem, the so-called dilaton runaway problem, is present in most formulations of gaugino condensation, in particular the so-called 'truncated superpotential' approach [10] , where the condensate field is assumed to be much heavier than the dilaton and therefore is integrated out below M_{cond}. In fact, the dilaton runaway problem is perhaps a more generic problem in string phenomenology where the underlying string theory is assumed to be weakly coupled. We shall return to the dilaton runaway later.

The second difficulty is the large cosmological constant that arises from the vacuum energy associated with gaugino condensation. An early attempt to remedy these difficulties was proposed by Dine *et al.* [10], in the context of no-scale supergravity whereby a constant term, c, is introduced in the superpotential which independently breaks supersymmetry and cancels the cosmological constant. The origin of c is traced to the vev of the 3 form in 10D supergravity, and is quantized in units of order M_{pl}. Therefore, this approach has the unsatisfactory feature of breaking supersymmetry at the scale of the fundamental theory.

Duality Symmetries (Modular Invariance and S-Duality)

Modular symmetry, with the group $SL(2,\mathcal{Z})$ subgroup of $SL(2,\mathcal{R})$ duality transformations, written in its simplest form:

$$T \to \frac{\alpha T - i\beta}{i\gamma T + \delta}, \tag{1}$$

where $\alpha\delta - \beta\gamma = 1$ and $\alpha, \beta, \gamma, \delta$ are integers,[2] is an exact invariance of the underlying string theory. However, this symmetry is anomalous in the LEEFT. Cancellation, or partial cancellation, of this anomaly in the effective theory can be achieved by the Green-Schwarz (GS) mechanism, which is especially clear in the linear-multiplet formulation of the LEEFT [11, 12, 13]. In the corresponding chiral formulation, the adding of GS counter-terms amounts to modifying the dilaton Kähler potential:

$$\ln(S+\bar{S}) \to \ln(S+\bar{S}-bG),$$

where $b = -\frac{2}{3}b_0$, and b_0 is the E_8 one-loop β-function coefficient. $G = \Sigma_i \ln(T^i + \bar{T}^i - \Sigma|\Phi|^2)$, and Φ is any untwisted sector (non-modulus) chiral field in the theory. For simplicity, here we only consider models where modular anomalies are completely cancelled by GS mechanism, for example, the (2,2) symmetric abelian orbifolds with

[1] These arguments are modified by, for instance, the requirement of modular invariance [9].

[2] There is, generally, one copy of the group per modulus field T^i.

no $N = 2$ fixed planes, like Z_3 or Z_7 [11, 12, 13]. The role of the gauge coupling and its renomalization in superstring effective theories, and the connection with modular anomaly cancellation are reviewed in Appendix.

Recently, another type of duality symmetry has been receiving much attention in string theories. In this case the group of duality transformations is $SL(2,\mathcal{Z})$, but acting on the field S instead of T^i, and is referred to as S-duality. Like its T-analogue, this group has a generator which is the transformation $S \rightarrow 1/S$, and since S is related to the gauge coupling, this duality transformation is also referred to as 'strong-weak' duality. Font *et al.* [14] have conjectured that S-duality, like T-duality is an exact symmetry of string theory. More recently, it has been mounting evidence that S-duality is a symmetry of certain string theories [15]. However, these theories all have $N = 4$ or $N = 2$ supersymmetries. At the level of string theory, there are two different types of S-duality, namely (i) those that map different theories into one another, and (ii) those that map strongly and weakly coupled regimes of the same theory into each another. Indeed, presently there is no evidence of an S-dual $N = 1$ theory, and it is therefore difficult to justify the use of S-duality as a true symmetry in the corresponding LEEFT. However, it has been shown that in the effective theory, the full $SL(2,\mathcal{R})$ duality transformation is a symmetry of the equations of motion of the gravity, gauge, and dilaton sector in the limit of weak gauge coupling [2, 3]. As in [3], we shall take S-duality as a guiding principle in constructing the Kähler potential for the gaugino condensate, which is, so far, the least understood element in the description of the effective theory for gaugino condensation. That is, we assume that S-duality invariance is recovered in limit of vanishing gauge coupling, $S + \bar{S} \rightarrow \infty$.

A Specfic Model

This model basically generalizes the momel of gaugino condensation with S-duality of ref. [3] to the case in which an intermediate scale is present. For details of the calculation and a more complete discussion, the reader is referred to ref [4].

The scheme of generating the intermediate scale considered here does not involve the spontaneous breaking of the hidden-sector gauge group. Here, we couple the hidden-sector gauge non-singlet fields Φ_i to a gauge singlet A. When A dynamically gets a vev, Φ_i become massive and the intermediate scale is thus generated. Since A is a singlet, the hidden-sector gauge group does not break. Such a scheme has interesting implications for gauge coupling unification [5]. For consistency, the pattern $M_{cond} \ll M_I \ll M_{string}$ is always assumed. Therefore, we shall integrate out the hidden-matter fields below M_I and the effective lagrangian at M_{cond} will consist of the moduli and the gauge composites only.

The superpotential for the hidden sector matter fields in our toy model is:

$$W_{HM} = \frac{1}{2}\lambda^{ij} A \Phi_i \Phi_j + \frac{1}{3}\lambda' A^3. \tag{2}$$

When constructing our model, two symmetry principles have been used to constrain the Lagrangian: First, the LEEFT must be T-modular invariant to all orders, according to all-loop string calculations. Second, S-duality is a symmetry in the weak-coupling limit $\langle S + \bar{S}\rangle \rightarrow \infty$. We will include the renormalization and intermediate-scale threshold corrections only in the dilatonic part of the Kähler potential. We simply write down the Kähler potential and for the full discussion we again refer the reader to ref. [4].

$$K = -\ln m - 3\ln(1 - m^{1/3}Q) + G \tag{3}$$

where[3],

$$m = 2/g_{eff}^2(M_{cond}) = S+\bar{S}-bG+3\left[b^{>}\ln\left(\frac{M_I^2}{M_{string}^2}\right)+b^{<}\ln\left(\frac{Q/(S+\bar{S}-bG)}{M_I^2}\right)\right], \quad (4)$$

is the renormalized coupling including the 1-loop threshold corrections at the canonically normalized, modular invariant intermediate mass M_I which can be computed to be:

$$M_I^2 = e^K(K^{\varphi\bar{\varphi}})^2|\lambda A|^2 = \frac{|\lambda A|^2 e^{G/3}}{9(s+\bar{s}-bG)}\left(1+\frac{b}{s+\bar{s}-bG}\right)^{-2}. \quad (5)$$

In these relations, G is the Green-Schwarz term,

$$G = -3\ln(T+\bar{T}-|A|^2), \quad (6)$$

and $Q = |H|^2 e^{G/3}$ (where H is the condensate superfield) is the modular invariant condensation scale. Various group theoretical factors are as foolws:

$$-3b = 2b_0 = \frac{1}{8\pi^2}C(E_8); \qquad b^{>} = (3C_G - C_M)/24\pi^2, \qquad b^{<} = C_G/8\pi^2, \quad (7)$$

where C_G and C_M are the quadratic Casimirs:

$$C_G = T(adj); \quad C_M = \sum_r n_r T(r); \quad T(r) = \mathrm{Tr}_r(T^2), \quad (8)$$

with r labelling the representations of the gauge group, and n_r being the number of fields in the r representation.

To summarize, our Kähler potential given in eq. (3) includes the one loop renormalization of the dilaton with the intermediate threshold corrections, as well GS counter terms that ensure modular anomaly cancellation. It is also constrained by approximate S-duality symmetry as discussed in references [3, 4].

The dynamical fields at the condensation scale in our model are S, H, and T. The scalar potential is given by:

$$V = e^K\left[K^{i\bar{j}}(K_iW+W_i)(K_{\bar{j}}\bar{W}+\bar{W}_{\bar{j}}) - 3|W|^2\right] \quad (9)$$

and the Kähler metric written in terms of $m = 2/g_{eff}^2(M_{cond})$ (eq. (5)), $Q = |H|^2e^{G/3}$, and their derivatives with respect to the scalar fields is given by:

$$\begin{aligned} K_{i\bar{j}} = m^{-2}\{m_i m_{\bar{j}}\tilde{x} &+ m(\xi-1)m_{i\bar{j}} + (\xi+\xi^2)(m_i q_{\bar{j}} + m_{\bar{j}} q_i) \\ &+ 3m^2[\xi q_{i\bar{j}} + (\xi+\xi^2)q_i q_{\bar{j}}] + m^2 G_{i\bar{j}}\}, \end{aligned} \quad (10)$$

where

$$x = m^{1/3}Q, \quad \xi = \frac{x}{1-x}, \quad \tilde{x} = 1 - 2\xi/3 + \xi^2/3,$$

and

$$m_i = \partial_i m, \quad q = \ln Q, \quad q_i = \partial_i \ln Q, \quad etc.$$

Notice that $G_{i\bar{j}} = 0$ unless $i = j = t$, $m_{h\bar{j}} = 0$, and $q_s = 0$. The nonperturbative part of the superpotential is of the form

$$W_{NP} = \alpha e^{-S/b}Y^n\left(\ln\frac{Y}{\mu}\right)^k, \quad Y = He^{S/3b^{<}}, \quad (11)$$

[3]Here, $M_{pl} = 1$; and notice that the UV cut-off is taken to be $M_{string} = (S+\bar{S}-bG)^{-1/2}$ meaning that the condensation scale is really in these units, $Q/(S+\bar{S}-bG)$.

with $n < 3$ (the Veneziano-Yankielowicz superpotential is the special case of $n = 3$ and $k = 1$). The reason the exponents n and k are introduced is because it is the Kähler potential (3) that already includes the gaugino condensate wave function renormalization, and so the superpotential should not.

We summarize the results of the numerical computation and analytic expansions of this model as follows. The scalar potential, V, is positive semi-definite and has a nontrivial minimum at finite values of the dilaton and the condensate field, at which the following relations are satisfied:

$$\langle W \rangle \simeq 0, \quad \text{and} \quad m = 2/g_{eff}^2(M_{cond}) \to 0. \tag{12}$$

This is in addition to the usual 'runaway' solution at $S \to \infty$ and $H = 0$. Notice that the second relation tells us that the coupling $g_{eff}(M_{cond})$ blows up (at a finite value of

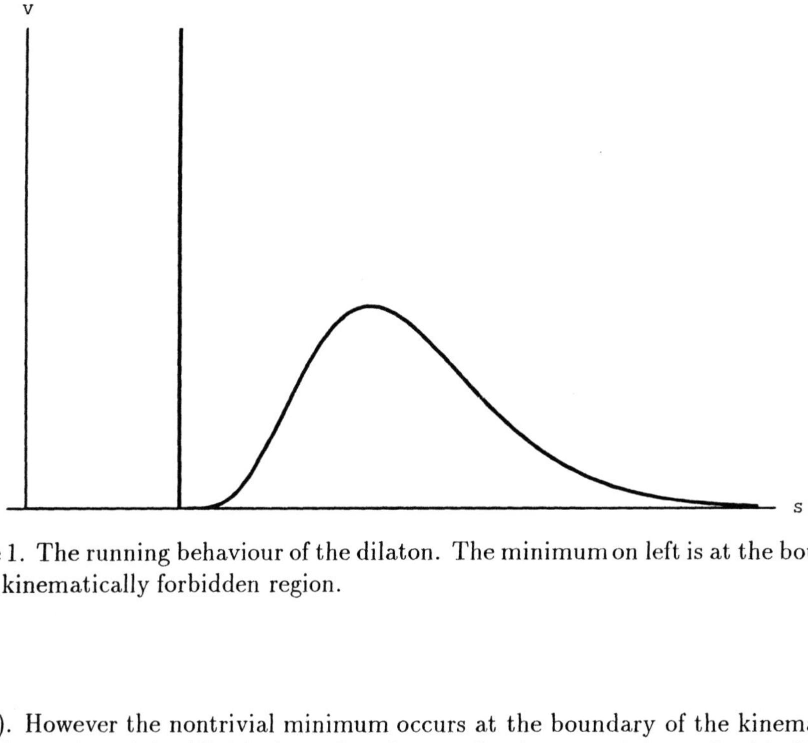

Figure 1. The running behaviour of the dilaton. The minimum on left is at the boundary of the kinematically forbidden region.

$S + \bar{S}$). However the nontrivial minimum occurs at the boundary of the kinematically forbidden region of the (S, H) plane. In other words, the potential runs in this direction as well! But this the correct direction, as the value that it runs to corresponds to strong coupling at the condensation scale, with a nonzero value of the condensate. The similar running behaviours (see Fig. 1) in both strong and weak coupling directions is attributed to S-duality. Notice however that the relations $\langle V \rangle = \langle W \rangle = 0$ imply that supersymmetry remains unbroken.

As for the effect of the intermediate mass, the two independent conditions $m = \langle W \rangle = 0$ imply that [4] the parameter μ of the nonperturbative superpotential in eq. (11) is 'locked' to M_I; and that the parameters of the superpotential which generates M_I allows for a phenomenologically sensible hierarchy between the condensation and the string scales.

Conclusion

We have discussed hidden sector gaugino condensation as a possible mechanism for supersymmetry breaking. In the model which was presented in the last section, which included some perturbbative corrections to the Kähler potential, as well as a nonperturbative constraint (S-dulaity), we saw that suprsymmetry remains unbroken. Perhaps the most peculiar feature of our model is the running behaviour of the dilaton, which is schematically shown in Fig. 1. Because the 'minimum' on the left hand side is at the boundary of the kinematically forbidden region, we hesitate to call this stabilization of the dilaton. The perturbative breakdown of supersymmetry and stabilization of moduli of string theory may require the full 1-loop corrections to the effective supergravity theory which have been recently calculated [16]. On the nonperturbative side, perhaps other stringy nonperturbative effects are more crucial as pointed ot in ref. [17]. A realization of this proposal in the context of linear multiplet formulation of gaugino condensate appears in ref [18]. Of course, the exact form of these nonperturbative corrections are not yet understood. But one can perhaps expect that the recent developments in string dualities can shed some light on the latter, and on the stabilization of string moduli and supersymmetry breaking.

Acknowledgements

I wish to thank Prof. Mary K. Gaillard for useful discussions, and Yi-Yen Wu for collaboration on parts of the work that this talk was based on. I also wish to thank the organizers of Cargèse summer school "Masses of Fundammental Perticles" (August 1996). This work was supported in part by the Director, Office of Energy Research, Office of High Energy and Nuclear Physics, Division of High Energy Physics of the U.S. Department of Energy under the Contract DE-AC03-76SF00098, and by the National Science Foundation under grant PHY-95-14797.

Appendix - The Role of the Gauge Coupling

In this appendix, we recall a few facts about the perturbative corrections of the gauge coupling function in the superstring effective field theories, as well as the connection with modular invariance of the effective theory.

As mentioned earlier, in our approach, the one-loop renormalization of the gauge coupling is completely included in the Kähler potential K, *i.e.,* the renormalization effects are completely absorbed into K by replacing the tree-level gauge coupling $S+\bar{S}$ in K by the one-loop renormalized gauge coupling. Therefore, it is worthwhile to discuss the renormalization of gauge couplings in supestring LEEFTs.

Let us first recall the Lagrangian for supergravity plus super-YM. In the Kähler covariant formalism [20] the classical superfield Lagrangian is given by:

$$\mathcal{L} = \int d^4\theta \{-3E + \frac{E}{8R} f_{ab}(Z) W^{\alpha a} W^b_\alpha + \frac{E}{2R} e^{K/2} W(Z)\} + h.c., \tag{A.1}$$

where $E = \mathrm{Sdet} E_M{}^A$, R is the curvature scalar of the superspace, and Z stands for the chiral fields in the theory. The first term in eq. (A.1) corresponds to the kinetic energy for the gravity sector as well as the chiral fields. The chiral fields enter through the dependence of the spinorial derivatives of E on the Kähler potential, $K(Z, \bar{Z})$. The second term describes the super-YM coupling to the theory, with the (holomorphic) gauge coupling function $f_{ab}(Z)$ and the YM 'field-strength' superfield

$$W_\alpha = W_{\alpha a}T^a = (\frac{1}{8}\bar{\mathcal{D}}^2 - R)e^{-2V}\mathcal{D}_\alpha e^{2V},$$

where V is the vector multiplet containing the YM gauge potential. We shall take $f_{ab} = f\delta_{ab} = S\delta_{ab}$ corresponding to the bare coupling of the effective superstring theories where S is the dilaton/axion chiral superfield. The component form of the second term contains:

$$\int d^4x\sqrt{g}\left(-\frac{1}{4}\text{Re}f\,\text{Tr}(F^2) - \frac{1}{4}\text{Im}f\,\text{Tr}(F\tilde{F})\right),$$

and thus Ref is the YM gauge coupling, while Imf gives the axionic coupling.

Finally in the last term of eq. (A.1), $W(Z)$ is the superpotential which is a holomorphic function of the chiral matter fields (independent of S and other internal moduli, until supersymmetry is broken nonperturbatively).

In discussing the gauge couplings in effective theories, it is important to to distinguish between the *Wilsonian* couplings, and the physical, or *effective* couplings. In particular in the effective supersymmetric theories that we are considering, there are powerful statements that can be made about the two types of gauge coupling. The (holomorphic) Wilsonian gauge couplings in supersymmetric YM theories, which appear in the Wilson effective action, do not renormalize beyond one loop. These are funcions that appear in the Wilson effective action, $S_W(\mu)$, the local functional of quantum operators. In $S_W(\mu)$, only momenta between the scale μ and the UV cut-off contibute to loops. The physically measurable 'effective' (or running) couplings appear in the c-number valued generating functional of 1PI graphs, Γ; this is in general a nonlocal functional of background fields that contain the IR momenta $p < \mu$ running through loops, as well. Right at the UV cut off, the Wilsonian couplings, *i.e.,* the coefficients appearing in front of the operator terms in S_W are the bare couplings of the theory. The relation between the two effective actions may formally be written as [21]

$$e^{i\Gamma[\Phi_{cl},\mu]} = \langle e^{iS_W[\Phi,\mu]}\rangle,$$

where the expectation value on the right hand side is taken in the the presence of background fields. In the supersymmetric YM theories, it is known that, unlike the Wilsonian gauge coupling, the effective coupling renormalizes perturbatively at all orders, and that, indeed, higher order corrections introduce nonholomorphicities [21]. The generalizations of these results to supergravity effective theories of superstrings have been carried out more recently [1, 13, 19, 22].

The gauge coupling in all $N = 1$ effective heterotic string constructions is given at tree level by:

$$g_\alpha^{-2} = k_\alpha \text{Re}S = k_\alpha g_{string}^{-2}. \tag{A.2}$$

ReS is the 'universal' gauge coupling at string scale, and k_α is the level of the affine Lie algebra associated with the factor G_α of the product gauge group. Subsequently, we shall set $k_\alpha = 1$, and throughout the analysis G_α refers to the IR strong group with gaugino condensation. The exact Wilsonian coupling is given by the holomorphic function: $f_W = S + f^{(1)}$, and the moduli dependent one-loop (*i.e.,* all-loop) correction $f^{(1)}(T^i)$ has been determined [22] (see below). The effective gauge coupling, with

LEEFT-loop corrections to all orders is given by [19, 21]:

$$g_{eff}^{-2}(p^2) = \mathrm{Re}S \;+\; b_0 \ln\frac{\Lambda^2}{p^2} + cK + \frac{T(adj)}{8\pi^2}\ln g_{eff}^{-2}(p^2)$$
$$- \;\frac{1}{8\pi^2}\sum_r T(r)\ln\det Z_{eff}^{(r)}(p^2), \qquad (A.3)$$

where, $b_0 \equiv (-3T(adj) + \sum_r n_r T(r))/16\pi^2$ (the YM β-function coefficient), and $c \equiv (-T(adj) + \sum_r n_r T(r))/16\pi^2$, and Z is the kinetic normalization matrix. To one-loop order, one has to evaluate the r.h.s. of the above equation at tree level, at 2-loop the r.h.s. is evaluated to one loop, etc. The one-loop result has also been obtaind in [13]. Threshold corrections due to integrating out the heavy string modes have been calculated in reference [1]. These corrections are only dependent on the moduli T^i, and not on the dilaton. All the perturbative dilaton dependences in the effective gauge coupling arise from *field-theoretical* loop effects. We have seen in section (3) that threshold corrections in the effective field theory also introduce dilaton-dependent terms in the running coupling.

Let us now turn to the question of modular invariance. As inputs from string theory, for general fields Φ^I (ignoring for the moment the GS counter terms), we have the normalisation matrix for the kinetic term, and the Kähler function. The former is given by:

$$Z_{I\bar{J}} = \delta_{I\bar{J}}\prod_i (T^i + \bar{T}^i)^{-q_I^i} + \mathcal{O}(\Phi^2), \qquad (A.4)$$

where the rational numbers q_I^i are the modular weights of the field Φ^I. They depend on the twist sector of the orbifold which gives rise to the matter fields Φ^I, and the modulus field T^i. The Kähler function at the tree level is given by $K = -\ln(S+\bar{S}) - \sum_i \ln(T^i + \bar{T}^i) + \mathcal{O}(\Phi^2)$. For the modular transformation given in eq. (1) of the text, K transforms by the usual transformation law:

$$K \to K + F + \bar{F}, \qquad F = \sum_i \ln(i\gamma_i T^i + \delta_i). \qquad (A.5)$$

Under a modular transformation, the non-modulus chiral fields, transforms as:

$$\Phi^I \to C_J^I(T^i)\Phi^J. \qquad (A.6)$$

Hence, the kinetic matrix $Z_{I\bar{J}}$ transforms according to:

$$Z \to (C^\dagger)^{-1} Z C^{-1}. \qquad (A.7)$$

It follows from eq. (A.5 – A.7) that the reparametrization induced on the matter fields by modular tranformations is given by:

$$C_J^I = C_{0J}^I \prod_i (i\gamma_i T^i + \delta_i)^{q_I^i}, \qquad (A.8)$$

where C_{0J}^I is moduli independent.

For a generic supergravity theory with super-YM, under the combined transformations: $K \to K + F + \bar{F}$ and $\Phi^I \to C_J^I \Phi^J$ with C_J^I holomorphic function of the moduli Φ^I, the Kähler invariance of the (exact) integral of the RGE's, *i.e.*, eq. (A.3) imply that:

$$f_W \to f_W + cF - \frac{1}{2\pi^2}\sum_r T(r)\mathrm{tr}\ln C^{(r)}, \qquad (A.9)$$

where c is the group theoretical factor given after eq. (A.3) above. For C^I_J and F corresponding to modular transformations, eq's (A.5) and (A.8), this gives:

$$\mathrm{Re}f_W \to \mathrm{Re}f_W - \frac{1}{16\pi^2}\sum_i 2\alpha^i \ln |(i\gamma_i T^i + \delta_i)|^2, \tag{A.10}$$

with

$$\alpha^i = \sum_I T(\Phi^I)(1 - 2q^i_I) - T(adj); \quad T(\Phi^I) = \sum_r \mathrm{tr}(T^2(r)), \tag{A.11}$$

and $T_a(r)$ are the generators of the representations of the fields Φ^I.

Furthermore, the transformation law (A.10) corresponds, up to a modular invariant function, to the transformation of the logarithm of Dedekind function. In fact it will give the complete modular dependent perturbative correction, $f^{(1)}$ to the Wilsonian coupling [1, 12, 13] :

$$\mathrm{Re}f^{(1)} = \delta \mathrm{Re}f_W = -\frac{1}{4\pi^2}\sum_i \alpha^i \ln |\eta(iT^i)|^2, \tag{A.12}$$

modulo a moduli independent part which has been argued to be a constant in most orbifold models [22]. These equations are interpreted as a parametrization of the string threshold corrections to the gauge couplings [1].

Modular invariance is restored by including factors of $\eta(iT^i)$ in the superpotential (see eq (2)), and in the definition of the fields, so as to cancel the above modular dependent correction of the gauge coupling, as well as by introducing GS counter term as discussed in the text. However, the inclusion of the η factors tends to spoil the boundedness from below of the scalar potential. To avoid this, we may restrict ourselves to the orbifold models which do not receive string threshold corrections. These models have been classified [1, 12, 22]. For such models, the modular anomaly is solely cancelled by the GS counter term.

References

[1] L. Dixon, V. Kaplunovsky, and J. Louis *Nucl. Phys.* **B355:** 649 (1991).

[2] M. K. Gaillard and B. Zumino, *Nucl. Phys.* **B193:** 221 (1981); S. Ferrara, J. Scherk., and B. Zumino, *Nucl. Phys.* **121:** 393 (1977).

[3] P. Binétruy and M.K. Gaillard, *Phys. Lett.* **B365:** 87 (1996).

[4] K. Saririan, *preprint* LBL-381995, UCB-PTH-96/03 (To appear).

[5] M. K. Gaillard, and Rulin Xiu, Phys. Lett. **B296:** 71 (1992) ; R. Xiu, Phys. Rev. **D49:** 6656 (1994) ; S. P. Martin and P. Ramond, Phys. Rev. **D51:** 6515 (1995); K.R. Dienes and A.E. Farraggi, Phys. Rev. Lett. **75:** 2646 (1995).

[6] G. Veneziano and S. Yankielowicz, *Phys. Lett.* **B113:** 231 (1982).

[7] E. Witten, *Nucl. Phys.* **B202:** 253 (1982).

[8] H. P. Nilles, *Phys. Lett.* **B115:** 193 (1982); S. Ferrara, L. Girardello, H. P. Nilles, *Phys. Lett.* **B125** 457 (1983).

[9] P. Binétruy and M.K. Gaillard, *Phys. Lett.* **B253:** 119 (1991).

[10] M. Dine, R. Rhom, N. Seiberg, and E. Witten, *Phys. Lett.* **B156:** 55 (1985).

[11] G.L. Cardoso and B.A. Ovrut, *Nucl. Phys.* **B369:** 351 (1992); *ibid.*, *Nucl. Phys.* **B392:** 315 (1993); *ibid.*,*Nucl. Phys.* **B418:** 535 (1994); S. Ferrara abd M. Villasante, *Phys. Lett.* **B186:** 87 (1987); P. Binétruy, G. Girardi, R. Grimm and M. Müller, *Phys. Lett.* **B265** 111 (1991); P. Adamietz, P. Binétruy, G. Girardi and R. Grimm, *Nucl. Phys.* **B401:** 257 (1993); J.P. Derendinger, F. Quevedo, M. Quiros, *Nucl. Phys.* **B428:** 228 (1994).

[12] I. Antoniadis, K.S. Narain, and T.R. Taylor, *Phys. Lett.* **B267:** 37 (1991).

[13] M.K. Gaillard and T.R. Taylor, *Nucl. Phys.* **B381:** 577 (1992).

[14] A. Font, L. I. Ibanez, D. Lust, F. Quevedo, *Phys. Lett.* **241:** 35 (1990).

[15] A. Sen, *Int. J. Mod. Phys.* **A9:** 3707 (1994), J. H. Schwarz and A. Sen, *Phys. Lett.* **B312:** 105 (1993); *Phys. Lett.* **B411:** 105 (1994); I. Girardello, A. Giveon, M. Porratti, A. Zaffaroni, *Nucl. Phys.* **B448** 127 (1995).

[16] M.K. Gaillard, V. Jain, and K. Saririan, *preprint* LBL-34984, hep-th/9606052 to be published in *Phys. Rev. D*; Also, *Phys. Lett.***B387:** 520 (1996).

[17] T. Banks and M. Dine, *Phys. Rev.* **D:50** 7454 (1994).

[18] P. Binétruy, M. K. Gaillard and Y-Y. Wu, *preprint* LBNL-38590, UCB-PTH 96/13; hep-th/9605170 (to be published in *Nucl. Phys.* **B**).

[19] V. Kaplunovsky and J. Louis, *Nucl. Phys.* **B422:** 57 (1994).

[20] P. Binétruy, G. Girardi, R. Grimm and M. Muller, *Phys. Lett.* **189B:** 83 (1987); P. Binétruy, G. Girardi and R. Grimm, Annecy preprint LAPP-TH-275-90, (1990).

[21] M. A. Shifman and A. I. Vainshtein, *Nucl. Phys.* **B277:** 456 (1986); and *Nucl. Phys.* **B359:** 571 (1991).

[22] V. Kaplunovsky and J. Louis, *Nucl. Phys.* **B444:** 191 (1995).

SOME NEW SALIENT FEATURES OF DEEP-INELASTIC SCATTERING AT HERA

J. Feltesse

DSM/DAPNIA, Centre d'Etudes de Saclay
F-91191 Gif-sur-Yvette Cedex (France)
email-address : J.Feltesse@cea.fr

INTRODUCTION

At HERA the lepton probes the structure of the nucleon with a spatial resolution $\lambda \sim \bar{h}c/\sqrt{Q^2}$ where Q^2, minus the squared 4-momentum of the exchanged virtual photon, can reach 90000 GeV^2, thus providing a probe transverse size as low as ~ 0.001 fm, i.e. 1000 times smaller than the size of the proton. At HERA x can be as low as $5\ 10^{-5}$ at Q^2 values $\lesssim 4\ \text{GeV}^2$. The low x and high Q^2 regimes are new domains in which QCD can be tested.

The large range of DIS data on inclusive and final state distributions at HERA provides a very rich source of information on perturbative QCD and on the transition between perturbative and non-perturbative QCD. In this contribution we concentrate on some new salient features. In the following section we give a brief overview of the theoretical expectations at low x. In the third section we present the most recent results on the structure function $F_2(x, Q^2)$ of the proton together with the first determination of the longitudinal structure function $F_L(x, Q^2)$ and of $F_2^{c\bar{c}}$, the charm contribution to the structure function F_2 at very low x. The last section is devoted to the most recent findings on the diffractive DIS events, the special class of events where the proton stays almost intact in the collision. We show how the new data can be interpreted in terms of the debatable structure function of the Pomeron.

THEORETICAL EXPECTATIONS

In the case of an elastic collision W, the outgoing invariant hadronic mass (Fig.1), is equal to M the mass of the nucleon and ν the energy lost by the lepton in the nucleon frame is simply

$$\nu = \frac{Q^2}{2M} \tag{1}$$

Masses of Fundamental Particles: Cargèse 1996
Edited by M. Lévy *et al.*, Plenum Press, New York, 1997

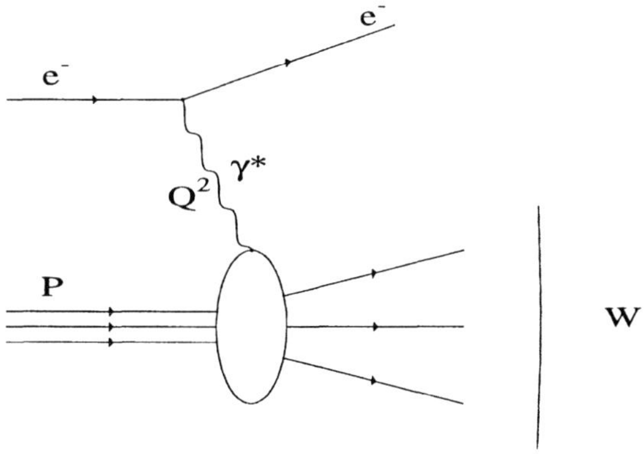

Figure 1. Basic Feynman diagram for electron proton deep inelastic scattering. Q^2 is minus the squared four momentum transferred or minus the squared mass of the virtual photon. W is the virtual photon-proton centre of mass energy.

In other words, using the Bjorken variable $x = \frac{Q^2}{2M\nu}$ the elastic cross section should have a peak around $x = 1$. Similarly when the lepton scatters elastically off valence quarks of the nucleon, the cross section should have a peak around $x = \frac{1}{3}$ and go to zero when x goes to zero.

More precisely, the inclusive differential cross section for DIS of a charged lepton off a nucleon is given in terms of structure functions as :

$$\frac{d^2\sigma(l^\pm N)}{dxdQ^2} = \frac{2\pi\alpha^2}{Q^4 x}\left[\left(2\,(1-y) + \frac{y^2}{1+R}\right) F_2(x,Q^2) \mp \left(2y - \frac{y^2}{2}\right) xF_3(x,Q^2)\right] \quad (2)$$

with

$$R = \frac{F_2 - 2xF_1}{2xF_1} = \frac{F_L}{2xF_1} \quad (3)$$

where F_L is the longitudinal structure function. The contribution from F_L is negligible except at large y, where y is the fractional energy transferred to the photon in the nucleon centre of mass. In the highest Q^2 range accessed by the present data ($Q^2 \approx 5000\ \mathrm{GeV}^2$) the contribution of the structure function xF_3 is still a small correction.

In the Quark Parton Model, the structure function F_2 is simply related to $q_i(x)$ the probability of finding the quark i in the nucleon carrying a fraction x of the proton momentum :

$$F_2(x) = \sum_i e_i^2\, x(q_i(x) + \bar{q}_i(x)) \quad (4)$$

where e_i is the electric charge of the quark i.

In the naive approach of quark constituents in the nucleon, the structure function is expected to have a peak around $x = \frac{1}{3}$ and to go to zero when x goes to zero or to one. This naive approach of the x dependence has been well supported by the lepton-nucleon data on fixed targets accumulated at SLAC, Fermilab and CERN up to the late eighties.

Although the full x and Q^2 dependence of the structure functions is not given by perturbative QCD (PQCD), there are prescriptions which can be compared to the data :

- Q^2 evolution
- x evolution of the gluon density at very low x
- asymptotic behaviour when $x \to 0$
- $F_L(x, Q^2)$

Q^2 evolution

The Q^2 evolution of the nucleon structure functions is one of the basic pre-

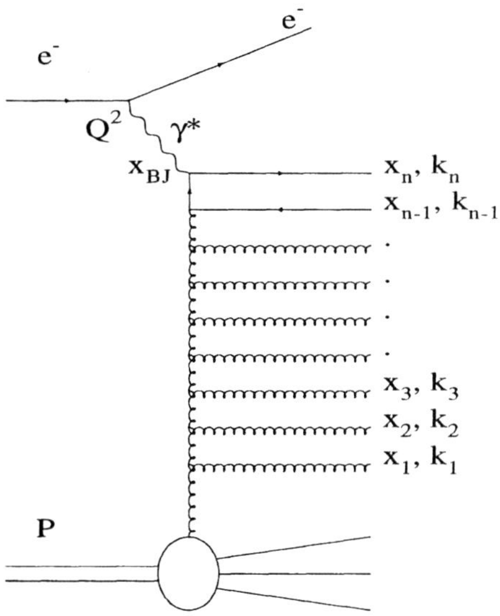

Figure 2. Diagrammatic representation of the gluon rungs contributing to deep inelastic scattering.

dictions of QCD, it has been initially derived in the language of renormalization group equations for coefficients of operator product expansions [1, 2]. An alternative derivation is based on sum of quark and gluon ladder diagrams of the type shown in Fig. 2. In the summation, gluons have ordered longitudinal momenta (momentum conservation) but also strongly ordered transverse momenta

$$x \leq x_n \leq \ldots \leq x_2 \leq x_1$$
$$Q^2 \gg k_n^2 \gg \cdots \gg k_2^2 \gg k_1^2 \gg Q_0^2 \tag{5}$$

The summation can be used in simulation programs to provide prediction on the hadronic final states and provides the so-called Dokshitzer Gribov Lipatov Altarelli Parisi (DGLAP) differential equations on Q^2 evolution of structure functions [3].

In Leading Order (LO), for Non-Singlet [1] structure functions $F^{NS}(x,Q^2)$, the Q^2 derivative at fixed x depends only on the quark densities at momentum fractions above x, but for Singlet [2] structure functions $F^S(x,Q^2)$, the Q^2 derivative depends on both the quark and the gluon densities :

$$\frac{\partial F^{NS}(x,Q^2)}{\partial \log(Q^2)} = \frac{\alpha_s(Q^2)}{2\pi}\int_x^1 \frac{dy}{y} F^{NS}(y,Q^2)P_{qq}^{NS}(\frac{x}{y}) \tag{6}$$

$$\frac{\partial F^{S}(x,Q^2)}{\partial \log(Q^2)} = \frac{\alpha_s(Q^2)}{2\pi}\int_x^1 \frac{dy}{y}\left[F^S(y,Q^2)P_{qq}^S(\frac{x}{y}) + 2n_f g(y,Q^2)P_{qg}^S(\frac{x}{y})\right] \tag{7}$$

$$\frac{\partial g(x,Q^2)}{\partial \log(Q^2)} = \frac{\alpha_s(Q^2)}{2\pi}\int_x^1 \frac{dy}{y}\left[F^S(y,Q^2)P_{gq}^S(\frac{x}{y}) + g(y,Q^2)P_{gg}^S(\frac{x}{y})\right] \tag{8}$$

where the P_{qq}, P_{gq} and P_{gg} are known splitting functions, n_f is the number of flavours and $xg(x,Q^2)$ is the gluon distribution. In parton language, the equations predict that when Q^2 increases there are more partons at low x ($x < 0.1$) and less partons at large x ($x > 0.2$).

The differential equations arise from the resummation of a series of $(\alpha_s \log(Q^2/Q_0^2))^n$ which is a priori only valid in the kinematical domain of the Leading Log Approximation (LLA(Q^2)) defined by (see Fig.3) :

$$\begin{aligned} \alpha_s(Q^2)\log(Q^2/Q_0^2) &\sim 1 \\ \alpha_s(Q^2)\log(1/x) &\ll 1 \\ \alpha_s(Q^2) &\ll 1 \end{aligned} \tag{9}$$

The validity of the DGLAP differential equations has been very precisely checked in fixed target experiments and has provided one of the best determinations of α_s [4]. Before data from HERA became available at very low x it had been commonly guessed that these equations would only be valid at about $x > 10^{-2}$ and $Q^2 > 4$ GeV^2.

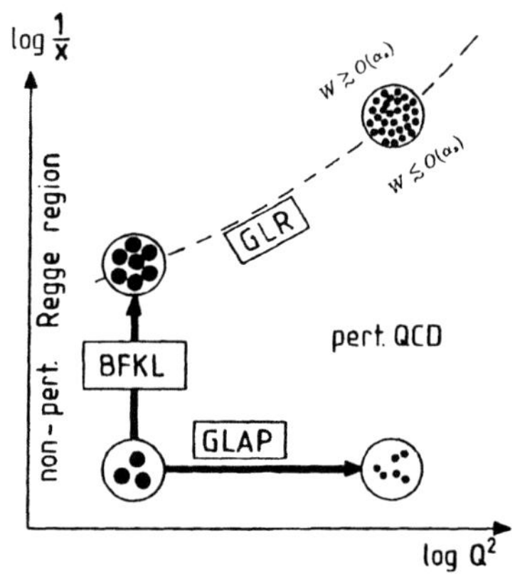

Figure 3. Schematic evolution of the quark densities in various (x, Q^2) regions according to the dominant dynamical effects. The dashed line is the theoretical limit of validity of perturbative QCD.

[1]Non-Singlet distribution of the quark flavour group is a combination of $q_i - q_j$ terms.
[2]The Singlet distribution is the combination of flavour symmetric terms $\sum_i(q_i + \bar{q}_i)$.

x evolution

At very low x the so-called Balitski Fadin Kuraev Lipatov (BFKL) equation [5] on the x evolution of the gluon distribution has been derived in the kinematical domain defined by (see Fig.3) :

$$\begin{aligned} \alpha_s(Q^2)\log(Q^2/Q_0^2) &\ll 1 \\ \alpha_s(Q^2)\log(1/x) &\sim 1 \\ \alpha_s(Q^2) &\ll 1 \end{aligned} \tag{10}$$

The differential equation is based on summation of $(\alpha_s \log(1/x))^n$ terms. It can be written in the form

$$x\frac{\partial f(x,k^2)}{\partial x} = \alpha_s \int_{Q_0^2} dk'^2 K(k^2,k'^2) f(x,k'^2) \tag{11}$$

with

$$f(x,k^2) = \frac{\partial xg(x,Q^2)}{\partial \log(Q^2)}|_{Q^2=k^2} \tag{12}$$

The kernel $K(k,k')$ is a known function [5]. Q_0 is an energy cut-off to prevent divergence of the kernel at $Q^2 = 0$. When x decreases, the equation evolution predicts a steep rise of the gluon density.

As in DGLAP evolution, the evolution at small x can be described as a branching process in the initial state, but with different orderings [6, 7, 8]. In the parton chain there is no strong ordering in k_n^2 but a strong ordering in x_n

$$\begin{aligned} x \ll x_n \ll \cdots \ll x_2 \ll x_1 \\ Q^2 \geq k_n^2 \geq \cdots \geq k_2^2 \geq k_1^2 \geq Q_0^2 \end{aligned} \tag{13}$$

The phase space available is larger than for strong transverse momentum ordering. There is "diffusion" or "random walk" in transverse momentum along the gluon chain. The difference of level of ordering in the gluon emission between DGLAP mechanism (strong order) and BFKL mechanism (soft order) has been exploited to look for signatures of the BFKL mechanism [9].

Asymptotic behaviour when $x \to 0$.

A long time ago, in 1974, it has been demonstrated that non-Abelian gauge theories are asymptotically free [10]. This is the fundamental property of QCD. A surprising consequence [11] of the asymptotic freedom is that, on the basis of certain uniformity assumptions, the structure function F_2 of the nucleon should grow faster than any power of $\log \frac{1}{x}$ and slower than any power of $\frac{1}{x}$ as $x \to 0$ and the growth should increase with increasing Q^2. This crucial property of QCD was forgotten for many years.

A different approach at low x is to resum the $(\alpha_s \log 1/x)^n$ terms as accomplished by the BFKL equation. When the effect of the running coupling constant is neglected, i.e. in a small range of Q^2, the BFKL evolution equation of the gluon can be solved analytically. The solution exhibits a Regge type $x^{-\lambda}$ dependence with

$$\lambda = \frac{12\alpha_s}{\pi}\log 2 \approx 0.5 \tag{14}$$

The quantity $1+\lambda$ is equal to the intercept of the so called BFKL bare Pomeron which is very different from the 1.08 intercept of the effective soft Pomeron which describes so well the energy dependence of all hadron-hadron and photon-hadron cross sections [13]. When the running of α_s and the strict conservation of energy is taken into account the exponent λ is expected to be a bit smaller [14, 15, 16] .

For completeness we should also mention that the rise of the gluon density has to be damped at some sufficiently small value of x by saturation (shadowing) effects when the parton density is so large that the partons can no longer be considered as independent. The Gribov Levin Ryskin (GLR) equation [17] is the first attempt to describe the shadowing correction. It is a perturbative QCD calculation applied in the transition region to the unknown domain of non-perturbative QCD (see Fig.3). Shadowing effects are however expected to be relatively weak in the HERA kinematical domain [16, 18, 19].

$F_L(x, Q^2)$

In the naive parton model and in leading order of QCD the longitudinal structure function F_L is zero. In Next-To-Leading Order of PQCD F_L gets contributions in α_s from both quarks and gluons [20]:

$$F_L(x,Q^2) = \frac{\alpha_s(Q^2)}{2\pi} x^2 \int_x^1 \frac{dy}{y^3} \left[\frac{8}{3} F_2(y,Q^2) + 2 \sum_{i=1}^{2n_f} e_i^2 (y-x) g(y,Q^2) \right] \tag{15}$$

F_L should vanish smoothly at large Q^2 as $\alpha_s(Q^2)$ and get sizeable contributions at low x ($x < 0.1$) due to the increase of the sea and of the gluon distributions. Higher orders of PQCD expressions of F_L, based on DGLAP evolution processes, have been recently calculated [21]. Estimates of F_L at low x and low Q^2, based on various assumptions on the dominant dynamics at this frontier of perturbative QCD, have been made this year [22, 23, 24]. We will see below what we can learn from the first experimental estimate of F_L at very low x by the H1 experiment.

STRUCTURE FUNCTION $F_2(x, Q^2)$

The first determination [25, 26] of the proton structure function $F_2(x, Q^2)$ at HERA in 1992 has been based on a recorded luminosity of about 30 nb^{-1} and has revealed the striking feature of a proton structure function rising as x decreases below 10^{-2}, for Q^2 values in the range $8 < Q^2 < 60$ GeV^2.

The published 1994 data [27, 28] from the H1 and ZEUS collaborations represent an increase of about a factor 100 in statistics at large Q^2 together with an extension of the accessible kinematic range towards $Q^2 = 1.5$ GeV^2 and $x = 0.00005$. The preliminary 1995 data [29] from the ZEUS collaboration provide a first glimpse into the $Q^2 < 1$ GeV^2 region.

At very high $Q^2 \geq 5000$ GeV^2 the number of recorded events is so far too meager to extract structure functions of neutral or charge currents. Data at high Q^2 are however very sensitive to new phenomena by comparing the measured differential cross section $d\sigma/dQ^2$ with the expected cross section from the standard model, the HERA experiments have provided new limits on masses and couplings of leptoquarks and on fermion compositeness scales and searches for supersymmetric particles [30, 31].

Scaling violations

The results on $F_2(x, Q^2)$ from H1 1994 data together with those of BCDMS [32], E665 [33] and NMC [34] are shown in fig. 4 as a function of Q^2 at fixed x values. In the HERA domain, to extract the structure function F_2 from the cross section (equation 2) assumptions based on the perturbative QCD expression (equation 15) have been made.

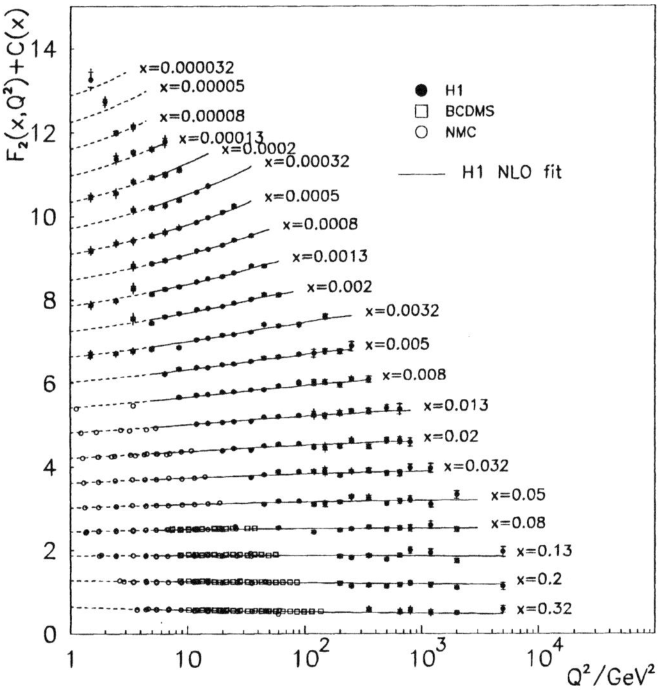

Figure 4. The proton structure function $F_2(x, Q^2)$ vs Q^2 for fixed values of x. The 1994 data from H1 (closed points) are compared to the fixed target data of BCDMS (open triangles) and NMC (open circles). The full lines represent the DGLAP NLO QCD fit using H1 data at $Q^2 > 5$ GeV^2. The dashed line represent the backward extension of the fit. For clarity the F_2 values are plotted adding a term $c(x) = 0.6(i_x - 0.4)$ to F_2, where i_x is the bin number starting at $i_x = 1$ for $x = 0.32$.

It is striking that in a large part of the kinematical domain at not too small nor too large Q^2 values, the systematics are at present below 5 %. The H1 data agree well with a smooth extrapolation from NMC and BCDMS data. As in the fixed target domain, there are no scaling violations at $x \sim 0.1$, but pronounced Q^2 scaling violations at low x, becoming steeper with x decreasing towards $x = 0.00005$.

The H1 collaboration has made a common fit of the 1994 data at $Q^2 > 5$ GeV^2 together with the new NMC data and the BCDMS measurements. The fit is based on Next-To-Leading order (NLO) DGLAP evolution equations. It gives a very good description of the data at $Q^2 > 5$ GeV^2 and evolving backwards down to $Q^2 = 1.5$ GeV^2 provides a prediction which is also in good agreement with the data (see fig. 4). The fit provides a determination of the gluon density with an accuracy a factor of two better than the previous determinations [35, 36]. A steep rise of the gluon density towards low x is observed (fig. 5). The rise is more pronounced when Q^2 increases.

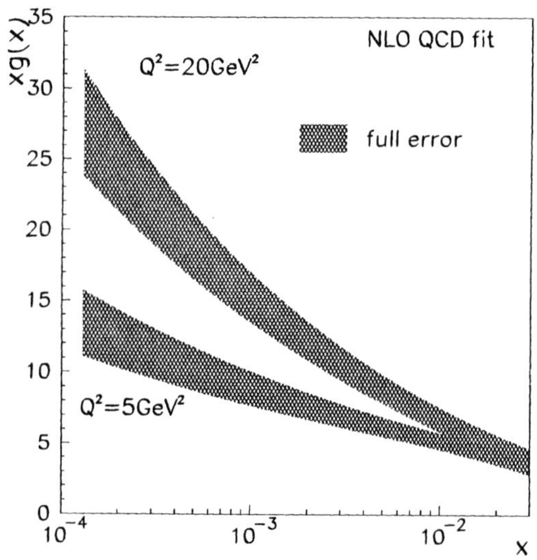

Figure 5. The gluon density $xg(x)$ at $Q^2 = 5$ GeV^2 and $Q^2 = 20$ GeV^2 extracted from a NLO QCD fit combining H1 1994 with NMC and BCDMS F_2 data. The band represents the full experimental errors.

The x dependence of F_2

The results on $F_2(x, Q^2)$ from H1 [27] and ZEUS [28] are shown in fig. 6 as a function of x at fixed Q^2 values. It is remarkable that the data sets of H1 and ZEUS agree well within the present small errors. It is striking that the steep rise of F_2 with

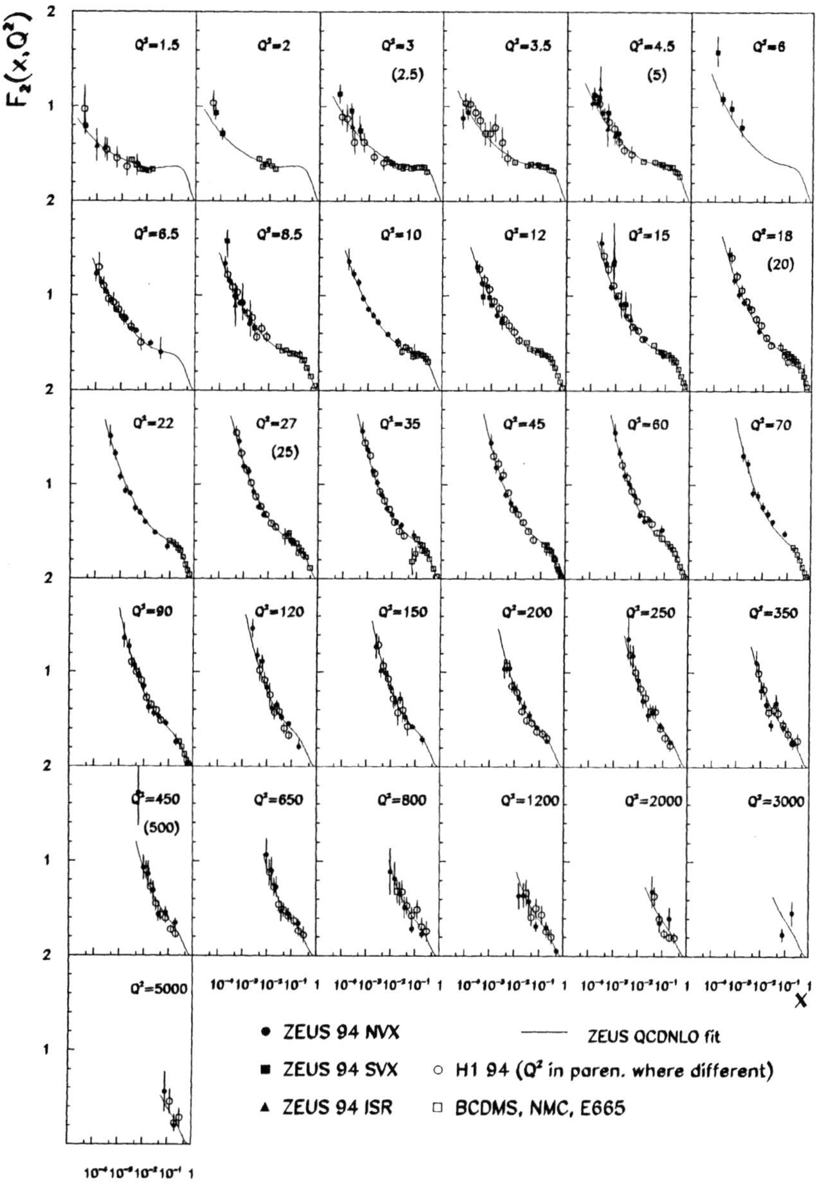

Figure 6. The proton structure function F_2. The 1994 data from H1 and ZEUS with Nominal Vertex (NVX), Shifted Vertex (SVX) and Initial State Radiation (ISR). The HERA data are compared to the fixed target data of BCDMS, NMC and E665. The full line represents the DGLAP NLO fit of all the ZEUS data points combined with the NMC data at $Q^2 > 4$ GeV^2.

decreasing x persists down to Q^2 values as low as 1.5 GeV^2 at $x < 10^{-2}$ and is already visible at $Q^2 = 2000$ GeV^2 at $x \sim 0.1$. To quantify the rise of F_2 the H1 collaboration has extracted the average derivative $< \frac{d \ln F_2}{d \ln Q^2} >$ by fitting the power λ characterising the rise of $F_2 \propto x^{-\lambda}$ for each Q^2 bin and $x < 0.1$. A rise of λ with $\log Q^2$ is observed from about 0.1 to 0.4 at Q^2 values from 1.0 to 800 GeV^2 (fig. 7).

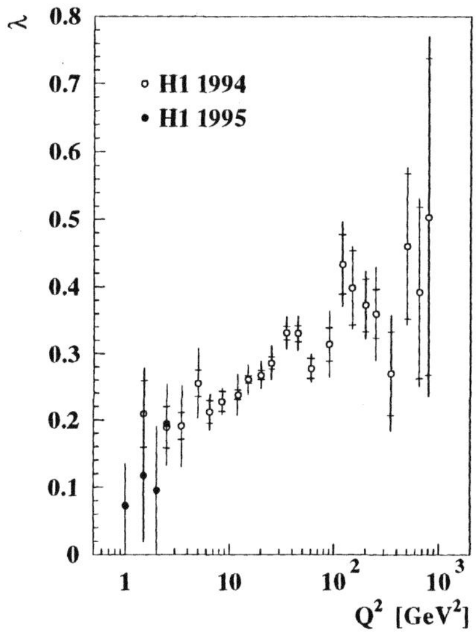

Figure 7. The exponent λ from fits of the form $F_2 \propto x^{-\lambda}$ of the measured structure function F_2 from the H1 experiment at fixed Q^2 values and for $x < 0.1$.

$F_2(x, Q^2)$ at very low Q^2

At low Q^2, in the transition region between photoproduction and DIS, there are two types of approach to generate the shape of the structure function :

- a model [37] where all parton densities are generated from intrinsic non-perturbative valence-like distributions at some low scale μ. The observed rise of F_2 at low x is generated by the perturbative QCD evolution starting at $\mu = 0.3$ GeV^2. This model has been so far very successful in predicting HERA data at low x and $Q^2 > 1.5$ GeV^2 [27, 28].

- models based on Regge theory which provide an economical parametrization of all hadron-hadron total cross sections. In these models [38, 39] the evolution at low Q^2 of F_2 is generated by empirical simple functions of Q^2 that vanish linearly with Q^2 as $Q^2 \to 0$. The Regge inspired models lie below the data at $Q^2 > 2$ GeV^2.

In 1995 the ZEUS collaboration [29] has added a small calorimeter close to the beam axis and obtained data at $0.15 < Q^2 < 0.6$ GeV^2. The preliminary results based on 1/4 of the statistics are shown in fig. 8. The data are in reasonable agreement with the Regge inspired parametrization of Donnachie and Landshoff [38] and in clear disagreement with the GRV parametrization at $Q^2 < 0.5$ GeV^2 [37].

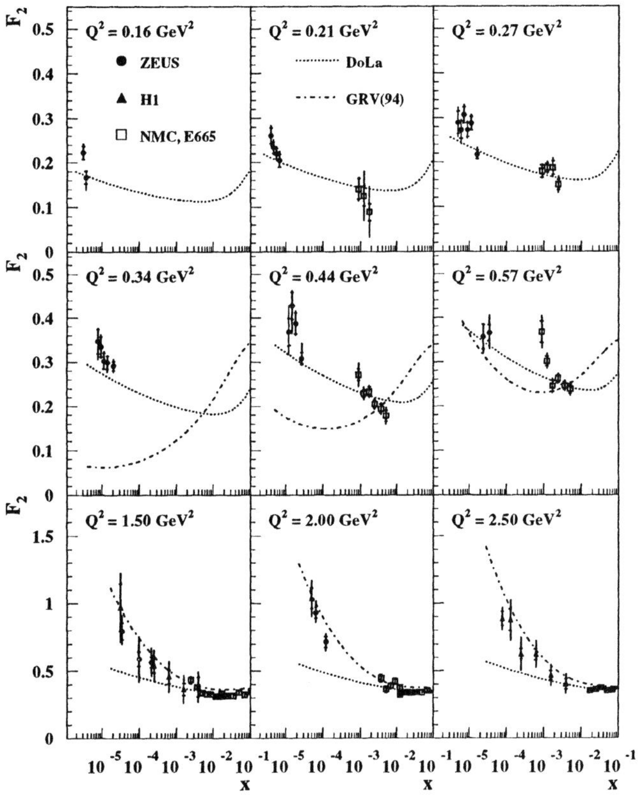

Figure 8. The low Q^2 structure function F_2 vs x. The data points of ZEUS, H1, NMC and E665 are compared to the predictions of the Donnachie and Landshoff (DoLa) Regge model and of the GRV parametrization.

Structure function $F_L(x, Q^2)$

The usual method to determine F_L is to measure the differential cross section at fixed (Q^2, x) bins but different y values by running at different beam energy settings. The H1 collaboration has proposed recently another method [40]. The structure function F_2 is extracted from the differential cross section at $y < 0.35$, a domain where F_L hardly contributes. Then at larger y, the value of F_2 is extrapolated with the DGLAP evolution equation. Finally from the measurement at large y together with the the extrapolated value of F_2 the structure function F_L is extracted in bins of x from the equation

$$\frac{d^2\sigma}{dydQ^2} = \frac{2\pi\alpha^2}{q^4 y}\left[\left(1+(1-y)^2\right)F_2(x,Q^2) - y^2 F_L\right] \tag{16}$$

which is a rewriting of equations 2 and 3. In contrast to the usual method it has to be assumed that the DGLAP evolution equations can be used to predict the values of F_2 at $Q^2 > 8$ GeV^2 and $x > 10^{-4}$ using as input values the result of the NLO fit of data at $Q^2 > 4$ GeV^2 and at $y < 0.35$. This assumption is reasonable because the HERA data so far have shown that the DGLAP evolution equations are still valid in this low x domain independently of the value of F_L. The resulting value based on a luminosity of 1.25 pb^{-1} is shown in fig. 9. Within the errors, the F_L values are consistent with the value of about 0.3 which can be inferred from perturbative QCD based on the DGLAP mechanism [21] and using the gluon density extracted from scaling violations [27].

Structure function $F_2^{c\bar{c}}$

The measurement of the charm production cross section in DIS is of importance for the understanding of the parton densities in the nucleon. There is some evidence from the EMC collaboration [41] that charm production should be dominated by the photon-gluon fusion process (Fig.10)

$$\gamma g \rightarrow c\bar{x}. \tag{17}$$

At HERA, in a sample of DIS events the H1 collaboration has observed nice signals of D^0 and D^* from the reconstruction of tracks of the hadronic final state [42]. For a recorded luminosity of 3 pb^{-1} H1 has obtained 103 ± 13 D^* events and 144 ± 19 D^0 events in the rapidity range $|\eta| < 1.5$. at $10 < Q^2 < 100$ GeV^2. Assuming charm tagging efficiencies

$$\begin{aligned} P(c \rightarrow D^0) * BR(D^0 \rightarrow k^-\pi^+) &= 0.0205 \pm 0.0011 \\ P(c \rightarrow D^*) * BR(D^{*+} \rightarrow D^0\pi^+) * BR(D^0 \rightarrow k^-\pi^+) &= 0.0069 \pm 0.0005 \end{aligned} \tag{18}$$

the signal has been converted into a charm cross section in each (Q^2, x) bin. The charm contribution $F_2^{c\bar{c}}$ to the proton structure function has been extracted after neglecting the F_L contribution at $y < 0.5$. The results are shown in fig.11 and compared to NLO calculations of the photon-gluon process [43] using the GRV-HO [37], MRSD0' [44] and MRSH [45] parametrizations of the gluon density for a charm quark mass $m_c = 1.5$ GeV. The data are also compared to the NLO calculations using the gluon density extracted from the H1 QCD fit to the F_2 data.

The only parametrization which fails to describe the data is the MRSD0' which was already exluded by inclusive F_2 measurements. The agreement between NLO QCD calculations and the data is remarkable for this very first measurement of $F_2^{c\bar{c}}$ at low x values.

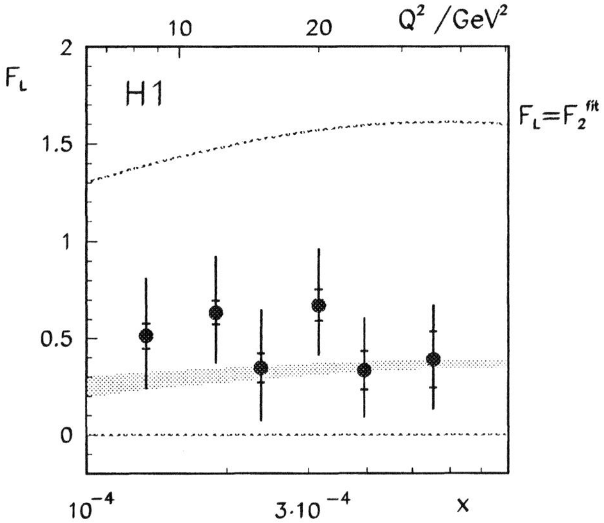

Figure 9. Longitudinal structure function F_L as a function of x or Q^2 for $y = 0.7$ by the H1 collaboration. The inner error bars are the statistical errors. The full error bars represent the statistical and systematic errors added in quadrature. The grey band represents the uncertainty of the calculation of F_L using the quark and the gluon determinations, as determined from the NLO QCD fit of of the H1 data [27] and the BCDMS data [32]. The dashed lines define the allowed range of F_L values from $F_L = 0$ to $F_L = F_2$.

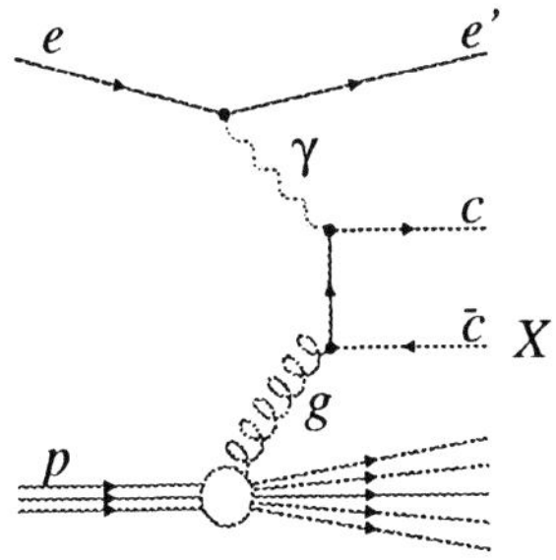

Figure 10. Feynman diagram of boson gluon fusion.

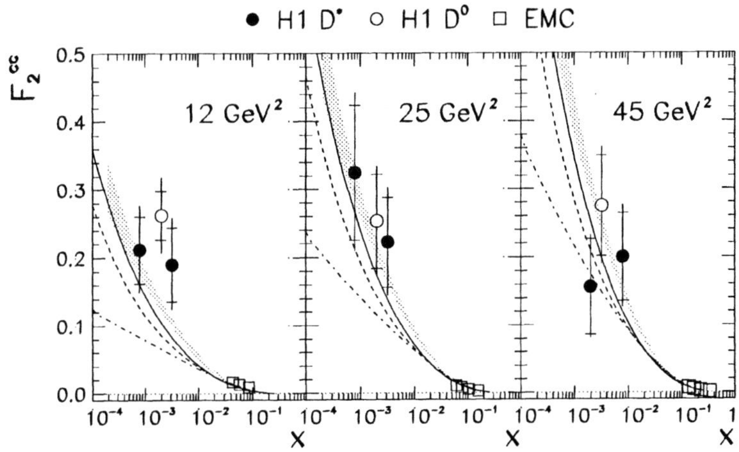

Figure 11. The charm contribution $F_2^{c\bar{c}}$ to the proton structure function as derived from the inclusive D^{*+} (full dots) and D^0 analysis (open circle) from the H1 collaboration. Data are compared with NLO calculations based on GRV (full lines), MRSH (dashed lines) and MRSD0' (dash-dotted line) parton distributions using a charm quark mass $m_c = 1.5$ GeV. The inner(outer) error bars refer to the statistical (total) errors. The shaded band represents the prediction using the parton distributions from the H1 NLO fit of the inclusive structure function F_2. The EMC data are also shown (open boxes).

DIFFRACTIVE DEEP-INELASTIC SCATTERING

It was anticipated by R.P. Feynman and J.D. Bjorken that deep-inelastic scattering at low x should provide a rather unique possibility to study diffractive dissociation at short distances [46] and that a large rapidity gap in the rapidity distribution of the final state would be a powerful criterion to select diffractive deep-inelastic events as shown in fig. 12 [47]. In a diffractive event the hadronic final state can be decomposed into two systems X an Y separated by a large rapidity gap and of invariant masses

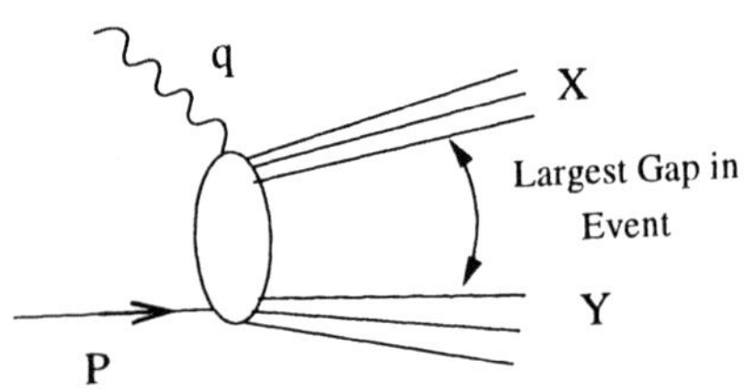

Figure 12. Basic Feynman diagram for diffractive deep-inelastic events.

M_X and M_Y small in comparison with the $\gamma^* p$ centre of mass energy W . Analysing the 1992 data, the ZEUS and H1 collaborations [48, 49] have reported observation of deep-inelastic events in which no energy flow is observed in a large region of rapidity close to the proton direction (fig. 12).

Independently of any assumption on the diffraction mechanism two variables can be defined :

$$x_P = \frac{Q^2 + M_X^2}{Q^2 + W^2} \tag{19}$$

$$\beta = \frac{Q^2}{Q^2 + M_X^2} \tag{20}$$

where M_X^2 and W^2 are the squared invariant masses of the hadronic system X and of the total hadronic system $(X + Y)$ respectively. The variables β and x_P are trivially related to x the usual Bjorken variable in DIS by

$$x = \beta\, x_P. \tag{21}$$

In a model where diffraction in DIS is produced by a Pomeron, an exchange object with the quantum numbers of the vacuum and with a partonic structure [50], x_P is the fraction of the proton energy carried by the Pomeron and β is the fraction of the Pomeron energy carried by the parton interacting with the virtual photon.

Structure function $F_2^{D(3)}(x_P, \beta, Q^2)$

In analogy to the proton structure function $F_2(x, Q^2)$ it is possible to define a 3-variable structure function $F_2^{D(3)}$ from the cross section

$$\frac{d^3\sigma^D}{dx_P d\beta dQ^2} = \frac{2\pi\alpha^2}{q^4\beta}\left[\left(1 + (1-y)^2\right) F_2^{D(3)}(x_P, \beta, Q^2)\right] \tag{22}$$

where the longitudinal structure function has been set to $F_L = 0$ and where the cross section has been averaged over t the squared 4-momentum transfer at the incident proton vertex. In models with partonic structure of the Pomeron, the structure function $F_2^{D(3)}(x_P, \beta, Q^2)$ should factorize into a Pomeron flux factor $f(x_P)$ and a Pomeron structure function $F_2^{D(2)}(\beta, Q^2)$:

$$F_2^{D(3)}(x_P, \beta, Q^2) = f(x_P) F_2^{D(2)}(\beta, Q^2) \tag{23}$$

The early measurements on 1993 data from the H1 (0.3 pb^{-1}) [51] and ZEUS (0.5 pb^{-1}) [52] experiments have indeed yielded a flux term $\propto x_P^{-n}$ which does not depend on β and Q^2. The two measurements are based on events selected by either a lack of activity in the forward detectors (H1) or by a large rapidity gap (ZEUS). The measured flux factors are perfectly consistent between the two experiments (table 1).

A recent measurement by the ZEUS collaboration of 1994 data based on events where the leading proton has been tagged [54] with $x_L = 1 - x_P > 0.97$ also yields a

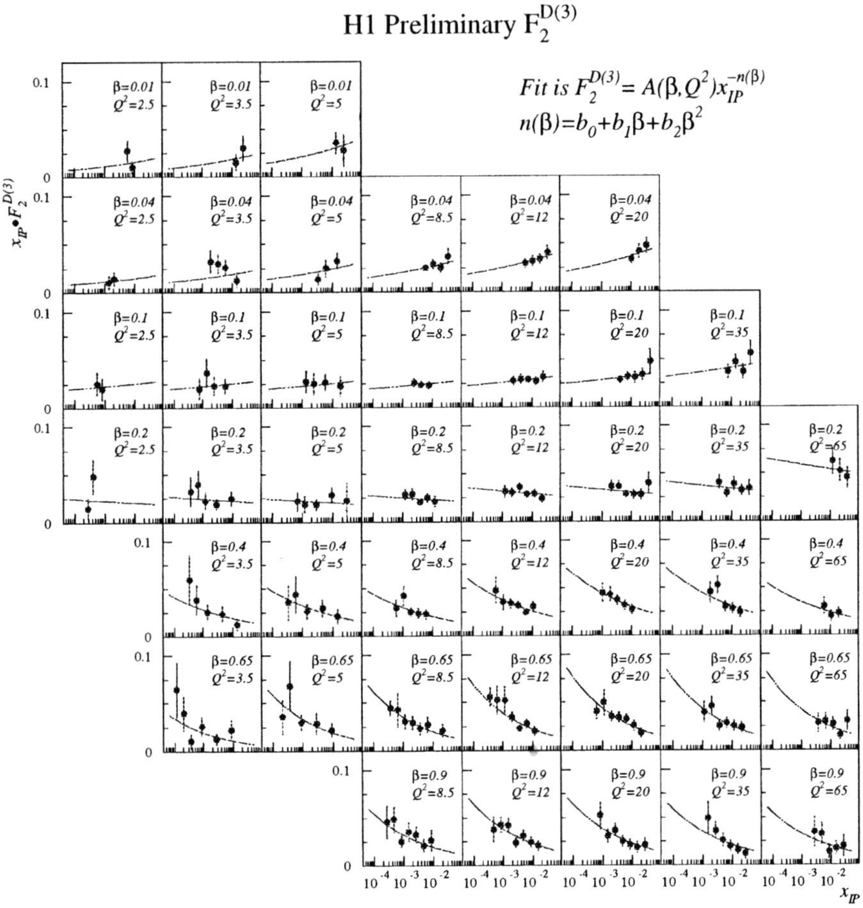

Figure 13. The 1994 diffractive contribution $F_2^{D(3)}(x_P, \beta, Q^2)$ to the proton structure function F_2 as a function of x_P for different β and Q^2 measured by the H1 collaboration. For clarity each data point is multiplied by x_P. Superimposed is the result of a fit establishing a breaking of factorization of the flux factor (see text).

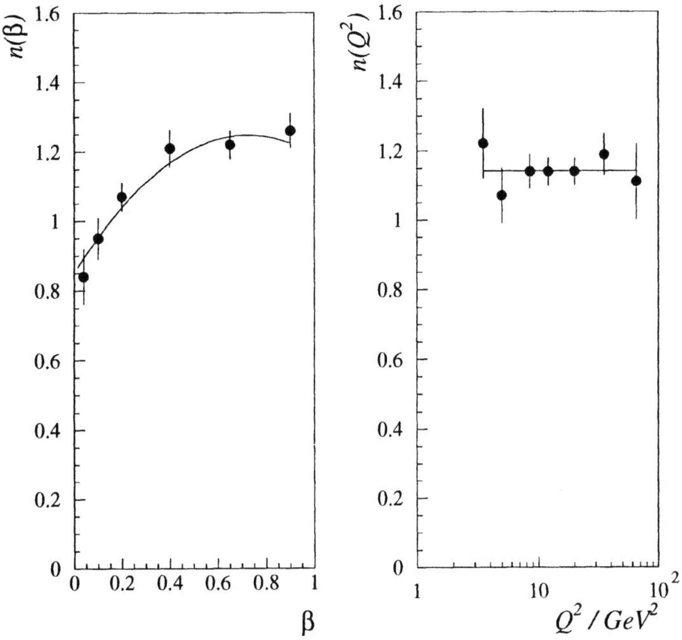

Figure 14. Preliminary result of a determination of the exponent n of the flux factor $x_{\mathbb{P}}^{-n(\beta,Q^2)}$ as fitted in each β or Q^2 interval of the measured $F_2^{D(3)}(x_{\mathbb{P}}, \beta, Q^2)$ by the H1 collaboration. Superimposed is the quadratic fit $n(\beta) = b_0 + b_1\beta + b_2\beta^2$.

flux term for $0.07 < \beta < 0.375$ in agreement with the two early measurements of H1 and ZEUS (table 1).

In the Regge formalism, the $x_{\mathbb{P}}$ dependence is described by a Reggeon flux term $x_{\mathbb{P}}^{-n}$ associated with the proton with

$$n \approx 2\alpha(0) - 1 \tag{24}$$

and where

$$\alpha(t) = \alpha(0) + \alpha' t \tag{25}$$

is the effective Regge Trajectory. To extract the intercept $\alpha(0)$ from the data, the measured n value has to be increased by a few percent to correct for the small t dependence. The resulting intercept $\alpha(0) = \frac{1+n}{2}$ (table 1) is a bit larger than the universal 1.08 intercept of the effective soft Pomeron in all hadron-hadron scattering (see Section on asymptotic behaviour when $x \to 0$).

Based on a luminosity of 2 pb^{-1} of good data recorded in 1994, the H1 collaboration has obtained a much more precise determination of the structure function $F_2^{D(3)}$

[55]. The results are shown in fig. 13. For clarity, each measured value is multiplied by x_P. The improved precision and extended kinematic range of these new data reveal a clear breaking of factorization. The flux factor is constant with Q^2 but varies with β. The flux power term can be parametrized as

$$n(\beta) = b_0 + b_1\beta + b_2\beta^2. \tag{26}$$

The result of the fit is superimposed on fig. 13 and shown as a function of β or Q^2 in fig. 14. The increase of n with increasing β (or equivalently with decreasing M_X) is at present clearly demonstrated by the new H1 data and is a challenge for all theoretical interpretations of the diffractive DIS events at HERA. Factorisation can be restored by assuming that two individually factorisable components contribute to $F_2^{D(3)}$. They may be identified with Pomeron and meson Regge trajectories with x_P dependencies $\propto x_P^{-n_P}$, $x_P^{-n_M}$ where $n_P = 1.29 \pm 0.03 \pm 0.07$ and $n_M = 0.3 \pm 0.3 \pm 0.6$.

Table 1. Flux power term in diffractive deep-inelastic scattering.

data sources	n	selection of events
H1 93 [51]	$1.19 \pm 0.06 \pm 0.07$	forward detectors
ZEUS 93,1 [52]	$1.30 \pm 0.08^{+0.08}_{-0.14}$	rapidity gaps
ZEUS 94 [54]	$1.28 \pm 0.07 \pm 0.15$	leading proton tagging
H1 94 [55]	*from* 0.8 *to* 1.3	forward detectors

Structure function

To investigate the Q^2 and β behaviour of $F_2^{D(3)}(x_P, \beta, Q^2)$ it is convenient to define

$$\tilde{F}_2^D(\beta, Q^2) = \int_{x_{P_l}}^{x_{P_h}} F_2^{D(3)}(x_P, \beta, Q^2) dx_P \tag{27}$$

where the integration limits $x_{P_l} = 0.003$ and $x_{P_h} = 0.05$ are chosen close to the experimentally accessed range.

The result of the integral of the structure function $F_2^{D(3)}(x_P, \beta, Q^2)$ over x_P measured by the H1 collaboration is shown in fig. 15. The structure function $\tilde{F}_2^D(\beta, Q^2)$ is almost constant with β but has strong scaling violations which persist to $\beta = 0.65$ as a rise in Q^2. If the scaling violations are interpreted as a putative partonic structure of $\tilde{F}_2^D$ which satisfies the DGLAP evolution equation, one obtains the striking result that at $Q^2 = 5$ GeV^2 the input parton distribution should be mainly gluonic with a gluon density peaking close to $\beta = 1$ (see fig. 16).

CONCLUSION

By summer 1996, most of the final results of the analysis of the 1994 data at HERA based on 3 to 5 pb^{-1} of good recorded data by the H1 and ZEUS experiments have been produced together with the first preliminary results from 1995 data. Precision measurements of the proton structure function F_2 at moderate Q^2 and first hints from the low Q^2 domain allow a thorough investigation on the evolution with Q^2 of the x dependence of F_2. The structure function F_2 exhibits a modest rise at $Q^2 = 0.2$ GeV^2 as $x \to 0$.The rise becomes rapidly steeper as Q^2 increases. At $Q^2 \approx 1$ GeV ^{2} there is a smooth transition between the Regge phenomenology and the perturbative QCD domain. First results on F_L, the longitudinal structure function and on $F_2^{c\bar{c}}$, the charm contribution to F_2 have been reported. The structure functions F_L and $F_2^{c\bar{c}}$ are in remarkable agreement with the prediction of perturbative QCD at very low x. In the precise diffractive DIS events a clear breaking of factorization has been observed showing that the naive model of exchange of a single Pomeron with a partonic structure is not tenable. These results represent some highlights of the physics message from HERA. Many other topics such as studies of the final states in DIS, α_s

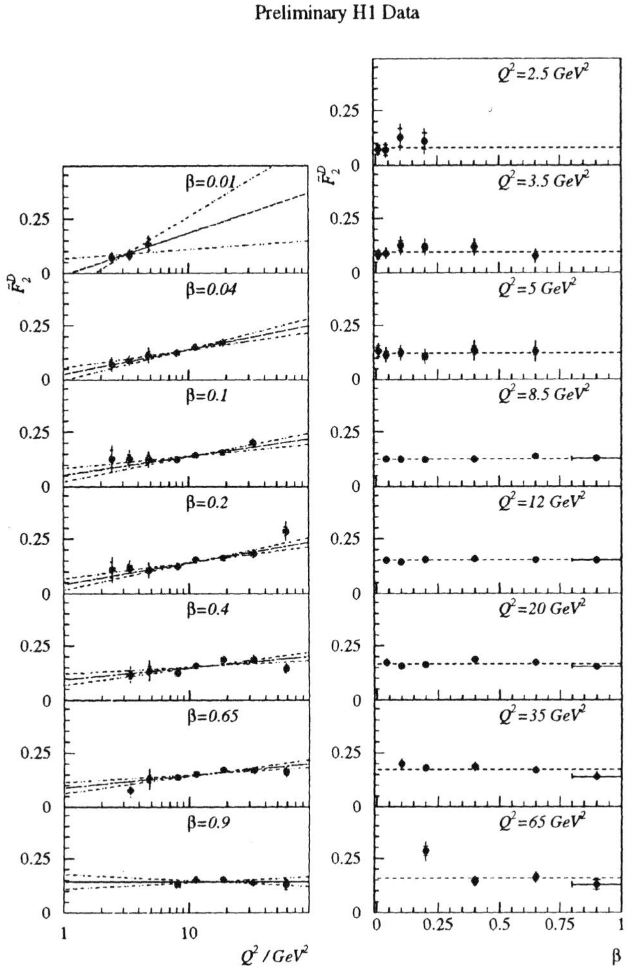

Figure 15. The structure function $\tilde{F}_2^D(\beta, Q^2)$ measured by the H1 collaboration. Superimposed are the results of fits which assume leading logarithm scaling violations. The best fit (continuous curve) and the curves corresponding to a change of ± 1 sigma (dashed curves) are shown.

measurement, photoproduction, heavy quark and vector meson production, searches beyond the standard model have been omitted in this contribution. Both experiments have recorded each another 15 pb^{-1} of data in 1995-1996 and more than 150 pb^{-1} per year are expected from the year 2000 onwards. After the present harvest of beautiful QCD results, the anticipated high luminosity should provide unique opportunities to the H1 and ZEUS experiments to discover signals beyond the standard model.

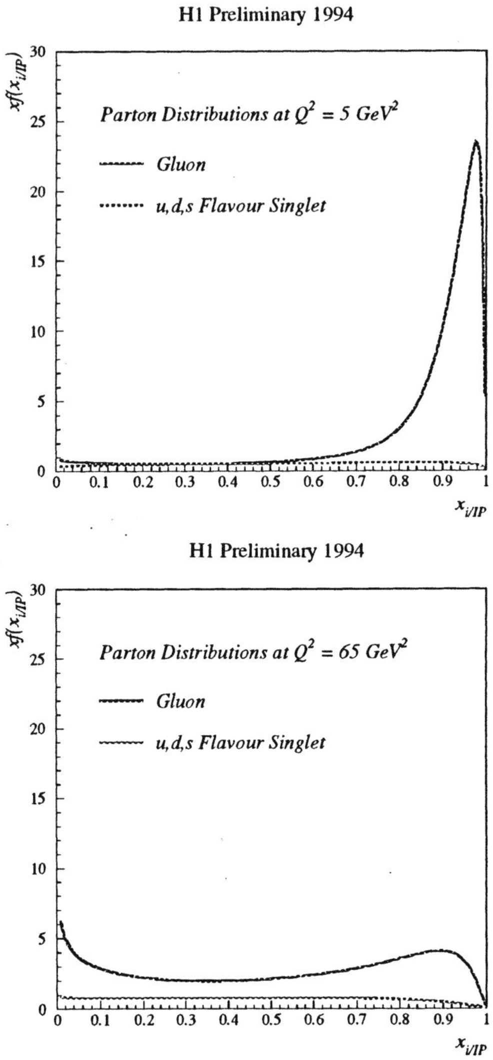

Figure 16. Parton distributions used as input values at $Q^2 = 5$ GeV^2 to the leading logarithm fits of $\tilde{F}_2^D(\beta, Q^2)$ from the H1 experiment. Also shown are the parton distributions after evolution to $Q^2 = 65$ GeV^2.

References

[1] H. Georgi and H.D. Politzer, *Phys. Rev.* D9 (1974) 416.

[2] D.J. Gross and F. Wilczek, *Phys. Rev.* D9 (1974) 980.

[3] V.N. Gribov and L.N. Lipatov, *Sov. J. Nucl. Phys.* 15 (1972) 438 and 675; G.Altarelli and G.Parisi, *Nucl. Phys.* B126 (1977) 298 ; Yu.L.Dokshitzer, *Sov. Phys. JETP* 46 (1977) 641.

[4] M. Virchaux, *QCD 20 years later, Proceedings of Aachen conf.*, eds P.M. Zerwas and H.A. Kastrup (World Scientific, 1993) page 205.

[5] E.A. Kuraev, L.N. Lipatov and V.S. Fadin, *Phys.Lett.* B60 (1975) 50 ;
Ya.Ya. Balitski and L.N. Lipatov, *Sov. J. Nucl. Phys.* 28 (1978) 822 and *Zh. Eksperiment. I. Teor.Fiz.* 72 (1977) 377.

[6] M. Ciafaloni, *Nucl. Phys.* B296 (1988) 49.

[7] S.Catani, F.Fiorani and G.Marchesini,*Phys. Lett.* B234 (1990) 339 and *Nucl. Phys.* B336 (1990) 18.

[8] K. Golec-Biernat et al., *Phys. Lett.* B335 (1994) 220.

[9] H1 Collaboration, S.Aid et al., *Phys. Lett.* B356 (1995) 118 and C.Adloff et al., DESY preprint 96-215 submitted to *Nucl. Phys. B.*

[10] D.J. Gross and F. Wilczek, *Phys. Rev. Lett.* 30 (1973) 1343.
H.D. Politzer, *Phys. Rev. Lett.* 30 (1973) 1346.

[11] A. De Rujula et al., *Phys. Rev.* D10 (1974) 1649.

[12] R.D. Ball and S. Forte, *Phys. Lett.* B351 (1995) 313.

[13] A. Donnachie and P.V. Landshoff, *Phys. Lett.* B296 (1992) 227.

[14] J.R. Forshaw and P.N. Harriman, *Phys. Rev.* D46 (1994) 3778.;
R.E. Hancock and D.A. Ross, *Nucl. Phys.* B383 (1992) 575.;
J.Kwiecinski, A.D. Martin and P.J.Sutton, *Phys. Rev.* D44 (1991) 2640. and *Phys. Lett.* B264 (1991) 199.;
J.C. Collins and P.V. Landshoff, *Phys. Lett.* B276 (1992) 196.;
A. Capella et al., *Phys. Lett.* B337 (1994) 358.;
R. Peschanski and S. Wallon, *Phys. Lett.* B349 (1995) 357.;
V.S. Fadin and L.N. Lipatov, *Proceedings of the Topical Conference on Hard Diffractive Processes*, Eilat (Israel), Feb. 1996, p.572.

[15] A.H. Mueller, *Nucl. Phys.* B415 (1994) 373.

[16] A.J. Askew et al., *Phys. Lett.* B325 (1994) 212; *Phys. Rev.* D49 (1994) 4402.

[17] L.V.Gribov, E.M.Levin and M.G.Ryskin, *Nucl. Phys.* B188 (1981) 555; *Phys. Rep.* 100 (1983).

[18] J. Bartels, K. Charchula and J. Feltesse, *Proceedings Workshop on Physics at HERA* (DESY, 1991) p.193.

[19] K. Golec-Biernat, M.W. Krasny and S. Riess, *Phys. Lett.* B337 (1994) 367.

[20] G. Altarelli and G. Martinelli, *Phys. Lett.* B76 (1978) 89.

[21] Zijlstra and W. van Neerven, *Nucl. Phys.* B383 (1992) 525;
S.A. Larin and J.A.M. Vermaseren, *Z.Phys.* C57 (1993) 93.

[22] H. Navelet et al., Preprint DESY 96-108 and Saclay DAPNIA/SPP 96-08, hep-ph/9605/389.

[23] C. López, F. Barreiro and F.J. Ynduráin, preprint DESY 96-087.

[24] B. Badelek, J. Kwieciński and A. Staśto, preprint Durham DTP 96/16.

[25] H1 Collaboration, I. Abt et al., *Nucl. Phys.* B407 (1993) 515.

[26] ZEUS Collaboration, M. Derrick et al., *Phys. Lett.* B316 (1993) 412.

[27] H1 Collaboration, S. Aid et al., *Nucl. Phys.* B470 (1996) 3.

[28] ZEUS Collaboration, M. Derrick et al., DESY preprint 96-076 and *Z. Phys.* C69 (1996) 607.

[29] ZEUS Collaboration, R. Yoshida and Q. Zhu, *Proceedings of the International Workshop on Deep Inelastic Scattering and Related Phenomena*, Roma (Italy), April 1996.

[30] H1 Collaboration, S. Aid et al., *Phys. Lett.* B353 (1995) 578 and *Phys. Lett.* B380 (1996) 462 and *Z. Phys.* C71 (1996) 211.

[31] ZEUS Collaboration, M. Derrick et al., *Z.Phys.* C65 (1995) 627.

[32] BCDMS Collaboration, A.C. Benvenuti et al., *Phys. Lett.* B223 (1989) 485.

[33] E-665 FNAL Collaboration, M.R. Adams et al., FNAL-Pub-95/396-E (1995).

[34] NMC Collaboration, M. Arneodo et al., *Phys. Lett.* B364 (1995) 107.

[35] H1 Collaboration, S. Aid et al., *Phys. Lett.* B354 (1995) 494.

[36] ZEUS Collaboration, M. Derrick et al., *Phys. Lett.* B345 (1995) 576.

[37] M. Glück, E. Reya and A. Vogt, *Z. Phys.* C67 (1995) 433.

[38] A. Donnachie and P.V. Landshoff, *Z. Phys.* C61 (1994) 139.

[39] A. Capella et al., *Phys. Lett.* B337 (1994) 358.

[40] H1 Collaboration, C.Adloff et al., DESY preprint 96-236, submitted to *Phys. Lett. B.*

[41] EMC Collaboration, J.J. Aubert et al., *Nucl. Phys.* B213 (1983) 31.

[42] H1 Collaboration, C. Adloff et al., DESY preprint 96-138 submitted to *Z. Phys. C.*

[43] E. Laenen et al., *Nucl. Phys.* B392 (1993) 162,229;
E. Laenen et al., *Phys. Lett.* B291 (1992) 325;;
S. Riemersma, J. Smith and W.L. van Neerven, *Phys. Lett.* B347 (1995) 143.

[44] A.D. Martin, R.G. Roberts and W.J. Stirling, *Phys. Lett.* B306 (1993) 145.

[45] A.D. Martin, W. J. Strirling amd R.G. Roberts, *Phys. Rev.* D50 (1994) 6743.

[46] R.P. Feynman, 'Photon Hadron Interactions', W. Benjamin Inc., 1972 p. 344 ; J.D. Bjorken in *Proceedings on Deep Inelastic Scattering and Related topics*, Eilat (Israel), 1994, p. 151 and earlier references therein;
see also J.Ellis, "Discussion session on large Q^2 and hadron structure", *Proceedings of the study of an ep Facility for Europe*, DESY 79-48.

[47] M.G. Ryskin and M. Besançon, *Proceedings of the Workshop on Physics at HERA* (DESY,1991) p.215 and earlier references therein.

[48] ZEUS Collaboration, M. Derrick et al., *Phys. Lett.* B315 (1993) 481.

[49] H1 Collaboration, T. Ahmed et al., *Nucl. Phys.* B429 (1994) 477.

[50] G. Ingelman and P. Schlein, *Phys. Lett.* B152 (1985) 256.

[51] H1 Collaboration, T. Ahmed et al., *Phys. Lett.* B348 (1995) 681.

[52] ZEUS Collaboration, M. Derrick et al., *Z. Phys.* C68 (1995) 569.

[53] ZEUS Collaboration, M. Derrick et al., *Z. Phys.* C70 (1996) 391.

[54] ZEUS Collaboration, R. Yoshida, E. Barberis, *Proceedings of the International Workshop on Deep Inelastic Scattering and Related Phenomena*, Roma (Italy), Apr. 1996.

[55] H1 Collaboration, A. Mehta, *Proceedings of the Topical Conference on Hard Diffractive Processes*, Eilat (Israel), Feb. 1996, p.710.

DETERMINATION OF THE GLUON STRUCTURE FUNCTION USING TAGGED CHARM EVENTS IN DEEP INELASTIC EP COLLISIONS

Catherine Adloff

Fachbereich Physik, Universität Wuppertal
Gaußstraße 20, D-42097 Wuppertal (Germany)
email-address : adloff@mail.desy.de

INTRODUCTION

In deep inelastic lepton nucleon scattering (DIS) the study of charm production is relevant for the understanding of the parton densities in the nucleon [1, 2]. In lowest order QCD the main processes in deep electron proton scattering are the quark parton model (QPM) process, the QCD-Compton (QCDC) process and the boson gluon fusion (BGF) process, which are represented in fig. 1.

As a confirmation of EMC collaboration measurements [3] from muon nucleon scattering 1994 data taken by the H1 detector at HERA electron proton collider have shown a clearly dominance of the BGF process in charm production. Additional contributions where the charm quarks may come from fragmentation or gluon splitting are supressed by heavy quark mass effects. Therefore tagged charm events are a unique probe for the gluon density $x_g\, g(x_g, Q^2)$ in the proton.

The easiest way to tag charm events is to reconstruct D^{*+} mesons via the decay chanel $D^{*+} \rightarrow D^0\pi_s^+ \rightarrow (K^-\pi^+)\pi_s^+$ [1]. In this case the leading order gluon density can be extracted directly by explicit reconstruction of x_g, the momentum fraction of the proton caried by the gluon. This enables a direct mesurement of $g(x_g, Q^2)$ at small x_g.

The aim of this contribution is to present the current analysis [4], to discribe the two main procedures for reconstructing x_g and to illustrate how feasible they are in case of charm production via the D^{*+} chanel.

1 $C\bar{C}$ PRODUCTION KINEMATICS AND x_g

In the BGF hard sub-process the momentum conservation is given by

$$x_g P + q = c + \bar{c} \tag{1}$$

[1]Henceforth, charge conjugate states are always implicitly included.

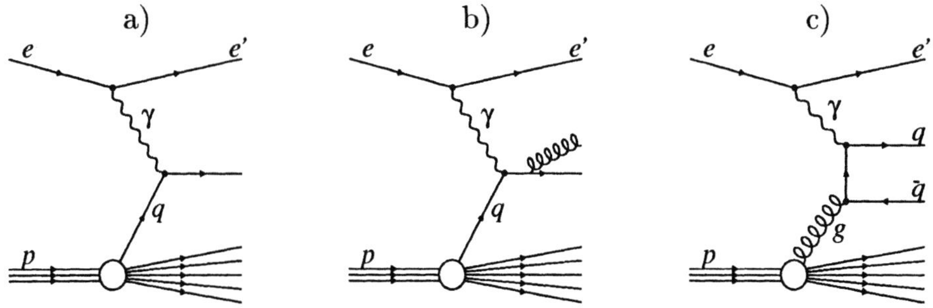

Figure 1. *Generic Feynman diagrams for deep inelastic ep scattering up to order* $\alpha\alpha_s$*: (a) ep interaction of the virtual probe* (γ, Z^0) *with a valence or a sea quark in the proton according to the quark parton model (QPM), (b) corrections to process (a) due to gluon radiation off the struck quark before or after interacting with the probe (QCD Compton scattering, QCDC), and (c) contribution due to boson gluon fusion (BGF).*

where P, q, c and $\bar{c}$ denote the proton, photon, c quark and $\bar{c}$ quark momenta respectively. Squaring equ. (1) gives in the approximation of small x_g

$$2x_g P\cdot q + q^2 = (c + \bar{c})^2. \tag{2}$$

From the momentum transfert squared Q^2 and the Bjorken scaling variable x definitions ($Q^2 = -q^2$, $x = -q^2/2P\cdot q$) we find

$$x_g = x(1 + \hat{s}/Q^2) \tag{3}$$

with $\hat{s} = (c + \bar{c})^2$, the invariant mass squared of the $c\bar{c}$ system. Experimentally the mesurements of the energy E'_e and the polar angel Θ_e of the scattered electron allow to determine x and Q^2 according to

$$Q^2 = 4E_e E'_e \cos^2\left(\frac{\Theta_e}{2}\right), \qquad x = \frac{\cos^2\left(\frac{\Theta_e}{2}\right)}{1 - \frac{E'_e}{E_e}\sin^2\left(\frac{\Theta_e}{2}\right)} \tag{4}$$

by neglecting the electron and proton masses.
Therefore the problem of reconstructing x_g is moved to the problem of measuring $\hat{s}$. This can be solved using the rapidity of the $c\bar{c}$ system or even by using the four momentum of only one of the charm quarks.

2 DATA ANALYSIS

From the data collected with the H1 detector at HERA in the 1994 running period ($\mathcal{L}_{int} = 2.97 \pm 0.04$ pb^{-1}) 103 ± 13 D^{*+} mesons where reconstructed via the decay $D^{*+} \to D^0\pi_s^+ \to (K^-\pi^+)\pi_s^+$ (see fig. 2) . The combinatorial background was reduced by making use of the hard fragmentation of charm quarks. In addition a better resolution and signal to background ratio is obtained in the $\Delta m = m(D^{*+}) - m(D^0)$ distribution than in the D^{*+} mass itself due to the small mass difference between D^{*+} and D^0 mesons.

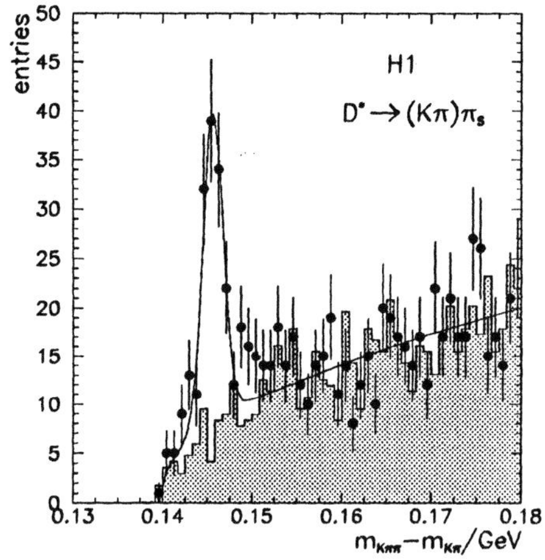

Figure 2. *Distribution of Δm from the D^{*+} analysis by requiring $|m_{K\pi} - m_{D^0}| < 90$ MeV. The shaded areas show the background expectations. The full lines are the results of fits to the data.*

2.1 Differential cross section and BGF dominance

The diferential cross section $1/\sigma\ d\sigma/dx_D$, where x_D is the scaled D^0 momentum in the γ^*p system given by

$$x_D = 2\ |\vec{p}_{D^0}^{\,\gamma^*p}|/W \approx |\vec{p}_{D^0}^{\,\gamma^*p}|/|\vec{p}_p^{\,\gamma^*p}|, \tag{5}$$

has been used to study the origin of charm production. Figure 3 show the corrected distribution $1/\sigma\ d\sigma/dx_D$ as well as the expectation of (a)BGF Monte Carlo generator (AROMA), (b)QPM Monte Carlo generator (LEPTO 6.1) and (c)νN scattering extrapolated to HERA energies. The closed points show the inclusive $D^{*+} \to D^0\pi_s^+ \to (K^-\pi^+)\pi_s^+$ analysis and the open points the inclusive $D^0\pi_s^+ \to (K^-\pi^+)\pi_s^+$ analysis, which was also performed for the 1994 data. A clear difference between data and QPM expectation can be seen whereas BGF describe the data fairly well. Finally, the QPM contribution is found to be less than 0.05 at 95% C.L. from a fit to $1/\sigma\ d\sigma/dx_D$ of a BGF plus a QPM contribution.

2.2 Charm Contribution $F_2^{c\bar{c}}(x, Q^2)$ to the Proton Structure Function

In the Callan-Gross approximation ($R = F_2/2xF_1 - 1 = 0$) the charm contribution $F_2^{c\bar{c}}(x, Q^2)$ to the proton structure function is obtained by the one photon exange cross section applied to the charm production

$$\frac{d^2\sigma^{c\bar{c}}}{dxdQ^2} = \frac{2\pi\alpha^2}{Q^4 x}\left(1 + (1-y)^2\right)\ F_2^{c\bar{c}}(x, Q^2)\,. \tag{6}$$

Since the data are found to be consistent with no charm sea in the proton it will be possible to extract $g(x, Q^2)$ from a QCD fit to $F_2^{c\bar{c}}$ with an implicit integration over x_g as it is done from the F_2 mesurement. At the present time the poor statistics in x and Q^2 bins prevents from extracting the gluon density with this method.

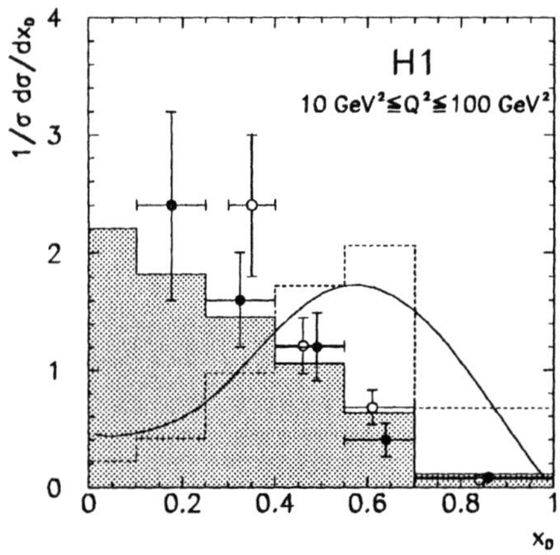

Figure 3. *Normalized x_D distribution in deep inelastic ep scattering at* $\langle W \rangle \approx 125$ GeV *for* $|\eta_D| < 1.5$ *compared to BGF, with* $m_c = 1.5$ GeV$/c^2$ *(shaded histogram), and to QPM expectations from LEPTO (dashed histogram), and from* νN *scattering evolved to HERA energies (full line).*

3 DIRECT x_g RECONSTRUCTION

3.1 Determination of x_g using $c\bar{c}$ pseudo-rapidity

In the hadronic center of mass system $\hat{s}$ can be mesured directly from the pseudo-rapidities $\eta_c^{\gamma^* p}$ and $\eta_{\bar{c}}^{\gamma^* p}$ of the charmed quarks and the invariant mass squared W^2 via

$$\hat{s} = W^2 e^{-(\eta_c^{\gamma^* p} + \eta_{\bar{c}}^{\gamma^* p})}. \qquad (7)$$

From the reconstructed D^{*+} we get the direction of one charm. The direction of the second charm can been obtained by applying a jet algorithm and therefore $\eta_c^{\gamma^* p} + \eta_{\bar{c}}^{\gamma^* p} \approx \eta_{D^{*+}}^{\gamma^* p} + \eta_{jet}^{\gamma^* p}$. Such a method has already been aplied but with two reconstructed jet [5] instead of one D^{*+} and one jet. The disavantage of the two jet analysis is the large contamination of QCDC events (30%) and the explicit cut on $\hat{s}$ at 100 GeV^2, which is needed to indentify the jets. The latter limits such an analysis to a $\langle x_g \rangle$ of 0.02. In the present analysis the minimum $\hat{s}$ is equal to $4m_c^2$ in the case where we could use double tagged charm events. This canot be applied because of the statistics (0.6 events with two reconstructed D^* mesons are expected for the 1994 data) so that we must used the D^{*+} plus one jet method which will bring the minimum $\hat{s}$ between $4m_c^2$ and 100 GeV^2.

3.2 Determination of x_g using the charm quark four momentum

An other method of extracting the gluon density uses the four momentum of a single charm quark produced by the BGF process. From this $\hat{s}$ may be reconstructed by the following relations

$$\hat{s} = \frac{(p_{c\perp}^{\gamma^* p})^2 + m_c^2}{z_c(1 - z_c)} \qquad z_c = \frac{E_c - p_{c\parallel}^2}{2yE_c}. \qquad (8)$$

Using reconstructed D^{*+} the four momentum of the charm quark may be approximate by applying a correction factor y_c to the D^{*+} momentum in order to take in account the effect of fragmentation, i.e. $\vec{p}_c = \vec{p}_{D^{*+}}/\langle y_c \rangle$ (the charm quark mass is fixed to 1.5 GeV^2). Figure 4 shows the $\Delta x_g/x_g$ distribution of events generated by AROMA program with $\langle y_c \rangle = 0.7$ which gives a resolution of 49%. A better resolution could be achieved by summing up all energies in a cone around the D^{*+} direction to get a better estimation of the energy of the charmed quark instead of using the mean fragmentation factor $\langle y_c \rangle$. This is presently investigated in the purpose to obtain a resolution of the order of 35%, which is the resolution on x_g achieved in case of an ideal detector.

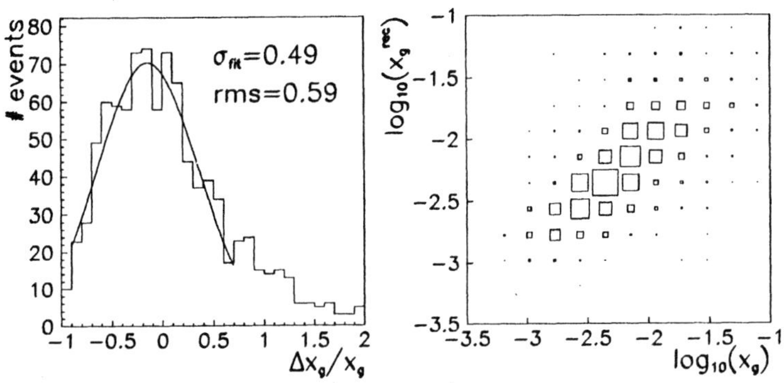

Figure 4. *Relative error in the reconstruction of the fractional gluon momentum* $\Delta x_g/x_g = (x_g^{rec} - x_g)/x_g$ *for BGF events generated by AROMA Monte-Carlo and correlation between the reconstructed and the true* x_g.

CONCLUSION

It has been shown that the D^{*+} analysis allows to select BGF events. The reconstruction of x_g is possible using a D^{*+} plus one jet or a single D^{*+} and enables a direct mesurement of $g(x_g, Q^2)$, which is essentially free from QCDC and QPM background. From the single D^{*+} we can already see that the sensitive $\langle x_g \rangle$ range is 5.10^{-3} (to compare with 2.10^{-2} for the two jets analysis). Although this method gives a bad resolution on x_g it profits from all the reconstructed D^{*+} statistics.

ACKNOWLEDGEMENTS

I would like to thank Karin Daum for the help she provided me in writing this contribution.

References

[1] A. Ali *et al.*, in *Proc. of the HERA Workshop*, Hamburg, Vol. 1,393 (1988), A. Ali and D. Wyler, in *Proc. of the Workshop "Physics at HERA"*, Hamburg, Vol. 2, 667 (1991), M.A.G. Aivazis *et al.*, *Phys. Rev.* D **50**, 3085 (1994), *Phys. Rev.* D **50**, 3102 (1994).

[2] S. Riemersma *et al.*, *Phys. Lett.* B **347**, 143 (1995).

[3] EMC Coll., J.J. Aubert *et al., Nucl. Phys.* B **231**, 31 (1983).

[4] H1 Coll., C. Adloff *et al.*, DESY 96-136, to appear in *Z. Phys.* C.

[5] H1 Coll., I. Abt *et al.*, DESY 95-086.

THE STRUCTURE OF THE PROTON PHENOMENOLOGY[1]

Laurent Schoeffel

DAPNIA - SPP, Centre d'Etudes de Saclay
F-91191 Gif-sur-Yvette Cedex (France)

INTRODUCTION

Deep inelastic scattering of an electron on a proton is a very interesting process to probe with high precision the structure of the proton. In deep inelastic scattering a highly virtual photon of mass Q^2 interacts with pointlike constituents (partons) of the proton. In the Breit frame the photon carries no energy and the proton has a momentum P proportionnal to $\sqrt{Q^2}$. For high Q^2, P is large and the proton looks like a highly contracted pancake. The time τ of interaction is proportional to $\frac{1}{\sqrt{Q^2}}$; for small τ we can safely suppose that the proton scatters incoherently on each parton. The cross section for deep inelastic scattering depends on the parton distribution seen when we look inside the proton with a resolution time τ.

The cross section for longitudinaly and transverse polarised photon (resp. σ_L and σ_T) can be written using two independent structure fonctions, $F_L(x, Q^2)$ and $F_2(x, Q^2)$ where x is the fraction of impulsion carried by the struck parton. In the quark-parton model

$$F_2(x, Q^2) = \sum_i e_i{}^2 x(q_i(x, Q^2) + \bar{q}_i(x, Q^2)) \tag{1}$$

where $q_i(x, Q^2)$ is the probability of finding a quark (of flavor i) localised within the transverse region $R_T \sim \frac{1}{Q}$ carrying the fraction x of the parton momentum.

HERA experiments, H1 and ZEUS, are particulary well suited for deep inelastic analysis and F_2 measurements. HERA provides an energy in the center of mass frame of $\sqrt{s} \sim 300 GeV$ by colliding an electron of energy $27.5 GeV$ and a proton of $820 GeV$. If we note p and q the quadrivectors of the proton and the photon the kinematic variables are defined as $Q^2 = -q^2$ and $x = \frac{Q^2}{2pq}$.

With HERA, we have accessed to a new kinematical range at small x where precise tests of the theory of stong interactions (perturbative QCD) can be done. The recently published 1994 results from HERA experiments [1] on the proton structure function F_2 have reached a high level of precision. It covers an extended kinematical range. In particular, it reaches very low values of x ($x \sim 10^{-5}$) and Q^2 ($Q^2 \sim 1.5\, GeV^2$). These data confirm with high statistics the strong rise of F_2 when x becomes very small, first noticed in 1992 experiments. In this new regime there are two main theoretical

---

[1])talk given at the CARGESE summer school on "Masses of fundamental particles" (5-17 August 1996) / [2])DGLAP stands for Dokshitzer, Gribov, Lipatov, Altarelli, Parisi

Masses of Fundamental Particles: Cargèse 1996
Edited by M. Lévy *et al.*, Plenum Press, New York, 1997

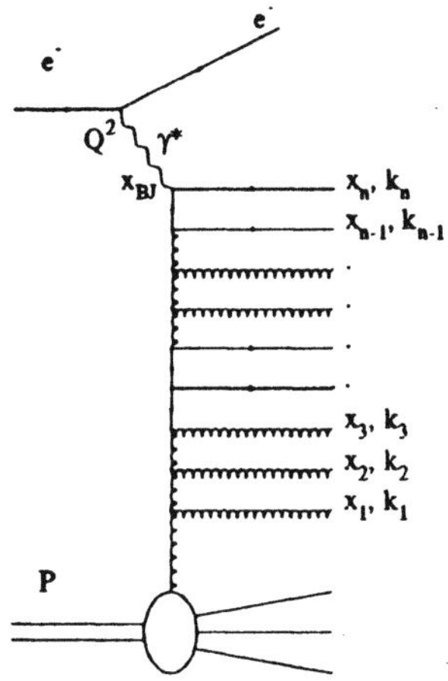

Figure 1. *Diagrammatic representation of quark and gluon rungs contributing to deep inelastic scattering.*

predictions to describe the evolution of F_2 with respect to x or Q^2, the DGLAP [2]and BFKL [3] evolution equations.

The purpose of this contribution is to present briefly the main aspects of these predictions and then comment experimental ideas which can be used to discriminate between these two alternatives. We will show that two observables are good candidates for this analysis, $\lambda = \frac{\partial \ln F_2}{\partial \ln 1/x}$ and $R = \frac{F_L}{F_2 - F_L} = \frac{\sigma_L}{\sigma_T}$. ($F_L$ is the longitudinal structure function and λ enables to quantify the stong rise of F_2 at small x).

1 THE DGLAP MODEL

Let us first notice that the full x and Q^2 dependence of the structure functions of the proton is not given by the theory, there are only predictions for evolution equations of these structure functions (according to which Feynman diagrams are summed up). The DGLAP model (leading order) is a sum of $(\alpha_S \ln Q^2)^n$ terms and this model describes an evolution in Q^2 [2]. (We will see later that the BFKL model which sums up $(\alpha_S \ln 1/x)^n$ terms corresponds to an evolution in $1/x$). We define the singlet and the non singlet structure functions $F_S(x, Q^2)$ and $F_{NS}(x, Q^2)$

$$F_2(x, Q^2) = F_S(x, Q^2) + F_{NS}(x, Q^2) \qquad (2)$$

with F_S completely symmetric in quark flavors and F_{NS} completely antisymmetric. For example $F_2 \sim F_{NS}$ at high x ($x > 0.2$) because the behaviour of F_2 at high x is dominated by the valence quarks and then by its non singlet part. On the contrary the low x regime is essentially driven by the singlet structure function (sea quarks) and the gluon distribution F_G, then $F_2 \sim F_S$ and the DGLAP evolution equations (in Q^2) exhibit couplings between F_S and F_G. These couplings are direct consequences of the elementary processes of perturbative QCD : $q \to q + g$, $\bar{q} \to \bar{q} + g$, $g \to q + \bar{q}$, $g \to g + g$, where the last process is typical of non abelian gauge theories. We will not give a complete derivation of the DGLAP equations but the idea is to sum up the $(\alpha_S \ln Q^2)^n$ contributions of diagrams like the one of Fig. 1. These calculations are performed

[3]BFKL stands for Balitskii, Fadin, Kuraev, Lipatov

with a strong ordering in ${k_T}^2$ and within the kinematic range $\alpha_S(Q^2)\ln 1/x << 1$ for $\alpha_S(Q^2) << 1$. DGLAP evolution equations can also be seen as a consequence of the renormalisation group equation and the operator product expansion.

At this stage we essentially deal with the predictions on $\lambda = \frac{\partial \ln F_2}{\partial \ln 1/x}$ at low x, we analyse in a next section the other observable $R = \frac{F_L}{F_2 - F_L}$. Thus, let us now determine the predictions on $\lambda(Q^2)$ in the DGLAP scheme. First, we consider the inverse Mellin transforms of the singlet structure function F_S, namely

$$F_S(x, Q^2, \mu^2) = \frac{1}{2i\pi}\int_{\omega_0 - i\infty}^{\omega_0 + i\infty} x^{-\omega} F_S(\omega, Q^2, \mu^2) d\omega \tag{3}$$

where the integration line $Re\omega = \omega_0$ is at the right of all singularities of $F_S(\omega, Q^2, \mu^2)$ and μ^2 defines the scale of the initial condition of the Q^2 evolution. In the DGLAP model $F_S(\omega, Q^2, \mu^2)$ and $F_G(\omega, Q^2, \mu^2)$ (the gluon distribution in Mellin-moment space) verify

$$\begin{pmatrix} F_S(\omega, Q^2, \mu^2). \\ F_G(\omega, Q^2, \mu^2) \end{pmatrix} = K(\omega, Q^2, \mu^2) \begin{pmatrix} F_S(\omega, \mu^2) \\ F_G(\omega, \mu^2) \end{pmatrix} \tag{4}$$

where $K(\omega, Q^2, \mu^2)$ is the Mellin transform of the DGLAP matrix kernel at the scale Q^2. The rightmost singularity of this kernel lies at $\omega_0 = 0$. Also $F_S(\omega, \mu^2)$ is the Mellin transform of the singlet input which is either regular or singular at the right of $\omega_0 = 0$, $F_G(\omega, \mu^2)$ is the Mellin transform of the gluon input. We then get two classes of models satisfying a DGLAP evolution.

(i) Either the rightmost ω-plane singularity is fixed by the DGLAP kernel singularity at $\omega = 0$, we obtain the Glück, Reya, Vogt (GRV) type parametrization [3]. In this case, at a given Q^2 near the $\omega = 0$ singularity we can show that

$$F_S(\omega, Q^2, \mu^2) \simeq \omega f(\omega) \exp\left(\frac{4N_c}{\omega} - a\right)\xi \tag{5}$$

where $f(\omega)$ is an input-dependent function regular at $\omega = 0$ by assumption and a is a constant ($a \sim 109/10$), $f(\omega)$ can be expanded as

$$f(\omega) = f(0)[1 + \Sigma_i \omega^i b_i] \tag{6}$$

Then we can determine $\frac{\partial \ln F_S}{\partial \ln \frac{1}{x}}$ using the following property of Bessel functions

$$\frac{1}{2i\pi}\int_{\omega_0 - i\infty}^{\omega_0 + i\infty} \omega^n e^{\omega \ln 1/x} e^{\frac{4N_c}{\omega}\xi} d\omega = \left(\frac{4N_c\xi}{\ln 1/x}\right)^{\frac{n+1}{2}} I_n[2(4N_c\xi \ln 1/x)^{\frac{1}{2}}] \tag{7}$$

Defining $\bar{\omega} = \left(\frac{4N_c\xi}{\ln 1/x}\right)^{\frac{1}{2}}$ and $v = 2(4N_c\xi \ln 1/x)^{\frac{1}{2}}$, it is straightforward to see that

$$\lambda \simeq \frac{\partial \ln F_S}{\partial \ln \frac{1}{x}} \simeq \bar{\omega}\frac{I_3(v)}{I_2(v)} \tag{8}$$

and the terms $b_i, i > 1$ are negligible (which means that λ is independent of the input). Moreover $\frac{\partial^2}{(\partial \ln Q^2)^2}\left(\frac{\partial \ln F_S}{\partial \ln \frac{1}{x}}\right) < 0$. This gives us an information on the concavity of the function $\lambda(Q^2)$ in the DGLAP model with the rightmost singularity imposed by the DGLAP kernel.

(ii) Or we have the López, Barreiro, Yndurain (LBY) type parametrization where the rightmost singularity lies at the right of $\omega = 0$ due to a singular input $F_s(\omega, Q^2, \mu^2)$. Moreover its location is essentially not modified by the perturbative evolution [4]. (Let us observe that the BFKL dynamics also leads to a rightmost singularity at ω greater than $\omega = 0$. We will come back to this later.) Starting from the LBY formulation of the singlet structure function

$$F_S(x, Q^2) = [B_s(Q^2)x^{-\lambda_s} + C_s(Q^2)](1-x)^{\nu(Q^2)} \tag{9}$$

where B_s, C_s and ν are Q^2-dependent functions [4]. $\lambda_s > 0$ defines the location of the rightmost ω–plane singularity and is Q^2–independent, but for charm and bottom thresholds. At small x

$$\lambda \simeq \frac{\partial \ln F_S}{\partial \ln \frac{1}{x}} \simeq \lambda_s, \tag{10}$$

up to the correction due to the phenomenological factor $C_s(Q^2)$.

2 THE BFKL MODEL

As already mentioned in this approach one sums up contributions of the type $(\alpha_S \ln 1/x)^n$ at low x [5]. There are two main features in this model : (1) BFKL evolution is an evolution from high longitudinal momentum partons to low longitudinal ones and (2) the evolution occurs at a fixed tranverse momentum (fixed Q^2). Then one can view the x evolution of the BFKL equation as creating the small x part of the wavefunction of the proton or simply as the "dressing" of a high momentum quark or gluon in the proton with low x gluons. In this model an interesting prediction (leading order) can be derived for the gluon distribution $F_G(x, Q^2)$

$$F_G(x, Q^2) \sim h(Q^2)\, x^{-\alpha_P} \tag{11}$$

$$\alpha_P = \frac{4\bar{\alpha} N_C \ln 2}{\pi} \leq 0.5 \tag{12}$$

showing a strong rise of F_G at low x and where α_P is the BFKL Pomeron intercept which is a constant, since the strong coupling constant $\bar{\alpha}$ is held fixed in this scheme. In this contribution we implement the BFKL dynamics using the recently developped QCD dipole model which is equivalent to BFKL for inclusive processes [6], [7]. We can then provide a prediction for $\lambda(Q^2)$. First we have for the singlet structure function

$$F_S = C a^{1/2} \frac{Q}{Q_0} e^{-\frac{a}{2} \ln^2 \frac{Q}{Q_0}}\, x^{-\alpha_P} \tag{13}$$

where $a = \left(\frac{\bar{\alpha} N_c}{\pi} 7\zeta(3) \ln \frac{1}{x}\right)^{-1}$. C and Q_0 are non-perturbative parameters to be determined by fitting with the data. We get

$$\lambda \simeq \frac{\partial \ln F_s}{\partial \ln \frac{1}{x}} = \alpha_p - 0.5 \frac{1}{\ln \frac{1}{x}} + \frac{1}{14 \frac{\bar{\alpha} N_c}{\pi} \zeta(3) \ln^2 \frac{1}{x}} \ln^2 \frac{Q}{Q_0} \tag{14}$$

Here we have, contrary to the GRV scheme $\frac{\partial^2}{(\partial \ln Q^2)^2} \left(\frac{\partial \ln F_S}{\partial \ln \frac{1}{x}} \right) > 0$.

Finally we can compare all the predictions with the data. As we can see on Fig. 2 the present accuracy of the data is not sufficient to distinguish between the different predictions but it appears a range of $Q^2 \in [1., 10.] GeV^2$ where new precise measurements could be very promissing [8].

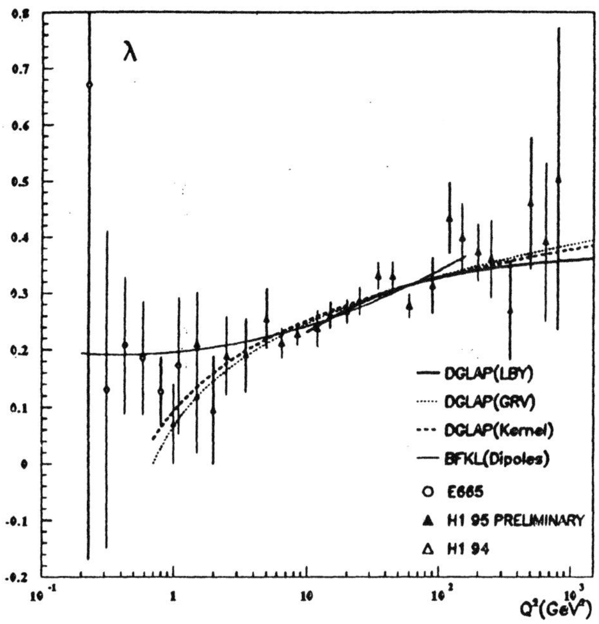

Figure 2. *Display of the data on λ compared with the prediction of the different parametrizations and evolution equations. DGLAP(Kernel) stands for $\lambda \simeq \bar{\omega}\frac{I_3(v)}{I_2(v)}$ and is verified to be in a good agreement with DGLAP(GRV).*

3 R MEASUREMENTS AND PREDICTIONS

$R = \frac{F_L}{F_2 - F_L}$ determination needs F_L measurement. There are several methods to get F_L, in the usual one we need at least two different beam energies. For example this can be achieved by decreasing the beam energy at the machine level or one can think of radiative events where a real photon is emitted from the incident electron which then losses one part of its initial energy. Other ideas are also in progress in the H1 collaboration and a precise measurement of F_L will become possible in a near future.

At this stage assuming $R \sim 0.3$ the total systematic error is expected to be of the order of 0.4. Statistical uncertainties depend on the luminosity of the machine and are expected to be of the order of 0.1 by running the experiment for a few weeks. Then we can compare with the theoretical predictions from DGLAP and BFKL (QCD dipole model) dynamics which are displayed on Fig. 3. We can not distinguish between them at this level of precision.

CONCLUSION

We have reminded the two types of QCD evolution equations describing the structure of the proton: the so called DGLAP and BFKL equations. These two descriptions cover two different physics images, thus it is an incentive question for experimentalists to find accurate observables (computable in the theory) which can afford to distinguish between them. We have shown that $\lambda = \frac{\partial \ln F_2}{\partial \ln 1/x}$ and $R = \frac{F_L}{F_2 - F_L}$ are good candidates for this analysis which is in progress in the H1 collaboration.

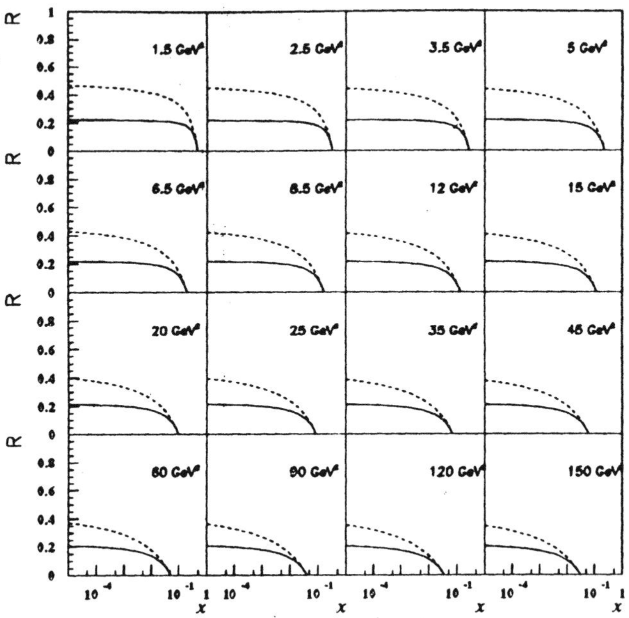

Figure 3. *Predictions on R, dashed line:DGLAP, continuous line:BFKL.*

ACKNOWLEDGEMENTS

The results described in the present contribution comes from a fruitful collaboration with J. Feltesse, H. Navelet, R. Peschanski, Ch. Royon and S. Wallon.

References

[1] H1 Coll., S. Aid et al., *Nucl. Phys.* **B470** (1996) 3.

[2] G. Altarelli and G. Parisi, *Nucl. Phys.* **B126** (1977) 298.

[3] M. Glück, E. Reya and A. Vogt, *Phys. Lett.* **B306** (1993) 391.

[4] C. López and F.J. Ynduráin, *Nucl. Phys.* **B183** (1981) 157: C. López, F. Barreiro and F.J. Ynduráin, preprint hep-ph/9605395.

[5] E.A. Kuraev, L.N. Lipatov and V.S. Fadin, *Phys. Lett.* **B60** (1975) 50; Ya.Ya. Balitskii and L.N. Lipatov, *Sov. J. Nucl. Phys.* **28** (1978) 822.

[6] A.H. Mueller, *Nucl. Phys.* **B415** (1994) 373: A.H. Mueller and B. Patel. *Nucl. Phys.* **B425** (1994) 471; A.H. Mueller. *Nucl. Phys.* **B437** (1995) 107.

[7] H. Navelet, R. Peschanski, Ch. Royon. *Phys. Lett.* **B366** (1995) 329; H. Navelet, R. Peschanski, C. Royon, S. Wallon, preprint desy 96-108 accepted by Phys. Lett. B.

[8] H. Navelet, R. Peschanski, Ch. Royon, L. Schoeffel, S. Wallon. subm. to Phys. Lett. B.

email-address : schoffel@hep.saclay.cea.fr

EXPERIMENTAL PHYSICS AT LHC pp COLLIDER

Jean-Jacques Aubert

CPPM IN2P3-CNRS Université de la Méditerranée
163, avenue de Luminy - Case 907
13288 Marseille Cedex 09 - France

ADVERTISEMENT

I have been involved with the LHC experiment only in the LHC committee, this means that I report mainly on work of the ATLAS, CMS and LHC-B groups[1,2,3]. They desserve all the credits for the splendid achievement of today.

Instead of showing the predictions of ATLAS and CMS for the same channel, I have choiced on purely non physical background either one or the other. There is no prejudice in these choices.

HIGGS PHYSICS

In the standard model, the particle masses come from the condensation of some field in vacuum. A simple way to get that, is to introduce a single elementary spin zero field, leading to a single physical Higgs boson. The Higgs mass should be smaller that 1 TeV to conserve unitarity.

But if the Higgs is an elementary particle, quantum corrections to its mass are very large. One way to cope with that is supersymmetry, then one gets in a minimum model three neutral Higgs plus a charged one.

Supersymmetry is not the only way, compositeness, like technicolor, may be an alternative.

Any how, at very high energy, the W_L W_L cross section should go up.

When the LHC will start, different scenarios are possible:

- LEP II has discover H^0 or h^0 (up to ~100 GeV), then LHC will have to look for the whole family;
- H^0 or h^0 is above 100 GeV. Then LHC explore the Higgs up to 1 TeV.

Standard Model Higgs Search at LHC

Predictions are given for CMS[1], but equivalent numbers are expected within ATLAS[2].

As a résumé of the following presentation, the figure 1 shows the standard model Higgs search possibilities at LHC.

H $\to \gamma\gamma$. It is a challenging measurement where one benefits from a very good γ energy measurement and a γ direction measurement at least at high luminosity.

In CMS, the signal cross-section, acceptance and mass resolution are given for m_H = 110, 130 GeV, in table 1.

Masses of Fundamental Particles: Cargèse 1996
Edited by M. Lévy *et al.*, Plenum Press, New York, 1997

Table 1. Signal, acceptance and mass resolution for $H^0 \rightarrow \gamma\gamma$

Signal	m_H 110 GeV	m_H 130 GeV	Background $d\sigma/dm_{\gamma\gamma}$[fb/GeV]	$m_{\gamma\gamma}$ 110 GeV	$m_{\gamma\gamma}$ 130 GeV
σ.B ($H \rightarrow \gamma\gamma$) [fb]	75.9	68.3	Quark annihilation	61.0	41.2
Acceptance	57%	63%	Gluon fusion	72.4	42.4
σ_m low L [MeV]	540	600	Isolated bremsstrahlung	84.4	50.4
σ_m high L [MeV]	870	960	TOTAL (isolated)	217.8	134.

This typically gives a significance of more than 7 standard deviations from 92 to 145 GeV.

Obviously, this measurement gives the same kind of information on $h^0 \rightarrow \gamma\gamma$.

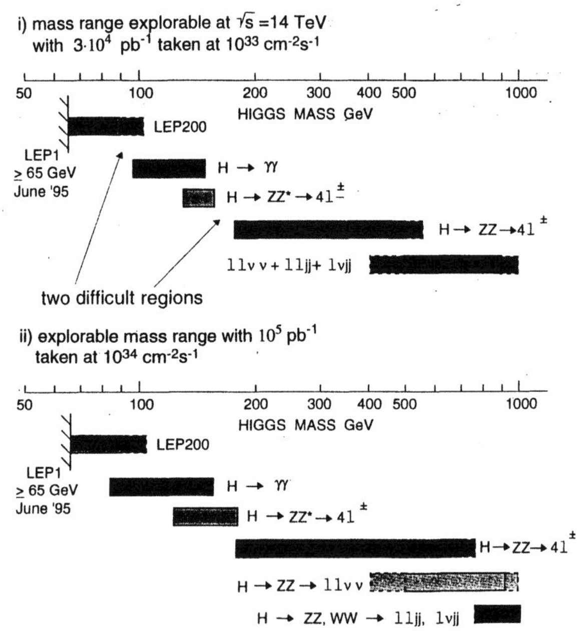

Figure 1. Standard model higgs search at LHC.

$H \rightarrow ZZ^* \rightarrow 4\ell^{\pm}$ and $H \rightarrow ZZ \rightarrow 4\ell^{\pm}$.

Again, the lepton measurement as well as the "transparency" of the tracker (small amount of matter) are important parameters. The ZZ* channel is more difficult because of a smaller cross section and also because the kinematic is less constraint.

The significance is shown in CMS detector on figure 2 for different luminosities.

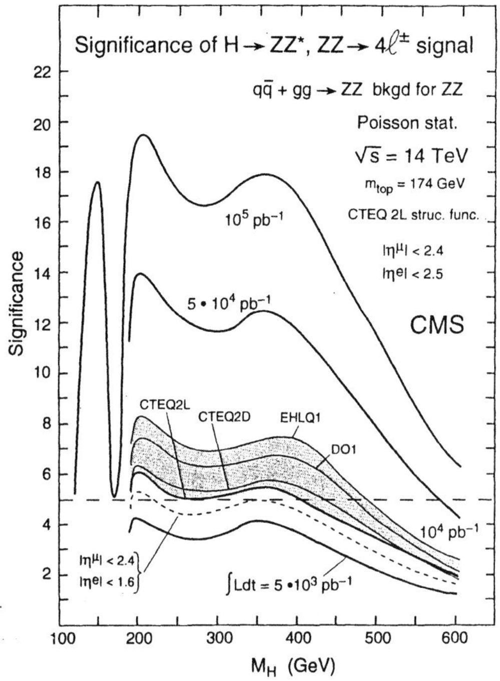

Figure 2. Standard model higgs search at LHC, significance for the decay channels $H \rightarrow ZZ^{*}, ZZ \rightarrow 4\ell^{\pm}$.

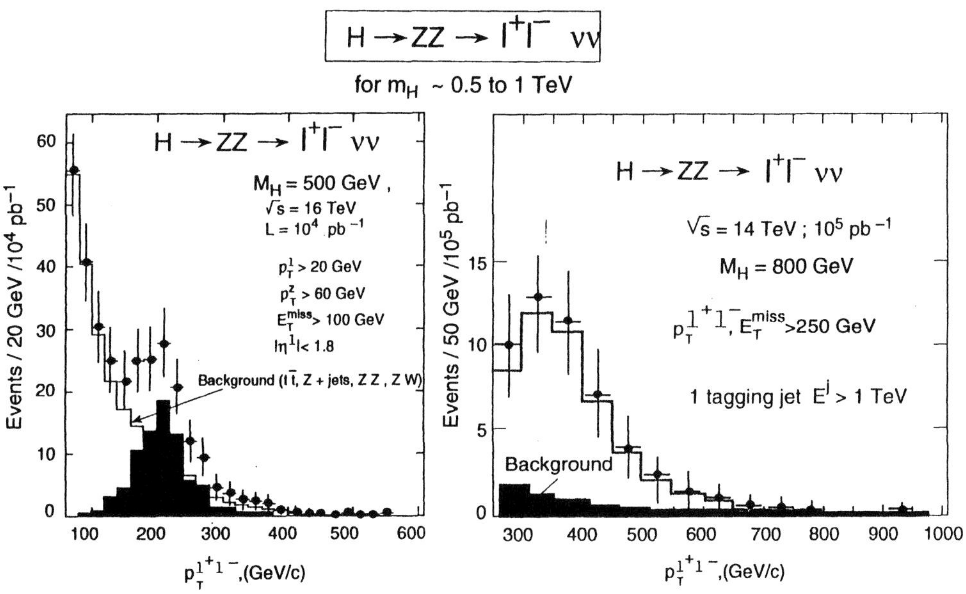

Figure 3. Standard model higgs search at LHC, significance for the decay channel $H \rightarrow ZZ \rightarrow \ell^{+}\ell^{-}\nu\bar{\nu}$.

H → ZZ → $\ell^{\pm}\, \nu\bar{\nu}$.

If one can also takes into account the Z → $\nu\bar{\nu}$ decay, the rate will be higher and the sensitivity will improve.

Figure 3 shows the sensitivity up to 800 GeV for m_H.

H → ZZ → $\ell\ell$ jj or WW → ℓ ν jj.

Even more rate is expected if one can address the Z decay in jet jet and the WW decay in leptons and jets.

Figure 4 shows the sensitivity up to 1000 GeV for M_H.

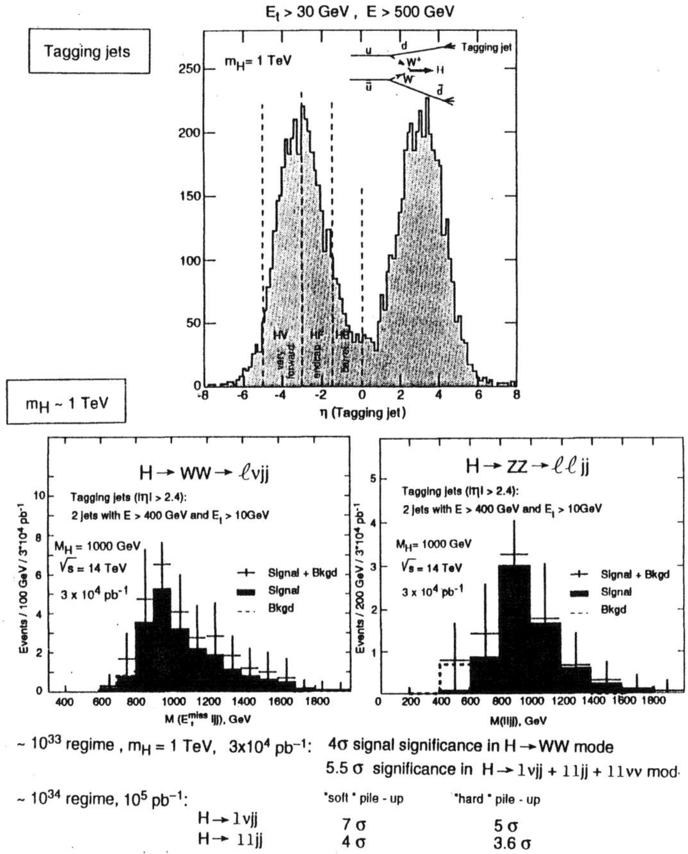

Figure 4. Standard model higgs search at LHC, significance for the decay channels $H^{*}ZZ \rightarrow \ell\ell jj$, $H \rightarrow WW \rightarrow \ell\nu jj$.

Minimum Supersymmetric Higgs Search

The channels h^0, H^0 → $\gamma\gamma$ and h^0, H^0 → ZZ^*, ZZ → $4\ell^{\pm}$ are analogue to the standard model search. Their sensitivity will be seen in the figure 8.

H_{susy} → $\mu^+\mu^-$.

It is a very clean signature, hopeless in the standard model where the cross section is too small, it is less efficient that H_{susy} → $\tau^+\tau^-$, but a very easy measurement (see figure 5).

Associated Production A$b\bar{b}$ or $H^0 b\bar{b}$.

This is an interesting channel where the ability to tag the B meson helps in reducing the natural background (see figure 6).

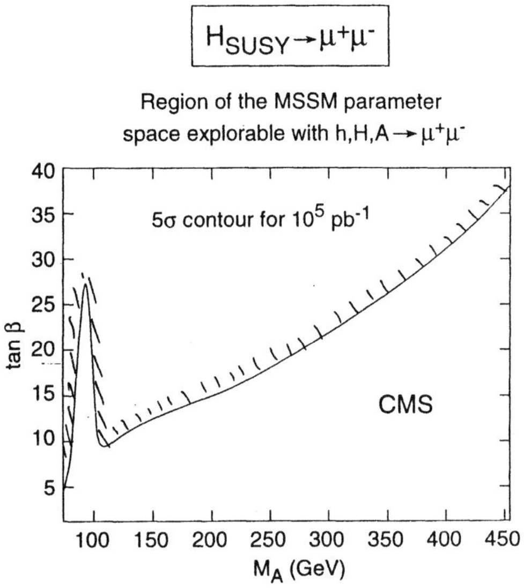

Figure 5. MSSM higgs search at LHC, significance for the decay channel $H_{SUSY} \rightarrow \mu^+\mu^-$.

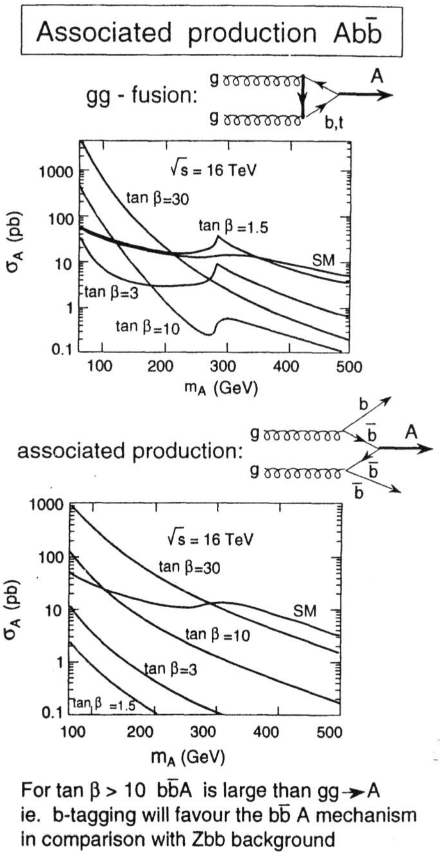

Figure 6. MSSM associated production cross section $Ab\bar{b}$ at LHC.

Charged Higgs $\rightarrow \tau^{\pm}\nu$.
A favorable scheme is to look through top decay :

$$t(E) \rightarrow H^{\pm}\, bWb$$
$$\downarrow$$
$$\tau^{\pm}\nu$$
$$\downarrow$$
$$\pi^{\pm}\nu$$

A good identification is favoured by the kinematic as well as the b tagging, see figure 7.

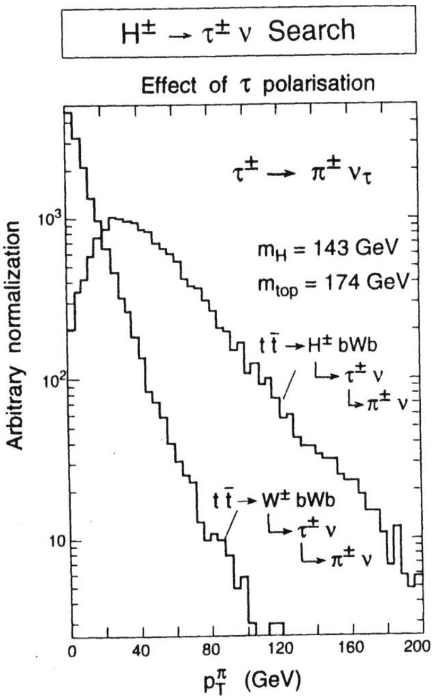

Figure 7. MSSM $H^{\pm} \rightarrow \tau^{\pm}\nu$ search, p_T^{π} spectrum.

Exclusion Plot

The MSSM can be characterized by two parameters, M_A and $\tan\beta$, the different channels exclude most of the plane M_A, $\tan\beta$, see figure 8.

SUPERSYMMETRY SEARCHES

The minimum supersymmetric model (MSSM) has five parameters, m_A, $\tan\beta$, μ, m(squark), m(gluino): there is a large parameter space. At LHC, one expects an important role since many theoretical advantages of SUSY require that the corresponding mass scale should not be larger than the electroweak scale breaking (EWSB).

We will explore the MSSM search with ATLAS.

Three signatures can explore a large part of the SUSY parameter space, namely:

- Jets + missing E_T signature for squarks and gluinos;
- Isolated same sign di-lepton signature for squarks and gluinos;
- Isolated tri-lepton signature for chargino/neutralino.

A susy search rely on R parity conservation, the lightest particle is stable and missing E_T is a characteristic signature.

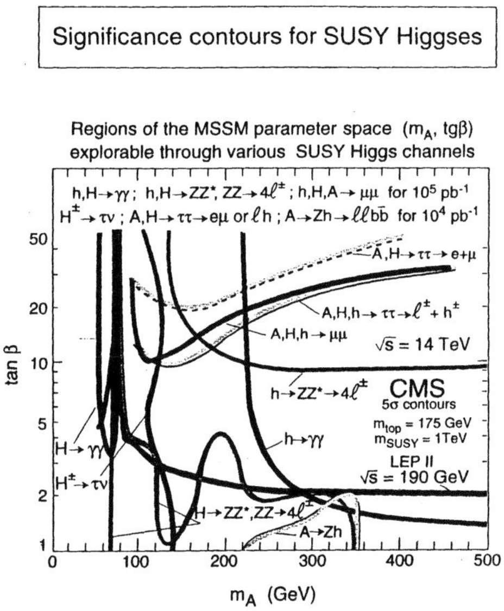

Figure 8. MSSM searches at LHC, significance contour in the m_A, $\tan\beta$ plane.

Missing E_T + jets Search

Search for squarks and gluinos. One requires 3 jets with $P_T > 200$ GeV, another jet with $P_T > 100$ GeV, and a missing transverse energy $E_T > 300$ GeV.

The backgrounds are physical from W+ jets, Z+ jets, $t\bar{t}$ events; instrumental from mis-measured jets. One can see on figure 9 that the instrumental background is small for $E_T > 300$ GeV. One gets a sensitivity to squarks and gluinos for masses above 300 GeV (figure 9).

More precisely, the 5σ limits are shown in the table 2.

Table 2. Sensitivity (5σ limit) for gluino and squark searches through missing E_T + jets

$\mathcal{L}$	m gluino if m squark = 2m gluinos	m gluino if m squark = m gluino	m gluino if m squark = m gluino /2
10^3 pb^{-1}	1050 GeV	1800 GeV	2600 GeV
10^5 pb^{-1}	1600 GeV	2300 GeV	3600 GeV

Same-sign di-lepton Search

Heavy squark or gluino decays to charginos or neutralinos will produce isolated same-sign leptons.

Requiring 2 leptons with P_T > 20 GeV, 4 jets with P_T > 110 GeV and E_T miss > 150 GeV, one gets rid off the main background coming from $t\bar{t}$.

A five sigmas mass reach are shown in the table 3.

The results are not too sensitive to μ or tanβ.

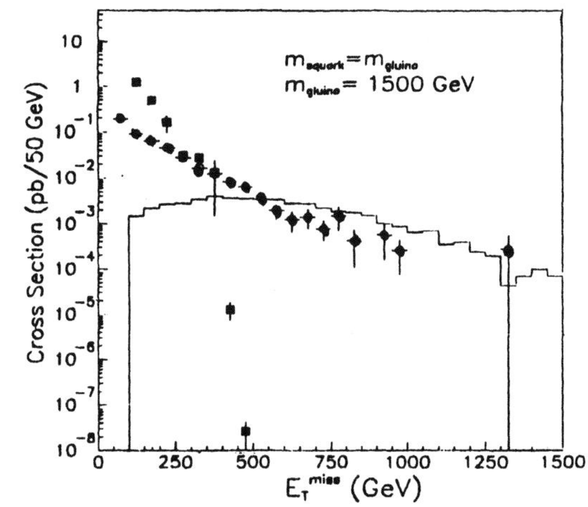

Figure 9. Missing E_T distribution for gluino production.
____ signal
● physical background
■ instrumental background

Chargino-Neutralino Searches

Heavy charginos and neutralinos can decay to the lightest supersymmetric particle (LSP) in the decay LSP + $\ell\nu$ or LSP + $\ell\ell$.

This will conduct to three isolated leptons. The dominant backgrounds are WZ, ZZ and $t\bar{t}$ productions.

This search technique is only usefull in limited region of parameters space where the purely leptonic branching ratios are significant.

With μ = -m (gluino), tanβ = 2, $\mathcal{L}$ = 10^4 pb^{-1}, the sensitivity extends up to ~500 GeV.

Table 3. Sensitivity (5σ limit) for gluino and squark searches through dileptons

$\mathcal{L}$	10^3 pb^{-1}	10^5 pb^{-1}
m gluino if m squark = 2m gluinos	900 GeV	1400 GeV
m gluino if m squark = m gluino	1100 GeV	1800 GeV
m gluino if m squark = m gluino /2	1200 GeV	2000 GeV

TOP PHYSICS AT LHC

LHC will be a nice top factory, already at a luminosity of 10^{32}cm^{-2}s^{-1}, 300 $t\bar{t}$ pairs are produced per hour and 10 clean isolated $e^{\pm}\,\mu^{\mp}$ pairs per day will be signed.

The $t\bar{t}$ decays will be mainly through $b\bar{b}$ $W^+ W^-$.

Top Mass Determination

$t\bar{t}$ with $t \to W^+ b \to \ell^+ \nu\, b$

$\bar{t} \to W^- \bar{b} \to j_1 j_2 \bar{b}$ will give the top mass, provide one tag b and calibrate properly the jets. The error will be dominated by the difficulty to taggle correctly a jet...

Many other channels exist but always the systematic error may be the dominating source of error.

Other Decays

We have seen already t to H^+ decay in MSSM searches.
Flavour changing neutral current decay can also be addressed.
A limit of BR ($t \to Zc$) $\leq 5.10^{-4}$ at 95% confidence level can be reached.

EXOTISMS

No Higgs Exists

There is room for non existing higgs, if no higgs resonance is observed at $m_{H^0} <$ 1 TeV, the higgs sector is strongly coupled. Then $W_L W_L$ high mass pair production cross-section will be higher than in the standard model.

It is not an easy measurement but one can see on figure 10 the magnitude of the signal expected in ATLAS for m_H = 1 TeV.

Strongly Coupled Electroweak Schemes

Some strongly coupled electroweak schemes predict resonances in some channels, W+Z or Z+γ resonance (ρ or ω like). LHC will be sensitive up to 1.5 to 2.0 TeV masses.

New Vector Bosons Searches

Detectors are sensitive to $Z' \to ee, \mu\mu, jj$ and to $W' \to e\nu, \mu\nu, jj$ up to 5.0, 5.5 TeV.

B PHYSICS ISSUES

Again, LHC is a very good B factory, a lot of potential physics is there in front of us, how much will be done before LHC is difficult to speculate but as we shall see, ultimate accuracy will be possible only with LHC in a foreseable future.

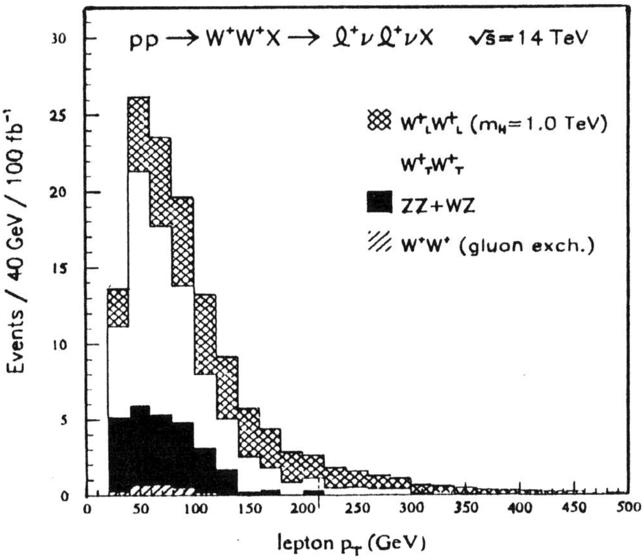

Figure 10. $W_L^+W_L^+$ event distribution as a function of p_T^{lepton} for m_H = 1.0 TeV.

B Physics and CP Violation

One possibility will be to determine the CKM parameters and so to determine the CP violation triangle.

Quarks Missing in the Standard Model.

The CKM matrix $\begin{pmatrix} V_{ud} & V_{us} & V_{ub} \\ V_{cd} & V_{cs} & V_{cb} \\ V_{td} & V_{ts} & V_{tb} \end{pmatrix}$ follows unitarity.

This means
$$V_{ud}^*V_{us} + V_{cd}^*V_{cs} + V_{td}^*V_{ts} = 0,$$
$$V_{us}^*V_{ub} + V_{cs}^*V_{cb} + V_{ts}^*V_{tb} = 0,$$
$$V_{ub}^*V_{ud} + V_{cb}^*V_{cd} + V_{tb}^*V_{td} = 0.$$

In the complexe plane, only the third relation represents a triangle called the unitarity triangle (see figure 11). This triangle defines three angles α, β, γ, and two sides related to the CKM matrix elements. Also shown on figure 11 the different B channels measurements which can help in measuring the triangle parameters.

Deviations from Unitarity

A fourth quark generation or additional box diagrams due to supersymmetry particles may affect the triangle which reflect the unitarity. Very precise measurements would then be requested.

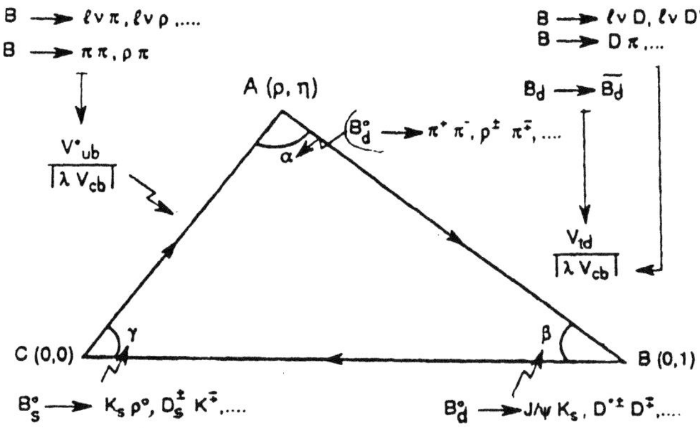

Figure 11. CP violation triangle.

Signatures of CP Violation and Measuring[4]

The semileptonic asymmetry, clean measurement, is in principle small of the order of 10^{-4}.

The decays of B_d, B_s to CP eigenstates is a fruitfull approach.

If f is a final state, CP eigenstate :

$$M = \left|\langle f | B^0 \rangle\right|$$

$$\overline{M} = \left|\langle f | \overline{B^0} \rangle\right|$$

$$\Gamma_{(t)}\left(B^0 \to f\right) = Ae^{-\Gamma t}[1 + B\cos\Delta mt - C\sin\theta\sin\Delta mt]$$

$$\Gamma_{(t)}\left(\overline{B}^0 \to f\right) = Ae^{-\Gamma t}[1 - B\cos\Delta mt + C\sin\theta\sin\Delta mt]$$

where $\Delta_m = m_{B1} - m_{B2}$ 2 eigenstates of the neutral B meson mass matrix

$$|B_{1(2)}\rangle = p|B^0\rangle + (-)q|\overline{B^0}\rangle$$

$$\theta \text{ phase of } \frac{q\langle f|\overline{B^0}\rangle}{p\langle f|B^0\rangle}$$

$$A = (M^2 + \overline{M^2})/2;\ B = \frac{M^2 - \overline{M}^2}{M^2 \to \overline{\overline{M}^2}}\ ;\ C = 2M\overline{M}/(M^2 + \overline{M}^2).$$

If $B^0 \to f$ involved only one CKM phase (ex.: K_s^0 J/Ψ decay), then

$$M = \overline{M},$$

and

$$A_{f(t)} = \frac{\Gamma_{(t)}(B^0 \to f) - \Gamma_{(t)}(\overline{B}^0 \to f)}{\Gamma_{(t)}(B^0 \to f) + \Gamma_{(t)}(\overline{B}^0 \to f)} = \sin 2\phi_f \sin(\Delta m t)$$

ϕ_f is one of the unitarity angle.

If $B^0 \to f$ involved more than one CKM phase (ex.: $\pi^+\pi^-$ decay), then

$$A_{f(t)} = B\cos\Delta mt + C\sin\theta\sin\Delta mt.$$

The measurement of the angle β can be done through the decay

$$B_d^0 \to J/\Psi\ K_s^0.$$

Measurement of α.

In the decay $B_d^0 \to \pi^+\pi^-$, one has two contributions from the tree diagramm and the penguin one. There are two isospin amplitudes :

$$A_{+-} = \langle \pi^+\pi^- | H_W | B^0 \rangle = \sqrt{\frac{2}{3}} \underbrace{\langle I=0 | H_W | B^0 \rangle}_{(a_t + a_p)e^{i\delta_0}} + \sqrt{\frac{1}{3}} \underbrace{\langle I=2 | H_W | B^0 \rangle}_{a_t e^{i\delta_2}}$$

where:

- δ_0 and δ_2 are strong phases,
- a_t tree amplitude $(\overline{b} \to \overline{u}(u\overline{d}))$ $\qquad \arg a_t = \arg V_{ub}^* = \alpha + \beta$,
- a_p penguin amplitude $(\overline{b} \to \overline{d}g)$ $\qquad \arg a_p = \pi + \beta$.

So

$$A(t) = \wedge_c \cos\Delta mt + \wedge_s \sin\Delta mt$$

where

$$\wedge_c = \frac{-2\sqrt{2}\sin\alpha\sin(\delta_2 - \delta_0)}{3 + 2\sqrt{2}\cos(\delta_2 - \delta_0)} \left|\frac{a_p}{a_t}\right|$$

$$\wedge_s = \sin 2\alpha \left[1 - \frac{4 + 2\sqrt{2}\cos(\delta_2 - \delta_0)\sin\alpha}{3 + \cos(\delta_2 - \delta_0)\sin 2\alpha} \left|\frac{a_p}{a_t}\right|\right].$$

Measurement of γ.

Different ways have been proposed:

- In decays of B_s mesons to non CP eigenstates[5], for the various $B_s \to D_s K$ decays channels to non CP eigenstates,

f final state $CP|f\rangle = |\overline{f}\rangle$

if $\langle f | B^0 \rangle$ and $\langle f | \overline{B}^0 \rangle$ make a unique weak phase, then

$$|\langle f|B^0\rangle| = |\langle \overline{f}|\overline{B}^0\rangle| = M \text{ and } |\langle \overline{f}|B^0\rangle| = |\langle f|\overline{B}^0\rangle| = \overline{M}$$

$$\Gamma(t)(B^0 \to f) = Ae^{-\Gamma t}\left[1 + B\cos\Delta mt - C\sin(\phi_f + \delta_s)\sin\Delta mt\right]$$

$$\Gamma(t)(B^0 \to \overline{f}) = Ae^{-\Gamma t}\left[1 - B\cos\Delta mt + C\sin(\phi_f - \delta_s)\sin\Delta mt\right]$$

where $A = (M^2 + \overline{M}^2)/2, \quad B = \dfrac{M^2 - \overline{M}^2}{M^2 + \overline{M}^2}, \quad C = 2M\overline{M}/(M^2 + \overline{M}^2),$

$\phi_f = \delta,$

δ_s = strong phase difference between $\langle f|B^0\rangle$ and $\langle f|\overline{B}^0\rangle$.

One can measure γ, δ_s

- in decays of charged B[6],

$B^\pm \to D_1^0 K^\pm,$

$B^\pm \to \overline{D^0} K^\pm$, with $D_1^0 = (D^0 + \overline{D^0})/\sqrt{2}$,

$B^\pm \to D^0 K^\pm.$

The amplitudes are:

$A(B^+ \to \overline{D^0} K^+) = A(B^- \to D^0 K^-),$

$A(B^+ \to D^0 K^+) = \exp(2i\gamma) A(B^- \to \overline{D^0} K^-),$

$A(B^+ \to D_1^0 K^+) = \dfrac{1}{\sqrt{2}}\left[A(B^+ \to D^0 K^+) + A(B^+ \to \overline{D^0} K^+)\right].$

This defines two triangles which will conduct to a two fold ambiguities.This can be, in principle, resolved using other channels as $B^\pm \to D^0 K^0 \pi^\pm, \overline{D}^0 K^0 \pi^\pm, \ldots$

Measuring xs (parameter of the B_s^0 oscillation).

Mixing parameters are defined such that

$$x_\alpha = \frac{\Delta m_\alpha}{\Gamma_\alpha}.$$

In practice, it is better to measure the ratio x_s / x_b which is less model dependent than x_b or x_s alone.

$$\frac{x_s}{x_b} = \frac{\tau_{B_s}}{\tau_{B_d}} \frac{m_{B_s}}{m_{B_d}} \frac{\eta_{QCD}^{B_s^0}}{\eta_{QCD}^{B_d^0}} \frac{f_{B_s}^2 B_{B_s}}{f_{B_d}^2 B_{B_d}} \left|\frac{V_{ts}}{V_{td}}\right|^2$$

using $B_s^0 \to \pi^+ D_s^-$ or J/Ψ $\overline{K}^{0*}$ or $K^- \pi^+$ and $D_s^\mp \pi^\mp \pi^+ \pi^-$

$$N(t) = N_0 A(t) e^{-t}\left[1 \pm D\cos x_s t\right],$$

where $A(t)$ is the acceptance,

D the dilution factor due to oscillation and mistag.

The Fourier transform (power structure) :

$$F(\omega)^2 = N_0^2 \frac{D^2}{4} \frac{1}{1 + (\omega - x_s)^2}$$

is sharply peaked around x_s.

In practice, x_s can be measured up to value of 50.

Rare decays

Flavor changing neutral current.

It can be looked for in the channel $B_s^0 \to \mu^+\mu^-$ and $B_d^0 \to \mu^-\mu^-$ (the e+e- decay is strongly suppressed) in the standard model $BR(B_s^0 \to \mu^+\mu^-)$= 2.10-9, this is likely to be reachable at LHC but $BR(B_d^0 \to \mu^+\mu^-)$= 8.10-11 is unlikely to be observed.

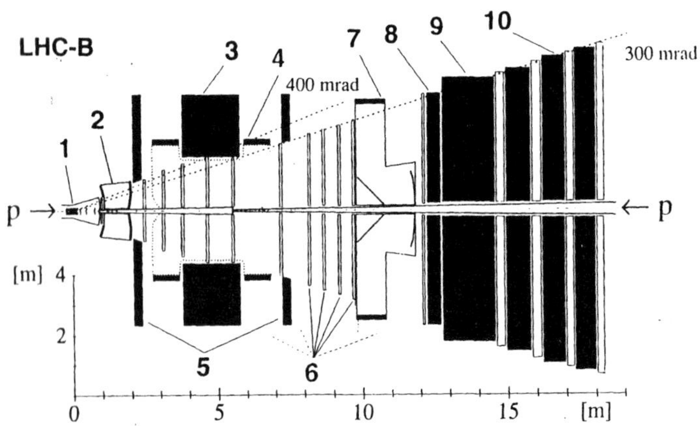

1) Vertex detector 2) Aerogel and Gas RICH's 3) Magnet yoke 4) Coils
5) Magnetic field shielding plates 6) Tracking chambers 7) Gas RICH
8) Electromagnetic calorimeter 9) Hadron calorimeter 10) Muon system

Figure 12. LHC-B detector.

$B^0 \to X_s^0, d\, \ell^+\ell^-$ decays

The inclusive decays of $B_s^0 \to X_s e^+e^-$ is around 1-2 10-5, this can be measured and is sensitive to non standard physics.

The exclusive decays ratio

$$\frac{BR(B^0 \to \rho^0 \ell^+\ell^-)}{BR(B^0 \to K^{*0} \ell^+\ell^-)} = \left|\frac{V_{td}}{V_{ts}}\right|^2 (1+\varepsilon_{QCD})$$

is another way of measuring $\left|\frac{V_{td}}{V_{ts}}\right|$ with little incertitude on ε_{QCD}.

Spectrosocopy of New States.

None of the B_c^0 has been observed but a whole spectroscopy is expected.

B_c^+ decays through $J/\Psi M^+$, $J/\Psi a_1^+$ or $B_s M^+$ may be accessible.

Measurement before LHC

HeraB[7], CdF[8] and D^0 [9], Stanford[10] and Japanese B[11] factories will achieve major steps.

HeraB and CdF depending on luck and usefull luminosity can measure at least $\sin 2\beta$ with few sigmas, dedicated B factories can measure many channels many years before LHC and overall it will depend strongly on the integrated luminosity.

At LHC, ATLAS and CMS will do their best and can convincely do a very good job on β and x_s measurements ($\delta(\sin 2\beta) = 0.01$).

The α angle measurement will benefit greatly from a K/π separation which can be obtained only in LHC-B[3], a dedicated detector (figure 12).

THE TECHNICAL CHALLENGE FOR DETECTORS

Most of the documents which have been presented in the Cargèse School have already been written for the Lepton-Photon Interactions Conference of August 95. I refer the reader to this paper[12]. For completeness, the ATLAS and CMS detectors are shown on figures 13 and 14.

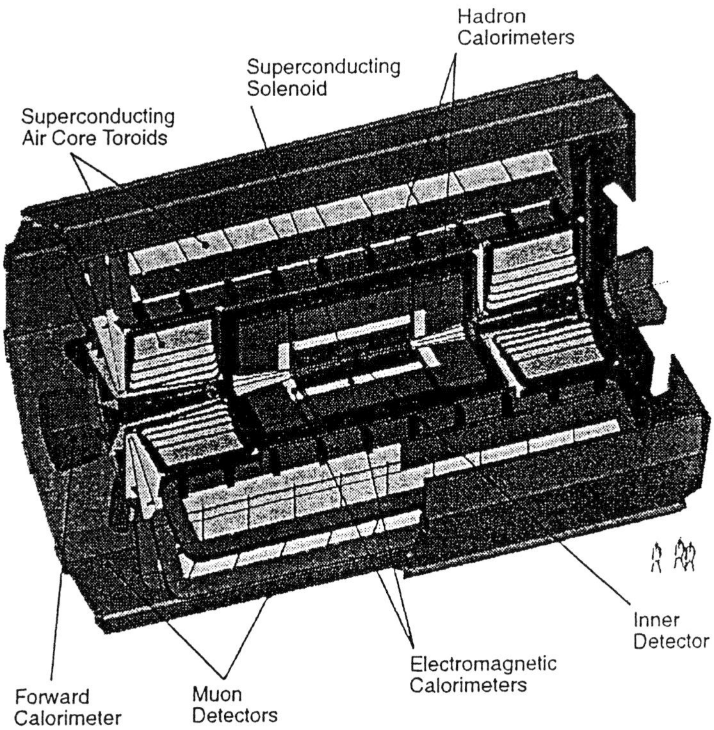

Figure 13. ATLAS detector.

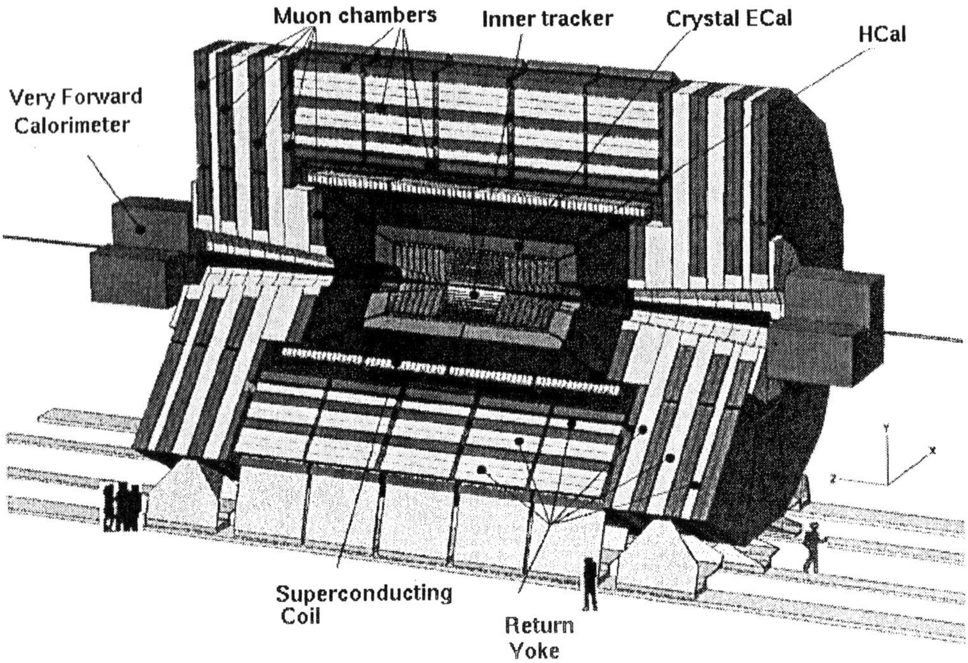

Figure 14. CMS detector.

ACKNOWLEDGMENTS

It is always a pleasure for me to come to the Cargèse School of Physics and I want to thank the students for their patience and the organizers for the warm atmosphere.

I have used extensively in the preparation, transparencies from experiments, talks from K. Einsweiler at the LHC-C, talks from D. Denegri, talks at the LHC-C, as well as many private discussions.

A special thank to Françoise Gonzalez which has typed this manuscript partly during Christmas time.

REFERENCES

1. CMS. *The Compact Muon Solenoïd Technical Proposal*, CERN/LHCC/94-38, 15-12-94.
2. ATLAS. *Technical Proposal*, CERN/LHCC/94-43, 15-12-94.
3. LHC-B. *Letter of Intent*, CERN/LHCC/95-5, 25-8-95.
4. COBEX. *Letter of Intent*, CERN/LHCC/93-50, 1993.
   GAJET. *Letter of Intent*, CERN/LHCC/93-54, 1993.
   LHB. *Letter of Intent*, CERN/LHCC/93-45, 1993.
   LHC-B. *Letter of Intent*, CERN/LHCC/95-5, 25-8-95.
   J. Ellis. *Referee work to the LHCC.*
5. R. Aleskan et al. Z. für Phys. C54, 653 (92).
6. Gronau and Wyler, Phys. Lett. B265, 172 (91), and Dunietz, Phys. Lett. B270, 75 (91).
7. HERA B. *Design Report*, DESY PRC 95/01, January 95.
8. The CDF Collaboration. *The CDF Upgrade,* CDF/Doc (CDF/Public/3171), June 95.
9. The D^0 Collaboration. *The D^0 Upgrade,* Fermilab (April 95).
10. BELLE. *Technical Design Report*, KEK (January 95).
11. BABAR. *Technical*, SLAC R95-457.
12. J.-J. Aubert, *New Detectors and Experimental Techniques, Proceedings of the 17th International Symposium on Lepton-Photon Interaction, August 95, Beijing*, p. 485.

INDEX